Basic Electric Machines

Vincent Del Toro

Professor Emeritus
Department of Electrical Engineering
City University of New York

Prentice Hall, Englewood Cliffs, New Jersey 07632

Del Toro, Vincent. (date)
  Basic electric machines / Vincent Del Toro.
    p. cm.
  Bibliography: p.
  Includes index.
  ISBN 0-13-060146-2
  1. Electric machinery.   2. Electromechanical devices.   I. Title.
TK2000.D44   1990
621.31'042—dc20

89-8714
CIP

Editorial/production supervision
and interior design: Sophie Papanikolaou
Cover design: Ben Santora
Manufacturing buyer: Bob Anderson

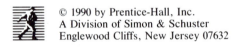

© 1990 by Prentice-Hall, Inc.
A Division of Simon & Schuster
Englewood Cliffs, New Jersey 07632

Printed in the United States of America

10 9 8 7 6 5 4 3 2 1

ISBN   0-13-060146-2

Prentice-Hall International (UK) Limited, *London*
Prentice-Hall of Australia Pty. Limited, *Sydney*
Prentice-Hall Canada Inc., *Toronto*
Prentice-Hall Hispanoamericana, S.A., *Mexico*
Prentice-Hall of India Private Limited, *New Delhi*
Prentice-Hall of Japan, Inc., *Tokyo*
Simon & Schuster Asia Ptc. Ltd., *Singapore*
Editora Prentice-Hall do Brasil, Ltda., *Rio de Janeiro*

To
Franca

Contents

PART II: TOPICS FOR FURTHER STUDY

Preface

This book is divided into two parts. Part I is devoted to an exposition of topics that can be considered to constitute the basic subject matter of a core course in electric machinery. Part II is concerned with a series of eclectic topics that easily builds on the foundation established by the basic topics of the first part. In most instances a suitable semester course can be fashioned by a syllabus that includes Chapters 1 to 9. Of course not all topics in each chapter need necessarily be part of such a syllabus. The treatment often lends itself to selective exclusion of specialty topics in each chapter should time be a pressing factor. Moreover, the Table of Contents was cast with an eye on flexibility in order to allow instructors, who are so inclined, to include in the syllabus such focused topics as stepper motors, synchros, and machine dynamics. This can be achieved by avoiding some of the more in-depth subject matter of the early chapters.

The first three chapters of Part I serve to establish the fundamental background which makes it possible to analyze the conventional electric machines—namely, single-phase and three-phase induction motors, dc generators and motors, and synchronous generators and motors. Reduced to simplest terms, one can say that the study of electric machinery involves two basic laws: Faraday's law of electromagnetic induction, which describes the generation of voltage in a winding by a moving or time-varying magnetic field; and Ampere's law, which describes the production of torque on a current-carrying conductor in a magnetic field. Because of the preeminence of the magnetic field in these laws, the book begins in Chapter 1 with a description of magnetic theory and their circuits. A primary goal here is to define the various magnetic quantities and to delineate their roles in magnetic circuit theory. After all, electric machines are constructed to an important degree to provide suitable paths for the magnetic field. The focus in Chapter 2 is on the application of Faraday's law as the basis for the construction of a very useful device—the transformer. The transformer fulfills many useful purposes in many fields of electrical engineering, and a comprehension of its principle of operation permits an easy understanding of the operation of all electrical machines. Here the treatment is heavily biased towards a *physical* understanding of transformer operation. This attitude is reflected even in the derivation of the equivalent circuit. However, to satisfy those individuals who prefer more mathematically oriented treatments, such are included as well. In Chapter 3 effort is directed to an analysis of torque production in an elementary machine by the use of the two fundamental laws and to a description of how the ensuing electrical quantities, e and i, are related to the associated mechanical quantities, T and ω, through the Law of Conservation of Energy. More detailed analyses of torque production and generation of voltage in situations associated with practical machine configurations appear, for the sake of simplicity, in the Appendix to Chapter 3 at the end of the book. However, the results found in this appendix are used in the last section of Chapter 3 to derive the *practical* forms of the torque and voltage equations as they apply to ac and dc machines. These latter forms are then often invoked throughout the remaining chapters of the book.

By drawing on the general background established from a study of Chapters 1 through 3, the study of the various types of electric machines becomes a routine matter. Attention is first directed in Chapter 4 to the three-phase induction motor. This choice is made not only because this motor is the most popular found in industry, but also because the analysis is very closely related to the static transformer despite the fact that a rotating element is involved. The equivalent circuit is exactly the same in form as that of the transformer. The significance of some of the circuit parameters, however, is different, and this distinction is drawn quite effectively by continuing with an emphasis on physical interpretation while simultaneously pursuing rigorous mathematical formulations. The power flow diagram is introduced in this chapter for the first time and its usefulness in helping to evaluate motor performance is amply illustrated. Speed control of these motors is an important application feature and, accordingly, receives a good deal of attention with special emphasis on electronic means of control. The chapter closes with a listing of operating features and areas of applications as well as with a discussion of controllers.

Chapters 5 and 6 deal with the subject matter of three-phase synchronous generators and motors, respectively. Both linear and nonlinear analysis by the general method and the synchronous reactance method are considered and these procedures are applied to cylindrical-rotor as well as salient-pole machines. The static stability of the synchronous motor under load, use of the synchronous motor for power factor control, and excitation of its field winding by rectified sources are topics that receive deserved attention. Again in these chapters the theme of stressing the physical interpretations of the mathematical results is continued.

The operation and performance of dc generators and motors are treated in Chapters 7 and 8. Because the dc machine is equipped with a mechanical rectifier (the commutator) to convert the induced ac coil voltages to unidirectional forms, the need for good commutation is imperative, and, for consistency, this topic is allotted corresponding attention. Also important to a more complete understanding of dc machine performance is an appreciation of the demagnetizing effect caused by the armature winding magnetomotive force. Hence, this topic too is singled out for special treatment. Moreover, because of the importance that the slope and shape of the speed-torque characteristics of the various types of dc motors have on their application suitability, an extensive analysis is given of these characteristics for the shunt, compound, and series motors. This study is also helpful in understanding the various methods of speed control of these motors. In addition to a treatment of speed control by the conventional schemes of field control, armature circuit resistance adjustment, and armature voltage adjustment, the use of electronic procedures based on thyristor and chopper drives is also discussed at length.

Chapter 9 deals with the single-phase induction motor, which is the most complex machine treated in Part I. Two procedures are described to analyze this motor. One is the double-revolving field theory, which requires no new theory beyond that appearing in preceding chapters. The other is the use of symmetrical components, which does call for new theory. However, in the interest of simplicity and continuity, the derivation of this new theory is placed in Chapter 10 but the results are used in Chapter 9 to derive an equivalent circuit which is shown to be identical to that obtained by the double-revolving field theory. The remainder of Chapter 9 is devoted to a description of the construction and operating features of the various types of single-phase motors that are frequently encountered in commercial and home environments.

The unbalanced two-phase motor is treated in Chapter 10, which is the first chapter of Part II. It contains a detailed analysis of four-phase symmetrical components which can be modified to furnish the results that are applicable to single-phase motors or to unbalanced two-phase motors such as the servomotor. The second half of this chapter is concerned exclusively with the servomotor, which is found in many servomechanisms.

Stepper motors are introduced in Chapter 11 through a detailed description of the permanent magnet version of these motors along with their drive amplifiers and translator logic devices. The reluctance-type stepper motor is also discussed briefly. Chapter 12 deals with another specialty topic—the synchro. The construction features of the various types are amply detailed along with their voltage relationships, applications, errors, and residual voltages.

All analysis of the machines of Part I is restricted to steady-state behavior. By contrast, in Chapter 13 the emphasis is placed on formulations that allow *dynamic* as well as steady-state behavior to be described. These results are achieved either through a transfer-function analysis or by a state-variable formulation. Both procedures are treated and applied to dc and ac machines.

1

Magnetic Theory and Circuits

An understanding of electromagnetism is essential to the study of electrical engineering because it is the key to the operation of a great part of the electrical apparatus found in industry as well as the home. All electric motors and generators, ranging in size from the fractional horsepower units found in home appliances to the 25,000-hp giants used in some industries, depend upon the electromagnetic field as the coupling device permitting interchange of energy between an electrical system and a mechanical system and vice versa. Similarly, static transformers provide the means for converting energy from one electrical system to another through the medium of a magnetic field. Transformers are to be found in such varied applications as radio and television receivers and electrical power distribution circuits. Other important devices—for example, circuit breakers, automatic switches, relays, and magnetic amplifiers—require the presence of a confined magnetic field for their proper operation. This chapter provides the reader with background so that he or she can identify a magnetic field and its salient characteristics and more readily understand the function of the magnetic field in electrical equipment.

The science of electrical engineering is founded on a few fundamental laws derived from basic experiments. In the area of electromagnetism it is Ampere's

law that concerns us and, in fact, serves as the starting point of our treatment.†
On the basis of the results obtained by Ampere in 1820, in his experiments on the
forces existing between two current-carrying conductors, such quantities as magnetic flux density, magnetic field intensity, permeability, and magnetic flux are
readily defined. Once this base is established, attention is then directed to a
discussion of the magnetic properties of certain useful engineering materials as
well as to the idea of a "magnetic circuit" to help simplify the computations
involved in analyzing magnetic devices.

1-1 AMPERE'S LAW—DEFINITION OF MAGNETIC QUANTITIES

Appearing in Fig. 1-1 is a simplified modification of Ampere's experiment. The
configuration consists of a very long conductor 1 carrying a constant-magnitude
current I_1 and an elemental conductor of length l carrying a constant-magnitude
current I_2 in a direction opposite to I_1. When taken together the elemental conductor and the current I_2 constitute a *current element* $I_2 l$. The elemental conductor 2 is actually part of a closed circuit in which I_2 flows, but for simplicity and
convenience the details of the circuit are omitted except for the length l. Moreover, it is assumed that conductors 1 and 2 lie in the same horizontal plane and are
parallel to each other. In accordance with Ampere's law it is found that with this
configuration there exists a force on the elemental conductor directed to the
right. Furthermore, the magnitude of the force is found to be directly proportional
to I_1, I_2, l, and the medium surrounding the conductors as well as inversely
proportional to the distance between them. In mks units‡ the magnitude of this
force can be shown to be given by

$$F = \frac{\mu I_1}{2\pi r} I_2 l \qquad \text{newtons (N)} \qquad (1\text{-}1)$$

where I_1 and I_2 are expressed in amperes, l and r in meters, and μ is a property of
the medium. A further interesting revelation about this experiment is that, if the
elemental conductor 2 is used as an exploring device to find those points in space
where the force is of constant magnitude and outwardly directed, the locus is
found to be a circle of radius r and centered along the axis of conductor 1. In other
words it is possible to identify a field having constant *lines of force*. In this
connection it is useful at this point to rewrite Eq. (1-1) as follows:

$$F = I_2 l B \qquad (1\text{-}2)$$

---

†It should not be here inferred that this is the only starting point in developing a quantitative
theory of the magnetic circuit. Faraday's law of induction is equally valid as a starting point and is
preferred when the goal is the development of an electromagnetic wave theory rather than a theory
leading to the treatment of electromechanical energy conversion.

‡Meter-kilogram-second system of units.

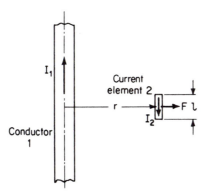

Figure 1-1 Illustrating the force existing between a current element $I_2 l$ and a very long conductor carrying current I_1 as described by Ampere's law.

where

$$B \equiv \frac{\mu I_1}{2\pi r} \tag{1-3}$$

Here for reasons that become clearer presently, B is defined† as the magnetic field or, still better, the *magnetic flux density* existing in the region where the elemental conductor 2 lies and has the units of *webers/meter*² or *teslas* (T). As Eq. (1-2) reveals, the magnetic flux density is defined as a measured force F per known current element $I_2 l$. Equation (1-3) is merely a means of identifying the manner in which the current I_1 influences the force field about the current element. It is significant to note that, as long as I_2 is not zero, the force field and the magnetic field have the same characteristics—both have a circular locus and both are vector quantities possessing magnitude and direction. However, because of the way B is defined, the magnetic field exists as long as I_1 is not zero irrespective of the value of I_2. Magnetic flux density thus represents the effects caused by the moving charges that make up I_1.

In our study of electrical machinery, which comes later, conductor 1 will be referred to as the *field winding* because it sets up the working magnetic field, whereas the total circuit of which the elemental conductor is a part is called the *armature winding*.

The direction of the magnetic field is readily determined by the *right-hand rule* which states that, if the field winding conductor (1 in this case) is grasped in the right hand with the thumb pointing in the direction of current flow, the lines of flux (or flux density) will be in the direction in which the fingers wrap around the conductor.

---

†This definition of B is correct for the configuration of Fig. 1-1. For other configurations the forms will be similar, but the constants will differ generally.

Permeability

A glance at Eq. (1-1) shows that all the factors are known in this equation with the exception of the proportionality factor μ, which is a characteristic of the surrounding medium. Upon repeating the experiment of Fig. 1-1 in iron rather than air, it is found that the force is many times greater for the same values of I_1, I_2, l, and r. Therefore, it follows that μ may be defined from Eq. (1-1) for various media because it is the only unknown quantity. Moreover, because of the way magnetic flux density was defined, Eq. (1-3) indicates that the effect of the surrounding medium may be described in terms of the degree to which it increases or decreases the magnetic flux density for a specified current I_1. Thus when iron rather than free space is the medium it can be said that the iron provides a greater penetration of the magnetic field in a given region, i.e., there is a greater flux density. This property of the surrounding medium in which the conductors are embedded is called *permeability*.

When the conductors of Fig. 1-1 are assumed placed in a vacuum (free space) and the force is measured for specified values of I_1, I_2, l, and r, the solution for the permeability of free space obtained from Eq. (1-1) and expressed in mks units comes out to be

$$\mu_0 = 4\pi \times 10^{-7} \ (\text{H/m}) \tag{1-4}$$

The unit of permeability also follows from Eq. (1-1). Thus

$$\mu_0 \propto \frac{\text{newtons}}{\text{ampere}^2}$$

where $\propto$ denotes "is proportional to." But

$$\text{newton-meter} = \text{joule} = \text{volt} \times \text{ampere} \times \text{second}$$

or

$$\mu_0 \propto \frac{\text{volt} \times \text{ampere} \times \text{second}}{\text{ampere}^2 \times \text{meter}} = \frac{\text{volt} \times \text{second}}{\text{ampere} \times \text{meter}}$$

However, volt $\times$ seconds/ampere is the unit of inductance expressed in henrys. Accordingly, permeability is expressed in units of henrys/meter (H/m).

In those cases where the surrounding medium is other than free space, the absolute permeability is again readily found from Eq. (1-1). A comparison with the result obtained for free space then leads to a quantity called *relative permeability*, μ_r. Expressed mathematically we have

$$\mu_r = \frac{\mu}{\mu_0} \tag{1-5}$$

Equation (1-5) clearly indicates that relative permeability is simply a numeric which expresses the degree to which the magnetic flux density is increased or decreased over that of free space. For some materials, such as Deltamax, the value of μ_r can exceed one hundred thousand. Most ferromagnetic materials, however, have values of μ_r in the hundreds or thousands.

Magnetic Flux Φ

Before considering the manner in which this quantity is defined, let us first investigate more closely the dimensions of the magnetic flux density B as described by its defining expression Eq. (1-2). Thus

$$B \propto \frac{\text{newtons}}{\text{amperes} \times \text{meter}}$$

$$B \propto \frac{\text{volts} \times \text{seconds} \times \text{amperes}}{\text{amperes} \times \text{meter}^2} = \frac{\text{volts} \times \text{seconds}}{\text{meter}^2} \tag{1-6}$$

A study of this expression reveals that B is in fact a density because it involves meters squared in the denominator. Accordingly, it is reasonable to expect that, since B is defined as a magnetic flux density, then multiplication by the effective area that B penetrates should yield the total *magnetic flux*. To illustrate this point refer to Fig. 1-2, which shows a coil of area ab lying in the same horizontal plane containing conductor 1. We already know that when a current I_1 flows through this conductor a magnetic field is created in space and specifically is described by Eq. (1-3). To find the total flux penetrating the coil it is necessary merely to perform an integration of B over the surface area involved. Of course if B were a constant over the area of concern, the flux would be simply the product of B and

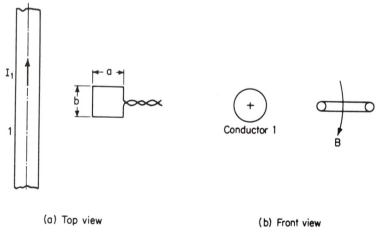

(a) Top view   (b) Front view

Figure 1-2   Associating an area with a magnetic field B to identify a magnetic flux.

Figure 1-3 Same as Fig. 1-2(b) except that the coil is tilted 60° relative to the horizontal plane.

the area ab. Next consider that the plane of the coil is tilted with respect to the plane of conductor 1 by 60°, as depicted in Fig. 1-3. Clearly now the total flux penetrating the coil is reduced by a factor of $\frac{1}{2}$. If the coil is oriented to a position of 90° with respect to the horizontal plane, no flux threads the coil.

On the basis of these observations, then, the magnetic flux through any surface is more rigorously defined as the surface integral of the normal component of the vector magnetic field B. Expressing this mathematically we have

$$\Phi = \int_s B_n \, dA \qquad (1\text{-}7)$$

where s stands for surface integral, A represents the area of the coil, and B_n is the normal component of B to the coil area. From expression (1-6) we know that magnetic flux must have the dimensions of *volt-seconds*. However, this is more commonly called *webers* and denoted by Wb. The volt-seconds unit of flux is better understood in terms of Faraday's law of induction (see Eq. (2-8)).

Magnetic Field Intensity H

Often in magnetic circuit computations it is helpful to work with a quantity representing the magnetic field which is independent of the medium in which the magnetic flux exists. This is especially true in situations such as are found in electrical machinery where a common flux penetrates several different materials, including air. A glance at Eq. (1-3) reveals that division of B by μ identifies such a quantity. Accordingly, magnetic field intensity is defined as

$$H \equiv \frac{B}{\mu} \qquad (1\text{-}8)$$

and has the units of

$$\frac{\dfrac{\text{newtons}}{\text{amperes} \times \text{meter}}}{\dfrac{\text{newtons}}{\text{ampere}^2}} = \frac{\text{amperes}}{\text{meter}} = \frac{A}{m}$$

Thus H is dependent upon the current that produces it and also on the geometry of the configuration but not the medium. For the system of Fig. 1-1 the value of the magnetic field intensity immediately follows from Eq. (1-3) and is given by

$$H = \frac{B}{\mu} = \frac{I_1}{2\pi r}$$

Because H is independent of the medium it is frequently looked upon as the intensity that is responsible for driving the flux density through the medium. Actually, though, H is a derived result.

More generally the units for H are ampere-turns/meter rather than amperes/meter. A little thought should make this apparent whenever the field winding is made up of more than just a single conductor.

Ampere's Circuital Law

Now that the magnetic field intensity has been defined and shown to have dimensions of ampere-turns per meter, we can develop a very useful relationship. Recall that H is a vector having the same direction as the magnetic field B. For the configuration of Fig. 1-1 H has the same circular locus as B. A line integration of H along any given closed circular path proves interesting. Of course the *line* integral is considered because H involves a per unit length dimension. Thus

$$\oint H \, dl = \int_0^{2\pi r} \frac{I_1}{2\pi r} \, dl = I_1 \qquad \text{amperes} \qquad (1\text{-}9)$$

(Again keep in mind that the units here would be ampere-turns if more than one conductor were involved in Fig. 1-1.) Equation (1-9) states that the closed line integral of the magnetic field intensity is equal to the enclosed current (or ampere-turns) that produces the magnetic field lines. This relationship is called *Ampere's circuital law* and is more generally written as

$$\boxed{\oint H \, dl = \mathcal{F}} \qquad (1\text{-}10)$$

where $\mathcal{F}$ denotes the ampere-turns enclosed by the assumed closed flux line path. The quantity $\mathcal{F}$ is also known as the *magnetomotive force* and frequently abbreviated *mmf*. This relationship is useful in the study of electromagnetic devices and is referred to in subsequent chapters.

Derived Relationships

In the preceding pages the fundamental magnetic quantities—flux density, flux, field intensity, and permeability—are defined starting with Ampere's basic experiment involving two current-carrying conductors. By the appropriate manipulation of these quantities additional useful results can be obtained.

Equation (1-8) is a vector equation describing the magnetic field intensity for a given geometry and current. If the total path length of a flux line is assumed to be l, then the total magnetomotive force (mmf) associated with the specified flux line is

$$\mathscr{F} = Hl = \frac{B}{\mu} l \qquad (1\text{-}11)$$

Now in those situations where B is a constant and penetrates a fixed, known area A, the corresponding magnetic flux may be written from Eq. (1-7) as

$$\boxed{\Phi = BA} \qquad (1\text{-}12)$$

Inserting Eq. (1-12) into Eq. (1-11) yields

$$\mathscr{F} = Hl = \Phi \left(\frac{l}{\mu A} \right) \qquad (1\text{-}13)$$

The quantity in parentheses in this last expression is interesting because it bears a very strong resemblance to the definition of resistance† in an electric circuit. Recall that the resistance in an electric circuit represents an impediment to the flow of current under the influence of a driving voltage. An examination of Eq. (1-13) provides a similar interpretation for the magnetic circuit. We are already aware that $\mathscr{F}$ is the driving mmf which creates the flux Φ penetrating the specified cross-sectional area A. However, this flux is limited in value by what is called the *reluctance* of the magnetic circuit, which is defined as

$$\boxed{\mathscr{R} = \frac{l}{\mu A}} \qquad (1\text{-}14)$$

No specific name is given to the dimension of reluctance except to refer to it as so many units of reluctance.

Equation (1-14) reveals that the impediment to the flow of flux which a magnetic circuit presents is directly proportional to the length and inversely proportional to the permeability and the cross-sectional area—results which are entirely consistent with physical reasoning.

Inserting Eq. (1-14) into Eq. (1-13) yields

$$\boxed{\mathscr{F} = \Phi \mathscr{R}} \qquad (1\text{-}15)$$

which is often referred to as Ohm's law of the magnetic circuit. It is important to keep in mind, however, that these manipulations are permissible as long as B and A are fixed quantities.

---

†The resistance of a coper wire of length l and cross-sectional area A is given by $R = \rho l / A$, where ρ is the resistivity of the copper.

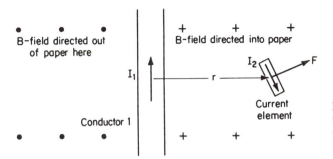

Figure 1-4 Showing the direction of the force when the current element is no longer located parallel to conductor 1 but remains in the same plane.

In most practical ferromagnetic materials the functional relationship between $\mathscr{F}$ and Φ (or the corresponding quantities $H = \mathscr{F}/l$ and $B = \Phi/A$) is nonlinear, which indicates that the reluctance $\mathscr{R}$ varies with the applied mmf. A glance at Fig. 1-10 makes this apparent. In some applications it is the change in reluctance about an operating point that is important rather than the total value of the reluctance. This leads naturally to the notion of *differential reluctance*, which is defined as

$$\mathscr{R}_d = \frac{d\mathscr{F}}{d\Phi} = \frac{d(Hl)}{d(BA)} = \frac{l}{A}\frac{dH}{dB} = \frac{l}{\mu_d A} \qquad (1\text{-}16)$$

Here μ_d is the differential permeability and is represented graphically as the slope of a tangent drawn to the B versus H curve.

Ampere's Law for Various Orientations of the Current Element

In Fig. 1-1 the assumption was made that the current element was located parallel to conductor 1 and lying in the same plane. Because this orientation was sufficient for the purpose at hand—to define the fundamental magnetic quantities—it was pursued as a matter of convenience. However, in the interest of furnishing a more complete picture of the experiment we shall now consider the effect on the force of placing the current element $I_2 l$ in two additional different orientations. Consider first that the current element is no longer placed parallel to conductor 1 but continues to be located in the same horizontal plane. Refer to Fig. 1-4. The dots in this figure indicate that the magnetic field is directed outward on the left side of conductor 1 and inward (with respect to the plane of the paper) on the right side of the conductor as revealed by the right-hand rule. The results of this experiment show that the magnitude of the force is the same as that found by using the configuration of Fig. 1-1.† This conclusion is not surprising because the value of the magnetic flux density as well as I_2 and l remain unchanged so that Eq. (1-2) is

---

†Because of the infinitesimal nature of the current element, the value of B is assumed to be constant about $I_2 l$ irrespective of its orientation.

still valid in describing the force. The only change is the direction of the force. However, as Fig. 1-4 indicates, the force continues to be normal to the current element. It is worthwhile to keep this point in mind. Presently a general rule for establishing the force direction for all configurations is described.

Next let us consider that orientation of the current element which places it parallel to conductor 1 but inclined at an angle $\theta = 30°$ with respect to the vertical. A side-view projection of the configuration is depicted in Fig. 1-5. Note that the magnetic field is directed downward along the vertical for this view. Actually, of course, the locus of B is circular, but in Fig. 1-5 we are looking at just that small portion of the B field about the plane containing conductor 1. With this configuration the force on the current element is found to have the same direction but one-half the magnitude of that obtained with the orientation of Fig. 1-1. It follows, then, that the angle between the current vector $I_2 l$ and the flux density B affects the magnitude of the force. As a matter of fact, further experimentation reveals that the general expression for the force is

$$\boxed{F = I_2 lB \sin \theta} \qquad \text{N} \qquad\qquad (1\text{-}17)$$

Equation (1-17) conveys information solely about the magnitude of the force and not its direction. It is possible, however, by employing the notation of vector analysis, to rewrite Eq. (1-17) so that information about magnitude as well as direction is present. This result is readily accomplished by the use of the *cross-product* notation between two vectors, yielding a third vector having magnitude and direction. Thus Eq. (1-17) is more completely expressed as

$$\boxed{\bar{F} = I_2 \bar{l} \times \bar{B}} \qquad \text{N} \qquad\qquad (1\text{-}18)$$

The cross notation must always be understood to involve the *sine* of the angle between the two vectors $\bar{B}$ and $\bar{l}$ (or the direction of I_2, which is determined by the orientation of l). Moreover, wherever the cross product is involved, the direction of the resultant vector is always normal to the plane containing the vectors $\bar{B}$ and $\bar{l}$ in the sense determined by the direction of advance of a right-hand screw as $\bar{l}$ is

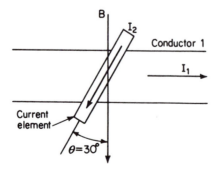

Figure 1-5 Side-view projection of Fig. 1-1 but with the current element tilted relative to the horizontal plane. The force is directed out of the paper.

turned into $\bar{B}$ through the smaller of the two angles made by the vectors. Accordingly, in the configuration of Fig. 1-5 the direction of the force is found by turning $I_2\bar{l}$ into $\bar{B}$ and then noting that this would cause a right-hand screw to advance out of the plane of the paper. Hence the force is directed outward, and this corresponds with the experimentally established result.

It is also possible to determine the direction of the force by means of another right-hand rule, which requires that the forefinger be put in the direction of the current and the middle finger in the direction of $\bar{B}$, with the two assumed to be lying in the same plane. The thumb of the right hand then points in the direction of the force when placed perpendicular to the other two fingers.

1-2 THEORY OF MAGNETISM

To understand the magnetic behavior of materials, it is necessary to take a microscopic view of matter. A suitable starting point is the composition of the atom, which Bohr described as consisting of a heavy nucleus and a number of electrons moving around the nucleus in specific orbits. Closer investigation reveals that the atom of any substance experiences a torque when placed in a magnetic field; this is called a *magnetic moment*. The resultant magnetic moment of an atom depends upon three factors—the positive charge of the nucleus spinning on its axis, the negative charge of the electron spinning on its axis, and the effect of the electrons moving in their orbits. The magnetic moment of the spin and orbital motions of the electron far exceeds that of the spinning proton. However, this magnetic moment can be affected by the presence of an adjacent atom. Accordingly, if two hydrogen atoms are combined to form a hydrogen *molecule*, it is found that the electron spins, the proton spins, and the orbital motions of the electrons of each atom oppose each other so that a resultant magnetic moment of zero should be expected. Although this is almost the case, experiment reveals that the relative permeability of hydrogen is not equal to 1 but rather is very slightly less than unity. In other words, the molecular reaction is such that when hydrogen is the medium there is a slight decrease in the magnetic field compared with free space. This behavior occurs because there is a precessional motion of all rotating charges about the field direction, and the effect of this precession is to set up a field opposed to the applied field regardless of the direction of spin or orbital motion. Materials in which this behavior manifests itself are called *diamagnetic* for obvious reasons. Besides hydrogen, other materials possessing this characteristic are silver and copper.

Continuing further with the hydrogen molecule, let us assume next that it is made to lose an electron, thus yielding the hydrogen ion. Clearly, complete neutralization of the spin and orbital electron motions no longer takes place. In fact, when a magnetic field is applied, the ion is so oriented that its net magnetic moment aligns itself with the field, thereby causing a slight increase in flux density. This behavior is described as *paramagnetism* and is characteristic of such

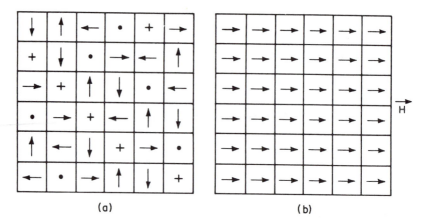

Figure 1-6 Representation of a ferromagnetic crystal: (a) unmagnetized and (b) fully magnetized by the field H.

materials as aluminum and platinum. Paramagnetic materials have a relative permeability slightly in excess of unity.

So far we have considered those elements whose magnetic properties differ only very slightly from those of free space. As a matter of fact the vast majority of materials fall within this category. However, there is one class of materials—principally iron and its alloys with nickel, cobalt, and aluminum—for which the relative permeability is very many times greater than that of free space. These materials are called *ferromagnetic*† and are of great importance in electrical engineering. We may ask at this point why iron (and its alloys) is so very much more magnetic than other elements. Essentially, the answer is provided by the *domain* theory of magnetism. Like all metals, iron is crystalline in structure with the atoms arranged in a space lattice. However, domains are subcrystalline particles of varying sizes and shapes containing about 10^{15} atoms in a volume of approximately 10^{-9} cubic centimeters. *The distinguishing feature of the domain is that the magnetic moments of its constituent atoms are all aligned in the same direction.* Thus in a ferromagnetic material, not only must there exist a magnetic moment due to a nonneutralized spin of an electron in an inner orbit, but also the resultant spin of all neighboring atoms in the domain must be parallel.

It would seem by the explanation so far that, if iron is composed of completely magnetized domains, then the iron should be in a state of complete magnetization throughout the body of material even without the application of a magnetizing force. Actually, this is not the case, because the domains act independently of each other, and for a specimen of unmagnetized iron these domains are aligned haphazardly in all directions so that the net magnetic moment is zero over the specimen. Figure 1-6 illustrates the situation diagrammatically in a simplified

---

†Derived from the Latin word for iron—*ferrum*.

fashion. Because of the crystal lattice structure of iron the "easy" direction of domain alignment can take place in any one of six directions—left, right, up, down, out, or in—depending upon the direction of the applied magnetizing force. Figure 1-6(a) shows the unmagnetized configuration. Figure 1-6(b) depicts the result of applying a force from left to right of such magnitude as to effect alignment of all the domains. When this state is reached the iron is said to be *saturated*— there is no further increase in flux density over that of free space for further increases in magnetizing force.

Large increases in the temperature of a magnetized piece of iron bring about a decrease in its magnetizing capability. The temperature increase enforces the agitation existing between atoms until at a temperature of 750°C the agitation is so severe that it destroys the parallelism existing between the magnetic moments of the neighboring atoms of the domain and thereby causes it to lose its magnetic property. The temperature at which this occurs is called the *curie point*.

1-3 MAGNETIZATION CURVES OF FERROMAGNETIC MATERIALS

If the experiment of Fig. 1-1 is repeated with iron or steel as the medium for increasing values of the field winding current I_1 and the corresponding values of μ computed, it is found that the relative permeability varies considerably with the magnetizing force that establishes the operating flux density. A typical variation of μ_r for cast steel appears in Fig. 1-7. Here μ_r is plotted versus flux density rather than the magnetizing force because it is the onset of the realignment of more and more domains that brings about the change in permeability. Unfortunately, the state of development of the theory of magnetism is not so far advanced that it allows the prediction of the magnetic properties of a material on a purely theoretical basis even though the exact composition of the material is known. For example, with the present theory it is not possible to say exactly what the flux density will be in a given specimen of iron for a specified value of the magnetizing force. Rather, it is customary to obtain this information by consulting technical and

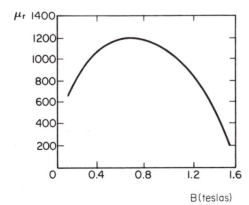

Figure 1-7 Graph of relative permeability versus flux density.

descriptive bulletins where the measured magnetic properties of a *representative sample* of the specimen are published. These bulletins are made available to users by the manufacturers of magnetic steels, and they include information on such varied shapes and forms as sheets, wires, bars, and even castings weighing up to hundreds of tons.

Usually the published magnetic characteristics of the various iron and steel samples are presented as plots of flux density B as a function of the magnetic field intensity H. In the interest of presenting a complete graphical picture of the functional relationship existing between these two quantities as well as to define additional terms used in this connection, refer to Fig. 1-8. Assume that the steel specimen is initially unmagnetized and is in the form of a toroidal ring with a coil of N turns wrapped around it. Assume too that the coil can be energized from a variable voltage source capable of furnishing current flow in either direction in the coil. As the current I is increased from zero in the positive direction (current flowing into the top terminal of the coil), an increasing magnetic flux Φ can be measured taking place within the body of the toroid in the clockwise direction. For any fixed value of I there is a specific value of flux. Then by Eq. (1-12) the corresponding flux density is determined since the toroidal cross-sectional area is known. Moreover, the magnetomotive force NI can be replaced by Hl_m in accordance with Ampere's circuital law [Eq. (1-10)], where l_m is the mean length of path of the toroid (as shown by the broken-line circle in Fig. 1-8). The two fundamental quantities involved in this arrangement then are B in webers/meter^2 and H expressed in ampere-turns per meter. H is the quantity we want to deal with rather than magnetomotive force because, for the same flux density, doubling the mean magnetic length will not change H but will require doubling the magnetomotive force. The conclusion to be drawn is that a plot of B versus H is a *universal* plot for the given material because it can be extended to any geometry of cross-sectional area and length. In contrast, a plot of Φ versus NI is limited to a single geometrical configuration. Therefore, a plot of the magnetic characteristics of a material always involves plotting B versus H. For the virgin sample of Fig. 1-8 the graph of B versus H follows the curve Oa of Fig. 1-9 for field intensities up to H_a. Take note of the nonlinear relationship existing between these two quantities.

Another interesting characteristic of ferromagnetic materials is revealed when the field intensity, having been increased to some value, say H_a, is subsequently decreased. It is found that the material opposes demagnetization and, accordingly, does not retrace along the magnetizing curve Oa but rather along a curve located above Oa. See curve ab in Fig. 1-9. Furthermore, it is seen that, when the field intensity is returned to zero, the flux density is no longer zero as was the case with the virgin sample. This happens because some of the domains remain oriented in the direction of the originally applied field. The value of B that remains after the field intensity H is removed is called *residual flux density*. Moreover, its value varies with the extent to which the material is magnetized. The maximum possible value of the residual flux density is called *retentivity* and results whenever values of H are used that cause complete saturation.

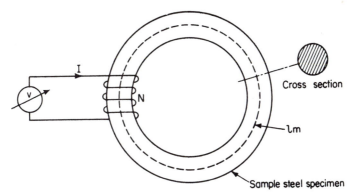

Figure 1-8   Obtaining the magnetization curve of a sample steel specimen.

Frequently, in engineering applications of ferromagnetic materials, the steel is subjected to cyclically varying values of H having the same positive and negative limits. As H varies through many identical cycles, the graph of B versus H gradually approaches a fixed closed curve as depicted in Fig. 1-9. The loop is always traversed in the direction indicated by the arrows. Since time is the implicit variable for these loops, note that *B is always lagging behind H*. Thus, when H is zero, B is finite and positive, as at point *b*, and when B is zero, as at *c*, H is finite and negative, and so on. This tendency of the flux density to *lag behind* the field intensity when the ferromagnetic material is in a symmetrically cyclically

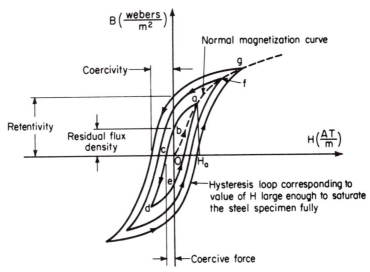

Figure 1-9   Typical hysteresis loops and normal magnetization curve.

magnetized condition is called *hysteresis*† and the closed curve *abcdea* is called a *hysteresis* loop. Moreover, when the material is in this cyclic condition the amount of magnetic field intensity required to reduce the residual flux density to zero is called the *coercive force*. Usually, the larger the residual flux density, the larger must be the coercive force. The maximum value of the coercive force is called the *coercivity*.

A glance at the hysteresis loops of Fig. 1-9 makes it quite evident that the flux density corresponding to a particular field intensity is not single-valued. Its value lies between certain limits depending upon the previous history of the ferromagnetic material. However, since in many situations involving magnetic devices this previous history is unknown, a compromise procedure is used in making magnetic calculations by working with a single-valued curve called the *normal magnetization curve*. This curve is found by drawing a curve through the tips of a group of hysteresis loops generated while in a cyclic condition. Such a curve is *Oafg* in Fig. 1-9. Typical normal magnetization curves of commonly used ferromagnetic materials appear in Fig. 1-10.

A final observation is in order at this point. By Eq. (1-8) the permeability of a material may be expressed as a ratio of B to H. Coupling this with the nonlinear variation existing between B and H (see Fig. 1-9) bears out the variation of permeability with flux density as already cited in connection with Fig. 1-7. As a matter of fact, for a material in a cyclic condition the permeability is nothing more than the ratio of B to H for the various points along the hysteresis loop.

1-4 THE MAGNETIC CIRCUIT: CONCEPT AND ANALOGIES

In general, problems involving magnetic devices are basically field problems because they are concerned with quantities such as Φ and B which occupy three-dimensional space. Fortunately, however, in most instances the bulk of the space of interest to the engineer is occupied by ferromagnetic materials except for small air gaps which are present either by intention or by necessity. For example, in electromechanical energy-conversion devices the magnetic flux must penetrate a stationary as well as a rotating mass of ferromagnetic material, thus making an air gap indispensable. On the other hand, in other devices an air gap may be intentionally inserted to mask the nonlinear relationship existing between B and H. But in spite of the presence of air gaps it happens that the space occupied by the magnetic field and the space occupied by the ferromagnetic material are practically the same. Usually this is because air gaps are made as small as mechanical clearance between rotating and stationary members will allow and also because the iron by virtue of its high permeability confines the flux to itself as copper wire confines electric current or a pipe restricts water. On this basis the three-dimensional field problem becomes a one-dimensional circuit problem and in accordance

---

†Derives from the Greek *hysterein* meaning to be behind or to lag.

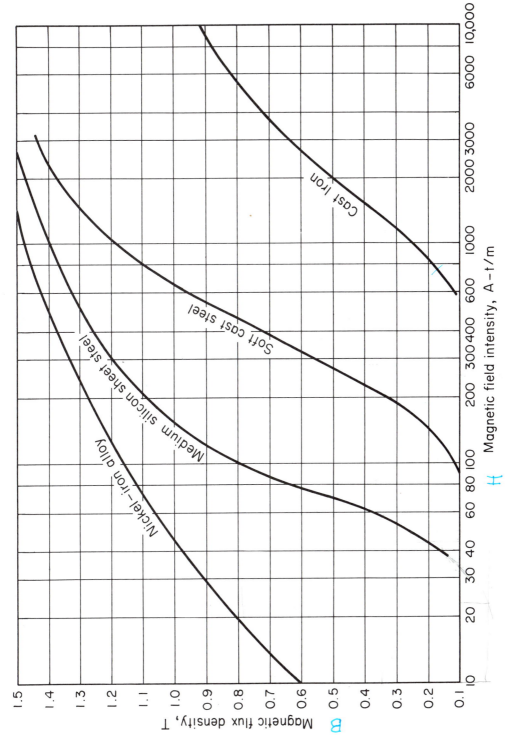

Figure 1-10 Magnetization curves of typical ferromagnetic materials.

17

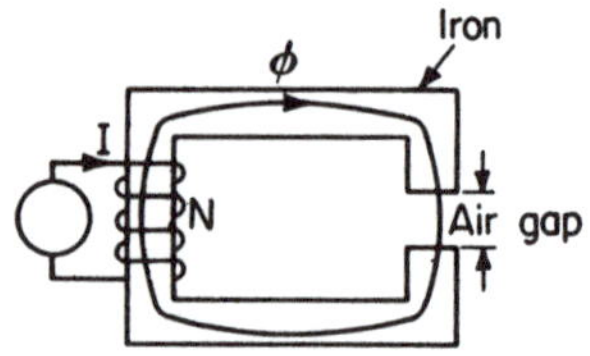

Figure 1-11 Typical magnetic circuit involving iron and air.

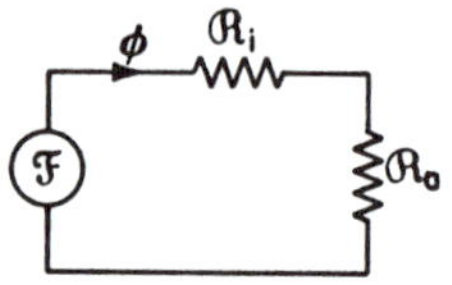

Figure 1-12 Single-line equivalent circuit of Fig. 1-11.

with Eq. (1-15) leads to the idea of a *magnetic circuit*. Thus we can look upon the magnetic circuit as consisting predominantly of iron paths of specified geometry which serves to confine the flux; air gaps may be included. Figure 1-11 shows a typical magnetic circuit consisting chiefly of iron. Note that the mmf of the coil produces a flux which is confined to the iron and to that part of the air having effectively the same cross-sectional area as the iron. Furthermore, a little thought reveals that this magnetic circuit may be replaced by a single-line equivalent circuit as depicted in Fig. 1-12. As suggested by Eqs. (1-14) and (1-15) the equivalent circuit consists of the magnetomotive force driving flux through two series-connected reluctances—$\mathcal{R}_i$, the reluctance of the iron, and $\mathcal{R}_a$, the reluctance of the air.

This analogy of the magnetic circuit with the electric circuit carries through in many other respects. For the sake of completeness these details are presented next for the case of a toroidal copper ring and a toroidal iron ring having the same mean radius r and cross-sectional area A.

Electric Case

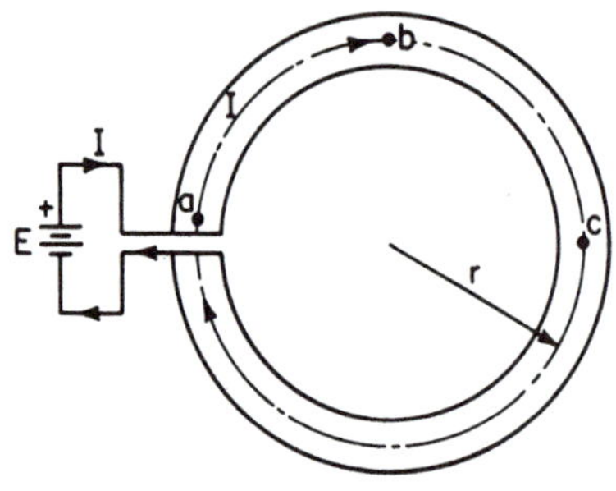

The toroidal copper ring is assumed open by an infinitesimal amount with the ends connected to a battery; a current of I amperes flows through the ring.

Magnetic Case

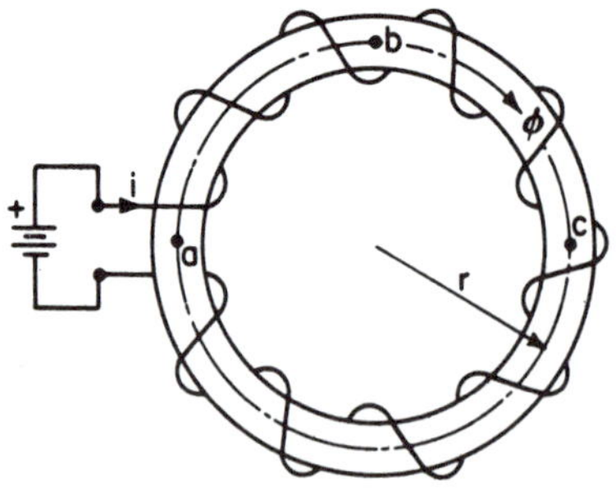

The toroidal iron ring is assumed to be wound with N turns of wire so that with a current i flowing through it the mmf creates the flux Φ.

Driving Force

applied battery voltage = E applied ampere-turns = $\mathcal{F}$

Response

current = $\dfrac{\text{driving force}}{\text{electric resistance}}$ flux = $\dfrac{\text{driving force}}{\text{magnetic reluctance}}$

or

$$I = \frac{E}{R} \qquad\qquad\qquad \Phi = \frac{\mathcal{F}}{\mathcal{R}} \qquad (1\text{-}19)$$

Impedance

Impedance is a general term used to indicate the impediment to a driving force in establishing a response.

resistance = $R = \rho\, \dfrac{l}{A}$ reluctance = $\mathcal{R} = \dfrac{l}{\mu A}$ (1-20)

where $l = 2\pi r$, the mean length of turn of the toroid, and A is the to-roidal cross-sectional area.

where $l = 2\pi r$, the mean length of turn of the toroid, and A is the to-roidal cross-sectional area.

Equivalent Circuit

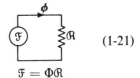

(1-21)

$$E = IR \qquad\qquad\qquad\qquad \mathcal{F} = \Phi\mathcal{R}$$

Electric Field Intensity

With the application of the voltage E to the homogeneous copper toroid, there is produced within the material an electric potential gradient given by

$$\varepsilon \equiv \frac{E}{l} = \frac{E}{2\pi r} \qquad \text{V/m}$$

This electric field must occur in a closed path if it is to be maintained. It then follows that the closed line integral of ε is equal to the battery voltage E. Thus

$$\oint \varepsilon\, dl = E$$

Magnetic Field Intensity

When a magnetomotive force is applied to the homogeneous iron toroid, there is produced within the material a magnetic potential gradient given by

$$H \equiv \frac{\mathcal{F}}{l} = \frac{\mathcal{F}}{2\pi r} \qquad \text{A-t/m} \quad (1\text{-}22)$$

As already pointed out in connection with Ampere's circuital law, the closed line integral of H equals the enclosed mmf. Thus

$$\oint H\, dl = \mathcal{F} \qquad (1\text{-}23)$$

Electric Potential Difference

If it is desired to find the voltage drop occurring between two points a and b of the copper toroid, we may write

$$V_{ab} = \int_a^b \varepsilon \, dl = \frac{E}{l} \int_a^b dl = \frac{IR}{l} \, l_{ab}$$

$$= \frac{I}{l} \rho \frac{l}{A} \, l_{ab} = I\rho \frac{l_{ab}}{A} = IR_{ab}$$

i.e.,

$$V_{ab} = IR_{ab}$$

where R_{ab} is the resistance of the copper toroid between points a and b.

Magnetic Potential Difference

This is the mmf drop appearing between two points when flux flows. Thus the portion of the total applied mmf appearing between points a and b is found similarly:

$$U_{ab} = \int_a^b H \, dl = \frac{\mathscr{F}}{l} \, l_{ab} = \frac{\Phi \mathscr{R}}{l} \, l_{ab}$$

$$= \frac{\Phi}{l} \frac{l}{\mu A} \, l_{ab} = \Phi \frac{l_{ab}}{\mu A} = \Phi \mathscr{R}_{ab}$$

$$U_{ab} = \Phi \mathscr{R}_{ab} \qquad (1\text{-}24)$$

where $\mathscr{R}_{ab}$ is the reluctance of the iron toroid between points a and b.

Current Density

By definition, current density is the number of amperes per unit area. Thus

$$J \equiv \frac{I}{A} = \frac{E}{AR} = \frac{\varepsilon l}{A\rho(l/A)} = \frac{\varepsilon}{\rho}$$

or

$$\varepsilon = \rho J$$

This last expression is often referred to as the *microscopic* form of Ohm's law.

Flux Density

Flux density is expressed as webers per unit area. Thus

$$B = \frac{\Phi}{A} = \frac{\mathscr{F}}{A\mathscr{R}} = \frac{Hl}{A(l/\mu A)} = \mu H$$

or

$$H = \frac{B}{\mu} \qquad (1\text{-}25)$$

It should not be inferred from the foregoing that electric and magnetic circuits are analogous in *all* respects. For example, there are no magnetic insulators analogous to those known to exist for electric circuits. Also, when a direct current is established and maintained in an electric circuit, energy must be continuously supplied. An analogous situation does not prevail in the magnetic case, where a flux is established and maintained constant.

1-5 UNITS FOR MAGNETIC CIRCUIT CALCULATIONS

Magnetic circuit calculations can be carried out by use of any one of several different systems of units. These various systems arose initially because it was thought that the phenomena of electricity and magnetism were unrelated—which lead to the development of a separate system of units for each—and, second, because there was a need to deal with practical values of the units once the relationship was discovered. Up to now attention has been given exclusively to the mks (meter-kilogram-second) system of units as developed by Giorgi about the turn of the twentieth century. This policy is prompted by the acceptance in 1960 of the mks system of units as the standard for scientific work and now referred to as SI units (Système International Unités). However, a good part of the past literature is written in the units of the cgs (centimeter-gram-second) system. Furthermore, many of the present-day computations are carried on in the *mixed* system employing such units as ampere-turns/inch, maxwells/inch², and ampere-turns because of the convenience they offer in dealing with dimensions that are expressed in inches. For these reasons the units of all three systems are shown in Table 1-1.

The weber, which is the unit of flux in the mks system, is equal to 10^8 *maxwells* (or lines), where the maxwell is the unit of flux in the cgs system. The *gilbert* is the cgs unit for mmf and is equal to 0.4π times the number of ampere-turns. The cgs unit for magnetic field intensity H is the *oersted* (or gilbert/cm) and the cgs unit for flux density B is the *gauss* (or lines/cm²). The relationships existing for the same quantity among the various systems of units are given in the last column of the table.

1-6 MAGNETIC CIRCUIT COMPUTATIONS

Basically magnetic circuit calculations involving ferromagnetic materials fall into two classes. In the first the value of the flux is known and it is required to find the mmf to produce it. This is the situation typical of the design of ac and dc electro-mechanical energy converters. On the basis of the desired voltage rating of an electric generator or the torque rating of an electric motor information about the required magnetic flux is readily obtained. Then with this knowledge and the configuration of the magnetic circuit the total mmf needed to establish the flux is determined straightforwardly. In the second case it is the flux for which we must solve, knowing the geometry of the magnetic circuit and the applied mmf. An engineering application in which this situation prevails is the magnetic amplifier, where it is often necessary to find the resultant magnetic flux caused by one or more control windings. Because the reluctance (or permeability) of the ferromagnetic material is not constant, the solution of this problem is considerably more involved than that of the first type as illustrated by the examples that follow.

TABLE 1-1 MAGNETIC UNITS

Quantity	Symbol	CGS Unit	CGS Relation	SI Unit	SI Symbol	SI Relation	Mixed English Unit	Mixed English Relation	Conversion factors
mmf	$\mathscr{F}$	gilberts	$\mathscr{F} = 0.4\pi NI$	ampere-turn	A-t	$\mathscr{F} = NI$	A-t	$\mathscr{F} = NI$	pragilbert = 10 gilberts A-t $= 0.4\pi$ gilberts $= 1.257$ gilberts
Permeability Free space Abs. norm. perm.	μ_0 μ		$\mu_0 = 1$ $\mu = \mu_0\mu_r$ $= \mu_r$			$\mu_0 = 4\pi10^{-7}$ $\mu = 4\pi10^{-7}\mu_r$		$\mu_0 = 3.19$ $\mu = 3.19\mu_r$	
Length	l	cm		meter	m		inch		1 centimeter = 0.01 meter 1 meter = 39.4 inches
Area	A	cm²		meter²	m²		inch²		1 meter² = 1550 inches²
Reluctance	$\mathscr{R}$		$\mathscr{R} = \dfrac{l}{\mu_r A}$			$\mathscr{R} = \dfrac{l}{4\pi10^{-7}\mu_r A}$		$\mathscr{R} = \dfrac{l}{3.19\mu_r A}$	
Flux	Φ	maxwells = lines	$\Phi = \dfrac{0.4\pi NI}{\dfrac{l}{\mu_r A}}$	weber	Wb	$\Phi = \dfrac{NI}{\dfrac{l}{\mu_r 4\pi10^{-7}A}}$	maxwells = lines	$\Phi = \dfrac{NI}{\dfrac{l}{3.19\mu_r A}}$	1 weber = 10^8 lines
Magnetic field intensity	H	gilb/cm = oersteds	$H = \dfrac{0.4\pi NI}{l}$	$\dfrac{\text{A-t}}{\text{meter}}$	$\dfrac{\text{A-t}}{\text{m}}$	$H = \dfrac{N\Phi}{l}$	$\dfrac{\text{A-t}}{\text{inch}}$	$H = \dfrac{NI}{l}$	1 oersted = 79.6 A-t/meter 1 praoersted = 100 oersteds $1\dfrac{\text{A-t}}{\text{inch}} = \dfrac{0.4\pi}{2.54} = 0.495$ oersted $1\dfrac{\text{A-t}}{\text{inch}} = \dfrac{2.02}{1000}$ praoersted
Flux density	B	lines/cm² = gausses	$B = \mu_r H$	$\dfrac{\text{webers}}{\text{meter}^2}$	T	$B = \mu_r\mu_0 H$	lines/inch²	$B = 3.19\mu_r H$	1 gauss = 6.45 lines/inch² $1\dfrac{\text{weber}}{\text{meter}^2} = 64{,}500$ lines/inch² $= 1$ tesla $1\dfrac{\text{weber}}{\text{meter}^2} = 10{,}000$ gausses $= 1$ tesla

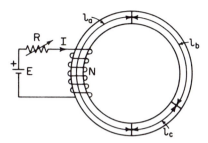

Figure 1-13   Toroid composed of three different materials.

Example 1-1

A toroid is composed of three ferromagnetic materials and is equipped with a coil having 100 turns as depicted in Fig. 1-13. Material a is a nickel-iron alloy having a mean arc length l_a of 0.3 m. Material b is medium silicon steel and has a mean arc length l_b of 0.2 m. Material c is of cast steel having a mean arc length equal to 0.1 m. Each material has a cross-sectional area of 0.001 m².

(a) Find the magnetomotive force needed to establish a magnetic flux of $\Phi = 6 \times 10^{-4}$ Wb = 60,000 lines.

(b) What current must be made to flow through the coil?

(c) Compute the relative permeability and reluctance of each ferromagnetic material.

Solution   (a) To obtain the total mmf of the coil all we need to do is to apply Ampere's circuital law. Thus

$$\mathscr{F} = U_a + U_b + U_c = H_a l_a + H_b l_b + H_c l_c$$

The unknown quantities here are H_a, H_b, and H_c. These can readily be found from a knowledge of the flux density, which here is the same for each section because the flux is common and the cross-sectional areas are the same. Hence

$$B_a = B_b = B_c = \frac{\Phi}{A} = \frac{0.0006}{0.001} = 0.6 \text{ T}$$

Now H_a is found by entering the B-H curve of the nickel-iron alloy of Fig. 1-10 corresponding to $B_a = 0.6$. This yields

$$H_a = 10 \text{ A-t/m}$$

Similarly

$$H_b = 77 \text{ A-t/m}$$

$$H_c = 320 \text{ A-t/m}$$

Accordingly, the total required mmf is

$$\mathscr{F} = H_a l_a + H_b l_b + H_c l_c$$

$$= 10(0.3) + 77(0.2) + 320(0.1)$$

$$= 3 + 15.4 + 32.0 = 50.4 \text{ A-t}$$

Note that, although the path length of cast steel is the smallest, it nonetheless requires the greatest portion of the mmf to force the specified flux through. This happens because of its much lower permeability as shown in part (c).

(b) In the SI system the mmf is equal to the number of ampere-turns. Hence

$$I = \frac{\mathcal{F}}{N} = \frac{50.4}{100} = 0.504 \text{ A}$$

(c) From Eq. (1-8)

$$\mu_a = \frac{B_a}{H_a} = \frac{0.6}{10} = 0.06 \text{ H/m}$$

Also

$$\mu_{ra} \mu_0 = \mu_a$$

$$\therefore \quad \mu_{ra} = \frac{\mu_a}{\mu_0} = \frac{0.06}{4\pi \times 10^{-7}} = 47{,}746$$

Furthermore, from Eq. (1-13) the reluctance is found to be

$$\mathcal{R}_a = \frac{\mathcal{F}_a}{\Phi} = \frac{3}{6 \times 10^{-4}} = 5000 \text{ rationalized mks units of reluctance}$$

Proceeding in a similar fashion for materials b and c leads to the following results:

$$\mu_{rb} = 6207 \qquad \mathcal{R}_b = 25{,}667$$

$$\mu_{rc} = 1492 \qquad \mathcal{R}_c = 53{,}333$$

Next we consider the more difficult problem: that of finding the flux in a given magnetic circuit corresponding to a specified mmf. The solution cannot be arrived at directly because, as a result of the nonlinear relationship between B and H, there are too many unknowns. The easiest way of finding the solution is to employ a cut-and-try procedure guided by a knowledge of the permeability characteristics of the materials such as appears in their magnetization curves. The following example illustrates the technique involved.

Example 1-2

For the toroid of Example 1-1, shown in Fig. 1-13, find the magnetic flux produced by an applied magnetomotive force of $\mathcal{F} = 35$ A-t.

The solution cannot be determined directly because to do so we must know the reluctance of each part of the magnetic circuit, which can be known only if the flux density is known—which means that Φ must be known right at the start. This is clearly impossible.

To obtain the solution by the cut-and-try procedure, we begin by first assuming that all of the applied mmf appears across the material having the highest reluctance. This yields an approximate value of Φ which can subsequently be refined. A glance at Fig. 1-10 shows that the poorest magnetic "conductor" is cast steel.

Hence by assuming the entire mmf to appear across material c we can find H, from which B follows, which in turn yields Φ. Thus

$$H_c = \frac{U_c}{l_c} = \frac{\mathcal{F}}{l_c} = \frac{35}{0.1} = 350 \text{ A-t/m}$$

From Fig. 1-10

$$B_c = 0.65 \text{ T}$$

$$\therefore \quad \Phi_1 = B_c A_c = 0.65(0.001) = 0.00065 \text{ Wb}$$

This value represents the first approximation for the flux as indicated by the subscript.

Also, since the cross-sectional area is the same for each material, it follows that

$$B_a = B_b = B_c = 0.65 \text{ T}$$

Reference to the nickel-iron magnetization curve reveals that the value of H_a corresponding to B_a is negligibly small compared with H_c. Hence for all practical purposes its effect can be neglected. However, note that for medium silicon steel the value of H_b is almost 81 A-t/m. This, coupled with the fact that $l_b = 2l_c$, indicates that material b takes about half as much mmf as material c in maintaining the flow of flux. In other words, at this point in our analysis we can make a refinement on our original assumption of assigning the entire mmf to material c. Now we see that about 50% of that assigned to c should be assigned to b. Thus

$$U_c + U_b = \mathcal{F} \qquad \text{(assumed)}$$

but

$$U_b = 0.5\,U_c \quad \text{(assumed)}$$

Hence

$$1.5\,U_c = \mathcal{F} = 35$$

$$\therefore \quad U_c = 23.3 \text{ A-t}$$

Accordingly a second approximation for the solution can be obtained. Therefore

$$H_c = \frac{23.3}{0.1} = 233 \text{ A-t/m}$$

which in turn yields

$$B_c = 0.4 \text{ T}$$

so that the value of the flux now becomes

$$\Phi_2 = B_c A_c = 0.0004 \text{ Wb}$$

To determine whether or not this is the correct answer we must at this point compute the mmf drops for each material and add to see whether they yield a value equal to the applied mmf. If not, the foregoing procedure must be repeated until Ampere's circuital law is satisfied. Making this check for the second approximation we have

$$H_b = 62 \text{ A-t/m} \qquad \text{corresponding to } B_b = 0.4 \text{ T}$$

and

$$H_a = 5.7 \text{ A-t/m} \qquad \text{for } B_a = 0.4 \text{ T}$$

Accordingly

$$mmf = H_a l_a + H_b l_b + H_c l_c$$
$$= 5.7(0.3) + 62(0.2) + 233(0.1)$$
$$= 1.7 + 12.4 + 23.3 = 37.4 \text{ A-t}$$

Obviously this is too high by about 7%. Hence, as a third try, reduce the biggest contributor to the mmf by a factor of 5%. That is, assume that now

$$U_c = 22 \text{ A-t}$$

Then

$$H_c = 220 \text{ A-t/m} \quad \text{and} \quad B_c = 0.375 \text{ T}$$
$$\therefore \quad \Phi_2 = 0.000375 \text{ Wb}$$

Corresponding to this flux we find

$$H_b = 59 \quad \text{and} \quad H_a = 5$$

Hence

$$mmf = 5(0.3) + 59(0.2) + 22 = 35.3 \text{ A-t}$$

Since this summation of mmf's agrees with the applied mmf of 35 A-t, the correct solution for the flux is

$$\Phi = 0.000375 \text{ Wb} = 37,500 \text{ lines}$$

When making magnetic circuit computations of the kind just illustrated, it is common practice to accept as valid any solution that comes within ±5% of the exact solution. The reason is that we are dealing with normal magnetization curves which neglect hysteresis and which are after all only *typical* of the material actually being used in the specified circuit. Deviations can and often do exist.

As a final example to illustrate magnetic circuit computations, we shall solve a problem involving parallel magnetic paths as well as the presence of an air gap. Moreover, we shall consider only the first type where the mmf needed to establish a specified flux is to be found. This is justified not only because the solution is straightforward but also because it is far more representative of the kind of magnetic circuit problem the engineer is likely to be concerned with.

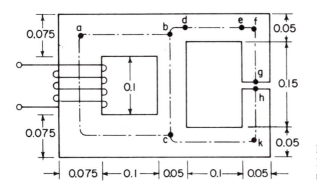

Figure 1-14 Magnetic circuit for Example 1-3. All dimensions are in meters.

Example 1-3

A magnetic circuit having the configuration and dimensions shown in Fig. 1-14 is made of cast steel having a thickness of 0.05 m and an air gap of 0.002 m length appearing between points g and h. The problem is to find the mmf to be produced by the coil in order to establish an air-gap flux of 4×10^{-4} Wb (or 40,000 lines).

The method of solution can be readily ascertained by referring to the equivalent circuit of this magnetic circuit as shown in Fig. 1-15. Knowledge of Φ_g enables us to find the mmf drop appearing across b and c. From this information the flux in leg bc can be determined and, upon adding it to Φ_g, we find the flux in leg cab. In turn, the mmf needed to maintain the total flux in leg cab can be computed, and when we add it to the mmf drop across bc we obtain the resultant mmf.

The computations involved for the various parts of the magnetic circuit are as follows:

Part gh. This is the air gap for which the flux is specified as 4×10^{-4} Wb. The cross-sectional area of the gap is $0.05(0.05) = 0.0025$ m². Normally, however, this area is slightly higher because of the tendency of the flux to bulge outward along the edges of the air gap—which is often referred to as *fringing*. For convenience this effect is neglected. Thus, the air-gap flux density is found to be

$$B_g = \frac{\Phi_g}{A_g} = \frac{4 \times 10^{-4}}{0.0025} = 0.16 \text{ T}$$

Since the permeability of air is practically the same as that of free space, we have

$$H_g = \frac{B_g}{\mu_0} = \frac{0.16}{4\pi \times 10^{-7}} = 127{,}300 \text{ A-t/m}$$

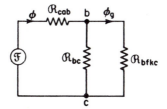

Figure 1-15 Equivalent circuit of Fig. 1-14.

and

$$U_g = H_g l_g = 127,300(0.002) = 255 \text{ A-t}$$

Part bg and hc. The length of ferromagnetic material involved here is

$$l_{bg} + l_{hc} = 2\left(l_{bd} + l_{de} + l_{ef} + \frac{l_{fk}}{2}\right) - l_g$$

$$= 2(0.025 + 0.1 + 0.025 + 0.1) - 0.002$$

$$= 0.498 \text{ m}$$

Moreover, corresponding to a flux density in the cast steel of 0.16 T, the field intensity H is found to be 125 A-t/m. Hence

$$U_{bg+hc} = 125(0.498) = 62.2 \text{ A-t}$$

Part bc. Because path *bfkc* is in parallel with path *bc*, the total mmf across path *bfkc* also appears across path *bc*. Hence

$$U_{bc} = 255 + 62.2 = 317.2 \text{ A-t}$$

Also

$$l_{bc} = 0.1 + 0.075 = 0.175 \text{ m}$$

$$\therefore \; H_{bc} = \frac{317.2}{0.175} = 1810 \text{ A-t/m}$$

And from Fig. 1-10 for cast steel the corresponding flux density is found to be

$$B_{bc} = 1.38 \text{ T}$$

Hence

$$\Phi_{bc} = 1.38(0.0025) = 0.00345 \text{ Wb}$$

Part cab. Accordingly, the total flux existing in leg *cab* is

$$\Phi_{cab} = \Phi_{bc} + \Phi_g = 0.00345 + 0.0004 = 0.00385 \text{ Wb}$$

Knowledge of this flux then leads to determination of the mmf needed in leg *cab* to sustain it. Thus

$$B_{cab} = \frac{0.00385}{0.00375} = 1.026 \text{ T}$$

From which

$$H_{cab} = 690 \text{ A-t/m}$$

Hence

$$U_{cab} = H_{cab} l_{cab} = 690(0.5) = 345 \text{ A-t}$$

Therefore the total mmf required to produce the desired air-gap flux is

$$\mathscr{F} = U_{cab} + U_{bc} = 345 + 317.2 = 662.2 \text{ A-t}$$

1-7 THE INDUCTANCE PARAMETER

Inductance is a characteristic of magnetic fields, and it was first discovered by Faraday in his renowned experiments of 1831. In a general way *inductance* can be characterized as that property of a circuit element by which energy is capable of being stored in a magnetic flux field. A significant and distinguishing feature of inductance, however, is that it makes itself felt in a circuit only when there is a *changing current* or flux. Thus, although a circuit element may have inductance by virtue of its geometrical and magnetic properties, its presence in the circuit is not exhibited unless there is a time rate of change of current. This aspect of inductance is particularly stressed when we consider it from the circuit viewpoint. However, for the sake of completeness, inductance is also treated from an energy and a physical view.

Circuit Viewpoint

The current-voltage relationship involving the inductance parameter is expressed by

$$v_L = L \frac{di}{dt} \tag{1-26}$$

In general both v_L and i are functions of time. Figure 1-16 depicts the potential difference v_L appearing across the terminals of the inductance parameter when a changing current flows into terminal c. Note that the arrowhead on the v_L quantity is shown at terminal c, indicating that this terminal is instantaneously positive with respect to terminal d. In turn, this means that current is increasing in the positive sense, i.e., the slope di/dt in Eq. (1-26) is positive. Any circuit element that exhibits the property of inductance is called an *inductor* and is denoted by the symbolism shown in Fig. 1-16. In the ideal sense the inductor is considered to be resistanceless, although practically it must contain the wire resistance out of which the inductor coil is formed.

It follows from Eq. (1-26) that an appropriate defining equation for inductance is

$$\boxed{L = \frac{v_L}{di/dt}} \qquad \text{volt-sec/ampere, or henrys (H)} \tag{1-27}$$

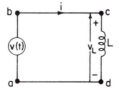

Figure 1-16   Circuit with pure inductance parameter.

Thus by recording the potential difference at a given time instant across the terminals of an inductor and dividing by the corresponding derivative of the current time function we determine the inductance parameter. Note that the units of inductance are volt-seconds per ampere. For simplicity this is commonly called the *henry*.

A *linear inductor* is one for which the inductance parameter is independent of current. As current flows through an inductor it creates a space flux. When this flux permeates air, strict proportionality between current and flux prevails so that the inductance parameter stays constant for all values of current. A plot of the potential difference across the coil as a function of the derivative of the current then appears as shown in Fig. 1-17, which is a plot of Eq. (1-27). When the flux is made to penetrate iron, however, it is possible for large currents to upset the proportional relationship between the current and the flux it produces. In such a case the inductor is called *nonlinear* and a plot of Eq. (1-27) will then no longer be a straight line.

For the resistance parameter Ohm's law can be written to express either voltage in terms of current or current in terms of voltage. The same procedure may be followed for the inductance parameter. Equation (1-26) already expresses the voltage as a function of the current. However, to express the current in terms of the potential difference across the inductor, Eq. (1-26) must be transposed to read as follows:

$$di = \frac{1}{L} v_L \, dt \qquad (1\text{-}28)$$

In integral form this becomes

$$\int_{i(0)}^{i(t)} di = \frac{1}{L} \int_0^t v_L \, dt \qquad (1\text{-}29)$$

or

$$\boxed{i(t) = \frac{1}{L} \int_0^t v_L \, dt + i(0)} \qquad (1\text{-}30)$$

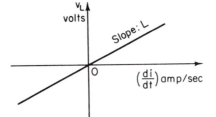

Figure 1-17 Graphical representation of the inductance parameter L, from the circuit viewpoint.

Equation (1-30) thus reveals that the current in an inductor is dependent upon the integral of the voltage across its terminals as well as the initial current in the coil at the start of integration.

An examination of Eqs. (1-26) and (1-30) reveals an important property of inductance: *The current in an inductor cannot change abruptly in zero time.* This is made apparent from Eq. (1-26) by noting that a finite change in current in zero time calls for an infinite voltage to appear across the inductor, which is physically impossible. On the other hand, Eq. (1-30) shows that in zero time the contribution to the inductor current from the integral term is zero so that the current immediately before and after application of voltage to the inductor is the same. In this sense, then, we may look upon inductance as exhibiting the property of inertia.

Example 1-4

An inductor has a current passing through it which varies in time in the manner depicted in Fig. 1-18(a). Find the corresponding time variation of the voltage drop appearing across the inductor terminals if it is assumed that the inductance of the coil is 0.1 H.

The solution appears in Fig. 1-18(b). Note that, in the interval from 0 to 0.1 s, $di/dt = 100$ A/s. Hence the voltage across the coil is then a constant given by

$$v_L = L \frac{di}{dt} = 0.1(100) = 10 \text{ V} \quad \text{for } 0 < t < 0.1$$

In the time range from 0.1 to 0.3 s the slope of the current curve is -50. Hence the voltage across the coil is -5 V. Finally, in the interval from 0.3 to 0.6 s the current wave shape is sinusoidal so that the corresponding voltage wave is a cosine. In drawing the cosine wave it is assumed that the maximum slope of the sine wave exceeds 100 A/s.

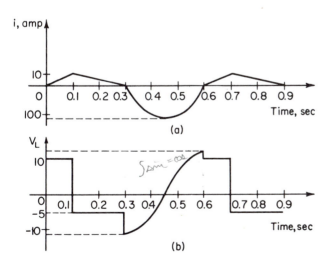

(a)

(b)

Figure 1-18 (a) Input current waveshape to an inductor; (b) corresponding voltage variation across the inductor terminal.

It is interesting to note in Fig. 1-18 that, unlike the current in an inductor, the voltage across the inductor is allowed to change discontinuously.

Energy Viewpoint

Assume that an inductor has zero initial current. Then, if a current i is made to flow through the coil across which appears the potential difference v_L, the total energy received in the time interval from 0 to t is

$$W = \int_0^t v_L i \, dt \quad \text{joules (J)} \tag{1-31}$$

Inserting Eq. (1-26) leads to

$$W = \int_0^t \left(L \frac{di}{dt} \right) i \, dt = \int_0^i Li \, di \tag{1-32}$$

or

$$\boxed{W = \tfrac{1}{2}Li^2} \quad \text{J} \tag{1-33}$$

Continuing with the assumption that the inductor has no winding resistance, Eq. (1-33) states that the inductor absorbs an amount of energy which is proportional to the inductance parameter L as well as the square of the instantaneous value of the current. Thus energy is stored by the inductor in a magnetic field. It is of finite value and retrievable. As the current is increased, so too is the stored magnetic energy. Note, however, that the energy is zero whenever the current is zero. Because the energy associated with the inductance parameter increases and decreases with the current, we can properly conclude that the inductor has the property of being capable of returning energy to the source from which it receives it.

A glance at Eq. (1-33) indicates that an alternative way of identifying the inductance parameter is in terms of the amount of energy stored in its magnetic field corresponding to its instantaneous current. Thus, in mathematical form we can write

$$\boxed{L = \frac{2W}{i^2}} \quad \text{H} \tag{1-34}$$

This is an energy description of the inductance parameter.

There is one final point worthy of note. It has already been demonstrated that for a potential difference to exist across the inductor terminals the current must be changing. A constant current results in a zero voltage drop across the ideal inductor. This is not true, however, about the energy absorbed and stored in the magnetic field of the inductor. Equation (1-33) readily verifies this fact. A constant current results in a fixed energy storage. Any attempt to alter this energy state is firmly resisted by the effects of the initial energy storage. This again reflects the inertial aspect of inductance.

Physical Viewpoint

The voltage across the terminals of an inductor may be expressed from a circuit viewpoint by Eq. (1-26). However, this same voltage may be described by Faraday's law in terms of the flux produced by the current and the number of turns N of the inductor coil. Accordingly, we may write

$$v_L = L\frac{di}{dt} = N\frac{d\phi}{dt} \tag{1-35}$$

It then follows that

$$L = N\frac{d\phi}{di} \tag{1-36}$$

In those cases, where the flux ϕ is directly proportional to current i for all values (i.e., for linear inductors), the last expression becomes

$$L = \frac{N\Phi}{i} \qquad \frac{\text{Wb-t}}{\text{A}} \text{ or } \text{H} \tag{1-37}$$

Here the inductance parameter has a hybrid representation because it is in part expressed in terms of the circuit variable i and in part in terms of the field variable Φ. To avoid this we replace flux by its equivalent, i.e.,

$$\Phi = \frac{\text{mmf}}{\text{magnetic reluctance}} = \frac{Ni}{\mathcal{R}} \tag{1-38}$$

where *mmf* is the magnetomotive force which produces the flux Φ in the magnetic circuit having a reluctance $\mathcal{R}$. Appearing in Fig. 1-19 is an inductor consisting of N turns wound about a circular iron core. If the core is assumed to have a mean length of l meters and a cross-sectional area of A_m meters^2, then the magnetic reluctance can be written as

$$\mathcal{R} = \frac{l}{\mu A_m} \tag{1-39}$$

where μ is the permeability, a physical property of the magnetic material.

Upon substituting Eqs. (1-38) and (1-39) in (1-37) there results the expression for the inductance parameter of the circuit of Fig. 1-19. Thus

$$L = \frac{N^2\mu A_m}{l} = \frac{N^2}{\mathcal{R}} \tag{1-40}$$

A study of Eq. (1-40) reveals some interesting and useful facts about the inductance parameter which are not readily available when this quantity is defined

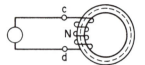

Figure 1-19   Linear inductor with iron core.

either from the circuit or the energy viewpoint. Most impressive is the fact that inductance, like resistance, is dependent upon the geometry of physical dimensions and the magnetic property of the medium. This is significant because it tells us what can be done to change the value of L. Thus, for the inductor illustrated in Fig. 1-19, the inductance parameter may be increased in any one of four ways: increasing the number of turns, using an iron core of higher permeability, reducing the length of the iron core, and increasing the cross-sectional area of the iron core.

It is interesting to note that neither the circuit viewpoint nor the energy viewpoint could tell us these things, because essentially they deal with the effects associated with a given inductor geometry. It is emphasized that all three viewpoints are needed to complete the picture of this circuit element.

1-8 HYSTERESIS AND EDDY-CURRENT LOSSES IN FERROMAGNETIC MATERIALS

The process of magnetization and demagnetization of a ferromagnetic material in a symmetrically cyclic condition involves a storage and release of energy which is not completely reversible. As the material is magnetized during each half-cycle, it is found that the amount of energy stored in the magnetic field exceeds that which is released upon demagnetization. The background for understanding this behavior was provided in Sec. 1-3. There the hysteresis loop was identified as the

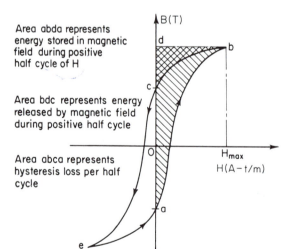

Area abda represents energy stored in magnetic field during positive half cycle of H

Area bdc represents energy released by magnetic field during positive half cycle

Area abca represents hysteresis loss per half cycle

Figure 1-20   Hysteresis loop and energy relationships per half-cycle.

variation of flux density as a function of the magnetic field intensity for a ferromagnetic material in a cyclic condition. The salient feature of the hysteresis loop is the delayed reorientation of the domains in response to a cyclically varying magnetizing force. A single hysteresis loop is depicted in Fig. 1-20. The direction of the arrows on this curve indicates the manner in which B changes as H varies from zero to a positive maximum through zero to a negative maximum and back to zero again, thus completing the loop. To appreciate the meaning of the various shaded areas shown in Fig. 1-20, let us look at the units associated with the product of B and H. Thus

$$\text{units of } HB = \frac{\text{amperes}}{\text{meter}} \times \frac{\text{newtons}}{\text{ampere-meter}} = \frac{\text{newtons}}{\text{meter}^2} = \frac{N}{m^2}$$

But

$$\text{newton-meter} = \text{joule}$$

Hence

$$\text{units of } HB = \frac{\text{joules}}{\text{meter}^3} = \frac{J}{m^3}$$

which is clearly recognized as an energy density. Therefore, in dealing with areas involving B and H in connection with a hysteresis loop we are really dealing with energy densities expressed on a per-cycle basis because the hysteresis loop is repeatable for each cyclic variation of H.

The foregoing is more precisely described by dealing with the energy as the integral of power with respect to time. Thus,

$$W(\text{energy in } J) = \int_{t_a}^{t_b} p \, dt$$

But,

$$p = ei = N \left(\frac{d\phi}{dt} \right) i$$

so that

$$W = \int_{\phi_a}^{\phi_b} Ni \left(\frac{d\phi}{dt} \right) dt = \int_{\phi_a}^{\phi_b} Ni \, d\phi \qquad (1\text{-}41)$$

Also,

$$i = \frac{\mathcal{F}}{N} = \frac{Hl}{N} \quad \text{and} \quad \phi = BA \quad \text{for } A \text{ fixed.}$$

Inserting these last expressions into Eq. (1-41) yields

$$W = \int_{B_a}^{B_b} N \left(\frac{Hl}{N} \right) A \, dB = Al \int_{B_a}^{B_b} H \, dB \qquad (1\text{-}42)$$

Hence the energy density is

$$w = \frac{W}{Al} = \int_{B_a}^{B_b} H \, dB \tag{1-43}$$

The energy stored in the magnetic field during that portion of the cyclic variation of H when it increases from zero to its positive maximum value (assuming the material is already in a cyclic state) is given specifically by

$$w_1 = \int_{B_a}^{B_b} H \, dB \qquad J/m^3 \tag{1-44}$$

when mks units are used; i.e., H must be expressed in A-t/m and B in webers per square meter. Note that the axes of Fig. 1-20 are so labeled. Moreover, during that portion of its cyclic variation when H decreases from its positive maximum value to zero (as it follows along curve bc of the hysteresis loop), energy is being released by the magnetic field and returned to the source, and this quantity can be represented as

$$w_2 = \int_{B_b}^{B_c} H \, dB \qquad J/m^3 \tag{1-45}$$

In this equation, since $B_b > B_c$, the quantity w_2 will be negative, indicating that the energy is being released rather than stored by the magnetic field.

A graphical interpretation of Eq. (1-44) leads to the result that the energy absorbed by the field when H is increasing in the positive direction can be represented by the area $abdca$. Similarly the energy released by the field as H varies from H_{max} to zero can be represented by area $bdcb$. The difference between these two energy densities represents the amount of energy which is not returned to the source but, rather, is dissipated as heat as the domains are realigned in response to the changing magnetic field intensity. This dissipation of energy is called *hysteresis loss*. Keep in mind that Fig. 1-20 depicts this energy density loss for a half-cycle variation of H. Hence area $abca$ represents the hysteresis loss per half-cycle. It certainly follows from symmetry that upon completion of the negative half-cycle variation of H an equal energy loss occurs. Therefore, as H varies over one complete cycle, the total energy loss per cubic meter is represented by the area of the hysteresis loop. More specifically this energy loss per cycle can be expressed mathematically as

$$\boxed{w_h = \text{(area of hysteresis loop)}} \qquad J/m^3/cycle \tag{1-46}$$

where mks units are used for H and B.

It is frequently desirable to express the hysteresis loss of ferromagnetic materials in watts—the unit of power. A little thought about the units of w_h in Eq. (1-46) shows how this can be directly accomplished. Thus

$$w_h = \frac{\text{energy}}{\text{vol} \times \text{cycles}} = \frac{\text{power} \times \text{seconds}}{\text{vol} \times \text{cycles}} = \frac{\text{power}}{\text{vol.} \times \text{cycles/second}} \quad (1\text{-}47)$$

Now let P_h = power loss, watts (W)
   v = volume of ferromagnetic material
   f = hertz (Hz) = frequency of variation of H

Then Eq. (1-47) becomes

$$w_h = \frac{P_h}{vf} \quad (1\text{-}48)$$

or

$$\boxed{P_h = w_h vf} \quad (1\text{-}49)$$

where w_h—the energy density loss—is determined from Eq. (1-46).

To obviate the need of finding the area of the hysteresis loop in order to compute the hysteresis loss in watts from Eq. (1-49), Steinmetz obtained an empirical formula for w_h based on a large number of measurements for various ferromagnetic materials. He expressed the hysteresis power loss as

$$\boxed{P_h = vf K_h B_m^n} \quad (1\text{-}50)$$

where B_m is the maximum value of the flux density and n lies in the range $1.5 \le n \le 2.5$ depending upon the material used. The parameter K_h also depends upon the material. Some typical values are: cast steel 0.025, silicon sheet steel 0.001, and permalloy 0.0001.

In addition to the hysteresis power loss, another important loss occurs in ferromagnetic materials that are subjected to time-varying magnetic fluxes—the *eddy-current* loss. This term is used to describe the power loss associated with the circulating currents that are found to exist in closed paths within the body of a ferromagnetic material and cause an undesirable heat loss. These circulating currents are created by the differences in magnetic potential existing throughout the body of the material owing to the action of the changing flux. If the magnetic circuit is composed of solid iron, the ensuing power loss is appreciable because the circulating currents encounter relatively little resistance. To increase significantly the resistance encountered by these eddy currents the magnetic circuit is invariably composed of very thin *laminations* (usually 14 to 25 mils thick) whenever the electromagnetic device is such that a varying flux permeates it in normal operation. This is the case with transformers and all ac electric motors and generators. An empirical equation for the eddy-current loss is

$$\boxed{P_e = K_e f^2 B_m^2 \tau^2 v} \qquad \text{W} \quad (1\text{-}51)$$

where K_e = a constant dependent upon the material
      f = frequency of variation of flux, Hz
    B_m = maximum flux density
     τ = lamination thickness
     ν = total volume of the material

A comparison of this equation with Eq. (1-50) reveals that eddy-current losses vary as the square of the frequency, whereas the hysteresis loss varies directly with the frequency.

Taken together the hysteresis and eddy-current losses constitute what is frequently called the *core losses* of electromagnetic devices that involve time-varying fluxes for their operation. More than just passing attention is devoted to these losses here because, as will be seen, core losses have an important bearing on temperature rise, efficiency, and rating of electromagnetic devices.

1-9 RELAYS—AN APPLICATION OF MAGNETIC FORCE

A *relay* is an electromagnetic device which can often be activated by relatively little energy, causing a movable ferromagnetic armature to open or close one or several pairs of electrical contact points located in another control circuit or a main circuit handling large amounts of energy. Alternating- and direct-current motor starters are equipped with relays designed to ensure proper operation of motors during starting and running conditions. These devices are found in many applications in all fields of engineering, especially in situations where control of a process or machine is involved. Our objective in this section is to describe the principles that underlie the operation of these electromagnetic devices. Besides the knowledge it offers, this treatment gives the motivation for studying the theory of magnetic fields and circuits.

The derivation of a magnetic force equation is our primary interest because it shows how it is possible to do mechanical work—moving a relay armature—by abstracting energy from that stored in the magnetic field. Moreover, since the emphasis is on the principles involved, the simplifying assumptions of no saturation and no losses are imposed; i.e., linear analysis is used throughout. Accordingly, the magnetization curve of the ferromagnetic material is assumed to be a straight line as depicted in Fig. 1-21(a). (Only the curve corresponding to positive values of H is shown.) Now, by appropriately modifying the axes of the curve plotted in Fig. 1-21(a), a significant and useful thing happens. First recall that the area between the B axis and the magnetization curve represents the energy absorbed from the source and stored in the magnetic field on a per-unit-volume basis. Equation (1-44) states this result mathematically.

$$w_f = \int_{B_a}^{B_b} H \, dB \qquad \text{J/m}^3 \qquad (1\text{-}52)$$

In the simplified situation of Fig. 1-21(a), Eq. (1-52) becomes merely the area of

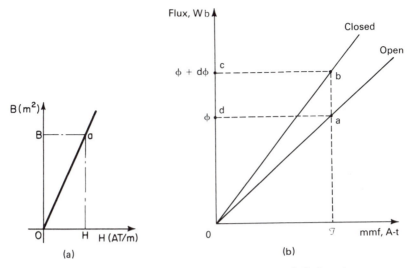

Figure 1-21   Linear magnetization curve of a relay. (a) Area OaB gives the energy density corresponding to fixed H. (b) Area OaB gives the total stored magnetic field energy corresponding to fixed $\mathcal{F}$.

triangle OaB. Thus

$$w_f = \tfrac{1}{2}BH \qquad J/m^3 \tag{1-53}$$

where H is assumed fixed at the value shown.

Moreover, since w_f is an energy density, the total energy stored in the magnetic field is found by multiplying Eq. (1-44) by the volume. Thus

$$W_f = w_f \nu = w_f Al \qquad J \tag{1-54}$$

where l is the length and A the cross-sectional area of the magnetic circuit. Inserting Eq. (1-53) into Eq. (1-54) yields

$$\boxed{W_f = \tfrac{1}{2}(BA)Hl = \tfrac{1}{2}\Phi\mathcal{F}} \qquad J \tag{1-55}$$

Note that in this equation A is combined with B to identify the flux Φ and H is combined with l to identify the mmf $\mathcal{F}$. A graphical representation of Eq. (1-55) appears in Fig. 1-21(b). It should be apparent that Fig. 1-21(b) derives from Fig. 1-21(a) by multiplying the ordinate axis by A and the abscissa axis by l, thus plotting Φ versus $\mathcal{F}$. Then for a fixed H (or $\mathcal{F}$) area OaB of Fig. 1-21(a) gives the energy density, whereas the corresponding area Oad of Fig. 1-21(b) gives the total energy stored in the magnetic field.

To understand how mechanical work can be done by the abstraction of energy stored in the magnetic field, consider the circuitry appearing in Fig. 1-22, which depicts the basic composition of an electromagnetic relay. It consists of an exciting coil placed on a fixed ferromagnetic core equipped with a movable ele-

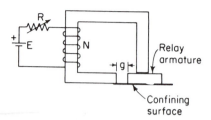

Figure 1-22 Basic composition of an electromagnetic relay.

ment called the *relay armature*. The relay is energized from a constant voltage source through an adjustable resistor R. To begin with, consider that R is fixed at that value which makes the coil mmf equal to $\mathscr{F}$ and produces the flux Φ as shown in Fig. 1-21(b) and that initially the relay armature is prevented from movement by being held fast. If the relay armature is now allowed to move to a smaller air gap position, the magnetization curve, *Oa*, moves to a steeper position, *Ob*, consistent with the lower reluctance of the magnetic circuit. It is important to understand that in the arrangement of Fig. 1-22, operation of the relay from the open to the closed position occurs under conditions of constant mmf because for a constant-source voltage *the coil current is determined by the circuit resistance and not by the reluctance of the magnetic circuit*. The lower reluctance of the magnetic circuit in the closed armature position simply means that the given mmf produces more flux. The increase in flux is indicated in Fig. 1-21(b) by the quantity $d\phi$.

The area of rectangle *abcd* in Fig. 1-21(b) represents an increase of energy (dW_i) to the relay as it goes from the open to the closed state. Quantitatively, it can be described by

$$dW_i = \mathscr{F}\, d\phi \tag{1-56}$$

Energy that is applied to the *relay* through the input circuit can be distributed in two forms: as stored magnetic field energy and as work done by the relay in moving from an open to a closed state. Of the total change in relay input energy, how much is expended to do the mechanical work in moving from the open to the closed position and how much is used to increase the initial field energy to a new total value? The latter quantity is readily determined by the area of triangle *Obc* in Fig. 1-21(b). Thus

$$(W_f)_{\text{relay closed}} = \tfrac{1}{2}\mathscr{F}(\phi + d\phi) = \tfrac{1}{2}\mathscr{F}\phi + \tfrac{1}{2}\mathscr{F}\, d\phi \tag{1-57}$$

Since the first term on the right side is the field energy corresponding to the open position, it follows that *half* of the energy represented by Eq. (1-56) goes to augment the stored field energy. Of course, the remaining half serves to do mechanical work which can be represented in differential form by ($F\,dx$) where x denotes distance. It is instructive to note that this mechanical work is represented in Fig. 1-21(b) by triangle *Oab*. An expression for the relay force can be readily derived by equating these two terms. Thus,

$$F \, dx = \tfrac{1}{2} \mathscr{F} \, d\phi = \tfrac{1}{2} \mathscr{F} \, d \left(\frac{\mathscr{F}}{\mathscr{R}} \right) = \tfrac{1}{2} \mathscr{F}^2 \, d \left(\frac{1}{\mathscr{R}} \right) \tag{1-58}$$

For convenience introduce

$$\mathscr{P} \equiv \frac{1}{\mathscr{R}} = \text{permeance of the magnetic circuit} \tag{1-59}$$

Then,

$$F \, dx = \tfrac{1}{2} \mathscr{F}^2 \, d\mathscr{P}$$

Or,

$$F = \tfrac{1}{2} \mathscr{F}^2 \frac{d\mathscr{P}}{dx} \qquad N \tag{1-60}$$

Equation (1-60) states that the force on the relay armature always acts to increase the permeance of the magnetic circuit, which is equivalent to saying that F acts to reduce the reluctance.

When the relay is energized from an ac source, the flux must remain essentially constant irrespective of the reluctance of the magnetic circuit (see Chap. 3). In such a situation the current automatically adjusts to meet the magnetomotive needs of the magnetic circuit. As the relay armature moves in circumstances that require operation at constant flux, the B-H curve for the magnetic circuit at the open and closed relay positions can now be depicted by Fig. 1-23. Here too the energy to move the relay armature from the open to the closed position is indicated by the area of triangle Oab in Fig. 1-23. Expressed mathematically we have

mechanical work on relay armature = reduced magnetic field energy

= area of triangle Oab

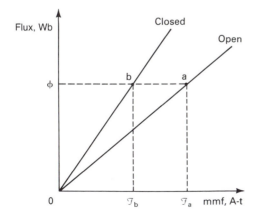

Figure 1-23 Flux and mmf conditions at open and closed positions of the relay armature when the relay is excited from an ac source that demands operation at constant flux.

Or

$$F \, dg = \tfrac{1}{2}\phi(\mathcal{F}_b - \mathcal{F}_a) \tag{1-61}$$

where

$$\mathcal{F}_a = \phi\mathcal{R}_a \quad \text{and} \quad \mathcal{F}_b = \phi\mathcal{R}_b \tag{1-62}$$

and where $\mathcal{R}_a$ is the reluctance of the magnetic circuit with the relay in the open position and $\mathcal{R}_b$ denotes the reluctance in the closed position. Moreover, $\mathcal{R}_a > \mathcal{R}_b$. Inserting Eq. (1-62) into Eq. (1-61) yields

$$F \, dg = \tfrac{1}{2}\phi(\phi\mathcal{R}_b - \phi\mathcal{R}_a) = \tfrac{1}{2}\phi^2(\mathcal{R}_b - \mathcal{R}_a) \tag{1-63}$$

We can now introduce

$$d\mathcal{R} = -(\mathcal{R}_a - \mathcal{R}_b) \tag{1-64}$$

as the change in reluctance from the open to the closed position. The negative sign is in recognition of the fact that a decrease in reluctance occurs. Substituting Eq. (1-64) into Eq. (1-63) yields finally

$$\boxed{F = -\tfrac{1}{2}\phi^2 \frac{d\mathcal{R}}{dg}} \qquad N \tag{1-65}$$

Equation (1-65) states that the action of the stored magnetic field energy is to reduce the reluctance of the magnetic circuit, a result which is consistent with the conclusion in Eq. (1-60).

Example 1-5

In the relay circuit of Fig. 1-22 assume the cross-sectional area of the fixed core and the relay armature to be A and the circuit mmf to be $\mathcal{F}$. Neglecting the reluctance of the iron, find the expression for the magnetic force existing on the relay armature.

Solution   The force is easily found by invoking Eq. (1-60). Thus,

$$F = \tfrac{1}{2}\mathcal{F}^2 \frac{d\mathcal{P}}{dx}$$

where

$$\mathcal{P} = \frac{1}{\mathcal{R}} = \frac{\mu_0 A}{g}$$

Hence,

$$F = \tfrac{1}{2}\mathcal{F}^2 \frac{d}{dx}\left(\frac{\mu_0 A}{g}\right)$$

Next introduce

$$dx = -dg$$

since movement of the relay armature reduces the gap length. Accordingly, the force on the relay becomes

$$F = -\tfrac{1}{2}\mathscr{F}^2 \frac{d}{dg}\left(\frac{\mu_0 A}{g}\right) = \tfrac{1}{2}\mathscr{F}^2 \frac{\mu_0 A}{g^2} \qquad N$$

An alternative expression for this force can be found by rewriting the last equation as

$$F = \tfrac{1}{2}\mathscr{F}^2\left(\frac{\mu_0 A}{g}\right)\frac{1}{g} = \tfrac{1}{2}\mathscr{F}^2 \frac{1}{\mathscr{R}}\frac{1}{g} = \tfrac{1}{2}\phi^2 \frac{\mathscr{R}^2}{\mathscr{R}}\frac{1}{g} = \tfrac{1}{2}\phi^2\mathscr{R}\frac{1}{g}$$

$$= \tfrac{1}{2}\phi^2 \frac{g}{\mu_0 A}\frac{1}{g} = \tfrac{1}{2}\frac{\phi^2}{\mu_0 A} \qquad N$$

Example 1-6

The cross-sectional view of a cylindrical plunger magnet appears in Fig. 1-24. The plunger (or armature) is free to move inside a nonferromagnetic guide around which the coil is surrounded by a cylindrically shaped steel shell. The plunger is separated from the shell by an air gap of length g.

(a) Derive the expression for the magnetic force exerted on the plunger when it is in the position shown in Fig. 1-24. The length of the plunger is at least equal to that of the shell. Also it has a radius of a meters. Neglect the reluctance of the steel.

(b) Find the magnitude of the force when the mmf is 1414 A-t and the plunger magnet dimensions are

$$x = 0.025 \text{ m} \qquad h = 0.05 \text{ m}$$

$$a = 0.025 \text{ m} \qquad g = 0.00125 \text{ m}$$

Solution   Before proceeding with the calculations let us review the principle that gives rise to the force. For the direction of coil current assumed, a typical flux path is that shown in Fig. 1-24. Note that the flux path must cross the air gap twice. It is particularly important to note too that the reluctance "seen" by the flux as it crosses the bottom gap is less than that seen by the same flux crossing the gap at the upper part of the plunger because the cross-sectional area of the magnetic circuit is smaller at the upper part than it is at the lower part of the plunger. As a matter of fact, at the lower part of the plunger the cross-sectional area seen by the flux is fixed for all

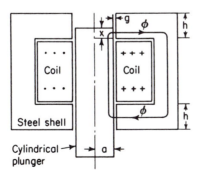

Cylindrical plunger

Figure 1-24   Plunger magnet for Example 1-6.

positions of the plunger lying in the range $0 \le x \le h$ and specifically is equal to $2\pi ah$ (assuming g is small compared to a). Accordingly, a force is created on the plunger so directed as to decrease the reluctance. This means that the magnetic force acts to move the plunger upward.

(a) The solution of this part is obtained from Eq. (1-65), but first we need to identify the correct expression for the reluctance as seen by the flux. The reluctance of the steel is negligible. Hence the reluctance associated with the typical flux shown in Fig. 1-24 is merely the sum of the reluctances associated with each air gap. Thus

$$\mathcal{R} = \frac{g}{\mu_0 2\pi ah} + \frac{g}{\mu_0 2\pi ax} = \frac{g}{2\pi a\mu_0}\left(\frac{1}{h} + \frac{1}{x}\right) \tag{1-66}$$

Then

$$\frac{d\mathcal{R}}{dx} = \frac{g}{2\pi a\mu_0}\left(-\frac{1}{x^2}\right) \tag{1-67}$$

Also

$$\Phi = \frac{\mathcal{F}}{\mathcal{R}} = \frac{\mathcal{F}}{g/\mu_0 2\pi a}\frac{xh}{h + x} \tag{1-68}$$

Inserting Eqs. (1-67) and (1-68) into Eq. (1-65) yields

$$F = \pi a\mu_0 \frac{\mathcal{F}^2 h^2}{g}\left(\frac{1}{x + h}\right)^2 = 3.94 \times 10^{-6}\frac{ah^2}{g}\mathcal{F}^2\left(\frac{1}{x + h}\right)^2 \quad \text{N} \tag{1-69}$$

(b) Inserting the specified dimensions into Eq. (1-69) gives the magnetic force as

$$F = 70 \text{ N} = 15.74 \text{ lb}$$

PROBLEMS

1-1. A long straight wire located in air carries a current of 4 A. Assume the relative permeability of air is unity.

(a) Find the value of the magnetic field intensity at a distance of 0.5 m from the center of the wire.

(b) A second long straight wire carrying a current of 2 A is placed parallel to the first one at a distance of 0.5 m with the current flowing in the same direction. Find the direction and magnitude of the force per meter existing between the wires.

(c) Repeat part (b) for wires that are embedded in iron having a relative permeability of 10,000 and a spacing of 0.05 m.

1-2. The wires shown in Fig. P1-2 are long, straight, and parallel and are completely embedded in iron having a relative permeability of 1000. Each wire carries a current of 10 A.

(a) Compute the magnitude and direction of the resultant force per meter on the wire in which I_2 flows.

(b) Compute the magnitude and direction of the force per meter on the wire in which I_1 flows.

(c) Repeat part (a) for the third wire.

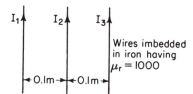

Figure P1-2

1-3. Repeat Prob. 1-2 for current I_2 flowing opposite to I_1 and I_3.

1-4. A uniform magnetic field of 0.7 Wb/m^2 in the iron is applied to the configuration of Fig. P1-2.

(a) Compute the direction and magnitude of the resultant force per meter when the magnetic field is directed perpendicularly into the plane of the paper.

(b) What is the value of this resultant force per meter when the magnetic field is applied in the plane of the wires and directed from right to left?

(c) Compute the resultant force per meter when the magnetic field is applied at an angle of 45° to the plane of the paper and directed into it from right to left.

(d) What is the resultant force per meter when the magnetic field is applied at an angle of 60° to the plane of the paper and directed into it from top to bottom?

1-5. In the configuration of Fig. P1-5, the current I_1 has a value of 40 A. Find the value of I_2 that causes the magnetic field intensity at point P to disappear.

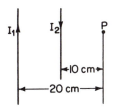

Figure P1-5

1-6. A circular loop of wire of radius r meters and consisting of a single turn carries the current I as shown in Fig. P1-6. Derive the expression for the magnetic field intensity at the center.

Figure P1-6

1-7. A magnetic circuit composed of silicon sheet steel has the square construction shown in Fig. P1-7.

(a) Find the mmf required to produce a core flux of 25×10^{-4} Wb.

(b) If the coil has 80 turns, how much current must be made to flow through the coil?

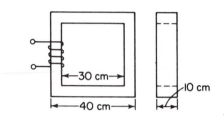

Figure P1-7

1-8. The magnetic circuit of Prob. 1-7 has an air gap of 0.1 cm cut in the right leg. For a coil having 100 turns find the current that must be allowed to flow in order for the core flux to be 0.0025 Wb.

1-9. In the magnetic circuit of Fig. P1-9 determine the coil mmf needed to produce a flux of 0.0014 Wb in the right leg. The thickness of the magnetic circuit is 0.04 m and is uniform throughout. Medium silicon steel is used.

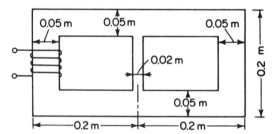

Figure P1-9

1-10. Repeat Prob. 1-9 for a coil placed on the center leg.

1-11. In the magnetic circuit of Prob. 1-7 find the core flux produced by a coil current of 200 A-t.

1-12. In the magnetic circuit of Prob. 1-8 find the core flux produced by a coil current of 600 A-t.

1-13. The core shown in Fig. P1-13 has a uniform cross-sectional area of 2 in.² and a mean length of 12 in. Also, coil A has 200 turns and carries 0.5 A, coil B has 400 turns and carries 0.75 A, and coil C carries 1.00 A. How many turns must coil C have for the core flux to be 120,000 lines? The coil currents have the directions indicated in the figure. The core is made of silicon sheet steel.

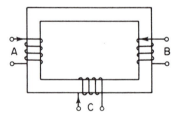

Figure P1-13

1-14. In the magnetic circuit shown in Fig. P1-14 the coil F_1 is supplied with 350 A-t in the direction indicated. Find the direction and magnitude of the mmf required in coil F_2 for the air-gap flux to be 180,000 lines. The core has an effective cross-sectional area of 9 in.^{2} and is made of silicon sheet steel. The length of the air gap is 0.05 in.

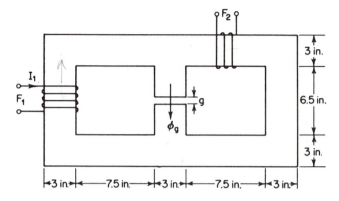

Figure P1-14

1-15. In the magnetic circuit of Fig. P1-14 the coil F_1 is supplied with 200 A-t in the direction shown. Find the direction and magnitude of the mmf required of coil F_2 for the air-gap flux to be 90,000 lines.

1-16. Find the inductance of a coil in which
   (a) A current of 0.1 A yields an energy storage of 0.05 J.
   (b) A current increases linearly from zero to 0.1 A in 0.2 s producing a voltage of 5 V.
   (c) A current of 0.1 A increasing at the rate of 0.5 A/s represents a power flow of $\frac{1}{2}$ W.

1-17. The coil in the configuration of Fig. 1-19 is equipped with 100 turns. Moreover, the mean length of turn of the magnetic core is known to be 0.2 m and the cross-sectional area is 0.01 m^2. The value of the permeability of the iron is 10^{-3}.
   (a) Find the inductance of the coil.
   (b) When a dc voltage is applied to the inductor it is found to draw a current of 0.1 A. How much energy is stored in the magnetic field?

1-18. When a dc voltage is applied to the coil of Prob. 1-17, the current is found to vary in accordance with $i = \frac{1}{10}(1 - \varepsilon^{-5t})$ expressed in amperes. Moreover, it is found that, after the elapse of 0.2 s, the voltage induced in the coil is 0.16 V. Find the inductance of the coil.

1-19. A sample of iron having a volume of 16.4 cm^3 is subjected to a magnetizing force sinusoidally varying at a frequency of 400 Hz. The area of the hysteresis loop is found to be 64.5 cm^3 with flux density plotted in kilolines per square inch and magnetizing force in ampere-turns per inch. The scale factors used are 1 in. = 5 kilolines/in.^{2} and 1 in. = 12 A-t/in. Find the hysteresis loss in watts.

1-20. A ring of ferromagnetic material has a rectangular cross section. The inner diameter is 7.4 in., the outer diameter is 9 in., and the thickness is 0.8 in. There is a coil of 600 turns wound on the ring. When the coil carries a current of 2.5 A, the flux produced in the ring is 1.2×10^{-3} Wb. Find the following quantities expressed in mks units:

(a) magnetomotive force;
(b) magnetic field intensity;
(c) flux density;
(d) reluctance;
(e) permeability;
(f) relative permeability.

1-21. In plotting a hysteresis loop the following scales are used: 1 cm = 10 A-t/in., and 1 cm = 20 kilolines/in.^{2} The area of the loop for a certain material is found to be 6.2 cm.^{2} Calculate the hysteresis loss in joules per cycle for the specimen tested if the volume is 400 cm^3.

1-22. The flux in a magnetic core is alternating sinusoidally at a frequency of 500 Hz. The maximum flux density is 50 kilolines/in.^{2}. The eddy-current loss then amounts to 14 W. Find the eddy-current loss in this core when the frequency is 750 Hz and the flux density is 40 kilolines/in^2.

1-23. The total core loss (hysteresis plus eddy current) for a specimen of magnetic sheet steel is found to be 1800 W at 60 Hz. If the flux density is kept constant and the frequency of the supply increased 50%, the total core loss is found to be 3000 W. Compute the separate hysteresis and eddy-current losses at both frequencies.

1-24. A flux ϕ penetrates the complete volume of the iron bars as shown in Fig. P1-24. When the bars are assumed separated by g meters, find the expression for the force existing between the two parallel plane faces. Neglect the reluctance of the iron.

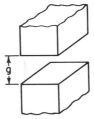

Figure P1-24

1-25. A magnetic flux ϕ penetrates a movable magnetic core which is vertically misaligned relative to the north and south poles of an electromagnet as depicted in Fig. P1-25. The depth dimension for the core and the electromagnet is b. Determine the expression for the force that acts to bring the core into vertical alignment. Express the result in terms of the air-gap flux density and the physical dimensions. Neglect the reluctance of the iron.

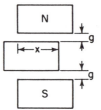

Figure P1-25

1-26. The magnetization curve of a relay is shown in Fig. P1-26.
  (a) Find the energy stored in the field with the relay in the open position.
  (b) Assume that the relay armature moves rapidly under conditions of constant flux. Compute the work done in going from the open to the closed position.
  (c) Calculate the force in newtons exerted on the armature in part (b).
  (d) Assume the relay armature moves slowly at constant mmf. Compute the work done in going from the open to the closed position.
  (e) Does the energy of part (d) come from the original stored field energy? Explain.

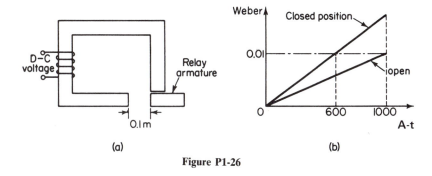

(a)   (b)

Figure P1-26

1-27. In the magnetic circuit of Fig. P1-27 the cross-sectional area of the air gap is 0.0025 m² and its length is 0.05 cm. Moreover, the cast steel core has a mean length of 0.2 m and it is equipped with 1000 turns. It is desirable to operate the circuit with an air-gap density of 1 T.
  (a) Determine the coil current needed to establish this gap density.
  (b) Compute the energy stored in the air gap.
  (c) Find the energy stored in the cast steel section of the circuit.
  (d) What is the inductance in henrys of this magnetic circuit?

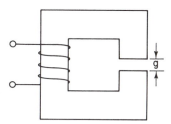

Figure P1-27

2

Transformers

An understanding of the transformer is essential to the study of electromechanical energy conversion. Although electromechanical energy conversion involves interchange of energy between an electrical system and a mechanical system, whereas the transformer involves the interchange of energy between two or more electrical systems, nonetheless the coupling device in both cases is the magnetic field, and its behavior in each case is fundamentally the same. As a result we find that many of the pertinent equations and conclusions of transformer theory have equal applicability in the analysis of ac machinery and some aspects of dc machinery. For example, the equivalent circuit of single-phase and three-phase induction motors is found to be identical in form to that of the transformer. Other examples can be cited, as a glance through the following chapters readily reveals.

In addition to serving as a worthwhile prelude to the study of electromechanical energy conversion, an understanding of transformer theory is important in its own right because of the many useful functions the transformer performs in prominent areas of electrical engineering. In communication systems ranging in frequency from audio to radio to video, transformers are found fulfilling widely varied purposes. *Input transformers* (to connect a microphone output to the first stage of an electronic amplifier), *interstage transformers*, and *output transformers* are to be found in radio and television circuits. Transformers are also used in communication circuits as impedance transformation devices which allow maximum transfer of power from the input circuit to the coupled circuit. Telephone lines and control circuits are two more areas in which the transformer is used extensively. In electric power distribution systems the transformer makes it possible to convert electric power from a generated voltage of about 15 to 20 kV (as determined by generator design limitations) to values of 380 to 750 kV, thus permitting transmission over long distances to appropriate distribution points (e.g., urban areas) at tremendous savings in the cost of copper as well as in power losses in the transmission lines. Then at the distribution points the transformer is the means by which these dangerously high voltages are reduced to a safe level (208 to 120 V) for use in homes, offices, shops, and so on. In short, the transformer is a useful device to be found in many phases of electrical engineering and therefore merits the attention given to it in this chapter.

2-1 THEORY OF OPERATION AND DEVELOPMENT
OF PHASOR DIAGRAMS

In the interest of clarity and motivation the theory of operation of the transformer is developed in five steps, starting with the ideal iron-core reactor and ending with the transformer phasor diagram under load. In the process the important transformer equations are developed along with a physical explanation of how the transformer operates. Emphasis on complete understanding of the physical behavior of a device is stressed, because, once this is grasped, the theory as well as the associated mathematical descriptions readily follow.

In its simplest form a transformer consists of two coils which are mutually coupled. When the coupling is provided through a ferromagnetic ring (circular or otherwise), the transformer is called an *iron-core transformer*. When there is no ferromagnetic material but only air, the device is described as an *air-core transformer*. In this chapter attention is confined exclusively to the iron-core type. Of the two coils the one that receives electric power is called the *primary winding*, whereas the other, which can deliver electric power to a suitable load circuit, is called the *secondary winding*.

Ideal Iron-Core Reactor

As the first step in the development of the theory of transformers refer to Fig. 2-1, which depicts an ideal iron-core reactor energized from a sinusoidal voltage source v_1. The reactor is ideal because the coil resistance and the iron losses are assumed to be zero and the magnetization curve is assumed to be linear.

Our interest at this point is to develop the phasor diagram that applies to the circuitry of Fig. 2-1, because thereby much is revealed about the theory. As v_1 increases in its sinusoidal variation, it causes a sinusoidal magnetizing current i_ϕ†

---

†Lowercase letters are used to represent instantaneous values,

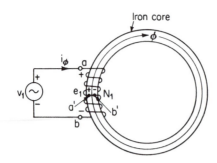

Figure 2-1 Ideal iron-core reactor: zero winding resistance and no core losses. Polarities are instantaneous.

to flow through the N_1 turns of the coil, which in turn produces a sinusoidally varying flux in time phase with the current. In other words, when the current is zero, the flux is zero and, when the current is at its positive maximum value, so too is the flux. Applying Kirchhoff's voltage law to the circuit we have

$$v_1 - e_1 = 0 \tag{2-1}$$

where e_1 is the *voltage drop* associated with the flow of current through the coil and can be written in terms of i_ϕ as

$$e_{ab} = e_1 = L \frac{di_\phi}{dt} \tag{2-2}$$

where L is the inductance of the coil and $i_\phi = i_{ab}$. Inserting Eq. (2-2) into Eq. (2-1) and rearranging yields

$$v_1 = e_1 = L \frac{di_\phi}{dt} = e_{ab} \tag{2-3}$$

We also know that the magnetizing current is sinusoidal because the applied voltage is sinusoidal and the magnetization curve is assumed linear. Consequently we can write for the magnetizing current the expression

$$i_\phi = \sqrt{2}\, I_\phi \sin \omega t \tag{2-4}$$

where I_ϕ is the rms value of the current.

Introducing Eq. (2-4) into Eq. (2-3) and performing the differentiation called for yields

$$v_1 = e_1 = \sqrt{2}\,(\omega L)I_\phi \cos \omega t = \sqrt{2}\,(\omega L)I_\phi \sin \left(\omega t + \frac{\pi}{2}\right) \tag{2-5}$$

The maximum value of this quantity occurs when the cos ωt has the value of unity. Then

$$E_{1\,max} = \sqrt{2}\,(\omega L)I_\phi = \sqrt{2}\, E_1 \tag{2-6}$$

where

$$E_1 \equiv \omega L I_\phi = \text{rms value of the inductive reactance drop} \tag{2-7}$$

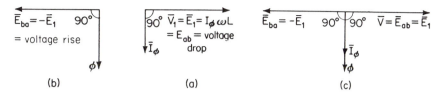

Figure 2-2 Phasor diagrams of the ideal iron-core reactor: (a) circuit viewpoint; (b) field viewpoint; (c) combined diagram.

A comparison of Eqs. (2-4) and (2-5) shows that the magnetizing current lags the voltage appearing across the coil by 90 electrical degrees. The phasor diagram of Fig. 2-2(a) depicts this condition. Since all the quantities involved here are sinusoidal, the phasors shown represent the rms value of the quantities.

Another approach that can be used in arriving at the phasor diagram for the ideal iron-core reactor is worth considering because of its use later. The approach just described can be referred to as a *circuit viewpoint* because it deals solely with the electric circuit in which I_ϕ flows as well as the circuit parameter L. No direct use is made of the magnetic flux ϕ, which is a space (or field) quantity. In contrast the second approach starts with the magnetic field as it appears in Faraday's law of induction. This is one of the fundamental laws upon which the science of electrical engineering is based. Faraday's law states that the emf induced in a coil is proportional to the number of turns linking the flux as well as the time rate of change of the linking flux. In applying Faraday's law to the circuit of Fig. 2-1 the expression may be written either as a voltage drop (*a* to *b*) or as a voltage rise (*b* to *a*). By the former description we can write

$$e_{ab} = e_1 = + N \frac{d\phi}{dt} \tag{2-8}$$

Expressed as a voltage rise, the induced emf equation becomes

$$e_{ba} = -e_1 = -N \frac{d\phi}{dt} \tag{2-9}$$

In either case, the sign is attributable to Lenz's law. The minus sign states that the emf induced by a changing flux is always in the direction in which current would have to flow to oppose the changing flux. On the other hand, the plus sign in Eq. (2-8) denotes that the polarity of *a* is positive whenever the time rate of change of flux is positive as determined by the direction of current flow.

Equation (2-9) then represents the starting point for the second approach in analyzing the circuit of Fig. 2-1. Corresponding to the sinusoidal variation of the magnetizing current as expressed by Eq. (2-4), there is also a sinusoidal variation of flux which may be expressed as

$$\phi = \Phi_m \sin \omega t \tag{2-10}$$

where Φ_m denotes the maximum value of the magnetic flux. Putting this expression into Eq. (2-9) and performing the required differentiation leads to

$$e_{ba} = -e_1 = -N_1 \frac{d\phi}{dt} = -N_1 \Phi_m \omega \cos \omega t = E_{1\,max} \sin\left(\omega t - \frac{\pi}{2}\right) \qquad (2\text{-}11)$$

where

$$E_{1\,max} \equiv N_1 \Phi_m \omega \equiv \sqrt{2}\, E_1 \qquad (2\text{-}12)$$

and

$$E_1 = \text{rms value of the induced emf}$$

A comparison of Eqs. (2-10) and (2-11) makes it clear that the reaction emf $-e_1$ *lags* the changing flux that produces it by 90°. This situation is depicted in Fig. 2-2(b). Basically, the emf $-e_1$ is a voltage rise (or generated emf) which has such a direction that, if it were free to act, would cause a current to flow opposing the action of I_ϕ. To verify this solely in terms of the action-reaction law consider any two adjacent points such as a' and b' on coil N_1 in Fig. 2-1. If the flux is assumed increasing in the direction shown, then the reaction in the coil must be such that current would have to flow from point b' to a' to oppose the increasing flux. Recall that by the *right-hand rule* any current assumed to be flowing from b' to a' produces flux lines opposing the increasing flux. For this condition to prevail, it is therefore necessary for point a' to be positive with respect to b' for the instant being considered. This line of reasoning when extended to the full coil, leads to the polarity markings shown in Fig. 2-1. The interesting thing to note here is that on the basis of the flux viewpoint the emf $-e_1$ in Fig. 2-1 is treated as a reaction (or generated) voltage rise in progressing from b' to a'. On the other hand, from the circuit viewpoint the voltage e_1, and its polarity when considered in the direction of flow of the current I_ϕ, appears as a voltage drop. The voltage e_1 remains the same; it is merely the point of view that changes.

Because I_ϕ and the flux ϕ are in time phase, it is customary to combine the two viewpoints into a single phasor diagram as depicted in Fig. 2-2(c). Since $E_{ab} = E_1$ represents the voltage drop, it follows that the same induced emf viewed on the basis of a reaction to the changing flux, being a voltage rise, is equal and opposite to E_1. That is, $E_{ba} = -E_1$. Another way of describing this is to say that the terminal voltage must always contain a component equal and opposite to the voltage rise.

It should be apparent up to this point that the reaction voltage viewed from b' to a' in Fig. 2-1 does not succeed in actually establishing a reverse current to oppose the increasing flux. This never happens in the primary winding of a two-winding transformer, but it does happen in the secondary winding as is presently described.

In the work that follows in this chapter preference is given to treating the induced emf as a voltage drop. That is, in applying Faraday's law, the version described by Eq. (2-8) will be used. Due regard is given to Lenz's law in this equation by noting that, when the time rate of change of flux in the configuration of Fig. 2-1 is positive, the polarity of point a is also positive.

Practical Iron-Core Reactor

As the second step in the development of transformer theory, let us consider how the phasor diagram of Fig. 2-2(a) must be modified to account for the fact that a practical reactor has a winding-resistance loss as well as hysteresis and eddy-current losses. A linear magnetization curve will continue to be assumed because the effect of the nonlinearity is to cause higher harmonics of fundamental frequency to exist in the magnetizing current—which, of course, cannot be included in the phasor diagram. Only quantities of the same frequency can be shown.

To develop the phasor diagram in this case we start with the flux phasor ϕ and then draw the induced emf E_1 90° leading. This much is shown in Fig. 2-3(a). Next we must consider the location of the current. Keep in mind that the iron is in a cyclic condition, i.e., the flux is sinusoidally varying with time. Moreover, this cyclic variation takes place along a hysteresis loop of finite area because the core losses are no longer assumed zero. Now, as pointed out in Chapter 1, it is characteristic of ferromagnetic materials in a cyclic condition to have the flux *lag* behind the magnetizing current by an amount called the *hysteretic angle* γ. See Fig. 2-3(b). Since the positive direction of rotation is counterclockwise in a phasor diagram, $\bar{I}_m$ is shown leading ϕ to bear out the physical fact. Note, too, that the magnetizing current $\bar{I}_m$ is now considered made up of two fictitious components $\bar{I}_\phi$ and $\bar{I}_c$. The quantity $\bar{I}_\phi$ is the purely reactive component and is the current that

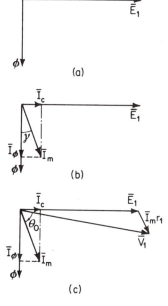

Figure 2-3   Development of the phasor diagram of a practical iron-core reactor: (a) building on the flux phasor; (b) accounting for core losses; (c) accounting for core losses and winding resistance.

would flow if the core losses were zero. The quantity I_c is that fictitious quantity which, when multiplied by the voltage E_1 produced by the changing flux, represents exactly the power needed to supply the core losses caused by the same changing flux. Thus we can write

$$I_c E_1 \equiv P_c \qquad \text{W} \tag{2-13}$$

or

$$I_c \equiv \frac{P_c}{E_1} \qquad \text{A} \tag{2-14}$$

where P_c denotes the sum of the hysteresis and eddy-current losses. This technique for handling the core loss of an electromagnetic device is a useful one and is frequently applied in the study of ac machinery.

To complete the phasor diagram for the practical iron-core reactor, it is only necessary to account for the winding-resistance drop. This is readily accomplished by applying Kirchhoff's voltage law to the circuit. Hence

$$\bar{V}_1 = \bar{I}_m r_1 + \bar{E}_1 \qquad \text{V} \tag{2-15}$$

where r_1 is the winding resistance in ohms. Since r_1 is a constant, it follows that the $\bar{I}_m r_1$ drop must be in phase with $\bar{I}_m$ or located parallel to $\bar{I}_m$ as depicted in Fig. 2-3(c). Note that the component of $\bar{I}_m$ that is in phase with the applied voltage $\bar{V}_1$ exceeds $\bar{I}_c$, as well it should because this component must supply not only the core losses but the winding copper losses too. Thus, the input power may be expressed as

$$P_{in} = V_1 I_m \cos \theta_0 = I_m^2 r_1 + P_c \qquad \text{W} \tag{2-16}$$

where θ_0 denotes the power-factor angle.

The Two-Winding Transformer

By placing a second winding on the core of the reactor of Fig. 2-1, we obtain the simplest form of a transformer. This is depicted in Fig. 2-4. The rms value of the induced emf E_1 appearing in the primary winding ab readily follows from Eq. (2-12) as

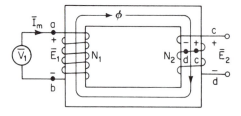

Figure 2-4   Two-winding transformer.

$$E_1 = \frac{N_1 \Phi_m \omega}{\sqrt{2}} = \frac{2\pi f}{\sqrt{2}} \Phi_m N_1 = 4.44 f \Phi_m N_1 \qquad (2\text{-}17)$$

Keep in mind that this result derives directly from Faraday's law and is very important in machinery analysis. One significant result deducible from this equation is that for transformers having relatively small winding resistance (i.e., $E_1 \approx V_1$), *the value of the maximum flux is determined by the applied voltage.*

The induced emf appearing across the secondary winding terminals is produced by the same flux that causes E_1. Hence the only difference in the rms values is brought about by the difference in the number of turns. Thus we can write

$$E_2 = 4.44 f \Phi_m N_2 \qquad (2.18)$$

Dividing Eq. (2-17) by (2-18) leads to

$$\frac{E_1}{E_2} = \frac{N_1}{N_2} \equiv a = \text{ratio of transformation} \qquad (2\text{-}19)$$

Another fact worth noting is that E_1 and E_2 are in time phase because they are induced by the same changing flux.

Transformer Phasor Diagram at No-Load

When a transformer is at no-load, no secondary current flows. Figure 2-5 depicts the situation. A study of Fig. 2-5 should make it apparent that the development of the phasor diagram follows from Fig. 2-3(c) by making two modifications. One is to identify the secondary induced emf in the diagram. But, as already observed in the preceding step, the emf induced in the secondary winding must be in phase with the corresponding voltage E_1 of the primary winding. For convenience call this secondary induced emf E_2. Then E_2 and E_1 must both lead the flux phasor by 90° as shown in Fig. 2-6. For convenience the turns ratio a is assumed to be unity. The second modification is concerned with accounting for the fact that not all of the flux produced by the primary winding links the secondary winding. The

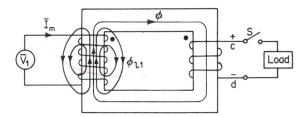

Figure 2-5 Two-winding transformer at no-load; S open.

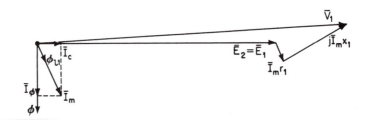

Figure 2-6   Phasor diagram of a transformer at no-load.

difference between the total flux linking the primary winding and the mutual flux linking both windings is called the *primary leakage flux* and is denoted by ϕ_{l1}. Note that some of the paths drawn to represent ϕ_{l1} encircle only several turns and not all N_1 turns. As long as such closed paths encircle an mmf different from zero, such a flux path does in fact exist. This is borne out by Ampere's circuital law.

An important distinction exists between the mutual flux ϕ and the primary leakage flux ϕ_{l1}. It is this. The mutual flux exists wholly in iron and so involves a hysteresis loop of finite area. The leakage flux, on the other hand, always involves appreciable air paths, and, although some iron is included in the closed path, the reluctance experienced by the leakage flux is practically that of the air. Consequently the cyclic variation of the leakage flux involves no hysteresis (or lagging) effect. Hence in the phasor diagram the primary leakage flux must be placed in phase with the primary winding current as shown in Fig. 2-6. Now this leakage flux induces a voltage E_{l1} which must lead ϕ_{l1} by 90°. This emf due to leakage flux may be replaced by an equivalent *primary leakage reactance drop*. Thus

$$\boxed{I_m x_1 \equiv E_{l1}}$$ (2-20)

This quantity must lead $\bar{I}_m$ by 90°. The phasor addition of this drop and the primary winding resistance drop as well as the voltage drop $\bar{E}_1$ associated with the mutual flux yields the primary applied voltage $\bar{V}_1$ as depicted in Fig. 2-6.

The quantity x_1 of Eq. (2-20) is called the *primary leakage reactance;* it is a fictitious quantity introduced as a convenience in representing the effects of the primary leakage flux.

Transformer under Load

In this last step in the development of the theory of operation of the transformer, we direct attention first to the phasor diagram representation of Kirchhoff's voltage law as it applies to the secondary circuit. For simplicity consider that the load appearing in the secondary circuit of Fig. 2-7 is purely resistive and further that the flux has the direction indicated with the flux increasing. By Lenz's law there is emf induced in the secondary winding which instantaneously makes terminal c positive with respect to terminal d. When switch S is closed, a current $\bar{I}_2$ flows

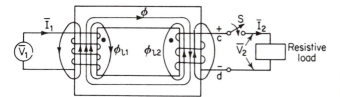

Figure 2-7 Two-winding transformer under load; S closed.

instantaneously from c through the load to d. For convenience the load resistor is assumed adjusted to cause rated secondary current to flow. $\bar{E}_2$ establishes the secondary load voltage besides accounting for other voltage drops existing in the secondary circuit. When equilibrium is established after the load switch is closed, the appropriate phasor diagram must show $\bar{V}_2$ and $\bar{I}_2$ in phase as depicted in Fig. 2-8. Since it is physically impossible for the secondary winding to occupy the same space as the primary winding, a flux is produced by the secondary current which does not link with the primary winding. This is called the *secondary leakage flux*. The voltage E_{l2} induced by this secondary leakage flux leads ϕ_{l2} by 90°. Again as a matter of convenience this secondary leakage flux voltage is replaced by a secondary leakage reactance drop. Thus

$$E_{l2} \equiv I_2 x_2 \tag{2-21}$$

where x_2 is the fictitious secondary leakage reactance which represents the effects of the secondary leakage flux. Also, because leakage flux has its reluctance predominantly in air, Eq. (2-21) is a linear equation. Thus, doubling the current doubles the leakage flux, which doubles $\bar{E}_{l2}$, and by Eq. (2-21) this is represented by a doubled reactance drop. The phasor sum of the terminal voltage, the secondary winding resistance drop, and the secondary leakage reactance drop adds up to the source voltage $\bar{E}_2$ as demonstrated in Fig. 2-8.

When a phasor diagram shows $\bar{V}_2$ and $\bar{I}_2$ in phase, the implication is that power is being supplied to the load equal to $V_2 I_2$. Clearly this power must come

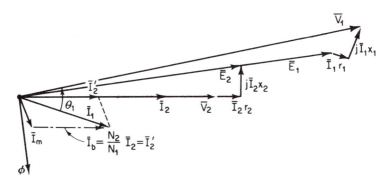

Figure 2-8 Complete phasor diagram of a transformer under load for the case where $a = N_1/N_2 > 1$.

from the primary circuit. But by what mechanism does this come about? To understand the answer to this question, consider the situation depicted in Fig. 2-7. With the flux increasing in the direction shown, the secondary emf has the polarity indicated, so that when the switch is closed the action of the secondary current is to cause a decrease in the mutual flux. This tendency for the flux to decrease owing to the action of $N_2 I_2$ causes the voltage drop E_1 [or $e_1 = N_1(d\phi/dt)$] to decrease. However, since the applied voltage $\bar{V}_1$ is constant, the slight decrease in $\bar{E}_1$ creates an appreciable increase in primary current. In fact the primary current is allowed to increase to that value which, when flowing through the primary turns N_1, represents sufficient mmf to neutralize the demagnetizing action of the secondary mmf. Expressed mathematically we have

$$\bar{I}_b N_1 - N_2 \bar{I}_2 = 0 \qquad (2\text{-}22)$$

where $\bar{I}_b$ denotes the increase in primary current needed to cancel out the effect of the secondary mmf to reduce the flux. $\bar{I}_b$ is often called the *balance* current. This is above and beyond the value $\bar{I}_m$ needed to establish the operating flux. Equation (2-22) reveals that $N_1 \bar{I}_b$ must be oppositely directed to $N_2 \bar{I}_2$. In short, then, it is through the medium of the flux and its small attendant changes that power is transferred from the primary to the secondary circuit of a transformer. Of course, the total primary current is the phasor sum of $\bar{I}_b$ and $\bar{I}_m$. In this case the primary leakage flux is in phase with $\bar{I}_1$ rather than $\bar{I}_m$, which is the case at no-load.

For transformers in which the winding resistance drop and the leakage reactance drop are negligibly small, the magnetizing mmf is the same at no-load as under load. In other words, the phasor sum of the total primary mmf $N_1 \bar{I}_1$ and the secondary mmf $N_2 \bar{I}_2$ must yield $N_1 \bar{I}_m$. Thus

$$\boxed{N_1 \bar{I}_1 - N_2 \bar{I}_2 = N_1 \bar{I}_m} \qquad (2\text{-}23)$$

Dividing through by N_1 leads to

$$\bar{I}_1 - \frac{N_2}{N_1} \bar{I}_2 = \bar{I}_m \qquad (2\text{-}24)$$

Inserting Eq. (2-22) then yields

$$\bar{I}_1 - \bar{I}_b = \bar{I}_m = \bar{I}_1 - I'_2 \qquad (2\text{-}25)$$

where I'_2 is equivalent to I_b. These equations are readily verified in the phasor diagram of Fig. 2-8.

Transformer Construction Features

Contrary to the schematic representation of Fig. 2-7 it should not be implied that the total number of primary winding turns are placed on one leg of the ferromagnetic core and all the turns of the secondary winding are placed on the other leg.

Such an arrangement leads to an excessive amount of leakage flux, so that for a given source voltage the mutual flux is correspondingly smaller. The usual procedure in transformer construction is to split the primary and secondary turns. This keeps the leakage flux within a few percent of the mutual flux. Of course still greater reduction in leakage flux can be achieved by further subdividing and sandwiching the primary and secondary turns, but this is clearly obtained at an appreciable increase in the cost of assembly.

The two principal types of construction used for single-phase transformers are the *core type* and the *shell type*. In the core type the windings are made to surround the iron in the fashion illustrated in Fig. 2-9. In the typical configuration one-half the turns of the low-voltage coil are placed on each leg nearest the iron separated by an appropriate layer of insulation. Then following another layer of suitable insulation half the turns of the high-voltage winding are placed over the low-voltage coil. The process is repeated for the second leg as well. The core-type construction is favored for high-voltage transformers because the problem of insulating is easier. In the shell-type transformer it is the iron that appears to surround the coils as depicted in Fig. 2-10. The coils in this configuration are generally of a pancake form as distinguished from the cylindrical forms used in the core-type transformer. Often the low-voltage winding is divided into three sections with the two outer sections each having a quarter of the turns and the

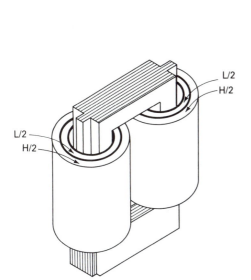

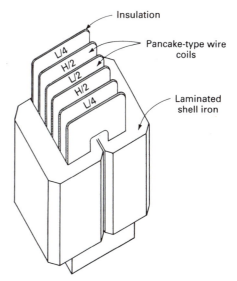

Figure 2-9 Construction features of the core-type transformer. H denotes the number of turns of the high voltage coil; L denotes the number of turns of the low voltage coil.

Figure 2-10 Construction features of the shell-type transformer. H denotes the number of turns of the high voltage coil; L denotes the number of turns of the low voltage coil.

remaining half is sandwiched in between the two halves of the high-voltage winding as indicated in the diagram. The shell-type construction is favored for power
transformers where large currents are made to flow. The iron shell provides
better mechanical protection to the windings in such circumstances.

Effect of Saturation on Magnetizing Current

All the quantities that appear in a phasor diagram such as the one depicted in Fig.
2-8 must necessarily be at the same frequency. Consequently, the magnetizing
component, $\bar{I}_m$, of the current $\bar{I}_1$ has to be treated as a quantity with a sinusoidal
shape when so used. However, the true shape of the actual magnetizing current is
other than sinusoidal chiefly because of the influence of the magnetic core. Two
factors are important in this consideration. One is the state of saturation of the
magnetic iron core and the other is the combined effect of hysteresis and core
losses.

Frequently, transformers function with time flux variations that place operation just above the knee of the saturation curve. It is instructive to examine the
consequences of such a mode of operation. This is best accomplished by referring
to Fig. 2-11. At the outset it is important to understand that the situation described in this diagram is a transitory one. We consider it in order to put into
perspective the forces that create the final shape of the magnetizing current.
The voltage waveshape that is applied to the primary winding of the transformer is assumed to be a pure sinusoid. Accordingly, this voltage source is
capable solely of supplying a sinusoidal magnetizing current to the primary coil of
the transformer. Figure 2-11 shows this magnetizing current drawn below the
magnetization curve, which in turn is depicted as possessing a fair degree of
saturation but no hysteresis. The effect of hysteresis in the magnetic core material is being neglected for the moment in the interest of simplicity. The projection
of the sinusoidally varying magnetizing current onto the magnetization curve
produces a flat-topped flux wave as shown to the right of the magnetization
curve. Moreover, this flat-topped flux wave contains primarily two harmonic
components: a large fundamental component, ϕ_1, and a substantial negative third
harmonic component, ϕ_3. The fundamental flux component induces a voltage, e_1,
in the primary coil which is equal and opposite to the applied primary voltage, v_1,
on the assumption of negligible primary winding resistance. However, the negative third harmonic flux component also induces a voltage in the primary coil, e_3.
But this voltage does not encounter an equal and opposite component in the
primary source voltage because it is a purely sinusoidal quantity. As a result, the
coil-generated third harmonic voltage is free to act in the primary circuit to produce a negative third harmonic current which then combines with the positive
fundamental component supplied by the source voltage. Thus, the resultant magnetizing current is the superposition of these two components. The result is a
waveshape for the magnetizing current that is characteristically *peaked* as depicted in Fig. 2-12. The peaking effect is associated with the fact that the third

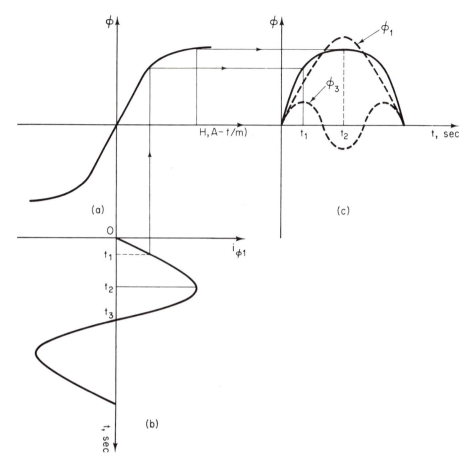

Figure 2-11   Illustrating the effect of a sinusoidal magnetizing current: (a) magnetization curve showing saturation at high magnetizing forces and no hysteresis; (b) sinusoidal magnetizing current assumed to flow from source voltage to primary coil; (c) resultant flat-topped flux wave produced by the current of (b).

harmonic component is made to flow in a direction opposite to the fundamental component because it originates from the primary coil.

It is interesting to observe that the action of the negative third harmonic current can also be interpreted as producing a primary-coil mmf that in turn acts to produce a negative third harmonic flux which serves to cancel the original third harmonic flux component of the flat-topped flux wave. Ultimately, of course, Kirchhoff's voltage law demands that the resultant core flux must be sinusoidal because only such a flux produces an induced coil voltage that balances the applied sinusoidal source voltage.

The peaked nature of the resultant magnetizing current can also be demonstrated by starting with the condition that the sinusoidal source voltage can be

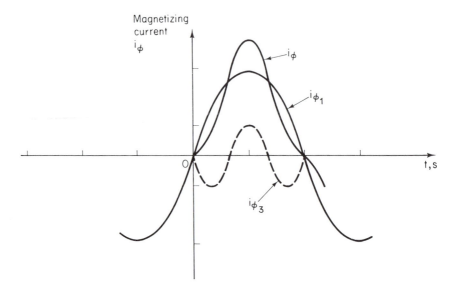

Figure 2-12 Transformer magnetizing current waveshape neglecting hysteresis and core losses. Here $i_{\phi 1}$ denotes the component supplied by the source voltage and $i_{\phi 3}$ is the component supplied by the transformer coil arising from saturation of the magnetic core.

balanced in the primary circuit only by the presence of a sinusoidal core flux. The subsequent projection of such a sinusoidal flux variation onto the saturated magnetization curve then leads directly to the peaked waveshape of the magnetization current. However, such an approach does not identify the source of the third harmonic current component, which we now know is not directly the sinusoidal source voltage.

The final waveshape of the magnetizing current is obtained by adding to the peaked curve of Fig. 2-12 a fundamental sinusoidal component (in phase with v_1 or

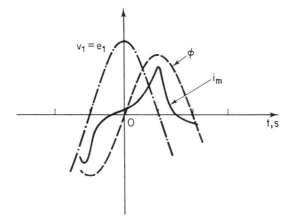

Figure 2-13 Final waveshape of magnetizing current i_m that results upon adding a sinusoidal fundamental component (in phase with the source voltage) to i_ϕ of Fig. 2-12. The in-phase component takes care of the hysteresis and eddy current losses.

e_1) to account for the hysteresis and eddy-current losses. Figure 2-13 shows the result.

2-2 THE EQUIVALENT CIRCUIT

Derivation

It is customary in analyzing devices in electrical engineering to represent the device by means of an appropriate equivalent circuit. In this way further analysis and design as well as computational accuracy is facilitated by the direct application of the techniques of electric circuit theory. This procedure is followed whenever new devices are investigated. Accordingly, equivalent circuits are derived not only for the transformer but for all ac and dc motors as well as important electronic devices such as the transistor. Generally, the equivalent circuit is merely a circuit interpretation of the equation(s) that describe the behavior of the device. For the case at hand there are two equations—Kirchhoff's voltage equations for the primary winding and for the secondary winding. Thus, for the primary winding we have

$$\bar{V}_1 = \bar{I}_1 r_1 + j\bar{I}_1 x_1 + \bar{E}_1 \tag{2-26}$$

and for the secondary winding

$$\bar{E}_2 = \bar{V}_2 + \bar{I}_2 r_2 + j\bar{I}_2 x_2 \tag{2-27}$$

Both equations are represented in the phasor diagram of Fig. 2-8, and the corresponding circuit interpretation appears in Fig. 2-14(a). The dot notation is used to indicate which sides of the coils instantaneously carry the same polarity. The schematic diagram of Fig. 2-14(a) can be replaced by the equivalent circuit of Fig. 2-14(b) by recognizing that $\bar{I}_1$ is composed of $\bar{I}_b$ and $\bar{I}_m$. Moreover, since by definition $\bar{I}_m$ consists of two components $\bar{I}_c$ and $\bar{I}_\phi$, it can be considered splitting into two parallel branches. One branch is purely resistive and carries the current $\bar{I}_c$, which was defined as in-phase with $\bar{E}_1$. The other branch must be purely reactive because it carries $\bar{I}_\phi$, which was defined as 90° lagging $\bar{E}_1$, as depicted in Fig. 2-3(b). The ideal transformer is included in Fig. 2-14(b) to account for the transformation of voltage and current that occurs between primary and secondary windings. Thus, with the ratio of transformation $a = 2$, the ideal transformer conveys the information that $\bar{E}_2 = \frac{1}{2}\bar{E}_1$ and that $\bar{I}_2 = 2\bar{I}_b$. The fact that the actual transformer has primary winding resistance and leakage flux and must carry a magnetizing current is taken care of by the circuitry preceding the symbol for the ideal transformer in Fig. 2-14(b).

The resistive element r_c must have such a value that, when the voltage E_1 appears across it, it permits the current I_c to flow. Thus

$$r_c \equiv \frac{E_1}{I_c} = \frac{P_c}{I_c^2} \tag{2-28}$$

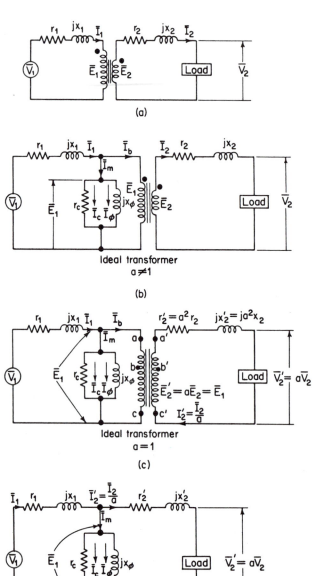

Figure 2-14 Development of the exact equivalent circuit: (a) schematic diagram; (b) two-winding equivalent circuit; (c) replacing the actual secondary winding with an equivalent winding having N_1 turns; (d) final complete form.

The second form of this equation is obtained by replacing E_1 with the expression of Eq. (2-13). Hence, r_c may also be looked upon as that resistance which accounts for the transformer core loss. Similarly, the magnetizing reactance of the transformer may be defined as that reactance which when E_1 appears across it causes the current I_ϕ to flow. Thus

$$x_\phi \equiv \frac{E_1}{I_\phi} \tag{2-29}$$

A single-line equivalent circuit of the transformer is a desirable goal. However, as long as $E_2 \neq E_1$ this is impossible to accomplish because the potential distribution along the primary and secondary windings can never be identical. Only when $a = 1$ is this condition possible. Therefore, in the interest of achieving the desired simplicity in the equivalent circuit, let us investigate replacing the actual secondary winding with an equivalent winding having the same number of turns as the primary winding. A little thought should make it reasonable to expect that, if an equivalent secondary winding having N_1 turns is made to handle the same kilovolt-amperes, the same copper loss and leakage flux, and the same output power as the actual secondary winding, then insofar as the applied voltage source is concerned it can detect no difference between the actual secondary winding of N_2 turns and the equivalent secondary winding of N_1 turns. That is, one winding can replace the other without causing any changes in the primary quantities.

Because the equivalent secondary winding is assumed to have N_1 turns, the corresponding induced emf must be different by the factor a. We have

$$E_2' = aE_2 = E_1 \tag{2-30}$$

where the prime notation is used to refer to the equivalent winding having N_1 turns. For the kilovolt-amperes of the equivalent secondary winding to be the same as that of the actual secondary winding the following equation must be satisfied:

$$E_2' I_2' = E_2 I_2 \tag{2-31}$$

Inserting Eq. (2-30) into the last equation shows that the current rating of the equivalent secondary winding must be

$$I_2' = \frac{I_2}{a} \tag{2-32}$$

Note that this current is the balance current I_b.

Since in general $a \neq 1$, it follows that the winding resistance of the equivalent secondary must be adjusted so that the copper loss is the same in both. Accordingly

$$(I_2')^2 r_2' = I_2^2 r_2 \tag{2-33}$$

This equation, together with Eq. (2-32), establishes the value of the winding resistance of the equivalent secondary. Thus

$$r_2' = \left(\frac{I_2}{I_2'}\right)^2 r_2 = a^2 r_2 \tag{2-34}$$

Often the quantity r_2' is described as the secondary resistance referred to the primary. The phrase "referred to the primary" is used because we are dealing with an equivalent secondary winding having the same number of turns as the primary winding.

By proceeding in a similar fashion the secondary leakage reactance referred to the primary can be expressed as

$$x_2' = a^2 x_2 \tag{2-35}$$

Moreover, since the kilovolt-ampere load must be the same for the two windings we have

$$V_2' I_2' = V_2 I_2 \tag{2-36}$$

from which it follows that the load terminal voltage in the equivalent secondary winding is

$$V_2' = \frac{I_2}{I_2'} V_2 = a V_2 \tag{2-37}$$

Appearing in Fig. 2-14(c) are all of the foregoing results in addition to an ideal transformer having a unity turns ratio. Keep in mind that Fig. 2-14(c) is drawn with the equivalent secondary winding having N_1 turns, whereas Fig. 2-14(b) is drawn with the actual secondary winding. As far as the primary is concerned it cannot distinguish between the two. For the ideal transformer of Fig. 2-14(c) note that joining point a to a', b to b', and c to c' causes no disturbance because instantaneously these points are at the same potential. Accordingly, this ideal transformer may be entirely dispensed with, in which case Fig. 2-14(c) becomes Fig. 2-14(d), which is called the *complete equivalent circuit* of the transformer. This is the single-line diagram we set out to derive.

Approximate Equivalent Circuit

In most constant-voltage, fixed-frequency transformers such as are used in power and distribution applications, the magnitude of the magnetizing current I_m is only a few percent (2 to 5) of the rated winding current. This happens because high-permeability steel is used for the magnetic core and so keeps I_ϕ very small. Furthermore, steel with a few percent of silicon added is often used and this helps to reduce the core losses, thereby keeping I_c small. Consequently, little error is introduced and considerable simplification is achieved by assuming I_m so small as to be negligible. This makes the equivalent circuit of Fig. 2-14(d) take on the configuration shown in Fig. 2-15. Note that the primary current $\bar{I}_1$ is now equal to

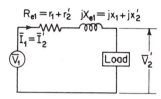

Figure 2-15   Approximate equivalent circuit referred to the primary.

the referred secondary current $\bar{I}_2'$. Also we can now speak of an equivalent resistance referred to the primary

$$R_{e1} \equiv r_1 + r_2' = r_1 + a^2 r_2 \tag{2-38}$$

and an equivalent leakage reactance referred to the primary

$$X_{e1} \equiv x_1 + x_2' = x_1 + a^2 x_2 \tag{2-39}$$

The subscript 1 denotes the primary.

The corresponding phasor diagram drawn for the approximate equivalent circuit appears in Fig. 2-16. Note the considerable simplification over the phasor diagram of the exact equivalent circuit which appears in Fig. 2-8.

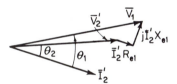

Figure 2-16   Phasor diagram of the approximate equivalent circuit for a lagging power factor angle θ_2.

The analysis of the preceding pages concerns itself with replacing the actual secondary winding by an equivalent winding having the same number of turns as the primary winding. A little thought should make it apparent that a similar procedure may be employed to draw a single-line equivalent circuit with all quantities referred to the secondary winding. In other words, the actual primary winding can be replaced by an equivalent primary winding having N_2 turns. The resulting equivalent circuit then assumes the form depicted in Fig. 2-17. It is interesting to note that, to refer the primary winding resistance r_1 to the secondary, it is necessary to divide by a^2. Thus

$$r_1' = \frac{r_1}{a^2} \tag{2-40}$$

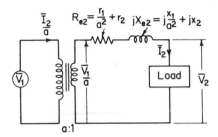

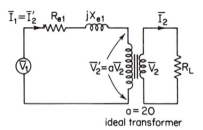

Figure 2-17 Approximate equivalent circuit referred to the secondary.

Figure 2-18 Approximate equivalent circuit drawn with the ideal transformer.

Furthermore, the equivalent winding resistance referred to the secondary is

$$R_{e2} = r_1' + r_2 = \frac{r_1}{a^2} + r_2 \qquad\qquad (2\text{-}41)$$

and the equivalent leakage reactance referred to the secondary is

$$X_{e2} = x_1' + x_2 = \frac{x_1}{a^2} + x_2 \qquad\qquad (2\text{-}42)$$

Note too that the ideal transformer is included in Fig. 2-17 merely to emphasize the transformation that takes place from the primary side to the secondary side. Normally, the primary circuit is omitted entirely leaving only the single-line secondary circuit.

Meaning of Transformer Nameplate Rating

A typical transformer carries a nameplate bearing the following information: 10 kVA 2200/110 V. What are the meanings of these numbers in the terminology and analysis developed so far? To understand the answer refer to Fig. 2-18, which is really the approximate equivalent circuit drawn to include the ideal transformer to bring into evidence the transformation action existing between the primary and secondary circuits. The number 110 refers to the rated secondary voltage. It is the voltage appearing across the load when rated current flows at rated kVA. We call this V_2. The 2200 number refers to the primary winding voltage rating, which is obtained by taking the rated secondary voltage and multiplying by the turns ratio a between primary and secondary. In our terminology this is V_2'. By specifying the ratio $V_2'/V_2 = 2200/110$ on the nameplate, information about the turns ratio is given. Finally, the kilovolt-ampere figure always refers to the *output* kilovolt-amperes appearing at the secondary load terminals. Of course the input kilovolt-amperes are slightly different because of the effects of R_{e1} and X_{e1}.

Transfer Function

The transfer function of the transformer may be specified with respect either to voltage or to current. The voltage transfer function is defined as the ratio of the output voltage $\bar{V}_2$ to the input voltage $\bar{V}_1$ with the load connected. Thus

$$\bar{T}_V = \frac{\bar{V}_2}{\bar{V}_1} = \frac{\bar{V}_2}{a\bar{V}_2 + \bar{I}_2' R_{e1} + j\bar{I}_2' X_{e1}} \approx \frac{1}{a} \qquad (2\text{-}43)$$

Equation (2-43) indicates that when the winding resistances and leakage fluxes are negligibly small the voltage transfer function $\bar{T}_V$ is approximately the reciprocal of the ratio of transformation. When these quantities cannot be neglected, $\bar{T}_V$ will be a complex number which varies with the load.

The current transfer function $\bar{T}_I$ is defined as the output current divided by the input current. For the approximate equivalent circuit of Fig. 2-18 we have

$$\bar{T}_I = \frac{\bar{I}_2}{\bar{I}_2'} = a \qquad (2\text{-}44)$$

Here the current transfer function is exactly equal to a because the magnetizing current is assumed negligible. When $\bar{I}_m$ is not negligible, $\bar{T}_I$ will be somewhat smaller than a and may be complex.

Input Impedance

The input impedance of a transformer is the impedance measured at the input terminals of the primary winding. In the case of the approximate equivalent circuit shown in Fig. 2-18 we have

$$\bar{Z}_i = \frac{\bar{V}_1}{\bar{I}_1} = R_{e1} + a^2 R_L + jX_{e1} \qquad (2\text{-}45)$$

A glance at this result indicates that the input impedance is dependent upon the load impedance. In the case of power applications the input impedance is of no particular concern because the internal impedance of the source is negligibly small. In communication applications, however, where $\bar{V}_1$ is obtained from a transistor, input impedance is a useful quantity because it describes the effect that the transformer has on the voltage source.

Frequency Response

One of the things we are interested in, when referring to the frequency response of a transformer, is the manner in which the voltage transfer function varies for a fixed load as the frequency of the source voltage $\bar{V}_1$ is varied. The amplitude of $\bar{V}_1$ is assumed fixed. For power transformers frequency response is of no serious interest because these units are operated at a single fixed frequency—usually 60

Hz and sometimes 50 Hz. However, in communication circuits the frequency of the source voltage is likely to vary over a wide range. For example, in the case of an output transformer which couples the power stage of an audio amplifier to a loudspeaker the frequency may vary over the entire audio range. What, then, is the effect of this varying frequency on the voltage transfer function? The answer follows from an inspection of Fig. 2-14(d). Keep in mind that inductive reactance is directly proportional to the frequency, i.e.,

$$x = \omega L = 2\pi f L \tag{2-46}$$

Hence at low frequencies the reactance x_ϕ, which normally is so high that it may be removed from the circuit, can now no longer be neglected. As a matter of fact this low magnetizing reactance can effectively shunt the fixed load impedance thereby causing a severe drop-off of output voltage as depicted in Fig. 2-20. Figure 2-19(a) shows the configuration of the equivalent circuit as it applies at very low frequencies. Note that x_1 and x_2 are so small at these frequencies that they may be omitted entirely from the diagram.

In the intermediate frequency range the design of these transformers for communication systems is such that x_1 and x_2' are still quite small while x_ϕ is sufficiently large that it may be omitted. The appropriate circuitry appears in Fig. 2-19(b). Note that for a resistive load the equivalent circuit consists solely of resistive elements so that the transfer function remains constant over the band of intermediate frequencies. Refer to Fig. 2-20.

At very high frequencies x_1 and x_2' are no longer negligible and therefore must be included in the circuit. However, x_ϕ may continue to be omitted. The equivalent circuit now takes the form depicted in Fig. 2-19(c). A little thought reveals that, as the frequency is allowed to get higher and higher, more and more of the

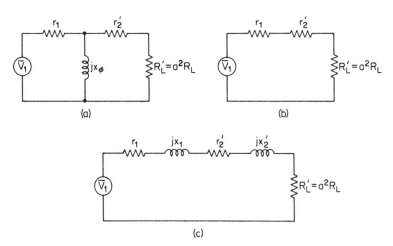

(a)   (b)

(c)

Figure 2-19   Equivalent circuit for various portions of the frequency spectrum: (a) low frequencies; (b) intermediate frequencies; (c) high frequencies.

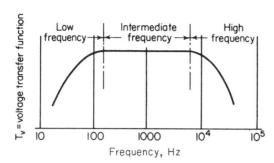

Figure 2-20 Voltage transfer function $\bar{V}_2/\bar{V}_1$ versus frequency.

source voltage $\bar{V}_1$ appears across x_1 and x_2' and less and less appears across the fixed load resistance. Therefore again a drop-off occurs in the value of the voltage transfer function as depicted in Fig. 2-20. In communication transformers a suitable frequency response such as that shown in Fig. 2-20 is very often the most important characteristic of the device.

Example 2-1

A distribution transformer with the nameplate rating of 100 kVA, 1100/220 V, 60 Hz, has a high-voltage winding resistance of 0.1 ohm (Ω) and a leakage reactance of 0.3 Ω. The low-voltage winding resistance is 0.004 Ω and the leakage reactance is 0.012 Ω. The source is applied to the high-voltage side.

(a) Find the equivalent winding resistance and reactance referred to the high-voltage side and the low-voltage side.

(b) Compute the equivalent resistance and equivalent reactance drops in volts and in percent of the rated winding voltages expressed in terms of the primary quantities.

(c) Repeat (b) for quantities referred to the low-voltage side.

(d) Calculate the *equivalent leakage impedances* of the transformer referred to the primary and secondary sides.

Solution From the nameplate data the turns ratio is $a = 1100/220 = 5$.

(a) Hence the equivalent winding resistance referred to the primary by Eq. (2-38) is

$$R_{e1} = r_1 + a^2 r_2 = 0.1 + (5)^2(0.004) = 0.2 \ \Omega$$

Also

$$X_{e1} = x_1 + a^2 x_2 = 0.3 + 25(0.012) = 0.6 \ \Omega$$

The corresponding quantities referred to the secondary are

$$R_{e2} = \frac{0.1}{25} + 0.004 = 0.008 \ \Omega$$

$$X_{e2} = \frac{0.3}{25} + 0.012 = 0.024 \ \Omega$$

These computations reveal that $r_1 = r_2'$ and $x_1 = x_2'$. This is a very common occurrence in well-designed transformers. A well-designed transformer is one that uses a minimum amount of iron and copper for a specified output. Note, too, that the resistance of the secondary winding is considerably smaller than that of the primary winding. This is consistent with the larger current rating of the former.

(b) The primary winding current rating in this case is

$$I_2' = \frac{100 \text{ kVA}}{1.1 \text{ kV}} = 91 \text{ A}$$

Hence

$$I_2' R_{e1} = 91(0.2) = 18.2 \text{ V}$$

Expressed as a percentage of the rated voltage of the high-tension winding this is

$$\frac{I_2' R_{e1}}{V_2'} 100 = \frac{18.2}{1100} 100 = 1.65\%$$

Also

$$I_2' X_{e1} = 91(0.6) = 54.6 \text{ V}$$

and

$$\frac{I_2' X_{e1}}{V_2'} 100 = \frac{54.6}{1100} = 4.96\%$$

(c) The secondary winding current rating is

$$I_2 = aI_2' = 5(91) = 455 \text{ A}$$

Hence

$$I_2 R_{e2} = 455(0.008) = 3.64 \text{ V}$$

and

$$\frac{I_2 R_{e2}}{V_2} 100 = \frac{3.64}{220} 100 = 1.65\%$$

Note that the percentage value of the winding resistance drop on the secondary side is identical with the value on the primary side. Similarly, for the leakage reactance,

$$I_2 X_{e2} = 455(0.024) = 10.9 \text{ V}$$

$$\frac{I_2 X_{e2}}{V_2} 100 = \frac{10.9}{220} 100 = 4.96\%$$

(d) The equivalent leakage impedance referred to the primary is

$$\bar{Z}_{e1} \equiv R_{e1} + jX_{e1} = 0.2 + j0.6 = 0.634\underline{/71.6°}$$

Referred to the secondary this is

$$\bar{Z}_{e2} = \frac{\bar{Z}_{e1}}{a^2} = 0.0253\underline{/71.6°}$$

2-3 PARAMETERS FROM NO-LOAD TESTS

The exact equivalent circuit of the transformer has a total of six parameters as Fig. 2-14(d) shows. A knowledge of these parameters allows us to compute the performance of the transformer under all operating conditions. When the complete design data of a transformer are available, the parameters may be calculated from the dimensions and properties of the materials involved. Thus, the primary winding resistance r_1 may be found from the resistivity of copper, the total winding length, and the cross-sectional area. In a similar fashion a parameter such as the magnetizing reactance x_ϕ may be determined from the number of primary turns, the reluctance of the magnetic path, and the frequency of operation. The calculation of the leakage reactance is a bit more complicated because it involves accounting for partial flux linkages. However, formulas are available for making a reliable determination of these quantities.

A more direct and far easier way of determining the transformer parameters is from tests that involve very little power consumption, called *no-load tests*. The power consumption is merely that which is required to supply the appropriate losses involved. The primary winding (or secondary winding) resistance is readily determined by applying a small dc voltage to the winding in the manner shown in Fig. 2-21. The voltage appearing across the winding must be large enough to cause approximately rated current to flow. Then the ratio of the voltage drop across the winding as recorded by the voltmeter, *VD*, divided by the current flowing through it as recorded by the ammeter, *AD*, yields the value of the winding resistance.

Open-Circuit Test

Information about the core-loss resistor r_c and the magnetizing reactance x_ϕ is obtained from an *open-circuit test,* the circuitry of which appears in Fig. 2-22. Note that the secondary winding is open. Therefore, as Fig. 2-14(d) indicates, the input impedance consists of the primary leakage impedance and the magnetizing impedance. In this case the ammeter and voltmeter are ac instruments and the second letter *A* is used to emphasize this. Thus *AA* denotes ac ammeter. The

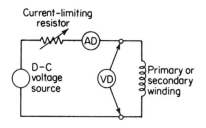

Figure 2-21 Circuitry for finding winding resistance.

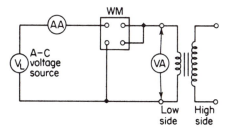

Figure 2-22 Wiring diagram for the transformer open-circuit test.

wattmeter shown in Fig. 2-22 is used to measure the power drawn by the transformer. The open-circuit test is performed by applying voltage either to the high-voltage side or the low-voltage side, depending upon which is more convenient. Thus, if the transformer of Example 2-1 is to be tested, the voltage would be applied to the low-tension side because a source of 220 V is more readily available than one of 1100 V. The core loss is the same whether 220 V is applied to the winding having the smaller number of turns or 1100 V is applied to the winding having the larger number of turns. The maximum value of the flux, upon which the core loss depends, is the same in either case as indicated by Eq. (2-17). Now for convenience assume that the instruments used in the open-circuit test yield the following readings:

$$\text{wattmeter reading} = P$$

$$\text{ammeter reading} = I_m$$

$$\text{voltmeter reading} = V_L$$

The voltmeter reading carries the subscript L to emphasize that the test is performed with the source and instruments on the low-voltage side.

Neglecting instrument losses, the wattmeter reading may be taken entirely equal to the core loss, i.e., $P_c = P$. This is because the attendant copper loss is negligibly small. Moreover, since I_m is very small, the primary leakage impedance drop may be neglected, so that for all practical purposes the induced emf is equal to the applied voltage, i.e., $E_1 = V_L$. In accordance with these simplifications the open-circuit phasor diagram is represented by Fig. 2-3(b). The no-load power-factor angle θ_0 is computed from

$$\theta_0 = \cos^{-1} \frac{P}{V_L I_m} \tag{2-47}$$

It then follows that

$$I_c = I_m \cos \theta_0 \tag{2-48}$$

and

$$I_\phi = I_m \sin \theta_0 \tag{2-49}$$

It is important to keep in mind that these currents are referred to the low-tension side. The corresponding values on the high-voltage side would be less by the factor $1/a$. The low-side core-loss resistor is

$$r_{cL} = \frac{P}{I_c^2} = \frac{P}{(I_m \cos \theta_0)^2} \tag{2-50}$$

The corresponding high-side core-loss resistor is

$$r_{cH} = a^2 r_{cL} \tag{2-51}$$

The magnetizing reactance referred to the low side follows from Eq. (2-29). Thus

$$x_{\phi L} = \frac{E_1}{I_\phi} = \frac{V_L}{I_m \sin \theta_0} \qquad (2\text{-}52)$$

On the high side this becomes

$$x_{\phi H} = a^2 x_{\phi L} \qquad (2\text{-}53)$$

Although the manner of finding the magnetizing impedance is treated in the foregoing, the motivation is not so much necessity as the desire for completeness. Invariably, the use of the approximate equivalent circuit is sufficient for all but a few special applications of transformers. Accordingly, the useful information obtained from the open-circuit test is the value of the core loss, which helps to determine the efficiency of the transformer.

Short-Circuit Test

Of the six parameters of the exact equivalent circuit two remain to be determined: the primary and secondary leakage reactances x_1 and x_2. This information is obtained from a short-circuit test which involves placing a small ac voltage on one winding and a short-circuit on the other as depicted in Fig. 2-23(a). Appearing in

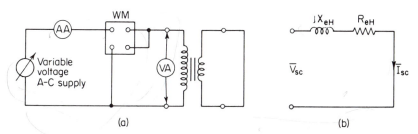

Figure 2-23 Transformer short-circuit test: (a) wiring diagram; (b) equivalent circuit.

Fig. 2-23(b) is the approximate equivalent circuit with the secondary winding short-circuited. Note that the input impedance is merely the equivalent impedance.

When performing the short-circuit test, a reduced ac voltage is applied to the high side usually because it is more convenient to do so. From the computations of Example 2-1 recall that the leakage impedance drop is only about 5% of the winding voltage rating. Hence 5% of 1100 V, which is 55 V, is easier and more accurate to deal with than 5% of 220 V, which is 11 V. Also, the wattmeter reading in the circuitry of Fig. 2-23(a) can be taken entirely equal to the winding copper losses. This follows from the fact that the greatly reduced voltage used in the short-circuit test makes the core loss negligibly small. Calling the wattmeter reading P_{sc} and the ammeter reading I_{sc}, we can determine the equivalent winding resistance as

$$R_{eH} = \frac{P_{sc}}{I_{sc}^2} \qquad (2\text{-}54)$$

The subscript H is used in place of L since it is known from the wiring diagram that the primary side is the high-voltage side. The equivalent resistance referred to the low side is then

$$R_{eL} = \frac{R_{eH}}{a^2} \qquad (2\text{-}55)$$

where

$$a = \frac{N_H}{N_L} = \frac{\text{high-side turns}}{\text{low-side turns}} \qquad (2\text{-}56)$$

Before the equivalent reactance can be determined, it is first necessary to find the equivalent impedance, which a glance at Fig. 2-23(b) reveals to be

$$Z_{eH} = \frac{V_{sc}}{I_{sc}} \qquad (2\text{-}57)$$

During the test the applied voltage V_{sc} is adjusted so that I_{sc} is at least equal to the rated winding current. The equivalent reactance follows from Eqs. (2-54) and (2-57). Thus

$$X_{eH} = \sqrt{Z_{eH}^2 - R_{eH}^2} \qquad (2\text{-}58)$$

It is important to note that this computation provides the _sum of the primary and secondary leakage reactances_. It gives no information about the individual breakdown between x_1 and x_2. Whenever the approximate equivalent circuit is used in analysis, such a breakdown is unnecessary. On the few occasions when the individual breakdown is needed, it is customary to assume that

$$x_1 = x_2' \qquad (2\text{-}59)$$

This statement is based on the assumption that the transformer is well-designed.

One final note. It should be apparent from the theory of transformer action that the establishment of rated current in the high-side winding requires a corresponding flow of rated current in the secondary circuit. It is for this reason that the _total_ winding copper loss is measured in the short-circuit test.

Example 2-2

$a = 20$

The following data were obtained on a 50-kVA, 2400/120-V transformer:

Open-circuit test, instruments on low side:

$$\text{wattmeter reading} = 396 \text{ W}$$

$$\text{ammeter reading} = 9.65 \text{ A}$$

$$\text{voltmeter reading} = 120 \text{ V}$$

Short-circuit test, instruments on high side:

$$\text{wattmeter reading} = 810 \text{ W}$$

$$\text{ammeter reading} = 20.8 \text{ A}$$

$$\text{voltmeter reading} = 92 \text{ V}$$

Compute the six parameters of the equivalent circuit referred to the high and low sides.

Solution   From the open-circuit test

$$\theta_0 = \cos^{-1} \frac{P}{V_L I_m} = \cos^{-1} \frac{396}{120(9.65)} = 70°$$

$$\therefore \quad I_{cL} = I_m \cos \theta_0 = 9.65(0.342) = 3.3 \text{ A}$$

$$I_{\phi L} = I_m \sin \theta_0 = 9.65 \sin 70° = 9.07 \text{ A}$$

Hence

$$r_{cL} = \frac{396}{3.3^2} = 36.4 \text{ }\Omega$$

This can also be found from

$$r_{cL} = \frac{E_1}{I_c} = \frac{V_L}{I_c} = \frac{120}{3.3} = 36.4 \text{ }\Omega$$

On the high side this quantity becomes

$$r_{cH} = a^2 r_{cL} = 400(36.4) = 14,560 \text{ }\Omega$$

$$\frac{N_1}{N_2} = a$$

The magnetizing reactance referred to the low side is

$$x_{\phi L} = \frac{E_1}{I_\phi} = \frac{V_L}{I_\phi} = \frac{120}{9.07} = 13.2 \text{ }\Omega$$

The corresponding high-side value is

$$x_{\phi H} = a^2 x_{\phi L} = 400(13.2) = 5280 \text{ }\Omega$$

From the data of the short-circuit test we get directly

$$\frac{Z_{eH}}{a^2}$$

$$Z_{eH} = \frac{92}{20.8} = 4.42 \text{ }\Omega, \qquad \therefore \quad Z_{eL} = \frac{4.42}{400} = 0.011 \text{ }\Omega$$

Also

$$R_{eH} = \frac{810}{(20.8)^2} = 1.87, \qquad \therefore \ R_{eL} = \frac{1.87}{400} = 0.0047$$

and

$$X_{eH} = \sqrt{Z_{eH}^2 - R_{eH}^2} = \sqrt{4.42^2 - 1.87^2} = 4 \ \Omega$$

$$\therefore \ X_{eL} = \frac{4}{400} = 0.01 \ \Omega$$

2-4 EFFICIENCY AND VOLTAGE REGULATION

The manner of describing the performance of a transformer depends upon the application for which it is designed. It has already been pointed out that in communication circuits the frequency response of the transformer is very often of prime importance. Another important characteristic of such transformers is to provide matching of a source to a load for maximum transfer of power. The efficiency and regulation is usually of secondary significance. This is not so, however, with power and distribution transformers, which are designed to operate under conditions of constant rms voltage and frequency. Accordingly, the material that follows has relevance chiefly to these transformers.

As is the case with other devices, the efficiency of a transformer is defined as the ratio of the useful output power to the input power. Thus

$$\eta = \frac{\text{output watts}}{\text{input watts}} \tag{2-60}$$

Usually the efficiency is found through measurement of the losses as obtained from the no-load tests. Therefore by recognizing that

$$\text{output watts} = \text{input watts} - \Sigma \text{ losses} \tag{2-61}$$

and inserting into Eq. (2-60), a more useful form of the expression for efficiency results. Thus

$$\boxed{\eta = 1 - \frac{\Sigma \text{ losses}}{\text{input watts}}} \tag{2-62}$$

where

$$\Sigma \text{ losses} = \text{core loss} + \text{copper loss} = P_c + I_2^2 R_{e2} \tag{2-63}$$

Equation (2-62) also leads to a more accurate value of the efficiency.

Power and distribution transformers very often supply electrical power to loads that are designed to operate at essentially constant voltage regardless of whether little or full-load current is being drawn. For example, as more and more light bulbs are placed across a supply line, which very likely originates from a distribution transformer, it is important that the increased current being drawn from the supply transformer does not cause a significant drop (more than 10%) in the load voltage. If this happens the illumination output from the bulbs is sharply reduced. Similar undesirable effects occur in television sets that are connected across the same supply lines. The reduced voltage means reduced picture size. An appreciable drop in line voltage with increasing load demands can also cause harmful effects in connected electrical motors such as those found in refrigerators, washing machines, and so on. Continued operation at low voltage can cause these units to overheat and eventually burn out. The way to prevent the drop in supply voltage with increasing load in distribution circuits is to use a distribution transformer designed to have small leakage impedance. The figure of merit used to identify this characteristic is the voltage regulation.

The *voltage regulation* of a transformer is defined as the change in magnitude of the secondary voltage as the current changes from full-load to no-load with the primary voltage held fixed. Hence in equation form,

$$\text{voltage regulation} = \frac{|\bar{V}_1| - |\bar{V}_2'|}{|\bar{V}_2'|} = \frac{\left|\dfrac{\bar{V}_1}{a}\right| - |\bar{V}_2|}{|\bar{V}_2|} \tag{2-64}$$

where the symbols denote the quantities previously defined, absolute signs are used in order to emphasize that it is the *change in magnitude* that is important. Equation (2-64) expresses the voltage regulation on a per unit basis. To convert to percentage, it is necessary to multiply by 100. The smaller the value of the voltage regulation, the better suited is the transformer for supplying power to constant-voltage loads (constant voltage is characteristic of most commercial and industrial loads).

In determining the voltage regulation it is customary to assume that $\bar{V}_1$ is adjusted to that value which allows rated voltage to appear across the load when rated current flows through it. Under these conditions the necessary magnitude of $\bar{V}_1$ is found from Kirchhoff's voltage law as it applies to the approximate equivalent circuit. Refer to Figs. 2-15 and 2-16. For the assumed lagging power-factor angle θ_2 of the secondary circuit the expression for $\bar{V}_1$ is

$$\bar{V}_1 = V_2'\underline{/0^\circ} + I_2'\underline{/-\theta_2}\, Z_{e1}\underline{/\theta_z} \tag{2-65}$$

where

$$\theta_z = \tan^{-1}\frac{X_{e1}}{R_{e1}} \tag{2-66}$$

Equation (2-65) is written using the referred value of rated secondary terminal voltage as the reference.

Example 2-3

For the transformer of Example 2-2(a) find the efficiency when rated kilovolt-amperes are delivered to a load having a power factor of 0.8 lagging; (b) compute the voltage regulation.

Solution   (a) The losses at rated load current and rated voltage are

$$\Sigma \text{ losses} = 396 + 810 = 1206 \text{ W} = 1.2 \text{ kW}$$

Also

$$\text{output kW} = 50(0.8) = 40 \text{ kW}$$

$$\text{input kW} = 40 + 1.2 = 41.2 \text{ kW}$$
$$\scriptstyle out + \Sigma \, loss$$

Hence

$$\eta = 1 - \frac{\Sigma \text{ losses}}{\text{input kW}} = 1 - \frac{1.2}{41.2} = 1 - 0.029 = 0.971$$

or

$$\eta = 97.1\%$$

$$\scriptstyle from \; pf = cos^{-1} \theta_z$$
$$\scriptstyle + \theta_z \; lead$$
$$\scriptstyle - \theta_z \; lag$$

(b) From Eq. (2-65) we have

$$\bar{V}_1 = \bar{V}_H = 2400 + 20.8\underline{/-37°} \; 4.42\underline{/65°}$$

where

$$\theta_z = \tan^{-1}\frac{X_{eH}}{R_{eH}} = \tan^{-1}\frac{4}{1.87} = 65°$$

Therefore

$$\bar{V}_1 = \bar{V}_H = 2400 + 92\underline{/28°} = 2400 + 81.1 + j43.1$$

$$\bar{V}_H \approx 2481.1$$

Note that the _j-component is negligible in comparison to the in-phase component_. This calculation indicates that, in order for rated voltage to appear across the load when rated current is drawn, the primary voltage $\bar{V}_1$ must be slightly different from the primary winding voltage rating of 2400 V. The percentage of voltage regulation is accordingly found to be

$$\text{percent of voltage regulation} = \frac{2481.1 - 2400}{2400} \; 100 = 3.38\%$$

Example 2-4

In this example our aim is to find the voltage regulation by assuming that the transformer of Example 2-3 is operated at the _input terminals_ with the rated voltage of 2400 V. To facilitate the comparison of the ensuing result with that obtained in Example 2-3, use is made of the same impedance which yields the rated output current when rated voltage appears at the secondary terminals; in other words, the load impedance has the value $\bar{Z}'_L = 115.4\underline{/30°} = 100 + j57.8$ ohms referred to the high side.

Solution   By Kirchhoff's voltage law for the primary circuit we have

$$\bar{V}_1 = 2400\underline{/0°} = \bar{I}_1[R_{e1} + R'_L + j(X_{e1} + X'_L)]$$

$$= \bar{I}_1[1.87 + 100 + j(4 + 57.8)]$$

$$\therefore \quad \bar{I}_1 = \frac{2400}{101.87 + j61.8} = \frac{2400}{119.15\underline{/-31.24°}} = 20.14\underline{/-31.24°}$$

$$\therefore \quad \bar{V}'_2 = \bar{I}_1\bar{Z}'_L = 20.14\underline{/-31.24°}\,(115.4\underline{/30°}) = 2324.465\underline{/-1.24°}$$

$$\text{percent voltage regulation} = \frac{2400 - 2324.465}{2324.465}100 = 3.25\%$$

2-5 MUTUAL INDUCTANCE

Analysis of the two-winding transformer by the methods of electric circuit theory involves the use of self- and mutual-inductance. So far, Kirchhoff's voltage equation applied to the primary winding has led to Eq. (2-26). In the material that follows this result will be shown to be identical with that obtained by the classical approach.

For simplicity assume the ferromagnetic material has a linear magnetization curve with no core losses. Refer to Fig. 2-24 where for the indicated current directions Kirchhoff's equation becomes

$$v_1 = i_1 r_1 + L_1 \frac{di_1}{dt} - M \frac{di_2}{dt} \qquad (2\text{-}67)$$

where lowercase letters for current and voltage denote instantaneous values. Here the quantity $L_1\, di_1/dt$ denotes the voltage drop associated with the total flux produced by i_1 and linking the primary turns N_1. Similarly, the quantity $-M\, di_2/dt$ refers to the *voltage rise* associated with the flux produced by i_2 and linking the primary turns. The existence of a voltage rise in the primary circuit for the assumed direction of i_2 is easily verified. Note that the secondary mmf $N_2 i_2$, if allowed to act freely, produces a flux which threads counterclockwise in the configuration of Fig. 2-24. Then the reaction in the primary coil between two points such as a and b by Lenz's law requires that point b should be at a higher potential than a. Accordingly, this induced emf becomes a voltage rise with respect to the specified direction of i_1.

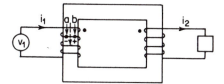

Figure 2-24   Elementary transformer for analysis by the classical approach.

Assuming that the applied primary voltage is a sinusoid, Eq. (2-67) may be written in rms quantities as

$$\bar{V}_1 = \bar{I}_1 r_1 + j\omega L_1 \bar{I}_1 - j\omega M \bar{I}_2 \tag{2-68}$$

Inserting the relationship

$$\bar{I}_2 = a\bar{I}_2' \tag{2-69}$$

wherein a is the ratio of transformation, we get

$$\bar{V}_1 = \bar{I}_1 r_1 + j\omega L_1 \bar{I}_1 - j\omega a M \bar{I}_2' \tag{2-70}$$

But by Eq. (2-24)

$$\bar{I}_2' = \bar{I}_1 - \bar{I}_m = \bar{I}_1 - \bar{I}_\phi$$

so that

$$\bar{V}_1 = \bar{I}_1 r_1 + j\omega(L_1 - aM)\bar{I}_1 + j\omega a M \bar{I}_\phi \tag{2-71}$$

Equation (2-71) is now in a form where correspondence with Eq. (2-26) is possible after some algebraic manipulation.

The expression for the self-inductance of the primary winding may be written as

$$L_1 = N_1 \frac{d\phi_1}{di_1} = N_1 \frac{\Phi_1}{i_1} \tag{2-72}$$

The equivalence of the differential form with the total quantities in this last expression is allowed by the linearity condition. Recalling that the total flux linking the primary coil consists of the mutual flux Φ_m plus the primary leakage flux Φ_{l1}, Eq. (2-72) can be rewritten as

$$L_1 = N_1 \frac{\Phi_m + \Phi_{l1}}{i_1} \tag{2-73}$$

Also,

$$\Phi_m = \frac{N_1 i_1}{\mathcal{R}_m} \quad \text{and} \quad \Phi_{l1} = \frac{N_1 i_1}{\mathcal{R}_{l1}} \tag{2-74}$$

where $\mathcal{R}_m$ is the reluctance experienced by the mutual flux and consists exclusively of iron and $\mathcal{R}_{l1}$ is the net reluctance experienced by the leakage flux of the primary in air. Accordingly, Eq. (2-73) becomes

$$L_1 = \frac{N_1^2}{\mathcal{R}_m} + \frac{N_1^2}{\mathcal{R}_{l1}} = L_m + L_{l1} \tag{2-75}$$

where L_m denotes an inductance associated with the mutual flux and L_{l1} is the inductance associated with the primary leakage flux. It is important to note here that L_m is not the mutual inductance.†

---

†L_m is in fact the mutual inductance referred to the primary winding.

The classical definition of mutual inductance is given in this case by

$$M_{21} = N_2 \frac{d\phi_{m1}}{di_1} = N_2 \frac{\Phi_{m1}}{i_1} \tag{2-76}$$

Thus the mutual inductance is related to the number of turns of winding 2 and the change of flux with respect to the current in winding 1 which produces the mutual flux Φ_{m1}. Upon inserting $N_1 i_1 = \Phi_{m1} \mathcal{R}_m$ into Eq. (2-76), there results

$$M_{21} = \frac{N_2 N_1}{\mathcal{R}_m} \tag{2-77}$$

The expression for mutual inductance may also be found by considering a change in mutual flux corresponding to a change in secondary current i_2 linking N_1 turns. Thus

$$M_{12} = N_1 \frac{d\phi_{m2}}{di_2} = N_1 \frac{\Phi_{m2}}{i_2} \tag{2-78}$$

But

$$\Phi_{m2} \mathcal{R}_m = N_2 i_2 \tag{2-79}$$

Inserting this into Eq. (2-78) yields

$$M_{12} = \frac{N_1 N_2}{\mathcal{R}_m} = M \tag{2-80}$$

A comparison of Eq. (2-80) with Eq. (2-77) shows them to be the same, and from here on either expression will be denoted by M.

Returning to Eq. (2-71) we are now in a position to evaluate the significance of the term in parentheses. Thus, by Eqs. (2-75) and (2-77)

$$L_1 - aM = \frac{N_1^2}{\mathcal{R}_m} + \frac{N_1^2}{\mathcal{R}_{l1}} - \frac{N_1}{N_2} \frac{N_2 N_1}{\mathcal{R}_m} = \frac{N_1^2}{\mathcal{R}_{l1}} = L_{l1} \tag{2-81}$$

Clearly this quantity is the primary leakage inductance, and it represents the effect of the primary leakage flux. Of course the total quantity

$$\omega(L_1 - aM)I_1 = \omega L_{l1} I_1 = x_1 I_1 \tag{2-82}$$

is nothing more than the primary leakage reactance drop and is the counterpart of the second term of Eq. (2-26).

Let us now examine the last term of Eq. (2-71). Thus

$$\omega a M I_\phi = \omega \frac{N_1}{N_2} \frac{N_2 N_1}{\mathcal{R}_m} I_\phi = \omega N_1 \frac{N_1 I_\phi}{\mathcal{R}_m} \tag{2-83}$$

The fraction quantity on the right side of the last equation is the expression for mutual flux. Specifically, however, since it involves the reluctance and the rms value I_ϕ of the magnetizing current, this expression represents the magnitude of the maximum value of the mutual flux divided by $\sqrt{2}$. That is,

$$\frac{N_1 I_\phi}{\mathcal{R}_m} = \frac{\Phi_m}{\sqrt{2}} \tag{2-84}$$

where Φ_m denotes the maximum value of the mutual flux. Accordingly, Eq. (2-83) becomes

$$\omega a M I_\phi = \omega N_1 \frac{\Phi_m}{\sqrt{2}} = \frac{2\pi}{\sqrt{2}} f N_1 \Phi_m = E_1 \tag{2-85}$$

It therefore follows from Eq. (2-85) that the mutual inductance parameter so commonly encountered in linear circuit analysis is implied in the expression for the induced emf. Clearly both the induced emf and the mutual inductance are related by the same mutual flux.

2-6 THE EQUIVALENT CIRCUIT: COUPLED CIRCUIT VIEWPOINT

In Sec. 2-2, the equivalent circuit of the transformer is derived from the field viewpoint based on interacting mmf's and fluxes. This approach is often favored because it cultivates a better physical understanding of the mechanism that underlies the operation and behavior of this device. However, it is entirely feasible to arrive at the same equivalent circuit by applying the techniques of circuit analysis to the coupled circuit, which is what the transformer is, after all. The usual starting point in such an analysis is Eq. (2-67), which is repeated here for convenience. Thus

$$v_1 = i_1 r_1 + L_1 \frac{di_1}{dt} - M \frac{di_2}{dt} \tag{2-86}$$

Keep in mind that L_1 denotes the self-inductance of the primary winding that is associated with N_1^2; on the other hand, M represents the mutual inductance between the primary and secondary windings and involves the product of N_1 and N_2 as indicated by Eq. (2-80). Figure 2-25 illustrates the schematic arrangement from which it is obvious that Eq. (2-86) describes the situation that prevails in the primary circuit in a manner consistent with Kirchhoff's voltage law. For ease of

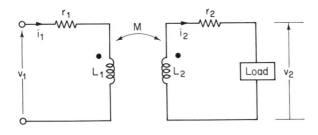

Figure 2-25 Schematic diagram of the two-winding transformer expressed in terms of the mutual and self-inductances.

handling, let us rewrite the last equation by replacing differentiation with the derivative operator s. Accordingly,

$$v_1 = i_1 r_1 + L_1 s i_1 - s M i_2 \tag{2-87}$$

Moreover, because the secondary emf is related to the primary emf by the turns ratio a and because power is preserved assuming negligible core losses, it follows that

$$i_2 = a i_b \tag{2-88}$$

where i_b is the required balance current needed in the primary winding to ensure power balance. In other words,

$$e_2 i_2 = \frac{e_1}{a} (a i_b) \tag{2-89}$$

where e_1 and e_2 are, respectively, the instantaneous voltages induced in the primary and secondary windings. Inserting Eq. (2-88) into Eq. (2-87) yields

$$v_1 = i_1 r_1 + L_1 s i_1 - a M s i_b \tag{2-90}$$

Next, we add zero to the right side of this equation in the form: $a M s i_1 - a M s i_1$. Hence we get

$$v_1 = i_1 r_1 + (L_1 - a M) s i_1 + a M s (i_1 - i_b) \tag{2-91a}$$

For sinusoidal sources the differential operator s can be replaced by $j\omega$ so that the last equation can be rewritten as

$$v_1 = i_1 r_1 + j\omega (L_1 - a M) i_1 + j\omega a M (i_1 - i_b) \tag{2-91b}$$

Before proceeding with a circuit interpretation of this equation, let us repeat the foregoing analysis for the secondary circuit in Fig. 2-25.

In accordance with the dot notation, when the current i_1 takes the direction shown in Fig. 2-25, the transformer action is such that the dotted side of the secondary coil is positive and the current i_2 assumes the clockwise direction indicated. Kirchhoff's voltage law then leads to

$$M \frac{di_1}{dt} = i_2 r_2 + v_2 + L_2 \frac{di_2}{dt} \tag{2-92}$$

In this expression the voltage drops appear on the right side. The use of the derivative operator then allows us to rewrite this equation as

$$M s i_1 = i_2 r_2 + v_2 + L_2 s i_2 \tag{2-93}$$

Now multiply both sides of Eq. (2-93) by the turns ratio a to yield

$$a M s i_1 = a v_2 + a i_2 r_2 + a L_2 s i_2 \tag{2-94}$$

Inserting Eq. (2-88) then leads to

$$a M s i_1 = a v_2 + i_b (a^2 r_2) + (a^2 L_2) s i_b \tag{2-95}$$

It is now necessary to manipulate this expression in such a manner that the ensuing interpretation will be entirely consistent with Eq. (2-91), which after all is part of the same equipment with the difference that Eq. (2-91) provides a description in the primary circuit while the latter does so in the secondary circuit. Because the mutual inductance is the circuit parameter that links the primary and secondary circuits, it should play the key role in providing a linkage between Eqs. (2-91) and (2-95). A study of Eq. (2-91) suggests that the last term on the right, namely, $aMs(i_1 - i_b)$, is the term that represents this linkage. This conclusion becomes even more apparent when $(i_1 - i_b)$ is replaced by i_m as called for by Eq. (2-25) because the magnetizing current and the mutual inductance go hand in hand. Therefore, to establish this term in Eq. (2-95), it is necessary merely to add zero to the left side in the form: $aMsi_b - aMsi_b$. Upon so doing and then rearranging, there results

$$aMs(i_1 - i_b) = av_2 + i_b(a^2 r_2) + (a^2 L_2 - aM)si_b \qquad (2\text{-}96a)$$

Then by replacing s by $j\omega$ we get

$$a\omega M(i_1 - i_b) = av_2 + a^2 r_2 i_b + j\omega(a^2 L_2 - aM)i_b \qquad (2\text{-}96b)$$

The circuit interpretation of Eqs. (2-91b) and (2-96b) taken together leads to the T configuration depicted in Fig. 2-26. Several important points are worth noting here. First, the quantity aM is really the mutual inductance referred to an equivalent winding having N_1 turns. This is apparent from

$$aM = \frac{N_1}{N_2}\left(\frac{N_1 N_2}{\mathcal{R}_m}\right) = \frac{N_1^2}{\mathcal{R}_m} \qquad (2\text{-}97)$$

Moreover, the quantity $(L_1 - aM)$ must necessarily represent the primary leakage flux. Recall that L_1 represents the total flux that links the primary coil which includes the leakage flux as well as the mutual flux whereas aM represents the mutual flux referred to primary turns. It then follows that $\omega(L_1 - aM)$ denotes the primary leakage reactance. A similar situation prevails in the secondary part of this circuit except that all quantities are expressed in terms of an equivalence based on replacing the actual secondary turns, N_2, by the primary turns N_1.

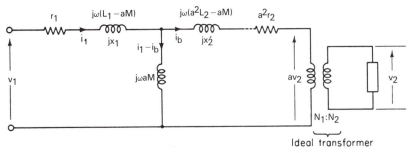

Figure 2-26 Equivalent circuit of the two-winding transformer derived from the coupled-circuit viewpoint.

Thus, although the load voltage v_2 is derived from a secondary winding having N_2 turns, multiplication by a in Eq. (2-96) means that the level is raised to that of a winding having N_1 turns. Similarly, the self-inductance of the secondary winding, which is generally expressed as $L_2 = N_2^2/\mathcal{R}_m$, is converted to an equivalent inductance of N_1 turns when it is multiplied by a^2 as shown in Eq. (2-96). It follows then that the quantity $(a^2L_2 - aM)$ can be considered to be an inductance that represents the secondary leakage flux referred to the primary. The resistance parameter too is modified in a like manner. The secondary winding resistance appears in Eq. (2-96) as well as in the circuit of Fig. 2-26 raised to a level of a^2r_2. This is consistent with the fact that the current is reduced by $1/a$, i.e., $i_b = i_2/a$, as required by the changed voltage level, av_2. The result is to keep the copper loss the same for power cannot be changed by the transformation of a winding of N_2 turns to one of N_1 turns. An ideal transformer is included in the equivalent circuit of Fig. 2-26 to provide the proper transition to the load where the voltage v_2 appears. Once the configuration of Fig. 2-26 is obtained, the matter of accounting for the core losses can readily be resolved by inserting a resistor in series with the mutual inductance parameter, aM, of such a value that it equals the core losses when multiplied by the square of the rms value of magnetizing current for sinusoidal sources.

2-7 THE AUTOTRANSFORMER

One of the principal features of the ordinary configurations of the transformer (see Fig. 2-24) is the electrical isolation that exists between the primary and secondary windings. Thus in the case of a distribution transformer that transforms power from 4400 V to 220 V, there does not exist the danger of the 4400 V appearing at the secondary terminals. Whenever large transformation ratios are involved, safety demands that ordinary transformers be used. However, in instances where the transformation ratio is close to unity (which minimizes the hazard), a transformer employing a single tapped winding may be used with notable advantages.

To understand some of these advantages, consider the arrangement of the conventional transformer appearing in Fig. 2-27(a). Assume the transformer is ideal for simplicity. A resistive load is shown to draw 12 A at a terminal voltage of 100 V. The corresponding primary current is 10 A. An important observation to make in this arrangement is that, all along the secondary winding between points b' and c' and all along that section of the primary winding between b and c, it is possible to identify points having the same instantaneous potential. If b and b' are two such points, they may be joined without adverse effect. Extending this procedure throughout the entire length of windings bc and $b'c'$ leads to the simplified configuration appearing in Fig. 2-27(b). This arrangement is called the *autotransformer*.

Several things are worth noting here. First, the secondary winding $b'c'$ of the ordinary transformer can be completely eliminated, and in its place an appro-

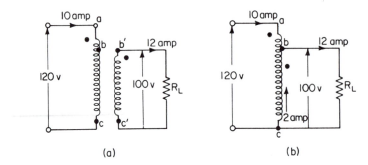

Figure 2-27 (a) Ordinary transformer arrangement; (b) derived autotransformer arrangement. Comparison is on the basis of same output kVA.

priate tap point, i.e., *b*, can be used to provide the required secondary voltage. Since the output volt-amperes remain intact in the arrangement of Fig. 2-27(b), it follows that the use of the autotransformer to provide the needed kVA at the load leads to a saving in cost. Less copper is required for the same output. As a matter of fact the saving in copper is even more pronounced than already indicated. The current that flows in the common winding of the autotransformer is always the difference between the primary and secondary currents of the ordinary transformer. In the case at hand this current is 2 A. Accordingly, the cross-sectional area of the copper of winding *bc* of the autotransformer can be much smaller than that of the primary of the ordinary transformer for the same design value of current density. Second, the efficiency of the autotransformer for the same output is greater. This result immediately follows from the smaller copper losses. Third, the voltage regulation of the autotransformer is superior because of the reduced resistance drop and lower leakage reactance drop. Keep in mind that one winding has been completely eliminated. Hence the resistance and leakage flux of this winding are zero. Fourth, the autotransformer is smaller in size for the same output. This too follows from the reduced copper need.

There is another way in which the autotransformer of Fig. 2-27(b) may be derived from an ordinary transformer. The second approach is predicated on the fact that it uses, completely, the primary and secondary windings of an appropriate ordinary transformer. The first approach is based on the same output for the ordinary transformer and the autotransformer. In contrast the second approach is based on the use of the same copper and iron; we should expect therefore that the autotransformer will have a greater output rating than the corresponding ordinary transformer. (The latter in this case is not the same as the ordinary transformer used in the first approach.)

A close examination of Fig. 2-27(b) reveals that the autotransformer can be derived by placing the winding voltages of two separate coils *ab* and *bc* in series aiding. The coils *ab* and *bc* can then be treated as the primary and secondary windings, respectively, of an *equivalent ordinary transformer* such as that illustrated in Fig. 2-28(a). Note that the volt-ampere rating of the equivalent ordinary transformer is 200, compared with 1200 for the derived autotransformer. In a practical situation coils *ab* and *b'c* appear on the same magnetic core. Refer to Fig. 2-28(b). The series-aid connection of the coils can be achieved by first energizing one of the coils, say *ab*, with rated voltage, then joining one terminal of the first coil, say *b*, to one terminal of the second coil, say *b'*, and finally checking that a voltmeter placed between the remaining two terminals *a* and *c* yields a reading that is the sum of the voltage ratings of the individual coils. If the voltmeter reads the difference, the connection to the second coil must be reversed. For the configuration of Fig. 2-28(b) the sense of the winding *b'c* has been so taken that the series-aid connection results when *b* is joined to *b'*.

Current and Voltage Relations for Step-Down Autotransformer

A step-down autotransformer is one in which the primary voltage is larger than the secondary voltage. By applying a source voltage to the full winding *abc* in Fig. 2-28(b) and placing the load across *bc*, operation as a step-down autotransformer results. In this configuration the applied primary voltage causes a magnetizing current to flow through the full winding *ac*. In view of its small magnitude compared with the rated input current, for the sake of ease and simplicity of analysis, this current will be considered equal to zero from here on.

The same basic principle that underlies the operation of the ordinary transformer is found to govern the operation of the autotransformer, i.e., the need for balancing mmf's. We use this as the starting point in calculating the current and voltage relationships for the autotransformer. To simplify the notation, use is made of the symbols depicted in Fig. 2-29. When the load switch *S* is closed in the

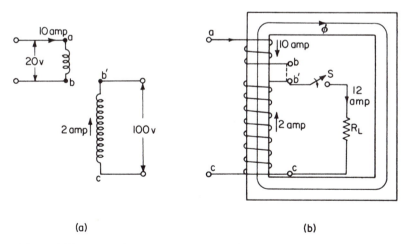

(a)   (b)

Figure 2-28 (a) Equivalent ordinary transformer; (b) its derived autotransformer. Comparison is on the basis of the same copper and iron.

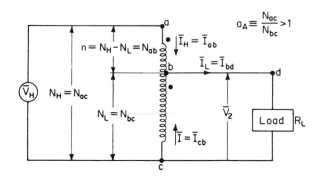

Figure 2-29 Symbolism for analysis of
the step-down autotransformer.

secondary of the autotransformer of Fig. 2-28(b), a current $\bar{I}$ begins to flow in the
common coil bc. It attempts to reduce the operating flux, but as soon as this
begins to happen there occurs an increase in primary current $\bar{I}_H$ to such a value
that the mmf in winding ab neutralizes the mmf in winding bc. That is,

$$n\bar{I}_H = N_L\bar{I} \tag{2-98}$$

or

$$(N_H - N_L)\bar{I}_H = N_L\bar{I} \tag{2-99}$$

Hence,

$$\boxed{\frac{\bar{I}}{\bar{I}_H} = \frac{I}{I_H} = \frac{N_H}{N_L} - 1 = a_A - 1} \tag{2-100}$$

where

$$a_A \equiv \frac{N_H}{N_L} = \frac{N_{ac}}{N_{bc}} > 1 \tag{2-101}$$

Since the right side of Eq. (2-100) is a pure number, it follows that the currents in
coils ab and cb are in-phase. The opposing action of the mmf's is derived rather
from the sense of the windings.

Applying Kirchhoff's current law at point b yields

$$\bar{I}_L = \bar{I}_H + \bar{I} \tag{2-102}$$

But since $\bar{I}_H$ and $\bar{I}$ are both in time phase, it follows that Eq. (2-102) is valid also
when only magnitudes are used. Thus,

$$\boxed{I_L = I_H + I} \tag{2-103}$$

Inserting $I = I_L - I_H$ into Eq. (2-100) yields the relationship between the low- and
the high-side currents. Hence

$$\frac{I_L - I_H}{I_H} = a_A - 1$$

or

$$\frac{I_L}{I_H} = a_A \tag{2-104}$$

Finally, inserting $I_H = I_L/a_A$ into Eq. (2-100) gives

$$\frac{I}{I_L} = \frac{a_A - 1}{a_A} \tag{2-105}$$

The induced voltages in windings ab and bc are also in time phase because they are caused by the same flux. Letting $E_1 = E_{ac} = E_H$, $E_2 = E_{bc} = E_L$, and $E_n = E_{ab}$, we then have

$$\frac{E_n}{E_L} = \frac{E_H - E_L}{E_L} = a_A - 1 \tag{2-106}$$

An inspection of Eqs. (2-100) and (2-106) shows that they bear out mathematically the equivalence described in Fig. 2-28. These equations indicate that the ratio of the voltages and currents in the coils ab and bc are the same as if the turns $n = N_{ab}$ formed the *primary* and the turns $N_2 = N_{bc}$ formed the *secondary* of an ordinary transformer having a ratio of transformation of $a_A - 1$. The value of a_A for the autotransformer of Fig. 2-28(b) is 1.2. Hence the equivalent ordinary transformer should have a turns ratio of $a_A - 1$, or 1/5. A glance at Fig. 2-28(a) corroborates the result.

One advantage of treating the autotransformer as an equivalent ordinary transformer is that use may be made immediately of previously established results. For example, to get the equivalent resistance of the autotransformer referred to the high side all that is necessary is to write the equivalent resistance of the equivalent ordinary transformer, treating winding ab as the primary and winding bc as the secondary. From our previous knowledge of the ordinary transformer we can then write

$$R_{eH} = r_{ab} + (a_A - 1)^2 r_{bc} = r_n + (a_A - 1)^2 r_2 \tag{2-107}$$

where $r_n = r_{ab}$, the resistance of winding ab, and $r_2 = r_{bc}$, the resistance of winding bc. Similarly the equivalent reactance referred to the high side becomes

$$X_{eH} = x_{ab} + (a_A - 1)^2 x_{bc} = x_n + (a_A - 1)^2 x_2 \tag{2-108}$$

The result expressed in Eq. (2-107) can be verified by writing the expression for the total copper loss when the secondary load terminals are short-circuited. If the current flowing through ab is I_H and that through bc is I, the copper loss at short circuit is then given by

$$P_{Cu} = I_H^2 r_{ab} + I^2 r_{bc} \qquad (2\text{-}109)$$

An equivalent resistance as seen from the high side may be derived from Eq. (2-109) provided I can be expressed in terms of I_H. This is readily accomplished by Eq. (2-100). Thus,

$$P_{Cu} = I_H^2 r_{ab} + (a_A - 1)^2 I_H^2 r_{bc}$$

or

$$P_{Cu} = I_H^2 [r_{ab} + (a_A - 1)^2 r_{bc}] \qquad (2\text{-}110)$$

The quantity in brackets can be interpreted as that equivalent high-side resistance which when multiplied by the high-side current squared yields the copper loss.

Power and Volt-Ampere Relations

Unlike the conventional transformer, the autotransformer has two kinds of volt-ampere power associated with it. It is the result of the direct copper connection existing between the source and the load. Refer to Fig. 2-29 to distinguish the two. Since the transformer action present in the autotransformer takes place in windings ab and bc, it follows that the number of volt-amperes in winding ab is equal to the number of volt-amperes in winding bc and specifically represents the transformed apparent power. Expressing this mathematically we have

$$\text{transformed VA} = E_{ab} I_{ab} = E_{bc} I_{bc} \qquad (2\text{-}111)$$

where VA denotes volt-amperes. If for simplicity we assume negligible winding resistance and leakage reactance, Eq. (2-111) can be written in terms of the terminal voltages. Thus

$$\text{transformed VA} = V_{ab} I_{ab} = V_{bc} I_{bc} \qquad (2\text{-}112)$$

For the values specified in Fig. 2-28 note that the transformed volt-amperes are $20(10) = 2(100) = 200$.

However, the input corresponding to the specified transformer current is $V_{ac} I_{ab} = 120(10) = 1200$ VA. This means that the number of volt-amperes that is not transformed is

$$V_{ac} I_{ab} - V_{ab} I_{ab} = (V_{ac} - V_{ab}) I_{ab} = V_{bc} I_{ab} \qquad (2\text{-}113)$$

The specific value for this example is then

$$V_{bc} I_{ab} = 100(10) = 1000 \text{ VA}$$

The quantity represented by Eq. (2-113) is called the *conducted volt-amperes*. The reason for the description follows from the fact that the input current I_{ab} flows *conductively* to the load at the potential V_{bc}. Although I_{ab} starts at the potential V_{ac}, it experiences a drop V_{bc} through transformation and then proceeds to the load at the reduced potential. However, it is important to keep in mind that the reduced potential at which I_{ab} flows to the load is compensated for exactly by an increase in load current originating from the transformer action.

In the notation of Fig. 2-29 we can write for the step-down autotransformer

$$\boxed{\text{transformed VA} = V_n I_H = V_L I} \qquad (2\text{-}114)$$

$$\boxed{\text{conducted VA} = V_L I_H} \qquad (2\text{-}115)$$

An interesting comparison to make at this point is the ratio of the output volt-amperes of the autotransformer to the transformed volt-amperes. The latter quantity in essence is the output of an ordinary transformer having the same current density and magnetic density ratings. Accordingly,

$$\frac{\text{output VA of autotransformer}}{\text{transformed VA}} = \frac{\text{output VA of autotransformer}}{\text{output VA of equiv. ord. transformer}}$$

$$= \frac{V_L I_L}{V_L I} = \frac{a_A}{a_A - 1} \qquad (2\text{-}116)$$

where a_A is the ratio of transformation of the autotransformer. The last term in the preceding equation follows from Eq. (2-105). Applying Eq. (2-116) to the configuration of Fig. 2-28 yields a factor of 6, which emphasizes that for many autotransformer applications (a_A close to unity) most of the power delivered to the load occurs through direct conduction and just a small portion is derived through transformation.

Phasor diagram. The phasor diagram of the step-down autotransformer can be conveniently developed by applying Kirchhoff's voltage law to the high-side circuit and making use of the equivalent ordinary transformer in establishing the quantity $V_{ab} = V_n$. Accordingly, in Fig. 2-27 Kirchhoff's voltage law yields

$$\bar{V}_H = \bar{V}_{ab} + \bar{V}_{bc} = \bar{V}_n + \bar{V}_L \qquad (2\text{-}117)$$

By treating winding ab as the primary of an equivalent ordinary transformer, it follows that

$$\bar{V}_{ab} = \bar{V}_n = (a_A - 1)\bar{V}_L + \bar{I}_H[r_{ab} + (a_A - 1)^2 r_{bc}] + j\bar{I}_H[x_{ab} + (a_A - 1)^2 x_{bc}]$$

$$(2\text{-}118)$$

or

$$\bar{V}_{ab} = \bar{V}_n = (a_A - 1)\bar{V}_L + \bar{I}_H(R_{eH} + jX_{eH}) = (a_A - 1)\bar{V}_L + \bar{I}_H\bar{Z}_{eH} \qquad (2\text{-}119)$$

Inserting Eq. (2-119) into Eq. (2-117) yields

$$\bar{V}_H = (a_A - 1)\bar{V}_L + \bar{I}_H\bar{Z}_{eH} + \bar{V}_L \qquad (2\text{-}120)$$

or more simply

$$\bar{V}_H = a_A\bar{V}_L + \bar{I}_H\bar{Z}_{eH} \qquad (2\text{-}121)$$

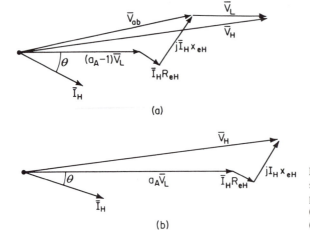

(a)

(b)

Figure 2-30   Phasor diagram of the step-down autotransformer: (a) interpretation in terms of Eq. (2-120); (b) interpretation in terms of Eq. (2-121).

Appearing in Fig. 2-30(a) is the phasor diagram representation of Eq. (2-120). The simplified version is depicted in Fig. 2-30(b).

Step-Up Autotransformer

The step-up autotransformer may be analyzed in a manner similar to that used for the step-down configuration. In the step-up case the source is applied to winding bc and the load is placed across the full winding ac. Refer to Fig. 2-31. The functional relationships between the input, output, and common winding currents can be derived by starting with the balancing mmf condition that must prevail in coils ab and bc. If this is done and if the ratio of transformation of the step-up case is defined as greater than unity, i.e., $a_A \equiv N_{ac}/N_{bc} = V_H/V_L$, the following results are obtained:

$$\frac{I}{I_H} = a_A - 1 \tag{2-122}$$

$$\frac{I_L}{I_H} = a_A \tag{2-123}$$

$$\frac{I}{I_L} = \frac{a_A - 1}{a_A} \tag{2-124}$$

$$I_L = I_H + I \tag{2-125}$$

and

$$E_{ab} = E_n = E_{bc}(a_A - 1) = E_L(a_A - 1) \tag{2-126}$$

A comparison of these equations with Eq. (2-100) and Eqs. (2-103) to (2-104) reveals that the results for the step-up case can be obtained from those of the step-

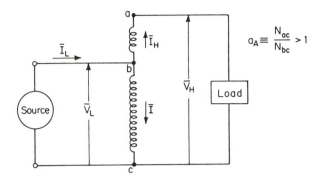

Figure 2-31 Configuration and symbolism for the step-up autotransformer.

down case by merely interchanging the roles of the currents I_L and I_H *provided that a_A* is defined the same way in both cases.

Example 2-5

A step-down autotransformer is to be used to start an induction motor in order to limit the magnitude of the starting current to an acceptable level which is known to be 34 A for a line voltage rating of 230 V. At starting, the autotransformer is to be set at an 80% tap point.

   (a) Show the schematic diagram of the autotransformer starter arrangement with voltages clearly indicated.

   (b) Find the autotransformer input current.

   (c) What is the value of the current in the common coil?

   (d) Determine the transformed VA.

   (e) What is the value of the conducted VA?

   (f) Find the VA rating of the autotransformer for the conditions specified.

   (g) Identify the rating of an equivalent ordinary transformer that can be used as the autotransformer.

Solution   (a) $V_H = 230$, $V_L = 0.8(230) = 184$ V, $I_L = 34$ A (See Fig. 2-32.)

   (b) First find $a_A = N_{ac}/N_{bc} = 1/0.8 = 1.25$. Then by Eq. (2-104)

$$I_H = \frac{I}{a_A} = \frac{34}{1.25} = 27.2 \text{ A}$$

   (c)

$$I = I_L - I_H = 34 - 27.2 = 6.8 \text{ A}$$

   (d)

$$\text{Transformed VA} = V_L I = 184(6.8) = 1251.2$$
$$= (V_H - V_L)I_H = 46(27.2) = 1251.2$$

   (e)

$$\text{Conducted VA} = I_H V_L = 27.2(184) = 5004.8$$

   (f) VA rating of autotransformer $= V_H I_H = 230(27.2) = 6256$

   (g) VA rating of equivalent ordinary transformer: $46/184$ V, 1251.2 VA

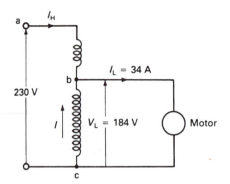

Figure 2-32 Solution to part (a) of Example 2-5.

2-8 PER-UNIT CALCULATIONS

Computations in per-unit values first found wide use in transmission-line calculations often involving several generators and many transformers. One feature of
the per-unit notation is that it makes it unnecessary to refer quantities on the high
or low side of transformers. The per-unit value of a circuit element such as
primary winding resistance is the same on either side. The bookkeeping procedure is therefore appreciably simplified by such a notation. However, the per-unit
notation bears another noteworthy advantage. It reveals important information
about the design and performance characteristics of the equipment to which it is
applied. It is this feature which is emphasized in this section.

Application of the per-unit system of notation depends upon the establishment of a set of specific base quantities by which other similar quantities can be
expressed. For example, all voltage drops in the low-voltage circuit of a transformer can be expressed as a ratio of the rated low-side voltage where the latter
quantity is identified as the base, or reference, voltage. In fact it is customary to
use the *rated quantities* of a device as the base values whenever per-unit calculations are to be performed. There is nothing that restricts this selection. Rather,
the choice of using the rated quantities as the base values is made solely because
of the significance that can be attached to the resulting numbers. Directing attention now specifically to the two-winding transformer we identify the following
base quantities:

base voltage on high side   = rated voltage on high side = V_R'

base current on high side   = rated current on high side = I_R'

base impedance on high side $= \dfrac{\text{rated high-side voltage}}{\text{rated high-side current}} = \dfrac{V_R'}{I_R'} = Z_{BH}$

base admittance on high side $= \dfrac{\text{rated high-side current}}{\text{rated high-side voltage}} = \dfrac{I_R'}{V_R'} = Y_{BH}$

base output VA $= $ rated output VA $= V_R'I_R' = V_RI_R$

base voltage on low side $= $ rated voltage on low side $= V_R$

base current on low side $= $ rated current on low side $= I_R$

base impedance on low side $= \dfrac{\text{rated low-side voltage}}{\text{rated low-side current}} = \dfrac{V_R}{I_R} = Z_{BL} = \dfrac{Z_{BH}}{a^2}$

base admittance on low side $= \dfrac{\text{rated low-side current}}{\text{rated low-side voltage}} = \dfrac{I_R}{V_R} = Y_{BL} = a^2 Y_{BH}$

It is worthwhile to note that *base* quantities have units. This is not true of the *per-unit values* of these quantities because by definition the per-unit value of a quantity is a ratio, or the quotient, of the actual value divided by the corresponding base value. Thus to obtain the per-unit value of the rated low-side voltage of a transformer we write

$$\text{p-u } V_R \equiv \frac{V_R \text{ (volts)}}{\text{base voltage on low side (volts)}} = \frac{V_R}{V_R} = 1 \qquad (2\text{-}127)$$

Clearly, if the rated quantities of a device are chosen to be the base values, then the per-unit values of all base quantities are equal to unity. A similar procedure is followed to find the per-unit value of other quantities. To illustrate further, let us find the per-unit value of the equivalent resistance referred to the high side of a two-winding transformer, R_{eH}. Here we write

$$\text{p-u } R_{eH} = \frac{R_{eH} \text{ (ohms)}}{Z_{BH} \text{ (ohms)}} \qquad (2\text{-}128)$$

The result is a numeric value—a quantity without units. However, considerable significance can be attached to this number by multiplying Eq. (2-128) by unity in the form of I_R'/I_R'. Equation (2-128) then becomes

$$\text{p-u } R_{eH} = \frac{R_{eH}I_R'}{Z_{BH}I_R'} = \frac{R_{eH}I_R'}{(V_R'/I_R')I_R'} = \frac{I_R'R_{eH}}{V_R'} \qquad (2\text{-}129)$$

Recalling that I_R' and V_R' denote the rated values of the current and voltage, respectively, it follows from the right side of Eq. (2-129) that the per-unit value of R_{eH} gives information about the portion of the applied voltage that appears across the equivalent winding resistance as a voltage drop when rated current flows.

Example 2-6

A 10-kVA, 200/100-V transformer has an equivalent high-side resistance of 0.1 Ω and an equivalent high-side leakage reactance of 0.16 Ω. Determine (a) the per-unit value of R_{eH} and (b) the percent of rated voltage absorbed by the equivalent winding resistance.

Solution   (a) From the transformer rating

$$I_R' = \frac{10}{0.2} = 50 \text{ A}$$

$$Z_{BH} = \frac{V_R'}{I_R'} = \frac{200}{50} = 4 \ \Omega$$

Hence by Eq. (2-128)

$$\text{p-u } R_{eH} = \frac{R_{eH}}{Z_{BH}} = \frac{0.1}{4} = 0.025$$

(b) By Eq. (2-129) it follows that

$$0.025 V_R' = 0.025(200) = 5 \text{ V}$$

or 2.5% of the rated voltage appears across the equivalent winding resistance.

Still more significance can be attached to the per-unit value of R_{eH} by multiplying Eq. (2-129) once again by unity in the form of I_R'/I_R'. Thus

$$\text{p-u } R_{eH} = \frac{I_R' R_{eH}}{V_R'} \frac{I_R'}{I_R'} = \frac{I_R'^2 R_{eH}}{V_R' I_R'}$$

$$= \frac{\text{total copper loss at rated value}}{\text{rated volt-amperes}} \qquad (2\text{-}130)$$

This last expression reveals that the per-unit value of R_{eH} also conveys information about the portion of the rated volt-amperes that is consumed as a copper loss at rated current. Thus for the transformer of Example 2-4 it can be said that, when rated current is delivered to the load, 2.5% of the transformer rating is dissipated as copper loss. In watts, this represents a loss of $0.025(10,000) = 250$ W. Note how much more meaningful it is to deal in the per-unit value of R_{eH} than merely to cite the absolute value.

If the equivalent resistance of a transformer is specified with reference to the low side, the per unit value is found by using the low-side base impedance. Thus

$$\text{p-u } R_{eL} = \frac{R_{eL}}{Z_{BL}} = \frac{R_{eL}}{V_R/I_R} = \frac{I_R R_{eL}}{V_R} = \frac{I_R^2 R_{eL}}{V_R I_R} \qquad (2\text{-}131)$$

A glance at this expression reveals that the per-unit value of R_{eL} carries the same interpretation as the per-unit value of R_{eH}. Moreover, the numerical values are identical for percent of voltage drop and percent of copper loss.

Computing the performance of a transformer—regulation and efficiency—can be accomplished quite conveniently in per-unit values. For example, in finding the voltage regulation for a lagging power factor, the expression for the required primary voltage we know to be

$$\bar{V}_1 = \bar{V}_R' + \bar{I}_R' \bar{Z}_{e1} = \bar{V}_R' + I_R'(\cos \theta - j \sin \theta)(R_{e1} + jX_{e1})$$

$$= V_R'\underline{/0°} + I_R' R_{e1} \cos \theta + I_R' X_{e1} \sin \theta + j(I_R' X_{e1} \cos \theta - I_R' R_{e1} \sin \theta) \qquad (2\text{-}132)$$

Dividing through both sides of this equation by V_R' yields

$$\frac{\bar{V}_1}{V_R'} = 1 + \frac{I_R' R_{e1}}{V_R'} \cos \theta + \frac{I_R' X_{e1}}{V_R'} \sin \theta + j \left(\frac{I_R' X_{e1}}{V_R'} \cos \theta - \frac{I_R' R_{e1}}{V_R'} \sin \theta \right) \qquad (2\text{-}133)$$

Keeping in mind that I_R' and V_R' denote rated quantities, we recognize the coefficients of the trigonometric terms as the per-unit values of the equivalent resistance and equivalent reactance. For ease of handling let

$$\frac{I_R' R_{e1}}{V_R'} = \text{p-u } R_{e1} \equiv \varepsilon_r \qquad (2\text{-}134)$$

$$\frac{I_R' X_{e1}}{V_R'} = \text{p-u } X_{e1} \equiv \varepsilon_x \qquad (2\text{-}135)$$

and

$$\frac{\bar{V}_1}{V_R'} = \text{p-u } \bar{V}_1 \qquad (2\text{-}136)$$

Equation (2-133) can then be expressed in a shorthand per-unit notation as

$$\text{p-u } \bar{V}_1 = 1 + \varepsilon_r \cos \theta + \varepsilon_x \sin \theta + j(\varepsilon_x \cos \theta - \varepsilon_r \sin \theta) \qquad (2\text{-}137)$$

The corresponding phasor diagram for this equation appears in Fig. 2-33. Since for practical transformers the per-unit values of ε_r and ε_x rarely exceed 0.1, the contribution of the j part of Eq. (2-137) in determining the per-unit value of $\bar{V}_1$ is negligibly small. Hence for all practical purposes Eq. (2-137) may be more simply expressed as

$$\text{p-u } \bar{V}_1 = 1 + \varepsilon_r \cos \theta + \varepsilon_x \sin \theta \qquad (2\text{-}138)$$

The equation for the voltage regulation in per-unit notation then becomes

$$\varepsilon = \frac{\left| \text{p-u } \bar{V}_1 \right| - \left| \text{p-u } \bar{V}_R' \right|}{\left| \text{p-u } \bar{V}_R' \right|} = \frac{1 + \varepsilon_r \cos \theta + \varepsilon_x \sin \theta - 1}{1}$$

or

$$\boxed{\varepsilon = \varepsilon_r \cos \theta + \varepsilon_x \sin \theta} \qquad (2\text{-}139)$$

Equation (2-139) is significant not only because it provides an accurate means of computing voltage regulation but even more because it emphasizes the factors

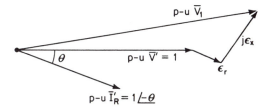

Figure 2-33   Phasor diagram of a transformer for lagging power factor, with all quantities expressed in per-unit notation.

that are responsible for its existence, i.e., the winding resistances, the leakage reactances, and the influence of the power-factor angle.

When the transformer delivers power to a leading power-factor load, θ is introduced as a negative quantity in Eq. (2-139). The expression for the per-unit voltage regulation is then found to be

$$\boxed{\varepsilon = \varepsilon_r \cos\theta - \varepsilon_x \sin\theta} \qquad (2\text{-}140)$$

A comparison with Eq. (2-139) makes it clear that a transformer is more capable of maintaining constant load voltage in the presence of a leading power-factor than a lagging one.

The efficiency of a transformer can also be expressed in the per-unit notation. However, it is helpful first to identify the core loss as a per-unit quantity. Equation (2-28) shows the core loss to be given by $P_c = I_c^2 r_c$. If the core-loss current I_c is replaced by $E_1/r_c \approx V_R'/r_c$, the core loss may be rewritten as

$$P_c \approx \frac{(V_R')^2}{r_c} \approx (V_R')^2 g_c \qquad (2\text{-}141)$$

where

$$g_c \equiv \frac{1}{r_c} = \text{core-loss conductance} \qquad (2\text{-}142)$$

The core-loss conductance can then readily be expressed as a per-unit quantity by

$$\text{p-u } g_c = \frac{g_c}{Y_{BH}} = \frac{g_c}{I_R'/V_R'} = \frac{V_R' g_c}{I_R'} = \frac{(V_R')^2 g_c}{V_R' I_R'} = \frac{\text{core loss}}{\text{rated VA}} \qquad (2\text{-}143)$$

Equation (2-143) indicates that the per-unit conductance representing core loss may be found either from a knowledge of the core loss itself expressed in watts or from a knowledge of the circuit parameter (g_c or r_c) associated with the core loss.

The efficiency of a transformer is affected by the presence of core losses and copper losses. Since we now know how to express both losses in the per-unit notation, let us return to the general expression for the efficiency, Eq. (2-62), and appropriately modify it so that the per-unit notation is brought into evidence. Thus

$$\eta = 1 - \frac{\text{core loss} + \text{copper loss}}{\text{rated output} + \text{core loss} + \text{copper loss}} \qquad (2\text{-}144)$$

Assuming operation at rated voltage, rated current, and a power-factor of $\cos\theta$, we have

$$\eta = 1 - \frac{(V_R')^2 g_c + (I_R')^2 R_{e1}}{V_R' I_R' \cos\theta + (V_R')^2 g_c + (I_R')^2 R_{e1}} \qquad (2\text{-}145)$$

Rearranging yields

$$\eta = 1 - \frac{\dfrac{(V_R')^2 g_c}{V_R' I_R'} + \dfrac{(I_R')^2 R_{e1}}{V_R' I_R'}}{\cos\theta + \dfrac{(V_R')^2 g_c}{V_R' I_R'} + \dfrac{(I_R')^2 R_{e1}}{V_R' I_R'}} \qquad (2\text{-}146)$$

and by Eqs. (2-130) and (2-143) this simplifies to

$$\eta = 1 - \frac{\text{p-u } g_c + \text{p-u } R_e}{\cos\theta + \text{p-u } g_c + \text{p-u } R_e} \qquad (2\text{-}147)$$

Note that the second subscript for equivalent resistance in the last equation was dropped because the per-unit value is the same irrespective of high or low side. It is important to understand that Eq. (2-147) is valid for finding the efficiency at *rated load only*. For any other load condition the value of per-unit R_e must be modified by the square of the per-unit value of the current. Thus to find the efficiency at load current I_2' Eq. (2-147) becomes

$$\eta = 1 - \frac{\text{p-u } g_c + (\text{p-u } R_e)(I_2'/I_R')^2}{\dfrac{I_2'}{I_R'}\cos\theta + \text{p-u } g_c + \left(\dfrac{I_2'}{I_R'}\right)^2 (\text{p-u } R_e)} \qquad (2\text{-}148)$$

The per-unit value of g_c remains intact because it is assumed that operation at rated voltage continues.

Example 2-7

For the transformer of Example 2-3 compute the efficiency when rated current is delivered to a 0.8 power-factor load.

Solution

$$\text{p-u } g_c = \frac{\text{core loss}}{\text{rated volt-amperes}} = \frac{396 \text{ W}}{50,000 \text{ VA}} = 0.00792$$

$$\text{p-u } R_e = \frac{\text{copper loss at rated current}}{\text{rated volt-amperes}} = \frac{810}{50,000} = 0.0162$$

Inserting these values into Eq. (2-147) yields the desired result:

$$\eta = 1 - \frac{0.00792 + 0.0162}{0.8 + 0.0241} = 1 - 0.0293 = 0.9707$$

Hence

$$\eta \text{ in percent} = 97.07 \approx 97.1$$

which compares favorably with the result computed in Example 2-3.

2-9 TRANSFORMERS FOR THREE-PHASE CIRCUITS

For reasons of efficiency and considerable economy in the use of copper, the generation, transmission, and distribution of electric power takes place on a three-phase basis rather than single-phase. The process begins with the electric generator, which is designed to develop a three-phase voltage. Characteristically, the three-phase voltage system is distinguished by the fact that each phase voltage has the same maximum value but is displaced in time from the other phase quantities by 120 electrical degrees. The original energy source may be steam originating from a boiler, water falling through an appropriate head, or the combustion of gasoline or diesel fuel. Frequently, the output voltage of the electric generator is limited by physical considerations to approximately 25 kV. The transmission of large amounts of electrical energy over considerable distances at such relatively low voltages would entail huge losses, thereby making it impractical to do so. A common solution is to introduce transformers between the generator and the transmission lines to boost the transmitting voltage to levels where the associated line losses are manageable. Often the transmitting line voltage is raised to a level of 230 kV and higher by means of step-up transformers. At the receiving end of the transmission line transformers are again needed to reduce the voltages at the substations to more reasonable levels. This is often followed by a further reduction at many places along a distribution network where the electrical energy is finally consumed. Throughout the process, transformers are used in a three-phase mode to provide the required transformations.

The three-phase mode of operation can be achieved either through use of a single three-phase transformer or by use of three single-phase transformers, which are in turn appropriately connected for three-phase operation. The single-unit three-phase transformer consists of a three-legged construction with each leg carrying the primary and secondary coils of one phase. When compared to three single-phase transformers equipped to deal with the same volt-amperes, the single three-phase unit is lighter, cheaper, requires less space, and is slightly more efficient. Some cost-conscious installations prefer them because of the smaller initial investment. However, a selection in favor of three single transformers has some notable advantages. The larger total surface area, for example, makes for better cooling and greater ease of repair. Moreover, in the initial development of a power system, it may be entirely possible to operate with just two of the transformers, albeit at reduced capacity. The third unit can be added at some future date when the full capacity of the system is needed. Also, if a reserve unit is needed to meet emergency conditions, only a single smaller unit is often sufficient to have on hand as compared to the need to have a complete single-unit three-phase transformer. In addition, the latter will require more space for storage and is more difficult to move around. In view of these pros and cons, it is obvious that the choice in a specific situation will be strongly influenced by the special circumstances surrounding that particular case. Because the operation of the transformers in either mode is virtually identical, the development of the subject matter of this section is continued by dealing with the use of three single transformers. The choice is made without prejudice.

The Δ-Δ Connection

When connecting three single transformers for three-phase operation, care must be taken to ensure that the secondary terminals provide a balanced system of voltages. The primary coils may be arbitrarily connected. To illustrate the point, refer to the arrangement depicted in Fig. 2-34. The objective is to arrange the secondary coil connections so that the line voltages will be balanced and no circulating current takes place in the secondary coils. To aid in distinguishing the primary windings from the secondary windings, uppercase letters are used for the former and lowercase letters for the latter. Moreover, the use of common letters (whether upper- or lowercase) is to be taken to mean that the coils are associated with the same core. Thus, in Fig. 2-34(a) coil AB denotes the primary winding and coil ab denotes the secondary winding of the same single-phase transformer. The generator that supplies the voltages to the primary coils of the three transformers is assumed to furnish the balanced three-phase set that is shown in Fig. 2-34(b). Observe that this set is represented by a phasor diagram in the closed form. Note, too, that in Fig. 2-34(a), the secondary coils are left unconnected. These connections must now be made in a prescribed manner.

Begin by joining terminal b of coil ab to terminal b of coil bc. The magnitudes of the primary phase voltages are each equal to one another, i.e., $V_{AB} = V_{BC} = V_{CA} = V_{p1}$, where V_{p1} denotes rms primary phase voltage. Similarly, the values of the secondary phase voltages are also equal to each other based on the assumption that the transformers are identical. Thus $V_{ab} = V_{bc} = V_{ca} = V_{p2}$, where V_{p2} denotes rms secondary phase voltage. It is helpful to be aware that although the magnitudes of these voltages are the same, they do differ in phase. Consequently, when a voltmeter is placed across the open terminals ac in the manner illustrated in Fig. 2-34(c), the reading will be one of two choices. Either it will be $\sqrt{3}\ V_{ab}$ or it will be equal in magnitude to V_{ab}. Keep in mind that the phasor voltage V_{bc} can be added to the phasor voltage V_{ab} at an angle of 60° (as shown by the dashed line in the phasor diagram of Fig. 2-34(c) or at an angle of −120° (as indicated by the solid line). These values are dictated by the relationship of V_{BC} to V_{AB} on the primaries. If the higher of the two values is found upon testing, it should be rejected and the coil connection reversed. It is important to keep in mind that in a Δ arrangement, the phase voltages are equal to the line voltages.

In the last step of this *phasing procedure,* terminal c of coil ac is joined to c of coil bc, with the gap in Fig. 2-34(d) left opened except for the insertion of a voltmeter. This voltage, too, presents one of two possibilities: either the reading will be twice V_{ab} or it will be zero. The result depends upon whether the voltage of coil ac is directed along a −60° direction or a direction of 120°. Clearly, that connection of coil ac must be chosen that yields a zero gap voltage. Surely, we

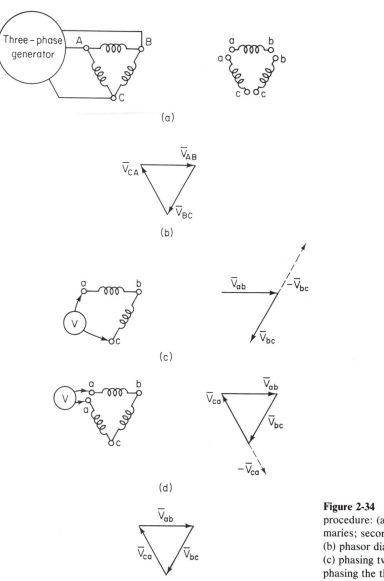

(a)

(b)

(c)

(d)

(e)

Figure 2-34 Transformer phasing procedure: (a) excitation of the primaries; secondaries unconnected; (b) phasor diagram of primary voltages; (c) phasing two secondary coils; (d) phasing the third coil for balanced voltages; (e) phasor diagram of the balanced secondary voltages.

would not want to connect the two *a* terminals except when the potentials of these terminals are the same. This avoids fireworks. Before leaving this matter, however, it is worthwhile to note that transformer manufacturers do use markings that lead to the proper phasing of three-phase transformers. But obviously the phasing procedure must be done by someone. Figure 2-34(e) shows the final phasor diagram as it pertains to the balanced secondary voltages.

The salient characteristics between the line and phase quantities from primary to secondary for the Δ-Δ connection may be summarized as described next. Introduce

$$a = \frac{V_{AB}}{V_{ab}} = \frac{V_{p1}}{V_{p2}}$$

where the quantities on the right side represent the primary and secondary phase voltages. Assume that a balanced three-phase load is applied to the secondary windings. Then

$$\text{Ratio of phase voltages:}\quad \frac{V_{AB}}{V_{ab}} = a$$

$$\text{Ratio of line voltages:}\quad \frac{V_{L1}}{V_{L2}} = \frac{V_{AB}}{V_{ab}} = a$$

where V_{L1} and V_{L2} represent the primary and secondary line voltages. The primary and secondary phase and line voltages are also respectively in phase with one another; i.e., $\bar{V}_{AB}$ is in phase with $\bar{V}_{ab}$, $\bar{V}_{BC}$ is in phase with $\bar{V}_{bc}$, and $\bar{V}_{CA}$ is in phase with $\bar{V}_{ca}$. A comparison of Fig. 2-34(b) with Fig. 2-34(e) makes this statement self-evident. Also,

$$\text{Ratio of phase currents:}\quad \frac{I_{AB}}{I_{ab}} = \frac{1}{a}$$

$$\text{Ratio of line currents:}\quad \frac{I_{L1}}{I_{L2}} = \frac{\sqrt{3}\,I_{p1}}{\sqrt{3}\,I_{p2}} = \frac{I_{AB}}{I_{ab}} = \frac{1}{a}$$

Here use is made of the fact that the line current in a balanced Δ-connection is $\sqrt{3}$ times the phase current.

It is important that the transformers that are used in these arrangements have equal transformation ratios in order to minimize the existence of circulating currents. Furthermore, the equivalent leakage impedances of the transformers should be as nearly identical as possible to ensure proper load sharing during operation. The use of a Δ secondary also makes only one voltage available with no possibility of including a neutral wire.

Finally, a word is in order concerning the magnetizing current in these arrangements. In the configuration of Fig. 2-34(a) a sinusoidal voltage is assumed to be applied to the primary coils of the three transformers. Because of its sinusoidal nature, the generator can deliver to the transformer a magnetizing current which is itself sinusoidal. Such a sinusoidal current, however, produces a flat-topped flux wave as a result of the saturation in the iron core of the transformer. In addition to the principal fundamental component, a flat-topped flux wave contains a large third harmonic part, which, in turn, serves to induce, by Faraday's law, a third harmonic voltage. This action takes place in each of the primary transformers. Furthermore, the third harmonic voltage induced in coil BC occurs at angle of $3(-120°) = -360°$ or in-phase with the component appearing in phase AB. Similarly, in coil CA, the phase of the third harmonic voltage is $3(120) = 360°$, so that it too is in-phase with the third harmonic voltage in the other two coils. Since the sinusoidal generator contains no such component in its voltage cycle, neutralizations of the third harmonic voltages cannot take place. Accordingly, the additive effect of these third harmonic voltages is to cause a third harmonic circulating current to exist in the three transformer primaries. This current, then, serves the useful purpose of changing the magnetizing current from a sinusoidal variation to one which is peaked to the extent that is necessary to ensure a sinusoidal flux variation. The sinusoidal flux variation in turn provides a sinusoidal voltage variation at the primary terminals of the transformers, thereby satisfying Kirchhoff's voltage law in the generator-transformer primary circuit.

Δ-Y Connection

The delta-wye arrangement of transformers is commonly found at the transmitting end of transmission lines as well as at the distribution end. At the transmitting end, the Δ-Y connection offers the advantage of an increased step-up ratio beyond that associated with the individual transformers. At the distribution end the inclusion of a neutral wire allows a choice of two voltages for the customer's use. A popular combination that is available from the electric power companies to homes and commercial establishments is the 208/120-V system. The larger figure is available between any two line terminals of the Y-connection, whereas the lower figure represents the voltage from line to neutral.

The phasing procedure for a Δ-Y connection proceeds differently from that of the Δ-Δ connection insofar as the secondary connections are concerned. The primary coils in each case can be wired arbitrarily in a closed delta arrangement. Because the wye secondaries involve a neutral terminal, a modified terminal marking is used for the individual transformers, as illustrated in Fig. 2-35(a). Of course, when the primary windings are joined to form a closed circuit, the n markings may be dropped as shown in Fig. 2-35(b). However, in the phasor representation of the primary voltages that is depicted in Fig. 2-35(c) the n notation is carried along because it makes it easier to understand what is happening at the secondaries.

We begin the phasing process at the secondaries by joining n of coil na to n of coil nb. A voltmeter placed across the free terminals will indicate one of two voltages, as was the case with the Δ-Δ connection. The specific reading depends upon the relative sense of the winding. The connection causes the voltage in coil an to be added to the voltage in coil nb along a $-60°$ direction in the phasor diagram or a $120°$ direction. These are the only two choices and the one that prevails depends upon the relative sense of the windings when the two terminals are connected. If the voltmeter reads a value for the line voltage V_{ab} that is equal to V_{nb}, it means the addition is occurring along the $120°$ direction [see the dashed-line component in Fig. 2-35(d)] and must be rejected. To correct this situation, it

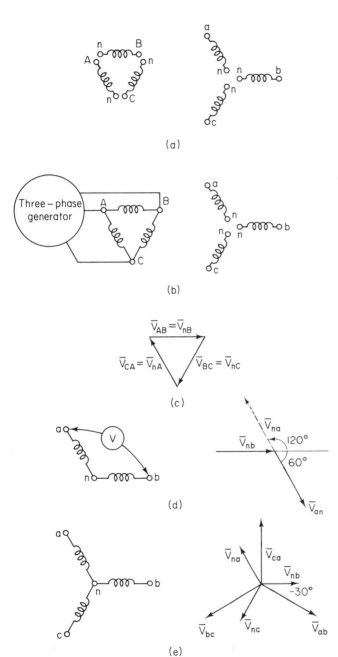

Figure 2-35 Phasing procedure for the Δ-Y connection: (a) terminal markings for the primary and secondary coils of transformers A, B, and C; (b) primaries in Δ mode with generator excitation; (c) phasor diagram of primary line and phase voltages; (d) phasing secondary coils *a* and *b*; (e) phasing coil *c* for balanced line-to-line and line-to-neutral secondary voltages.

is merely necessary to reverse one of the coil connections. Now the voltmeter will yield for V_{ab} a value of $\sqrt{3}\, V_{nb}$. This is the proper value for the line-to-line voltage of the wye connection. The line voltage must always be greater than the phase voltage by a factor of $\sqrt{3}$. When the third coil c is joined in the manner of Fig. 2-35(e), the correct phasing is determined when the voltages between each pair of terminals is equal in magnitude to $\sqrt{3}$ times the phase voltage. (What are the voltage readings when coil c is incorrectly phased?)

A comparison of the phasor diagram of a properly phased Δ-Y connection such as appears in Fig. 2-35(e) reveals that the line-to-line voltages at the secondaries are displaced by 30 electrical degrees from the line-to-line voltages at the primaries. Thus, for the situation depicted in Fig. 2-35, the line voltage between terminals a and b at the secondaries is $\bar{V}_{ab} = V_{ab}\underline{/-30°}$, whereas the line voltage between terminals A and B at the primaries is $\bar{V}_{AB} = V_{AB}\underline{/0°}$. Accordingly, there is a *lag* of 30 electrical degrees in this arrangement. It will be recalled that this condition did not occur in the Δ-Δ connection where it was found that the line-to-line voltages between primaries and secondaries were all in phase. The importance of this distinction lies in the conclusion that a bank of Δ-Δ connected transformers cannot be placed in parallel with a bank of Δ-Y transformers even if the voltage magnitudes correspond. The phase angle discrepancy which is inherent with these connections will result in wasteful circulating currents.

At the generator end of a transmission line, the Δ-Y connection is used to step up the generator voltage to a suitable transmission line voltage. Let the step-up voltage ratio be represented by

$$a = \frac{V_{nA}}{V_{na}} = \frac{V_{p1}}{V_{p2}} \qquad (2\text{-}149)$$

where a is less than unity for a step-up arrangement and V_{p1} denotes the rms primary value of the phase voltage and V_{p2} is the rms secondary value of the phase voltage. The ratio of the line-to-line voltages at the secondaries to the line-to-line voltages at the primaries can then be expressed as

$$\frac{(V_{L\text{-}L})_Y}{(V_{L\text{-}L})_\Delta} = \frac{V_{ab}}{V_{AB}} = \frac{\sqrt{3}\, V_{p2}}{V_{p1}} \qquad (2\text{-}150)$$

Upon replacing V_{p1} by its equivalence from Eq. (2-149), there results

$$\frac{(V_{L\text{-}L})_Y}{(V_{L\text{-}L})_\Delta} = \frac{V_{ab}}{V_{AB}} = \frac{\sqrt{3}\, V_{p2}}{a V_{p2}} = \frac{\sqrt{3}}{a} \qquad (2\text{-}151)$$

Thus, if a is assigned a value of one-tenth, the line-to-line voltage at the secondaries will be 17.32 times greater than the line-to-line voltages at the primary. The attractive feature about this arrangement, of course, is that insulation for the secondary windings need be provided only to the extent of a factor of 10 over that of the primary voltage rather than 17.32. Furthermore, in view of the fact that output volt-amperes at the secondaries of the Y must equal the input volt-amperes at the primaries of the Δ under the assumption of negligible losses, it follows that the ratio of the line current at the secondaries $(I_L)_Y$ to the line current at the primaries $(I_L)_\Delta$ for a balanced load is the reciprocal of Eq. (2-151). That is,

$$\frac{(I_L)_Y}{(I_L)_\Delta} = \frac{I_{p2}}{\sqrt{3}\, I_{p1}} = \frac{I_{p2}}{\dfrac{\sqrt{3}\, I_{p2}}{a}} = \frac{a}{\sqrt{3}} \tag{2-152}$$

where I_{p1} and I_{p2} denote the primary and secondary phase currents, respectively. Therefore, by raising the line voltage substantially, the corresponding transmission line currents can be markedly reduced, which in turn can serve to improve the efficiency of transmission appreciably.

The presence of a delta arrangement on the primary side, which allows a third harmonic magnetizing current component to flow, ensures that the line-to-neutral voltages on the wye side will be sinusoidal. Of course, the foregoing statement is based on the premise that the source voltage that energizes the Δ-connected primaries is itself purely sinusoidal.

Y-Y Connection

When equipped with a primary and secondary neutral, the wye-wye connection offers two advantages: (1) it provides a choice of two voltages at the secondary side; and (2) transformers need to be specified for an insulation that guards against the magnitude of the line-to-phase voltage rather than line-to-line voltage. Also, if the ratio of the primary phase voltage to the secondary phase voltage is a, this same ratio obtains for the line-to-line voltages. On the other hand, the ratio of the primary phase or line currents to the secondary phase or line currents is $1/a$. Moreover, on either side, the line voltage is phase displaced from the nearest phase voltage by a magnitude of 30 electrical degrees.

Any attempt to operate a Y-Y arrangement of transformers without the presence of a primary neutral connection will lead to difficulty and potential failure. In fact, trouble occurs even under no-load conditions. To illustrate the point, assume that the primary wye connection is energized from a sinusoidal three-phase generator. Since such a generator delivers a sinusoidal magnetizing current, the typical saturating magnetization curve of the transformer core causes the flux variation to be flat-topped. In turn, this flat flux wave contains a large third harmonic component which induces an appreciable third harmonic voltage in the transformer coils. The magnitude of this voltage depends upon the degree of saturation of the core and in most instances, this third harmonic voltage is not negligible. Recall that this same condition was noted to exist in the Δ-connected primaries. But with the Y-Y connection, there is an important difference. Whereas in the Δ-connected primaries the third harmonic voltage was presented with a closed path that allowed a circulating current to flow, this situation does not happen with the Y connection without a neutral. As a matter of fact, because the

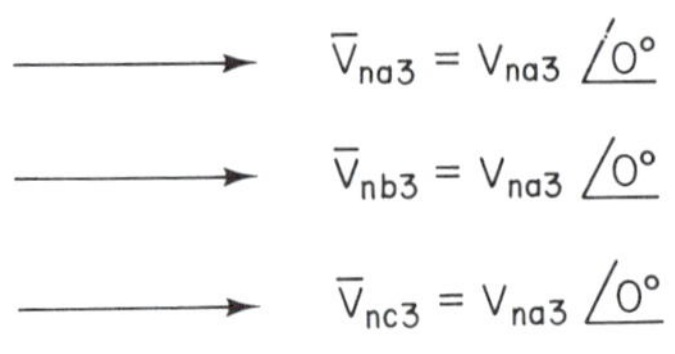

$\overrightarrow{}$ $\bar{V}_{na3} = V_{na3} \,\underline{/0°}$

$\overrightarrow{}$ $\bar{V}_{nb3} = V_{na3} \,\underline{/0°}$

$\overrightarrow{}$ $\bar{V}_{nc3} = V_{na3} \,\underline{/0°}$

Figure 2-36 The line-to-neutral voltages caused by the third harmonic flat-topped flux wave present in a Y-Y connection without a primary neutral.

third harmonic phase voltages are each in phase with one another, they fail to make their presence felt at the line-to-line terminals. Figure 2-36 shows the phase and magnitudes of the third harmonic rms voltages from line-to-neutral in each phase. Observe that their magnitudes and phase angles are identical. To obtain the effect of these voltages between line terminals, it is necessary to perform the phasor addition

$$(\bar{V}_{ab})_3 = \bar{V}_{an3} + \bar{V}_{nb3} = -\bar{V}_{na3} + \bar{V}_{nb3} = 0$$

which clearly adds to zero. The significance of this result in the generator-primary transformer circuit is that Kirchhoff's voltage law is satisfied despite the presence of a flat-topped (nonsinusoidal) flux wave in the transformer core. One consequence of this condition is that the resultant line-to-neutral voltage can now readily exceed the rated transformer voltage, so that harmful insulation stresses can take place. Of course, when a primary neutral connection is established, the third harmonic voltage that originates in the transformer from line-to-neutral is now provided with a path that allows the magnetizing current once again to become peak-shaped, thereby restoring a sinusoidal flux variation. In turn, this virtually eliminates the third harmonic voltages in the transformer coils. In those instances where the ensuing third harmonic neutral ground currents cause telephone or other related interference, it may be necessary to resort to an alternate transformer connection involving the Δ arrangement on the primary or secondary side or else to purchase transformers equipped with tertiary windings which can then be connected in the Δ mode.

    Another serious shortcoming of a Y-Y connection without a primary neutral is that it makes it practically impossible to supply power to a load that is placed from line-to-neutral on the secondary side. For energy to be delivered to the load in such an arrangement, the generator source, which supplies the primary windings, must do so through the two unloaded primaries. However, as we learned earlier in this chapter, an unloaded transformer presents an impedance which is essentially the magnetizing reactance. This high impedance mitigates against the flow of significant currents. In this regard it is interesting to note that even if a short-circuit were placed from one line-to-neutral on the secondary, very little current would flow. Because such a short-circuit has the effect of placing the neutral at the potential of one line terminal, it follows that the line-to-neutral voltages appearing across each of the unloaded transformers is increased almost by a factor of $\sqrt{3}$. Such a high overvoltage can readily cause damage to the insulation. In conclusion, it ought to be obvious that a Y-Y connection should not be used without the primary neutral connected to the source.

Y-Δ Connection

The wye-delta arrangement of single-phase transformers is often found at the receiving end of a transmission line. The reason is that it provides a large step-down voltage ratio from primary line voltage to secondary line voltage. If we let a denote the ratio of the primary phase voltage (Y side) to the secondary phase voltage (Δ), it follows that the ratio of the line voltage from primary to secondary is $\sqrt{3}\, a$. Correspondingly, the ratio of the primary line current to the secondary line current is $1/\sqrt{3}\, a$.

Third harmonic voltage stresses do not exist in the wye-delta arrangement because the closed path provided by the secondary Δ connection permits the third harmonic magnetizing current to exist. In turn, this current acts to virtually eliminate the third harmonic component in the flat-topped flux wave, thus ensuring a sinusoidal flux variation. Unbalanced secondary loads, however, can present some difficulties, as explained for the Y-Y connection, without a primary neutral. To avoid problems, it is best to establish the primary neutral to the source.

Example 2-8

A Δ-Y connection is used to connect a large three-phase generator to a loaded transmission line. The turns ratio of each single-phase transformer from the primary side to the secondary side is 0.1 (see Fig. 2-37). The three-phase load on the trans-

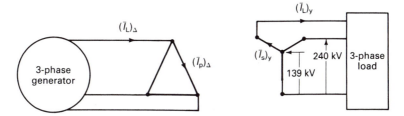

Figure 2-37 Schematic diagram for Example 2-8.

mission line is 10 MVA at a line-to-line voltage of 240 kV. The load power factor is 0.8 lagging.

(a) Find the magnitude of the transformer winding currents on both the high and low sides. Assume all losses are negligible.

(b) Determine the magnitude of the line currents supplied by the generator.

(c) If the line-to-line voltages on the high side are described by

$$V_{ab} = 240\,\underline{/0°}\ \text{kV}, \quad V_{bc} = 240\,\underline{/-120°}\ \text{kV}, \quad V_{ca} = 240\,\underline{/-240°}\ \text{kV}$$

find the corresponding expressions of the line-to-line voltages on the Δ side. Assume all leakage impedances are negligible.

Solution (a) Working with the expression for the volt-amperes on the high side of the transmission line we can write

$$(I_L)_y = \frac{10^7}{\sqrt{3}\,(V_L)_y} = \frac{10^7}{\sqrt{3}\,(240)10^3} = 24 \text{ A} = (I_s)_y$$

$$(I_p)_\Delta = \frac{(I_L)_\Delta}{a} = \frac{24}{0.1} = 240 \text{ A}$$

(b) Here,

$$(I_L)_\Delta = \sqrt{3}\,(I_p)_\Delta = \sqrt{3}\,(240) = 415.7 \text{ A}$$

Or, by Eq. (2-152),

$$(I_L)_\Delta = \frac{\sqrt{3}\,(I_L)_y}{a} = \frac{\sqrt{3}\,(24)}{0.1} = 415.7 \text{ A}$$

(c) As per the discussion relating to Fig. 2-35(e), the corresponding line voltages of the Δ side lead the respective line voltages on the Y side by 30°. Hence,

$$V_{AB} = 24\,\underline{/30°}\text{ kV}, \quad V_{BC} = 24\,\underline{/-90°}\text{ kV}, \quad V = 24\,\underline{/-210°}\text{ kV}$$

PROBLEMS

2-1. A reactor coil with an iron core has 400 turns. It is connected across the 115-V, 60-Hz power line.
(a) Neglecting the resistance voltage drop, calculate the maximum value of the operating flux.
(b) If the flux density is not to exceed 75 kilolines/in.^{2}, what must be the cross-sectional area of the core?

2-2. A fixed sinusoidal voltage is applied to the circuit shown in Fig. P2-2. If the voltage remains connected and the shaded portion of the iron removed, state what happens to the maximum value of the flux and the maximum value of the magnetizing current. Justify your answer. Assume negligible leakage impedance and winding resistance.

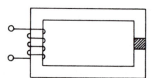

Figure P2-2

2-3. Repeat Prob. 2-2 assuming the voltage applied to the core is a fixed dc quantity. Assume dc current is limited by an external resistor.

2-4. A transformer coil rated at 200 V and having 100 turns is equipped with a 0.5 tap. If 50 V is applied to half the number of turns, find the value of the maximum flux. The frequency is 60 Hz.

2-5. Refer to the transformer coil of Prob. 2-4.
  (a) Determine the maximum value of the operating flux when 150 V is applied across half the number of turns.
  (b) What is the required change in magnetizing current in part (a)? Neglect saturation.
  (c) Briefly describe the effect on the magnetizing current in part (b) if saturation is not neglected.
  (d) What must be the frequency of the 150-V source in part (a) if the flux is to remain unchanged?

2-6. In an open-circuit test of a 25-kVA, 2400/240-V transformer made on the low side the corrected readings of amperes, volts, and watts are, respectively, 1.6, 240, and 114. In the short-circuit test the low side is short-circuited and the current, voltage, and power to the high side are measured to be, respectively, 10.4 A, 55 V, and 360 W.
  (a) Find the core loss.
  (b) What is the full-load copper loss?
  (c) Find the efficiency for a full-load of 0.8 power factor leading.
  (d) Compute the percent voltage regulation for part (c).

2-7. A transformer with a rating of 25 kVA, 2400/240 V has the following parameters: $R_{eH} = 3.33\ \Omega$ and $Z_{eH} = 5.28\ \Omega$. The core loss at rated voltage is 114 W.
  (a) Find the efficiency when the transformer delivers rated kVA at 0.8 pf lagging.
  (b) What is the value of the high-side voltage in part (a)?
  (c) Determine the percent voltage regulation in part (a).

2-8. A 50-kVA, 2300/230-V, 60-Hz transformer has a high-voltage winding resistance of 0.65 Ω and a low-voltage winding resistance of 0.0065 Ω. Laboratory tests showed the following results:

Open-circuit test:	V = 230 V,	I = 5.7 A,	P = 190 W
Short-circuit test:	V = 41.5 V,	I = 21.7 A,	P = none used

  (a) Compute the value of primary voltage needed to give rated secondary voltage when the transformer is connected *step-up* and is delivering 50 kVA at a power factor of 0.8 lagging.
  (b) Compute the efficiency under the condition of part (a).

2-9. A 100-kVA transformer was measured to have a core loss of 1250 W at rated voltage. A short-circuit test was performed at 125% of rated current and the input power was measured to be 2875 W. Calculate the efficiency of this transformer when it delivers rated kVA at a pf of 0.9 lagging.

2-10. The following test data were taken on a 15 kVA, 2200/440-V, 60-Hz transformer:

Short-circuit test:	P = 620 W,	I = 40 A,	V = 25 V
Open-circuit test:	P = 320 W,	I = 1 A,	V = 440 V

  (a) Calculate the voltage regulation of this transformer when it supplies full load at 0.8 pf lagging. Neglect the magnetizing current.
  (b) Find the efficiency at the load condition of part (a).

2-11. The following test data were taken on a 110-kVA, 4400/440-V, 60-Hz transformer:

Short-circuit test: $P = 2000$ W, $I = 200$ A, $V = 18$ V

Open-circuit test: $P = 1200$ W, $I = 0.2$ A, $V = 4400$ V

Calculate the voltage regulation of this transformer when it supplies rated current at 0.8 pf lagging. Neglect the magnetizing current.

2-12. A 50-kVA, 2300/230-V, 60-Hz distribution transformer takes 360 W at a power factor of 0.4 with 2300 V applied to the high-voltage winding and the low-voltage winding open-circuited. If this transformer has 230 V impressed on the low side with no load on the high side, what will be the current in the low-voltage winding? Neglect saturation.

2-13. A 25-kVA, 2400/240-V transformer draws 254 W at a pf of 0.15 when 240 V is impressed on the low-voltage side with the high-voltage side open-circuited. Find the current that is drawn from the line when 2400 V is applied to the high side with the low side open-circuited.

2-14. A 10-kVA, 2400/240-V transformer draws 165 W at a pf of 0.2 when 220 V is applied to the low side with the high side open. Find the current drawn by the high side when 2400 V is applied to the high side with the low side open. Neglect saturation.

2-15. A 200/100-V, 60-Hz transformer has an impedance of $0.3 + j0.8$ Ω in the 200-V winding and an impedance of $0.1 + j0.25$ Ω in the 100-V winding. What are the currents on the high and low side if a short-circuit occurs on the 100-V side with 200 V applied to the high side?

2-16. A 60-Hz, three-winding transformer is rated at 2300 V on the high side with a total of 300 turns. Of the two secondary windings, each designed to handle 200 kVA, one is rated at 575 V and the other at 230 V. determine the primary current when rated current in the 230-V winding is at unity power factor and the rated current in the 575-V winding is at 0.5 pf lagging. Neglect all leakage impedance drops and magnetizing current.

2-17. A transformer is equipped with three windings. The high side (primary) is rated at 2300 V and has a total of 3000 turns. Each of the low-side secondary windings has a kVA rating of 200 and voltage ratings of 460 V and 230 V. When rated current at unity pf flows through the 460-V winding and rated current simultaneously is made to flow through the 230-V winding, the primary current is found to be $122.98\underline{/45°}$ A. Describe the type of load that appears across the 230-V winding. Neglect all leakage impedance and magnetizing current.

2-18. An "ideal" transformer has a secondary winding tapped at point b. The number of primary turns is 100 and the number of turns between a and b is 300 and between b and c is 200. The transformer supplies a resistive load connected between a and c and drawing 7.5 kW. Moreover, a load impedance of $10\underline{/45°}$ Ω is connected between a and b. The primary voltage is 1000 V. Find the primary current.

2-19. Draw a neat phasor diagram of a transformer operating at rated conditions. Assume that $N_1/N_2 = 2$ and

$$I_1 r_1 = 0.1E_1, \qquad I_2 r_2 = 0.1V_R, \qquad I_m = 0.3I'_R$$

$$I_1 x_1 = 0.2E_1, \qquad I_2 x_2 = 0.2V_R, \qquad I_c = 0.1I'_R$$

Consider the load power factor to be 0.6 leading. Use V_R as the reference phasor and show all currents and voltages drawn to scale.

2-20. A 10-kVA, 500/100-V transformer has $R_{eH} = 0.3\ \Omega$ and $X_{eH} = 5.2\ \Omega$, and it is used to supply power to a load having a lagging power factor. When supplying power to this load an ammeter, wattmeter, and voltmeter placed in the high-side circuit read as follows:

$$I_1 = 20\ \text{A}, \qquad V_1 = 500\ \text{V}, \qquad \text{WM} = 8\ \text{kW}$$

For this condition calculate what a voltmeter would read if placed across the secondary load terminals. Assume the magnetizing current to be negligibly small.

2-21. A 10-kVA, 460/115-V transformer has a high-side winding resistance of 0.4 Ω and a low-side winding resistance of 0.02 Ω. The equivalent leakage reactance on the high side is 3.2 Ω. When this transformer is used to supply power to a passive load, it draws a lagging current of 21.7 A at 460 V and 8 kW. Determine the resistive and reactive components of the load impedance. Neglect the magnetizing current.

2-22. The following data are taken on a 30-kVA, 2400/240-V, 60-Hz transformer:

Short-circuit test:	$V = 70\ \text{V}$,	$I = 18.8\ \text{A}$,	$P = 1050\ \text{W}$
Open-circuit test:	$V = 240\ \text{V}$,	$I = 3.0\ \text{A}$,	$P = 230\ \text{W}$

(a) Determine the primary voltage when 12.5 A at 240 V is taken from the low-voltage side supplying a load of 0.8 pf lagging.
(b) Compute the efficiency in part (a).

2-23. A 30-kVA, 240/120-V, 60-Hz transformer has the following data:

$$r_1 = 0.14\ \Omega, \qquad r_2 = 0.035\ \Omega$$
$$x_1 = 0.22\ \Omega, \qquad x_2 = 0.055\ \Omega$$

It is desired to have the primary induced emf equal in magnitude to the primary terminal voltage when the transformer carries the full-load current. Neglect the magnetizing current. How must the transformer be loaded to achieve this result?

2-24. The two windings of a 48-kVA, 2400/240-V, 60-Hz transformer have resistances of 0.6 Ω and 0.025 Ω for the high- and low-voltage windings, respectively. This transformer requires that 238 V be impressed on the high-voltage coil in order that rated current be circulated in the short-circuited low-voltage winding.
(a) Calculate the equivalent leakage reactance referred to the high side.
(b) How much power is needed to circulate rated current on short-circuit?
(c) Compute the efficiency at full load when the power factor is 0.8 lagging. Assume that the core loss equals the copper loss.

2-25. A transformer coil is equipped with a 0.5 tap, has 100 turns, and is rated for 200 V. When this voltage is applied to the full winding, the magnetizing current is 1 A. If now 50 V is applied to half the number of turns, what will be the magnetizing current? Assume no saturation.

2-26. A short-circuit test is to be performed on a transformer with a rating of 50 kVA, 460/115 V. The high-side and low-side resistances are known to be 0.06 Ω and 0.004 Ω, respectively. Moreover, the power factor at short-circuit conditions is 0.3.

Compute the input voltage that is required to force rated current through the low-side coil at short-circuit.

2-27. A coil located on an iron core has a constant 60-Hz voltage applied to it. The total inductance is found to be 10 H. A second identical core is placed in parallel with the first one and is so wound that it produces aiding flux. What is the total inductance of the parallel combination? Neglect the winding resistance and leakage flux of each winding.

2-28. Repeat Prob. 2-27 for the case where the second identical coil is placed in series with the first coil. Assume that aiding flux is produced.

2-29. Show the derivation that allows Eq. (2-68) to be written from Eq. (2-67) for a sinusoidal forcing function and no saturation.

2-30. A two-winding transformer has a high-voltage coil equipped with 1000 turns and a low-voltage coil equipped with 100 turns. The self-inductance of the high voltage primary is known to be 20 H. Moreover, the primary winding has a leakage inductance of 1 H. The secondary leakage inductance is 0.01 H.
   (a) Find the mutual inductance in henrys.
   (b) Determine the value in henrys of the self-inductance of the secondary winding.

2-31. A conventional transformer with a rating of 10 kVA, 2000/1000 V is available.
   (a) Show how to arrange this transformer for use as a step-down autotransformer with a secondary voltage of 2000 V.
   (b) For a 0.8 pf lagging load, what is the maximum current that can be delivered without overloading the transformer coils?
   (c) At the load conditions of part (b), express the conducted kVA of the autotransformer as a percentage of the rating of the conventional transformer.
   (d) List the advantages which the autotransformer has over an ordinary transformer of equal kVA and voltage rating for doing the same job.

2-32. A 10-kVA, 2300/230-V transformer has an iron loss of 190 W and a copper loss of 196 W at rated voltage and rated current. The secondary winding is made up of two equal sections, each section series-connected for 230-V operation. It is desired to reconnect this conventional transformer as an autotransformer in order to raise the 2300-V supply line to 2415 V and to deliver maximum power without overheating the coils. Using *all* windings of the transformer, determine for rated current:
   (a) kVA delivered to the load.
   (b) kVA conducted.
   (c) Efficiency for a load pf of 0.8 leading.
   (d) The rating (voltage and kVA) of an ordinary transformer to do the same job. What are the advantages of using an autotransformer?

2-33. The following data were found during a short-circuit test on an autotransformer rated at 22 kVA, 440/330 V. With the low side short-circuited and voltage applied to the high side, the readings were:

$$P = 320 \text{ W}, \qquad V = 20 \text{ V}, \qquad I = 50 \text{ A}$$

   (a) Find the voltage regulation of this autotransformer at full-load output and unity power factor.

(b) If the noncommon portion of the winding has a resistance of 0.064 Ω, determine the resistance in ohms of the common portion.

(c) Find the transformed and conducted power for full-load output at unity power factor.

(d) What must be the rating (kVA and voltages) of an ordinary transformer that can be used to do the same job?

2-34. The expression for the equivalent resistance of an autotransformer looking into the high-side terminals is given by Eq. (2-107). Derive the expression for the equivalent resistance of the autotransformer looking into the low-side terminals, with the high side short-circuited.

2-35. When the transformer of Prob. 2-15 is used as a step-up autotransformer (200/300), what are the line currents on the high and low sides if a short circuit occurs on the 300-V side with 200 V applied to the low side?

2-36. An autotransformer has $N_{ac} = 100$ turns and $N_{bc} = 60$ turns, where N_{ac} represents the full winding. A short-circuit test is performed by short-circuiting winding ab and applying a reduced voltage to winding bc. The equivalent impedance looking into winding bc is found to be: $1.5 + j4.5$.

(a) Determine the equivalent impedance looking into winding ac for the condition where winding bc is short-circuited.

(b) Find the equivalent resistance looking into bc when ac is short-circuited and voltage is applied to bc.

2-37. A single-phase autotransformer rated at 40 kVA supplies a load impedance of $4\underline{/36.9°}$ Ω at a terminal voltage of 200 V from a 125-V supply. All power losses and leakage reactances are negligible. Calculate the magnitudes of the currents in the common and noncommon parts of the transformer winding, taking into account the magnetizing current of 0.075 per unit.

2-38. A 25-kVA, 2400/240-V single-phase transformer is connected to act as a booster from 2400 to 2640 V. With the maximum kVA load that it is possible to supply without overloading the transformer coils, determine:

(a) kVA received and delivered.

(b) kVA transformed.

(c) kVA conducted.

Neglect all losses, magnetizing current, and leakage impedance.

2-39. Repeat Prob. 2-38 for the case where the secondary coil is reversed so that the output (load) voltage is 2160 V.

2-40. A transformer rated at 40 kVA has a total copper loss of 250 W when operating at 50% of rated current. Determine the per-unit value of the equivalent resistance.

2-41. Refer to the configuration of Fig. P2-41.

(a) Assume that the current that enters terminal a is increasing in the positive direction. Identify the instantaneous polarity markings of each terminal.

(b) The coils have the number of turns indicated. A resistive load that draws 20 A is placed across terminals c and d. A purely capacitive load drawing 40 A is placed across ef. Find the source current. Neglect all leakage impedances and magnetizing current.

(c) Disregard winding *ef*. Assume winding *ab* has a voltage rating of 400 V and a
current rating of 50 A. A short-circuit test on this transformer yields the following data:

$$WM = 600 \text{ W}, \qquad VM = 50 \text{ V}, \qquad AM = 20 \text{ A}$$

Find the voltage regulation at unity pf.

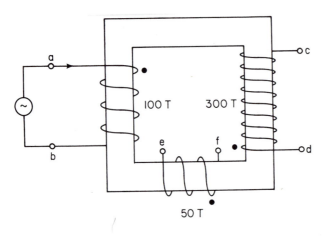

Figure P2-41

2-42. The high-side and low-side winding resistances of a 100-kVA, 440/110-V single-
phase transformer are 0.04 Ω and 0.0025 Ω, respectively.
   (a) Find the per-unit values of the high-voltage and low-voltage winding resistances.
   (b) If the core loss at rated voltage is 1200 W, find the transformer efficiency when it
   delivers half-rated kVA at unity power factor.

2-43. A transformer rated at 2500 kVA, 10,000/2000 V, 60 Hz is designed for maximum
efficiency at 80% of rated load. The per-unit equivalent impedance of this transformer is 0.02 + *j*0.06. For a resistive load and operation at maximum efficiency,
find the losses and the change in voltage from 80% load (at rated voltage) to no load.

2-44. A transformer rated at 1100 kVA has a short-circuit test conducted at a current equal
to 75% of rated value, and takes an amount of power equal to 7030 W. An open-
circuit test is also conducted and at rated voltage the input power is measured to be
9850 W.
   (a) Find the per-unit value of the equivalent resistance.
   (b) Find the per-unit value of the conductance.

2-45. If the per-unit value of the equivalent resistance of the conventional transformer of
Prob. 2-31 is 0.02, find the value of the equivalent resistance of the autotransformer
in ohms referred to the high side.

2-46. Describe the predominate waveshape of the primary magnetizing current of a Y-Δ
three-phase transformer connection employing single-phase transformers.

2-47. A three-phase transformer bank is connected with the primaries in delta and the
secondaries in wye. The line voltage on the primary side is 120 V and on the secondary side 240 V. Find the ratio of primary to secondary turns for each of the single-
phase transformers.

2-48. Three single-phase transformers with ratings of 2400/120 V are used to step down a distribution line voltage from 4160 V to a lower value.

    (a) Assuming the use of a Y-Δ arrangement, find the value of the secondary Δ-side voltage.

    (b) If the maximum value of the magnetic flux puts operation beyond the knee of the magnetization curve, does a third harmonic voltage exist between the line and neutral terminals on the Y-side? Explain.

2-49. Refer to the arrangement shown in Fig. P2-49. The single-phase transformers are rated for 2400/120 V. Consider that the three coils for the secondary delta connection are properly joined with regard to the fundamental induced voltage component. Assume further that the magnitude of the third harmonic induced voltage is one-third that of the fundamental.

    (a) Find the gap voltage when a sinusoidal voltage of 4160 V rms is applied to the line terminals of the wye.

    (b) Calculate the line-to-neutral voltage at the Y connection.

    (c) When the gap is closed, describe the role played by the voltage found in part (a) in the operation of these transformers. Be specific.

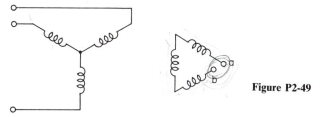

Figure P2-49

2-50. A Y-Y connection of transformers consists of single-phase transformers which carry a rating of 69/6.9 kV. A sinusoidal transmission line voltage of 120 kV is applied to the high side. The third harmonic voltage, when it exists, is known to be one-third the value of the fundamental voltage. Neither the high nor the low side is provided with a ground neutral.

    (a) Determine the rms value of the line-to-neutral voltage on each side.

    (b) What recommendation can you make to eliminate the third harmonic voltage from the line-to-neutral voltage?

3

Fundamentals of Electromechanical Energy Conversion

Electromechanical energy conversion involves the interchange of energy between an electrical system and a mechanical system through the medium of a coupling magnetic field. The process is essentially reversible except for a small amount which is lost as heat energy. When the conversion takes place from electrical to mechanical form the device is called a *motor*. When mechanical energy is converted to electrical energy the device is called a *generator*. Moreover, when the electrical system is characterized by alternating current the devices are referred to as *ac* motors and *ac* generators, respectively. Similarly, when the electrical system is characterized by direct current the electromechanical conversion devices are called *dc* motors and *dc* generators.

The same fundamental principles underlie the operation of both ac and dc machines, and they are governed by the same basic laws. Thus, in the computation of the developed torque of an electromechanical energy-conversion device, one basic torque formula (which derives directly from Ampere's law) applies whether the machine is ac or dc. The ultimate forms of the torque equations appear different for the two types of machines only because the details of mechanical construction differ. In other words, starting with the same basic principles for the production of electromagnetic torque, the final forms of the torque equations differ to the extent that the mechanical details differ. These comments apply equally in generating emf in the armature winding of a machine whether it is ac or dc. Once again a simple basic relationship (Faraday's law) governs the voltage induced. The final forms of the voltage equations differ merely as a reflection of the differences in the construction of the machines. It is important that this is understood at the outset, because a prime objective of the treatment of the subject matter as it unfolds in this chapter is to impress upon the reader the fact that ac machines are not fundamentally different from dc machines. They differ merely in construction details; the underlying principles are the same.

This chapter is arranged to give still greater emphasis to this theme. Attention first is given to a basic analysis of the production of electromagnetic torque as well as the generation of voltages starting with the classical fundamental laws. Then the construction features of the various types of electric machines are examined with the purpose of indicating how the conditions for the production of torque and voltage are fulfilled and also to point out why differences are necessary. Once this background is established, the fundamental equations for developed torque and induced voltage are modified to make these expressions consistent with the particular construction features of the machines involved. In this way the equations are put in a more useful form for purposes of analysis and design. Attention is next directed in the following chapters to an analysis of each of the major types of electromechanical energy-conversion devices to be found performing innumerable tasks throughout industry and in homes, shops, and offices everywhere.

3-1 ELECTROMAGNETIC TORQUE

Since electromechanical energy conversion involves the interchange of energy between an electrical and a mechanical system, the primary quantities involved in the mechanical system are torque and speed, while the analogous quantities in the electrical system are current and voltage respectively. Figure 3-1 represents the situation diagrammatically. *Motor action* results when the electrical system causes a current i to flow through conductors that are placed in a magnetic field. Then by Eq. (1-2) a force is produced on each conductor so that, if the conductors are located on a structure that is free to rotate, an electromagnetic torque T results, which in turn manifests itself as an angular velocity ω_m. Moreover, with the development of this motor action the revolving conductors cut through the magnetic field and thereby undergo an emf e, which is really a reaction voltage

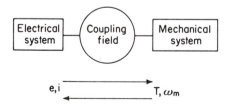

Figure 3-1 Block representation of electromechanical energy conversion.

analogous to the induced emf in the primary winding of a transformer. Note that the coupling field is involved in establishing the electromagnetic torque T as well as the reaction induced emf e. In the case of *generator action* the reverse process takes place. Here the rotating member—the *rotor*—is driven by a prime mover (steam turbine, gasoline engine, and so on), causing an induced voltage e to appear across the armature winding terminals. Upon the application of an electrical load to these terminals, a current i is made to flow, delivering electrical power to the load. Of course the flow of this current through the armature conductors interacts with the magnetic field to produce a reaction torque opposing the applied torque originating from the prime mover.

In the interest of emphasizing further the character of this motor and generator action and its relationship to the coupling magnetic field, a mathematical description is now undertaken. Consider that we have an electromechanical energy-conversion device which consists of a stationary member—the *stator*—and the rotating member, the rotor. Refer to Fig. 3-2(a).

Assume that the stator is equipped with a coil 1-2, which is energized to cause a constant current to flow in the direction shown by the dot-cross notation. By Flemmings' right-hand rule, the flux lines that are produced by the coil ampere-turns leave the stator from the upper half, cross their air gap, penetrate the rotor iron, and then once again cross the air gap to form the closed path illustrated in Fig. 3-2(a). If the rotor and stator surfaces were to be cut and laid out in a flat manner, the resulting configuration would appear as shown in Fig. 3-2(b). If the reluctance of the iron is neglected, half the coil mmf is used for each crossing of the air gap. Accordingly, the value of the flux density that appears between coil sides 1-2 in Fig. 3-2(b) is one-half the coil mmf divided by the air-gap reluctance and the cross-sectional area per pole. Since these quantities are fixed, it follows that the flux density, too, is constant from coil side 1 to 2 and 2 to 1. This situation is depicted in Fig. 3-2(c) in a manner that emphasizes that the flux that leaves the stator (in the upper section between coil sides 1-2) behaves like a north-pole flux, while the flux that enters the stator (in the lower section between coil sides 2-1) behaves like a south-pole flux.

With the uniformly distributed field now established, the focus of attention is shifted to the rotor, which is assumed to be equipped with a total of Z conductors, only two of which (a and b) are shown in Fig. 3-2(a). It is further assumed that conductor a is joined to conductor b to form a coil in the fashion illustrated in Fig. 3-2(d). Because the principle of electromechanical energy conversion is concerned with the movement of a coil relative to a magnetic field, the analysis now proceeds from the arrangement shown in Fig. 3-2(e), where only the presence of the field density B is given importance and not its source. It is helpful to observe that initially the flux that links rotor coil a-b is zero. In Fig. 3-2(e), this situation is represented by noting that coil a-b links one-half of the positive (north-pole) air-gap flux in addition to one-half of the negative (south-pole) air-gap flux for a resultant of zero. By Faraday's law, displacement of the coil by a differential

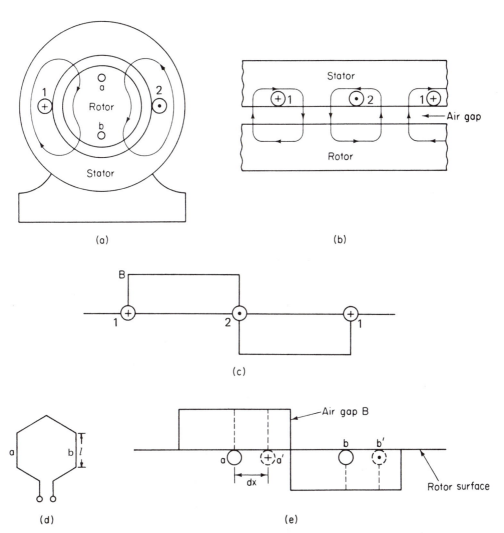

(a)

(b)

(c)

(d)

(e)

Figure 3-2 Illustrating some construction features of an electromechanical energy conversion device: (a) stator and rotor; (b) a plan view; (c) a uniform airgap flux density produced by coil 1-2; (d) a full-pitch coil with axial length l; (e) changing flux linkage of coil a-b with displacement dx.

distance dx to the new position $a'b'$ causes a change in the flux linking the coil that is expressed by

$$d\phi = 2B\,dA \qquad\qquad (3\text{-}1)$$

The factor 2 accounts for the two coil sides; B is the value of the flux density at the points where conductors a and b lie, and dA is the differential area through which the B-field penetrates. A little thought reveals this area to be equal to the axial length of the rotor, l, times dx. Thus

$$dA = l\, dx \qquad (3\text{-}2)$$

However, if it is further assumed that dx results from a linear velocity v imparted to the conductor for the differential time dt, it follows that

$$dx = v\, dt \qquad (3\text{-}3)$$

Insertion of Eqs. (3-2) and (3-3) into Eq. (3-1) then leads to

$$d\phi = 2Blv\, dt \qquad (3\text{-}4)$$

The corresponding emf induced in coil ab becomes, by Faraday's law,

$$e = \frac{d\phi}{dt} = 2Blv \qquad (3\text{-}5)$$

Because the coil a-b is assumed driven from left to right, the $\bar{v} \times \bar{B}$ rule indicates that the direction of the induced emf is into the paper for coil side a' and out of the paper for coil side b'. The magnetic field is considered to be directed downward on side a' and upward on side b'. Moreover, the application of the $\bar{I} \times \bar{B}$ rule discloses that a developed force exists on sides a' and b' in association with a coil current I that is directed from right to left. This produces a torque that opposes the torque that drives the coil.

The factor 2 appears in Eq. (3-5) because the coil involves two conductors. It then follows that, if the rotor is equipped with a total of Z conductors, which are all series-connected, the total induced voltage can be expressed as

$$e = ZBlv \qquad (3\text{-}6)$$

When a current i flows through conductors a and b in the configuration of Fig. 3-2(e), the direction of i and the direction of B are 90° apart. Accordingly, a force exists on each conductor which is given by Eq. (1-2). Thus

$$F_c = Bli \qquad (3\text{-}7)$$

where F_c denotes the force developed on an individual conductor. However, since there are now Z conductors on the rotor surface, it follows that the total developed force is†

$$F = ZBli \qquad (3\text{-}8)$$

Moreover, if this force is made to act through a moment arm r, which is the radius of the rotor, the resulting developed torque can be expressed as

$$T = Fr = ZBlri \qquad (3\text{-}9)$$

---

†This equation is valid provided that the conductors are paired to form series-connected coils that have a span of one *pole pitch*. A pole pitch is equal to 180 electrical degrees.

A glance at Eqs. (3-6) and (3-9) shows that both the induced voltage and the developed electromagnetic torque are dependent upon the coupling magnetic field. The factor that determines whether the induced voltage e and the developed torque T are actions or reactions depends upon whether generator or motor action is being developed.

The two mechanical quantities T and ω_m and the two electrical quantities e and i, which are coupled through the medium of the magnetic field B, are related further by the law of conservation of energy. This is readily demonstrated by dividing Eq. (3-6) by Eq. (3-9). Thus

$$\frac{e}{T} = \frac{ZBlv}{ZBlir} = \frac{\omega_m}{i} \tag{3-10}$$

where

$$\omega_m = \frac{v}{r} \quad \text{mechanical angular velocity of the rotor} \tag{3-11}$$

Rearranging Eq. (3-10) leads to

$$ei = T\omega_m \tag{3-12}$$

which states that the developed electrical power is equal to the developed mechanical power. This statement is valid for generator as well as motor action.

Finally, on the basis of the foregoing discussion, it should be apparent that a conventional electromechanical energy-conversion device must involve two components. One is the *field winding*, which by definition is that part of the machine which produces the coupling field B. The other is the *armature winding*, which by definition is that part of the machine in which the "working" emf e and the current i exist.

Example 3-1

A two-pole electric machine has a rectangular flux-density wave of height 0.4 T. It also has an axial length of $l = 0.2$m and a rotor with a radius of $r = 0.1$m. The rotor is equipped with a single coil of 20 full-pitch turns. A current of 5 A is supplied to the coil.

(a) Find the maximum flux in webers that can be made to link the coil.

(b) When the coil carries 5 A, find the developed torque in N-m.

(c) Assuming that the torque produced in part (b) results in a speed of 900 rpm, determine the coil induced emf.

(d) Obtain the result in part (c) by an alternative procedure.

(e) What is the mechanical power developed in watts?

Solution   (a) The maximum flux occurs when the coil is in the position relative the flux density wave illustrated in Fig. 3-2(c). The area is that associated with one-half the rotor diameter. Thus

$$\phi = BA = B\pi rl = 0.4\pi(0.1)(0.2) = 0.025133 \text{ Wb}$$

(b) By Eq. (3-9)

$$T = ZBlri = 2NBlri = 40(0.4)(0.2)(0.1)5 = 1.6 \text{ N-m}$$

(c) From the principle of conservation of energy

$$ei = T\omega_m$$

where

$$\omega_m = \frac{2\pi n}{60} = 0.10472(900) = 94.25 \text{ rad/s}$$

$$\therefore \quad e = \frac{T\omega_m}{i} = \frac{1.6(94.25)}{5} = 30.16 \text{ V}$$

(d) Invoking Eq. (3-9) we get

$$e = ZBlv = 2NBlv = 40(0.08)(0.1)\omega_m r = 0.32(94.25)0.1 = 30.16 \text{ V}$$

(e) Two alternative expressions provide the result.

$$T\omega_m = 1.6(94.25) = 150.8 \text{ W}$$

or

$$ei = 5(30.16) \quad = 150.8 \text{ W}$$

Important note. The foregoing notions regarding the development of electromagnetic torque can be applied in a general way to the actual field and ampere-conductor distributions that are found in real ac and dc machines. This treatment appears at the end of the book in the Appendix to Chapter 3. The principal results are then invoked in Sec. 3-4 as the starting point in deriving more readily useful forms of the torque formulas for these practical ac and dc machines.

3-2 INDUCED VOLTAGES

The electromechanical energy-conversion process involves the interaction of two fundamental quantities which are related through the law of conservation of energy and the coupling magnetic field. These are, of course, electromagnetic torque and induced voltage. The manner in which electromagnetic torque is developed is described in the preceding section. Here we consider the generation of voltage in the armature winding brought about by changing flux linkages. The polarity of this voltage depends upon the direction of the relative motion of the winding with respect to the magnetic flux and is independent of whether the machine is a motor or a generator, i.e., the direction of energy flow.

Consider the configuration shown in Fig. 3-3. The flux-density curve is assumed to be sinusoidally distributed along the air gap of the machine, and the armature winding is assumed to consist of a full-pitch coil having N turns. A full-pitch coil is one that spans π electrical radians. Just as Ampere's law serves as

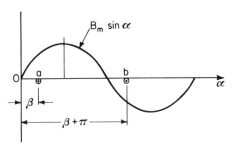

Figure 3-3 Configuration for derivation of the induced-emf equation.

the starting point in the development of the torque relationships, so Faraday's law serves as the starting point in establishing the induced-voltage relationships. Equation (3-5) is not the result we are seeking, because it applies when the flux-density curve is uniform over the pole pitch. It is more generally useful to deal with a sinusoidal distribution of the flux density, as explained more fully later. Since the energy-conversion process involves relative motion between the field and the armature winding, we know that the flux linking the coil changes from a positive maximum to zero to a negative maximum and back to zero again. This cycle repeats every time the coil passes through a pair of flux-density poles. It should be clear from Fig. 3-3 that maximum flux penetrates the coil when it is in a position corresponding to $\beta = 0°$ or $\beta = 180°$ or multiples thereof. Minimum flux linkage occurs when $\beta = 90°$ or $270°$ or multiples thereof. It should be apparent too that the maximum flux that links the coil is the same as the flux per pole, which for sinusoidal distributions is given by Eq. (3-64) of Appendix to Chapter 3. Moreover, as the coil moves relative to the field, it follows that the instantaneous flux linking the coil may be expressed as

$$\phi = \Phi \cos \beta = \frac{4}{p} B_m lr \cos \beta \tag{3-13}$$

The $\cos \beta$ factor is included because of the sinusoidal distribution. Furthermore, since β increases with time in accordance with

$$\beta = \omega t \tag{3-14}$$

where ω is expressed in electrical radians per second, the expression for the instantaneous flux linkage becomes

$$\phi = \frac{4}{p} B_m lr \cos \omega t = \Phi \cos \omega t \tag{3-15}$$

Then by Faraday's law the expression for the instantaneous induced voltage becomes

$$\boxed{e = -N \frac{d\phi}{dt} = \omega N \Phi \sin \omega t} \tag{3-16}$$

It is important to keep in mind that Φ represents the *maximum flux*, which is the same as the flux per pole.

Equation (3-16) is applicable to dc as well as ac machines. In a later section the induced emf per phase in an ac machine and the induced voltage appearing between brushes in a dc machine are derived starting with Eq. (3-16). The fact that in the dc machine the flux-density curve is nonsinusoidal (as Fig. 3-16 of Appendix to Chapter 3 shows) is relatively unimportant; what is of primary importance is the fundamental component of the nonsinusoidal distribution. It is for this reason that Eq. (3-16) is derived for sinusoidal distributions. Recall that by means of the Fourier series any nonsinusoidal periodic wave may be expressed entirely in terms of sinusoids.

Example 3-2

The purpose of this example is to draw a distinction in the character of voltage that is induced in a revolving coil when it finds itself under the influence of a *vertically directed* uniform field as illustrated in Fig. 3-4(a) and a *radially directed* uniform field as occurs in the configuration of Fig. 3-4(b). Assume that coil AA' is square and measures 30 cm on a side with a total of 40 turns. The coil is driven at a constant speed of 200 rad/s. The value of the uniform magnetic field in both cases is 0.1 T.

(a) Obtain an expression for the flux linkage through the coil as a function of time.

(b) Find the expression for the voltage induced in the coil as a function of time using Faraday's law.

(c) Determine the result found in part (b) by employing the *Blv* concept using Eq. (3-5).

(d) Now assume that coil AA is put under the influence of a radially directed field as shown in Fig. 3-4(b). Coil CC' is assumed to carry a fixed unidirectional

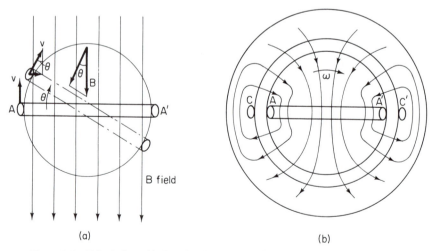

(a)   (b)

Figure 3-4 Calculation of induced voltages: (a) coil AA' operating in a vertically directed field; (b) coil AA' operating in a radially directed field.

current, thereby producing a constant field distribution over the entire periphery of the air gap. Find the coil emf in this case as a function of time.

Solution   (a) Let time be counted starting with the coil in the horizontal position in Fig. 3-4(a). As time elapses, the coil, which is rotating, takes a new position displaced by $\theta = \omega t$ degrees from its reference position. In this new position, the component of the vertical field that threads the coil in a quadrature fashion is

$$B \cos \theta = B \cos \omega t$$

The corresponding amount of flux that links with the coil is described by

$$BA \cos \theta = BA \cos \omega t$$

or

$$\phi_m \cos \theta = \phi_m \cos \omega t = \phi$$

where ϕ_m denotes the maximum flux that links with the coil. But <u>flux linkage is simply this flux multiplied by the number of turns in the coil.</u> Thus

$$\lambda = N\phi = N\phi_m \cos \omega t = 40(0.1)(0.3)(0.3) \cos \omega t$$

$$= 0.36 \cos 200t \quad \text{Wb-t}$$

(b) By Faraday's law, we know that the induced emf is equal to the rate of change of flux linkage with time. Hence

$$e = -\frac{d\lambda}{dt} = -\frac{d}{dt}(0.36 \cos \omega t) = (0.36)(200) \sin 200t$$

$$= 72 \sin 200t \quad \text{V}$$

This result states that the emf varies sinusoidally with time.

(c) With the Blv approach it is the cutting action of the flux lines by the conductors that receives the focus. When the coil is in the horizontal position, the conductors fail to be under the influence of the B field, so the instantaneous value of the voltage is zero. The coil sides are not cutting the flux lines. It is important to note that at this position, the velocity vector of the conductor is collinear with the field. Hence the component that is in quadrature with the field lines is zero and so no cutting occurs. As time is allowed to elapse and the coil takes a position θ degrees from its horizontal reference point, the quadrature component of the velocity vector is no longer zero. Rather, it is equal to $v \sin \theta = v \sin \omega t$, as can be seen from Fig. 3-4(a). Employing only the useful part of the velocity vector then permits the induced emf to be written as

$$e = 2NBlv \sin \omega t = 2NBl\omega r \sin \omega t$$

$$= 2(40)(0.1)(0.3)(200)(0.15) \sin \omega t$$

$$= 72 \sin \omega t = 72 \sin 200t$$

(d) Unlike the situation that occurs in the geometry of Fig. 3-4(a), the use of a radially directed field means that the velocity vector is at all times in a position of quadrature with the B-field. Hence the voltage induced in this case is

$$e = 2NBlv = 2NBl\omega r = 2(40)(0.1)(0.3)(200)(0.15) = 72 \text{ V}$$

This quantity is constant in magnitude with time and specifically it is equal to the peak value found in part (c). Another point of interest associated with the geometry of Fig. 3-4(b) compared to that of Fig. 3-4(a) is that flux linkage in the case of Fig. 3-4(b) is not described by the cosine function but rather in the manner already described in Sec. 3-1.

3-3 CONSTRUCTION FEATURES OF ELECTRIC MACHINES

The energy-conversion process usually involves the presence of two important features in a given electromechanical device. These are the field winding, which produces the flux density, and the armature winding, in which the working emf is induced. In this section the salient construction features of the principal types of electric machines are described to show the location of these windings, as well as to demonstrate the general composition of such machines.

The Three-Phase Induction Motor

This is one of the most rugged and most widely used machines in industry. Its stator is composed of laminations of high-grade sheet steel. The inner surface is slotted to accommodate a three-phase winding. In Fig. 3-5(a) the three-phase winding is represented by three coils, the axes of which are 120 electrical degrees apart. Coil aa' represents all the coils assigned to phase a for one pair of poles. Similarly, coil bb' represents phase b coils, and coil cc' represents phase c coils. When one end of each phase is commonly connected, as depicted in Fig. 3-5(b), the three-phase stator winding is said to be Y-connected. Such a winding is called a three-phase winding because the voltages induced in each of the three phases by a revolving flux-density field are out of phase by 120 electrical degrees—a distinguishing characteristic of a symmetrical three-phase system.

The rotor also consists of laminations of slotted ferromagnetic material, but the rotor winding may be either the squirrel-cage type or the wound-rotor type. The latter is of a form similar to that of the stator winding. The winding terminals are brought out to three slip rings as depicted in Fig. 3-6(a). This allows an

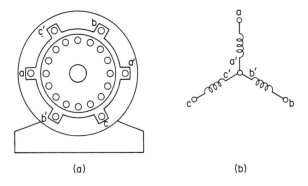

(a)

(b)

Figure 3-5 Three-phase induction motor: (a) showing a stator with three-phase winding and squirrel-cage rotor; (b) schematic representation of a three-phase Y-connected stator winding.

external three-phase resistor to be connected to the rotor winding for the purpose of providing speed control. As a matter of fact it is the need for speed control which in large measure accounts for the use of the wound-rotor type induction motor. Otherwise, the squirrel-cage induction motor would be used. The squirrel-cage winding consists merely of a number of copper bars imbedded in the rotor slots and connected at both ends by means of copper end rings as depicted in Fig.

(a)

(b)

Figure 3-6   (a) Wound rotor for three-phase induction motor; (b) showing details of rotor bars of squirrel-cage induction motor.

3-6(b). (In some of the smaller sizes aluminum is used.) The squirrel-cage construction is not only simpler and more economical than the wound-rotor type but more rugged as well. There are no slip rings or carbon brushes to be bothered with.

In normal operation a three-phase voltage is applied to the stator winding at points *a-b-c* in Fig. 3-5. As described in Sec. 4-1, magnetizing currents flow in each phase which together create a revolving magnetic field having two poles. The speed of the field is fixed by the frequency of the magnetizing currents and the number of poles for which the stator winding is designed. Figure 3-5 shows the configuration for two poles. If the pattern *a-c'-b-a'-c-b'* is made to span only 180 mechanical degrees and then is repeated over the remaining 180 mechanical degrees, a machine having a four-pole field distribution results. For a *p*-pole machine the basic winding pattern must be repeated $p/2$ times within the circumference of the inner surface of the stator.

The revolving field produced by the stator winding cuts the rotor conductors, thereby inducing voltages. Since the rotor winding is short-circuited by the end rings, the induced voltages cause currents to flow which in turn react with the field to produce electromagnetic torque—and so motor action results.

Accordingly, on the basis of the foregoing description, it should be clear that for the three-phase induction motor the field winding is located on the stator and the armature winding on the rotor. Another point worth noting is that this machine is singly excited, i.e., electrical power is applied only to the stator winding. Current flows through the rotor winding by induction. As a consequence both the magnetizing current, which sets up the magnetic field, and the power current, which allows energy to be delivered to the shaft load, flow through the stator winding. For this reason, and in the interest of keeping the magnetizing current as small as possible in order that the power component may be correspondingly larger for a given rating, the air gap of induction motors is made as small as mechanical clearance will allow. The air-gap lengths vary from about 0.02 in. for smaller machines to 0.05 in. for machines of higher rating and speed.

Synchronous Machines

The essential construction features of the synchronous machine are depicted in Fig. 3-7. The stator consists of a stator frame, a slotted stator core, which provides a low-reluctance path for the magnetic flux, and a three-phase winding imbedded in the slots. Note that the basic two-pole pattern of Fig. 3-5(a) is repeated twice, indicating that the three-phase winding is designed for four poles. The rotor either is cylindrical and equipped with a distributed winding or else has salient poles with a coil wound on each leg as can be seen in Fig. 3-7(b). The cylindrical construction is used almost exclusively for turbogenerators which operate at high speeds. On the other hand, the salient-pole construction is used exclusively for synchronous motors operating at speeds of 1800 rpm or less.

When operated as a generator the synchronous machine receives mechani-

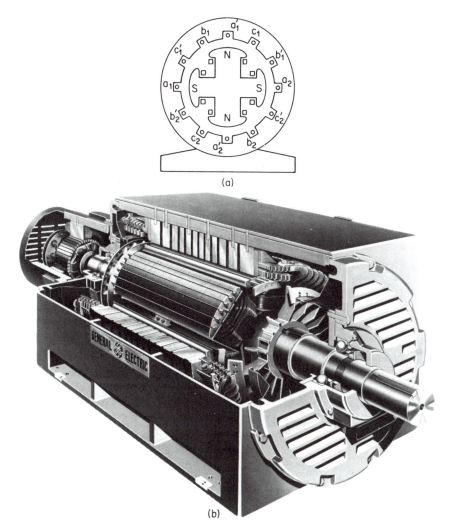

(a)

(b)

Figure 3-7   Salient-pole synchronous machine: (a) schematic diagram; (b) photograph showing field coils wrapped around salient poles. Note too the squirrel-cage winding embedded in the pole faces.

cal energy from a prime mover such as a steam turbine and is driven at some fixed speed. Also, the rotor winding is energized from a dc source, thereby furnishing a field distribution along the air gap. When the rotor is at standstill and direct current flows through the rotor winding, no voltage is induced in the stator winding because the flux is not cutting the stator coils. However, when the rotor is being driven at full speed, voltage is induced in the stator winding and upon application of a suitable load electrical energy can be delivered to it.

A little thought should make it apparent that for the synchronous machine the field winding is located on the rotor; the armature winding is located on the stator. This conclusion is valid even when the synchronous machine operates as a motor. In this mode ac power is applied to the stator winding and dc power is applied to the rotor winding for the purpose of energizing the field poles. Mechanical energy is then taken from the shaft. Note, too, that unlike the induction motor, the synchronous motor is a doubly fed machine. In fact it is this characteristic which enables this machine to develop a nonzero torque at only one speed—hence the name "synchronous." This matter is discussed further in Chapter 6.

Because the magnetizing current for the synchronous machine originates from a separate source (the dc supply or an exciter generator such as the one in the leftmost part of Fig. 3-7(b)), the air-gap lengths are larger than those found in induction motors of comparable size and rating. However, synchronous machines are more expensive and less rugged than induction motors in the smaller horsepower ratings because the rotor must be equipped with slip rings and brushes to allow the direct current to be conducted to the field winding.

Direct-Current Machines

Electromechanical energy-conversion devices that are characterized by direct current are more complicated than the ac type. In addition to a field winding and armature winding, a third component is needed to serve the function of converting the induced ac armature voltage into a direct voltage. Basically the device is a mechanical rectifier and is called a *commutator*.

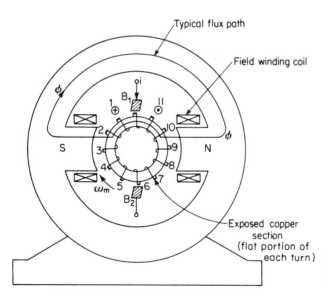

Figure 3-8 Construction features of the dc machine, showing the Gramme-ring armature winding.

Appearing in Fig. 3-8 are the principal features of the dc machine. The stator consists of a laminated ferromagnetic material equipped with a protruding structure around which coils are wrapped. The flow of direct current through the coils establishes a magnetic field distribution along the periphery of the air gap in much the same manner as occurs in the rotor of the synchronous machine. Hence in the dc machine the field winding is located on the stator. It follows then that the armature winding is on the rotor. The rotor is composed of a laminated core, which is slotted to accommodate the armature winding. It also contains the commutator—a series of copper segments insulated from one another and arranged in cylindrical fashion as shown in Fig. 3-9(a). Riding on the commutator are appropriately placed carbon brushes which serve to conduct direct current to or from the armature winding depending upon whether motor or generator action is taking place.

In Fig. 3-8 the armature winding is depicted as a coil wrapped around a toroid. This is merely a schematic convenience. An actual winding is wound so that no conductors are wasted by placing them on the inner surface of the rotor core where no flux penetrates. In Fig. 3-8 those parts of the armature winding which lie directly below the brush width are assumed to have the insulation removed, i.e., the copper is exposed. This allows current to be conducted to and from the armature winding through the brush as the rotor revolves. In a practical winding each coil is made accessible to the brushes by connecting the coils to individual commutator segments as indicated in Fig. 3-9(a) and then placing the brushes on the commutator.

For motor action direct current is made to flow through the field winding as well as the armature winding. If current is assumed to flow into brush B_1 in Fig. 3-8, then note that on the left side of the rotor for the *outside* conductors current flows into the paper while the opposite occurs for the conductors located on the outside surface of the right side of the rotor. By Eq. (1-2) a force is produced on each conductor, thereby producing a torque causing a clockwise rotation. Now the function of the commutator is to ensure that, as a conductor such as 1 in Fig. 3-8 revolves and thus goes from the left side of brush B_1 to the right side, the current flowing through it reverses, ensuring a continuous unidirectional torque for the entire armature winding. Recall that a reversed conductor current in a flux field of reversed polarity keeps the torque unidirectional. The reversal of current comes about because the commutator always allows current to be conducted in the same directions in either side of the armature winding whether or not it is rotating.

Another point of interest in Fig. 3-8 concerns the location of the brushes. By placing the brushes on a line perpendicular to the field axis all conductors contribute in producing a unidirectional torque. If, on the other hand, the brushes were placed on the same line as the field axis, then half of the conductors would produce clockwise torque and the other half counterclockwise torque, yielding a zero net torque.

Figure 3-9(a)   Armature winding of a dc machine, showing the coils connected to the commutator. (Courtesy of General Electric Company.)

Figure 3-9(b)   Complete stack of stator laminations for the two-pole dc machine equipped with interpoles.

Figure 3-9(c)   Complete stator structure of the two-pole dc machine of Fig. 3-9(b) with field and interpole windings in place.

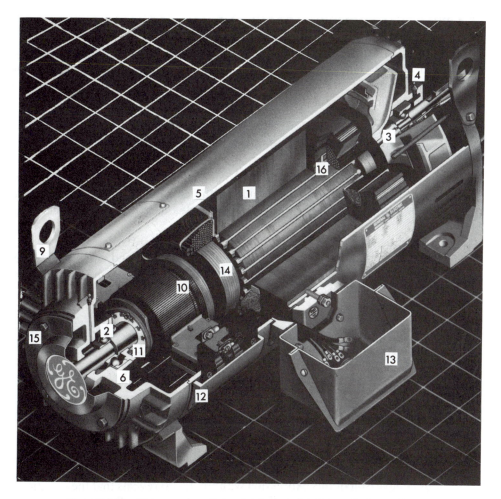

Figure 3-9(d)   Cutaway view of completely assembled two-pole dc machine with notations for various construction features. See text for descriptions. (Courtesy of General Electric Company.)

Additional construction features of the dc machine are displayed in the remaining parts of Fig. 3-9. Shown in Fig. 3-9(b) is the laminated stator stack of the two-pole machine. The small protrusions are called interpoles and are used to aid the change of current direction in an armature coil as it passes from one side of a brush to the other. Figure 3-9(c) depicts the field and interpole windings in place. Finally, Fig. 3-9(d) displays a cutaway view of the assembled machine. Here item 2 locates the vacuum degassed steel ball bearings; 5 denotes the steel enclosure; 11 is the commutator support which includes cooling air passages for improved cooling of the armature conductors; 12 identifies the brush-access port for maintenance; and 14 denotes contamination resistant armature insulation.

3-4 PRACTICAL FORMS OF TORQUE AND VOLTAGE FORMULAS

Our objective here is to modify the basic expressions for induced voltage and electromagnetic torque to forms that are more meaningful for the specific design of a particular electromechanical energy-conversion device, whether it is alternating or direct current.

Alternating-Current Machines

Voltage.   If N denotes the total number of turns per phase of a three-phase winding, the instantaneous value of the emf induced in any one phase can be represented by Eq. (3-16), which is repeated here for convenience. Thus

$$e = \omega N \Phi \sin \omega t \qquad (3\text{-}17)$$

Keep in mind that Φ denotes the total flux per pole and ω represents the relative cutting speed in electrical radians per second of the winding with respect to the flux-density wave. It is related to the frequency f of the ac device by

$$\omega = 2\pi f \qquad (3\text{-}18)$$

where f is expressed in cycles per second.

The maximum value of this ac voltage occurs when $\sin \omega t$ has a value of unity. Hence

$$E_{\max} = \omega N \Phi \qquad \text{V} \qquad (3\text{-}19)$$

The corresponding rms value is

$$E \equiv \frac{E_{\max}}{\sqrt{2}} = \sqrt{2}\,\pi f N \Phi = 4.44 f N \Phi \qquad \text{V} \qquad (3\text{-}20)$$

A comparison of this expression with Eq. (2-18) reveals that the equations have an identical form. There is a difference, however, and it lies in the meaning of Φ. In the transformer, Φ_m is the maximum flux in time corresponding to the peak magnetizing current consistent with the magnitude of the applied voltage. In the ac electromechanical energy-conversion device Φ is the maximum flux per pole (a space quantity) that links with a coil having N turns and spanning the full-pole pitch.

In any practical machine the total turns per phase are not concentrated in a single coil but rather are distributed over one-third of a pole pitch (or 60 electrical degrees for each of the three phases). In addition the individual coils that make up the total N turns are intentionally designed to span not the full-pole pitch but rather only about 80% to 85% of a pole pitch. Such a coil is called a *fractional-pitch coil*. The use of a *distributed* winding employing *fractional-pitch* coils has the advantage of virtually eliminating the effects of all harmonics that may be present in the flux-density wave while only slightly reducing the fundamental component.† The reduction in the fundamental component can be represented by a winding factor denoted by K_w. Usually K_w has values ranging from 0.85 to 0.95. Accordingly, the final practical version of the induced rms voltage equation for an ac machine is

$$\boxed{E = 4.44fNK_w\Phi} \qquad \text{V} \qquad\qquad (3\text{-}21)$$

Torque.   Analysis of the manner in which electromagnetic torque is produced by the interaction of the armature ampere-conductor distribution with the field distribution that prevails in actual ac machines can be shown (see Appendix to Chapter 3, Sec. 3A) to lead to the following basic expression:

$$T = \frac{\pi}{8} p^2 \Phi J_m \cos \psi \qquad \text{N-m} \qquad\qquad (3\text{-}22)$$

where p is the number of poles

Φ is the flux per pole in Wb

J_m is the equivalent current sheet that represents an idealized ampere-conductor distribution expressed in A/rad

ψ is the phase displacement angle between the start of the current sheet (i.e., ampere-conductor distribution) and the start of the flux density wave beneath a pole.

A study of Eq. (3-22) stresses the three conditions that must be satisfied for the development of torque in conventional electromechanical energy-conversion devices. There must be a field distribution represented by Φ. There must be an ampere-conductor distribution represented by J_m. Finally, there must exist a favorable space-displacement angle between the two distributions. Note that it is possible to have both a field and an ampere-conductor distribution and still the torque can be zero if the field pattern is such that equal and opposite torques are developed on the conductors, i.e., $\psi = 90°$.

There are two useful and practical forms to which this basic expression for developed electromagnetic torque may be converted. In one form it is the ampere-conductor distribution of the armature winding that is emphasized. In the other, it is the conservation of energy that is stressed. The mmf per pole, $\mathcal{F}_p$, is related to J_m by

$$J_m = \mathcal{F}_p \qquad\qquad (3\text{-}23)$$

(see Appendix to Chapter 3, Eq. (3-50)). Moreover, for any given ac machine the quantity $\mathcal{F}_p$ is entirely determined from the design data as they appear in the following expression:‡

†See Appendix B.
‡See Appendix C.

$$\mathcal{F}_p = \frac{2\sqrt{2}}{\pi} q \frac{N_2 K_{w2}}{p} I_2 = 0.9 q \frac{N_2 K_{w2}}{p} I_2 \tag{3-24}$$

where q = number of phases of the armature winding
N_2 = number of turns per phase of the armature winding
K_{w2} = armature-winding factor
I_2 = armature-winding current per phase
p = number of poles

Hence the expression for electromagnetic torque as described by Eq. (3-22) becomes

$$T = \frac{\pi}{8} p^2 \Phi \mathcal{F}_p \cos \psi = \frac{\pi}{8} 0.9 p \Phi N_2 K_{w2} q I_2 \cos \psi \tag{3-25}$$

Also, qN_2 is the *total* number of turns on the armature surface. This can be expressed in terms of the total number of conductors, Z_2, by making use of the fact that it requires two conductors to make one turn. Thus

$$N_2 = \frac{Z_2}{2q} \tag{3-26}$$

Inserting this into Eq. (3-25) yields

$$\boxed{T = 0.177 p \Phi (Z_2 K_{w2} I_2) \cos \psi} \qquad \text{N-m} \tag{3-27}$$

The quantity in parentheses emphasizes the role played by the ampere-conductor distribution of the armature winding. Of course to compute the torque by this equation we need information about the space-displacement angle ψ in addition to the design data, i.e., p, Φ, Z_2, K_{w2}. Normally, ψ can be determined by the same data that lead to the determination of I_2, which is described later in the next chapter.

A second approach to obtaining a useful form of the basic torque equation is to replace Φ with Eq. (3-21) and J_m with Eqs. (3-23) and (3-24) in Eq. (3-22). Thus

$$T = \frac{\pi}{8} p^2 \left(\frac{E_2}{\sqrt{2}\,\pi f N_2 K_{w2}} \right) \left(2 \frac{\sqrt{2}}{\pi} q \frac{N_2 K_{w2}}{p} I_2 \right) \cos \psi$$

or

$$T = \frac{p}{4\pi f} q E_2 I_2 \cos \psi \tag{3-28}$$

Moreover, since the mechanical angular velocity ω_m is related to the electrical angular velocity ω by

$$\omega_m = \frac{2}{p} \omega = \frac{2}{p} 2\pi f = \frac{4\pi f}{p} \tag{3-29}$$

it follows that Eq. (3-28) may be written as

$$T = \frac{1}{\omega_m} qE_2I_2 \cos \psi \qquad \text{N-m} \qquad (3\text{-}30)$$

where E_2 and I_2 are the rms induced voltage and current per phase of the armature winding.

It can be shown that the *space*-displacement angle ψ in this equation is identical with the time-displacement angle θ_2,† which is the phase angle existing between the two time phasors E_2 and I_2. Accordingly, the expression for the electromagnetic torque may be written as

$$\boxed{T = \frac{1}{\omega_m} qE_2I_2 \cos \theta_2} \qquad \text{N-m} \qquad (3\text{-}31)$$

Note the power balance, which states that the developed mechanical power is equal to the developed electrical power.

Direct-Current Machines

Voltage. The useful version of the expression for the emf induced in the armature winding of a dc machine and appearing at the brushes also derives from Eq. (3-16). The presence of the commutator, however, makes a difference in the manner in which this equation is to be treated to obtain the desired result. We are interested in the voltage as it appears at the brushes. If we follow coil 1 in Fig. 3-8 as it leaves brush B_2 and advances to brush B_1, we notice that the emf induced in this coil is always directed out of the paper beneath the south-pole flux. In fact this is true of any coil that moves from B_2 to B_1. Having established the direction of this voltage for the coils on the left side of the armature, we then note that upon tracing through the winding starting in the direction called for by the sign of the emf induced in coil 1 we terminate at brush B_2. When coil 1 rotates to a position on the right side of the armature winding, such as position 11, the direction of the induced emf is into the paper. If we start with this indicated direction and trace the armature winding, we again terminate at brush B_2. Therefore it is reasonable to conclude that brush B_1 is at a fixed *positive* polarity and brush B_2 is at a fixed *negative* polarity. In essence this means that, if we were to ride along as observers on coil 1 as it traverses through one revolution beginning at brush B_2, the fundamental component of the induced emf would exhibit the variation shown in Fig. 3-10. Note that *relative to brush B_1 the voltage induced in the coil as it moves from B_1 to B_2 appears rectified because the action of the commutator in conjunction with the brushes is to fix the armature winding in space in spite of its rotation.* Accordingly, whether we look at a coil moving from B_2 to B_1 on the left side

†See page 175.

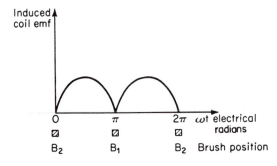

Figure 3-10 Fundamental component of the induced emf in a coil of the armature winding of the dc machine.

or from B_1 to B_2 on the right side, the directions of the induced emf's are such as to make B_1 assume one fixed polarity, and B_2 the opposite fixed polarity.

For a single coil having N_c turns the average value of the unidirectional voltage appearing between the brushes is obtained by integrating Eq. (3-16) over π electrical radians and dividing by π. Thus

$$E_c = \frac{1}{\pi} \int_0^\pi \omega N_c \Phi \sin \omega t \, d\omega t = 4fN_c\Phi \qquad (3\text{-}32)$$

It is customary to express the frequency in terms of the speed of the armature in *rpm* which is given by the relationship†

$$f = \frac{pn}{120} \qquad (3\text{-}33)$$

where p denotes the number of poles and n the speed in rpm. Inserting Eq. (3-33) into Eq. (3-32) and rearranging leads to

$$E_c = p\Phi 2N_c \frac{n}{60} = p\Phi z \frac{n}{60}$$

where z denotes the number of conductors per coil.

This last expression is the average value of the induced emf for a single coil. If many such coils are placed to cover the entire surface of the armature, the total dc voltage appearing between the brushes can be considerably increased. If the armature winding is assumed to have a total of Z conductors and a parallel paths, then the induced emf of the armature winding appearing at the brushes becomes

$$E_a = p\Phi \frac{Z}{a} \frac{n}{60} = \frac{pZ}{60a} \Phi n = K_E \Phi n \qquad (3\text{-}34)$$

†Rotating a coil in a two-pole field at a rate of 1 rps results in a frequency of 1 Hz. For a four-pole field each rps yields 2 Hz. For a p-pole distribution the frequency generated in hertz is

$$f = \frac{p}{2} \times \text{rps} = \frac{p}{2} \frac{\text{rpm}}{60} = \frac{pn}{120}$$

where K_E is a winding constant defined as

$$K_E \equiv \frac{pZ}{60a} \tag{3-35}$$

Whenever the speed is expressed in radians per second, the induced emf equation can then be expressed as

$$E_a = K_T \Phi \omega \tag{3-36}$$

where

$$K_T = \frac{pZ}{2\pi a} \tag{3-36a}$$

It is worth keeping in mind that the induced-emf equation for the ac machine and that for the dc machine originate from the same starting point. To further emphasize this common origin let it be said that we can derive Eq. (3-34) from Eq. (3-21). It merely requires introducing the appropriate winding factor as it applies to the dc armature winding. In fact, it is the quasi-annihilating effect of the winding factor on the harmonics that allows the derivation of Eq. (3-34) to proceed in the fundamental terms in spite of the nonsinusoidal field distribution characteristic of dc machines. It is assumed throughout, however, that the fundamental component of flux is virtually the same as the total flux per pole.

Torque.   Analysis of the manner in which electromagnetic torque is developed by the interaction of the ampere-conductor distribution with the flux density wave in the dc machine can be shown (see Appendix to Chapter 3, Sec. 3B) to lead to the following fundamental formula:

$$T = \frac{p}{2} J\Phi \qquad \text{N-m} \tag{3-37}$$

where p is the number of poles
   J is the current-sheet equivalent of the armature winding ampere-conductor distribution in A/rad
   Φ is the flux per pole in Wb

It is worthwhile to note again that electromagnetic torque is dependent upon a field distribution represented by Φ, an ampere-conductor distribution represented by J, and the angular displacement between the two distributions. Of course, ψ does not appear in Eq. (3-37) because for simplicity the current sheet was assumed to be in phase with the flux-density curve, thereby ensuring that all parts of the current sheet experience a unidirectional torque.

A practical version of this basic torque expression for dc machines results when the current sheet J is replaced by its equivalent expression involving the appropriate design data. In this connection refer to Fig. 3-11. Applying Ampere's

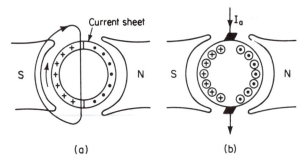

Figure 3-11 Direct-current machine: (a) armature with current sheet; (b) armature with finite ampere-conductor distribution.

(a)   (b)

circuital law to the typical flux path shown makes it apparent that the mmf per pole pair is given by

$$\mathcal{F} = \pi(\text{rad}) \, J \left(\frac{\text{A-t}}{\text{rad}}\right) \tag{3-38}$$

Hence the mmf per pole is

$$\mathcal{F}_p = \frac{\mathcal{F}}{2} = \frac{\pi J}{2} \tag{3-39}$$

This expression is applicable whenever the armature winding is represented by a current sheet. In a practical situation the armature winding is always represented by a finite ampere-conductor distribution as depicted in Fig. 3-11(b). In such cases the armature mmf per pole may be expressed in terms of the total current I_a entering or leaving a brush and the total conductors Z on the armature surface. Thus

$$\mathcal{F}_p = \frac{I_a}{a} \frac{Z}{2} \frac{1}{p} \qquad \text{mmf/pole, or A-t/pole} \tag{3-40}$$

Upon equating the last two expressions we obtain the equation for J. Thus

$$J = \frac{I_a Z}{\pi p a} \tag{3-41}$$

Insertion of Eq. (3-41) into Eq. (3-37) then yields the desired expression for the electromagnetic torque developed by the dc machine. Hence

$$T = \frac{p^2}{2} J\Phi = \frac{pZ}{2\pi a} \Phi I_a \tag{3-42}$$

or

$$\boxed{T = K_T \Phi I_a} \qquad \text{N-m} \tag{3-43}$$

where

$$K_T \equiv \frac{pZ}{2\pi a} = \text{torque constant} \tag{3-44}$$

The foregoing form of the torque formula stresses the ampere-conductor distribution taken in conjunction with the flux field. It is analogous to Eq. (3-27) for the ac machine. Of course in Eq. (3-43) ψ does not appear because it was made zero by placing the brush axis in quadrature with the field axis. It is important to note, however, that the fundamental quantities for the production of torque are there. The equations differ only to the extent that the mechanical details of construction differ.

An alternative expression for the electromagnetic torque results upon substituting in Eq. (3-42) the expression for Φ obtained from Eq. (3-34). Thus, from Eq. (3-34)

$$\Phi = \frac{60a}{pZn} E_a \tag{3-45}$$

so that

$$T = \frac{pZ}{2\pi a} \frac{60a}{pZn} E_a I_a = \frac{60}{2\pi n} E_a I_a \tag{3-46}$$

But

$$\omega_m = \frac{2\pi n}{60} \tag{3-47}$$

Therefore

$$\boxed{T = \frac{1}{\omega_m} E_a I_a} \tag{3-48}$$

A glance at this result again points out the power-balance relationship that underlies the operation of electromechanical energy-conversion devices. Equation (3-48) for the dc machine is analogous to Eq. (3-31) for the ac machine.

Example 3-3

An ac machine is equipped with four poles, three phases, and 36 armature winding turns per phase. The field winding is designed to produce a flux per pole of 0.015 Wb. The armature winding is designed to be essentially resistive over its operating range and the coil span is made equal 83.3% of a pole pitch.

(a) When the motor develops a torque of 115 N-m, find the armature current in amperes.

(b) Determine the armature mmf per pole.

(c) Find the developed torque when the armature current is 15 A.

(d) In part (a) the developed power is known to be equal to 21,677 W. Calculate the armature induced voltage per phase.

(e) Compute the mechanical speed of the motor in rpm for the condition of part (a).

Solution   (a) We can use Eq. (3-27) directly here to get the current I_2, which is the armature current. But first two matters need to be resolved. The first is the angle ψ.

By the problem statement this quantity is $0°$ because the inductive component of the armature winding is considered to be negligible. Second, we need to determine K_{w2}. This quantity is readily calculated by invoking Eq. (B-2) of Appendix B. Thus

$$K_{w2} = \sin (0.833) \frac{\pi}{2} = \sin 75° = 0.966$$

It therefore follows that the armature current is

$$I_2 = \frac{T}{0.177p\phi Z_2(0.966)} = \frac{115}{0.177(4)(0.015)(2 \times 3 \times 36)(0.966)} = 51.9 \text{ A}$$

(b) The armature mmf per pole is found from

$$\mathcal{F}_p = J_m = 0.9q \frac{N_2 K_{w2}}{p} I_2 = 0.9(3) \frac{36(0.966)}{4} 51.9 = 1218.3 \text{ A-t}$$

(c) Using the second form of Eq. (3-25) we get

$$T = \frac{\pi}{8} (0.9)p\phi N_2 K_{w2}qI_2 \cos \psi$$

$$= \frac{\pi}{8} (0.9)(4)(0.015)(\underset{36}{26})(0.966)(3)(15) = 33.2 \text{ N-m}$$

(d) From Eq. (3-31)

$$3E_2 I_2 = 21,677$$

$$E_2 = \frac{21,677}{3(51.9)} = 139.2 \text{ V per phase}$$

(e) By conservation of energy

$$T\omega_m = 21,677$$

$$\omega_m = \frac{21,677}{115} = 188.5 \text{ rad/s}$$

or, expressed in rpm, we write

$$W_M = \frac{2\pi n}{60} \qquad n_m = \frac{30}{\pi} \omega_m = 1800 \text{ rpm}$$

where n_m denotes the mechanical speed in revolutions per minute.

3-5 FINAL REMARKS

For those readers who are interested in a detailed and rigorous treatment of the derivation of the basic electromagnetic torque formulas for ac and dc machines, the subject matter is continued at the end of the book in the section entitled Appendix to Chapter 3. In the interest of continuity, both the equation numbers and figure numbers pick up from the last notations used in Sec. 3-4. A full complement of problems is also included.

PROBLEMS

3-1. A two-pole dc motor is equipped with a single ten-turn full-pitch coil. It has a rectangular field distribution of height 0.5 T. The radius of the rotor is 0.2 m and the axial length is 0.3 m.

    (a) Find the flux per pole in webers.

    (b) Current is supplied to the coil and the motor is found to reach a speed of 625 rpm. Find the emf induced in the coil.

    (c) What is the rate of change of flux with time?

3-2. The motor of Prob. 3-1 is used to supply a torque of 9 N-m to an attached load at a speed of 500 rpm.

    (a) Find the current drawn by the coil.

    (b) Calculate the coil emf.

    (c) What is the developed mechanical power?

3-3. At a particular load condition the motor of Prob. 3-1 delivers 400 W while drawing a coil current of 8 A. All losses are negligible.

    (a) Find the developed torque at the motor shaft.

    (b) What is the coil emf?

    (c) Determine the speed of rotation of the motor shaft in rpm.

3-4. At a given operating condition the machine of Prob. 3-1 runs at 350 rpm. Find the coil induced emf.

3-5. The instantaneous voltage generated in a coil revolving in a magnetic field can be calculated either by the flux-linkage concept or by the *Blv* concept. With this in mind consider the following problem. A square coil 20 cm on a side has 60 turns and is so located that its axis of revolution is perpendicular to a vertically directed uniform magnetic field in air of 0.06 Wb/m^2. The coil is driven at a constant speed of 150 rpm. Compute:

    (a) The maximum flux passing through the coil.

    (b) The maximum flux linkage.

    (c) The time variation of the flux linkage through the coil.

    (d) The maximum instantaneous voltage generated in the coil using both concepts referred to in the foregoing. Indicate by a sketch the position of the coil at this instant.

    (e) The average value of the voltage induced in the coil over one cycle.

    (f) The voltage generated in the coil when the plane of the coil is 30° from the vertical. Compute by both methods.

3-6. A square coil of 100 turns is 10 cm on each side. It is driven at a constant speed of 300 rpm.

    (a) The coil is placed so that its axis of revolution is perpendicular to a vertically directed *uniform* field of 0.1 Wb/m^2. At time $t = 0$, the coil is in a position of maximum flux linkages. Derive an expression for the instantaneous voltage generated. Sketch the voltage waveshape for one cycle.

    (b) The coil is now placed in a *radial* field. All other data remain the same as for part (a). Sketch the voltage waveshape for one cycle showing the numerical value of the maximum voltage.

3-7. In Figs. P3-7(a) and (b) the flux per pole is 0.02 Wb. The coil is revolving at 1800 rpm and consists of 2 turns.

(a) For the configuration of Fig. P3-7(a) derive an expression for the flux linking the coil in terms of the flux per pole and α_0.

(b) What is the instantaneous value of the coil voltage in Fig. P3-7(a)?

(c) Compute the maximum value of the voltage induced in the arrangement of Fig. P3-7(b).

(d) Assuming that the coil of Fig. P3-7(b) is connected to a pair of commutator segments, find the dc value of this voltage.

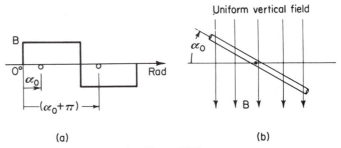

(a)   (b)

Figure P3-7

3-8. An electric machine has a field distribution as shown in Fig. P3-8. The coil spans a full 180° and has N turns. The machine has two poles and an axial length l and rotor radius r.

(a) Obtain the expression for the flux linkage per pole in terms of α expressed in electrical degrees.

(b) By using Faraday's law find the expression for the induced emf in terms of the flux per pole.

(c) Use the Blv form for the induced emf and verify the result of part (b).

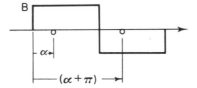

Figure P3-8

3-9. A two-pole field distribution has the sinusoidal shape shown in Fig. P3-9. The maximum value of the flux density is 0.4 T. A coil having full pitch, 50 turns, an axial length of 0.3 m, and a radius of 0.2 m is driven at 600 rpm.

(a) Derive the expression for the flux linkage of the coil when at position α_0.

(b) Using the flux linkage approach, find the instantaneous value of the coil voltage when $\alpha_0 = 120°$.

(c) Repeat part (b) using the Blv approach.

(d) Find the maximum voltage induced in the coil and the position at which it occurs.

(e) What is the average value of the induced coil voltage?

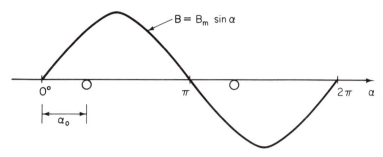

Figure P3-9

3-10. A two-pole machine has the field distribution shown in Fig. P3-10. The coil is full pitch and has N turns, radius r, and an axial length l.

(a) Show that the flux linkage (λ) for the coil position is given by

$$\lambda = \frac{NB_p lr}{\pi}\left(\frac{\pi^2}{2} - 2\alpha_0^2\right) \qquad \text{where } 0 \leq \alpha_0 \leq \frac{\pi}{2}$$

(b) Assume that the coil is driven at ω_0 rad/s. Find the expression for the induced coil voltage using the change-of-flux-linkage approach. Express the result in terms of the flux per pole and for the interval $0 \leq \omega_0 t \leq \pi/2$.

(c) Repeat part (b) using the Blv approach.

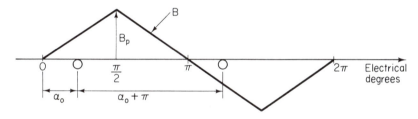

Figure P3-10

3-11. Derive Eq. (3-27) by starting with the conservation of energy [i.e., Eq. (3-31)].

3-12. A four-pole ac induction motor is characterized by a sinusoidal rotor mmf and sinusoidal flux density. Moreover, the flux per pole is 0.02 Wb. When operating at a specified load condition, the space displacement angle is found to be 53.3°. What is the amplitude of the rotor mmf wave required to produce a torque of 40 N-m? Assume two poles.

3-13. A 60-Hz, eight-pole ac motor has an emf of 440 V induced in its armature winding, which has 180 effective turns. When this motor develops 10 kW of power, the amplitude of the sinusoidal field mmf is equal to 800 A-t. $= F_M$

(a) Find the value of the resultant air-gap flux per pole.

(b) Find the angle between the flux wave and the mmf wave.

3-14. The induced emf per phase of a 60-Hz, four-pole induction motor is 120 V. The number of effective turns per phase is 100. When this motor develops a torque of

60 N-m the space displacement angle is 30°. Find the value of the total armature ampere-turns.

3-15. The three-phase armature winding of a 60-Hz, four-pole, ac induction motor is equipped with 320 effective turns per phase. When the motor develops 10 kW it has an armature current of 40 A, a space displacement angle of 37°, and a speed of rotation of 1700 rpm.

(a) Find the amplitude of the armature mmf per pole.

(b) What is the value of the developed torque?

(c) Compute the induced emf in the armature winding per phase.

(d) Find the flux per pole.

3-16. Identify the physical location and the function(s) performed by each of the following in the dc machine: shunt field winding, interpole winding, compensating winding, and series field winding.

3-17. A machine is built with a two-pole single-phase winding and equipped with a dc commutator winding as depicted in Fig. P3-17. The brush axis is displaced by 20° from the coil axis. If a single-phase voltage is applied to the stator winding, will a continuous, unidirectional torque be developed? Explain. Assume negligible armature leakage flux.

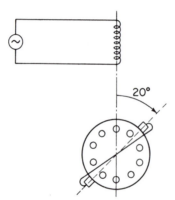

Figure P3-17

3-18. Show that the basic equation for electromagnetic torque in the form

$$T = \frac{\pi}{2} \left(\frac{p}{2}\right)^2 \Phi \mathscr{F} \sin \Delta$$

is correct dimensionally.

4

Three-Phase Induction Motors

One distinguishing feature of the induction motor is that it is a *singly excited* machine. Although such machines are equipped with both a field winding and an armature winding, in normal use an energy source is connected to one winding alone, the field winding. Currents are made to flow in the armature winding by induction, which creates an ampere-conductor distribution that interacts with the field distribution to produce a net unidirectional torque. The frequency of the induced current in the conductor is affected by the speed of the rotor on which it is located; however, the relationship between the rotor speed and the frequency of the armature current is such as to yield a resulting ampere-conductor distribution that is stationary in relation to the field distribution. As a result the singly excited induction machine is capable of producing torque *at any speed below synchronous speed*.† For this reason the induction machine is placed in the class of *asynchronous machines*. In contrast, *synchronous machines* are electromechanical en-

---

†Synchronous speed is determined by the frequency of the source applied to the field winding and the number of poles for which the machine is designed. These quantities are related by Eq. (3-33). Thus

$$\text{synchronous speed} = \frac{120f}{p} = n_s$$

ergy-conversion devices in which a net torque can be produced at only one† speed of the rotor. The distinguishing characteristic of the synchronous machine is that it is a *doubly excited* device except when it is being used as a reluctance motor.

The salient construction features of the three-phase induction motor are described in Sec. 3-3. Because the induction machine is singly excited, it is necessary that both the magnetizing current and the power component of the current flow in the same lines. Moreover, because of the presence of an air gap in the magnetic circuit of the induction machine, an appreciable amount of magnetizing current is needed to establish the flux per pole demanded by the applied voltage. Usually, the value of the magnetizing current for three-phase induction motors lies between 25% and 40% of the rated current. Consequently, the induction motor is found to operate at a low power factor at light loads and at less than unity power factor in the vicinity of rated output.

We are concerned in this chapter with a description of the theory of operation and of the performance characteristics of the three-phase induction motor. Our discussion begins with an explanation of how a revolving magnetic field is obtained with a three-phase winding. After all it is this field that is the driving force behind induction motors.

4-1 THE REVOLVING MAGNETIC FIELD

The application of a three-phase voltage to the three-phase stator winding of the induction motor creates a rotating magnetic field, which by transformer action induces a working emf in the rotor winding. The rotor induced emf is called a working emf because it causes a current to flow through the armature winding conductors. This combines with the revolving flux-density wave to produce torque in accordance with Eq. (3-27). Consequently, we can view the revolving field as the key to the operation of the induction motor.

The rotating magnetic field is produced by the contributions of space-displaced phase windings carrying appropriate time-displaced currents. To understand this statement let us turn attention to Figs. 4-1 and 4-2. Appearing in Fig. 4-1 are the three-phase currents which are assumed to be flowing in phases *a*, *b*, and *c* respectively. Note that these currents are time-displaced by the equivalent of 120 electrical degrees. Depicted in Fig. 4-2 are the stator structure and the three-phase winding. Note that each phase (normally distributed over 60 electrical degrees) for convenience is represented by a single coil. Thus coil *a-a'* represents the entire phase *a* winding having its flux axis directed along the vertical. This means that whenever phase *a* carries current it produces a flux field directed along the vertical—up or down. The right-hand rule readily verifies this statement. Similarly, the flux axis of phase *b* is 120 electrical degrees displaced from phase *a*,

---

†Theoretically, there are two rotor speeds at which a net torque different from zero can exist, but at the second speed enormous currents flow, which makes operation impractical.

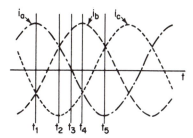

Figure 4-1 Balanced three-phase alternating currents.

and that of phase c is 120 electrical degrees displaced from phase b. The unprimed letters refer to the beginning terminal of each phase.

Let us consider the determination of the magnitude and direction of the resultant flux field corresponding to time instant t_1 in Fig. 4-1. At this instant the current in phase a is at its positive maximum value while the currents in phases b and c are at one-half their maximum negative values. In Fig. 4-2 it is arbitrarily assumed that, when current in a given phase is positive, it flows out of the paper

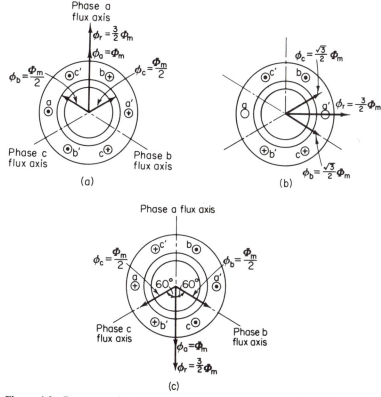

Figure 4-2 Representation of the rotating magnetic field at three different instants of time: (a) time t_1 in Fig. 4-1; (b) time t_3; (c) time t_5.

with respect to the unprimed conductors. Thus, since at time t_1 i_a is positive, a dot is used for conductor a. See Fig. 4-2(a). Of course a cross is used for a' because it refers to the return connection. Then by the right-hand rule it follows that phase a produces a flux contribution directed upward along the vertical. Moreover, the magnitude of this contribution is the maximum value because the current is a maximum. Hence $\phi_a = \Phi_m$, where Φ_m is the maximum flux per pole of phase a. It is important to understand that phase a really produces a sinusoidal flux field with the amplitude located along the axis of phase a as depicted in Fig. 4-3. However, in Fig. 4-2(a) this sinusoidal distribution is conveniently represented by the vector ϕ_a.

To determine the direction and magnitude of the field contribution of phase b at time t_1, we note first that the current in phase b is negative with respect to that in phase a. Hence the conductor that stands for the beginning of phase b must be assigned a cross while b' is assigned a dot. Hence the instantaneous flux contribution of phase b is directed upward along its flux axis and the magnitude of phase b flux is one-half the maximum because the current is at one-half its maximum value. Similar reasoning leads to the result shown in Fig. 4-2(a) for phase c. A glance at the space picture corresponding to time t_1, as illustrated in Fig. 4-2(a), should make it apparent that the resultant flux per pole is directed upward and has a magnitude $\frac{3}{2}$ times the maximum flux per pole of any one phase. Figure 4-3 depicts the same results as Fig. 4-2(a) but does so in terms of sinusoidal flux waves rather than flux vectors. Keep in mind that the resultant flux vector in Fig. 4-2 shows the direction in which flux crosses the air gap. Once across the air gap, the flux is confined to the iron in the usual fashion.

Next let us investigate how the situation of Fig. 4-2(a) changes as time passes through 90 electrical degrees from t_1 to t_3 in Fig. 4-1. Here phase a current

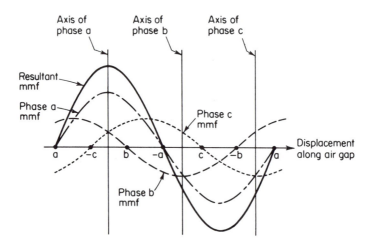

Figure 4-3 Component and resultant field distributions corresponding to t_1 in Fig. 4-1.

is zero, yielding no flux contribution. The current in phase b is positive and equal to $\sqrt{3}/2$ its maximum value. Phase c has the same current magnitude but is negative. Together phases b and c combine to produce a resultant flux having the same magnitude as at time t_1. See Fig. 4-2(b). It is important to note, too, that an elapse of 90 electrical degrees in time results in a rotation of the magnetic flux field of 90 electrical degrees.

A further elapse of time equivalent to an additional 90 electrical degrees leads to the situation depicted in Fig. 4-2(c). Note that again the axis of the flux field is revolved by an additional 90 electrical degrees.

On the basis of the foregoing discussion it should be apparent that the application of three-phase currents through a balanced three-phase winding gives rise to a rotating magnetic field that exhibits two characteristics: (1) it is of constant amplitude and (2) it is of constant speed. The first characteristic has already been demonstrated. The second follows from the fact that the resultant flux traverses through 2π electrical radians in space for every 2π electrical radians of variation in time for the phase currents. Hence, for a two-pole machine, where electrical and mechanical degrees are identical, each cycle of variation of current produces one complete revolution of the flux field. Hence this is a fixed relationship which is dependent upon the frequency of the currents and the number of poles for which the three-phase winding is designed. In the case where the winding is designed for four poles, it requires two cycles of variation of the current to produce one revolution of the flux field. Therefore it follows that for a p-pole machine the relationship is

$$\omega = 2\eta f \qquad f = \frac{\omega}{2\pi}$$

$$f = \frac{p}{2} \times \text{rps} = \frac{2\pi p}{2}\frac{n}{60} = \frac{\omega}{\partial\pi} \qquad \omega = \frac{\pi p n}{60} \qquad (4\text{-}1)$$

where f is in cycles per second and rps denotes revolutions per second. Note that Eq. (4-1) is identical with Eq. (3-33).

An inspection of the ampere-conductor distribution of the stator winding at the various time instants reveals that the individual phases cooperate in such a fashion as to produce a solenoidal effect in the stator. Thus in Fig. 4-2(a) the directions of the currents are such that they all enter into the page on the right side and leave on the left side. The right-hand rule indicates that the flux field is then directed upward along the vertical. In Fig. 4-2(b) the situation is similar except that now the cross and dot distribution is such that the resultant flux field is oriented horizontally toward the right. Therefore, it can be concluded that the rotating magnetic field is a consequence of the revolving mmf associated with the stator winding.

In the foregoing it is pointed out that the flow of balanced three-phase currents through a balanced three-phase winding yields a rotating field of constant amplitude and speed. If neither of these conditions is exactly satisfied, it is still possible to obtain a revolving magnetic field but it will not be of constant amplitude nor of constant linear speed. In general, for a q-phase machine a rotating field of constant amplitude and constant speed results when the following two conditions are satisfied: (1) there is a *space* displacement between balanced phase windings of $2\pi/q$ electrical degrees, and (2) the currents flowing through the phase windings are balanced and *time*-displaced by $2\pi/q$ electrical degrees. For the three-phase machine $q = 3$ and so the now familiar 120° figure is obtained. The only exception to the rule is the two-phase machine. Because the two-phase situation is a special case of the four-phase system, a value of q equal to 4 must be used.

One final point is now in order. The speed of rotation of the field as described by Eq. (4-1) is always given relative to the phase windings carrying the time-varying currents. Accordingly, if a situation arises where the winding is itself revolving, then the speed of rotation of the field relative to inertial space is different than it is relative to the winding.

Mathematical Analysis

The mmf distribution of the stator (field) winding of the three-phase induction motor is usually trapezoidal in shape, but it contains a very prominent fundamental component. In fact it is this component that is depicted in Fig. 4-3. Although some mmf harmonics do exist, we shall continue to confine our attention here to the important fundamental term.

The resultant stator mmf at any time instant is composed of the contributions of each phase. It is important to keep in mind that each phase winding makes a contribution that varies with time along a fixed space axis. An observer looking only at a single phase thus sees an alternating field. If the axis of phase *a* in Fig. 4-3 is taken as the zero reference point for the displacement angle α along the air gap, the alternating field created by this phase may be mathematically expressed by

$$\mathcal{F}_a = \mathcal{F}_m \cos \omega t \cos \alpha \qquad\qquad (4\text{-}2)$$

where $\mathcal{F}_m$ denotes the maximum value of the mmf per pole and is specified by Eq. (3-24) when q is set equal to unity (one phase). It is worthwhile to pause a moment to be sure that the meaning of Eq. (4-2) is understood. This equation states that an observer stationed in space, say, at $\alpha = 0°$ (i.e., on the axis of phase *a*) sees at time $t = 0$ the maximum value of phase *a* mmf. As time progresses the value of this mmf changes. Thus at $\omega t = 60°$ the observer standing at $\alpha = 0°$ now sees $\mathcal{F}_m/2$ as the phase contribution. At $\omega t = 120°$ the contribution becomes $-\mathcal{F}_m/2$, which means that the pole reverses polarity. At $\omega t = 180°$ the mmf contribution is found to be $-\mathcal{F}_m$. Accordingly, the observer placed at $\alpha = 0°$ sees the mmf contribution of phase *a* alternate between a positive $\mathcal{F}_m$ and a negative $\mathcal{F}_m$ as time progresses. If the point of observation is changed, Eq. (4-2) states that the observer continues to see an alternating field but now it varies between different maxima. Thus, if the observer is fixed at $\alpha = 30°$, then as time elapses phase *a* mmf is noted to alternate between $\pm(\sqrt{3}/2)\mathcal{F}_m$. An observer stationed at this position can never see the peak value of fundamental mmf of phase a. Of course an observer situated at $\alpha = 90°$ sees a zero value of phase a mmf for all time instants.

The expression for the mmf of phase b can be written by introducing two modifications in Eq. (4-2) relating to the time variable t and the space variable α. Recalling that a balanced three-phase voltage is applied to the balanced three-phase stator winding, it follows that the time-varying current in phase b winding is displaced by 120 electrical degrees from that in phase a as depicted in Fig. 4-1. Moreover, the axis of the winding of phase b is also displaced by 120 electrical degrees in space. Accordingly, the expression for the instantaneous mmf of phase b becomes

$$\mathscr{F}_b = \mathscr{F}_m \cos (\omega t - 120°) \cos (\alpha - 120°) \qquad (4\text{-}3)$$

where $\mathscr{F}_m$ carries the same meaning as for phase a. Proceeding in a similar fashion the mmf of phase c can be written as

$$\mathscr{F}_c = \mathscr{F}_m \cos (\omega t - 240°) \cos (\alpha - 240°) \qquad (4\text{-}4)$$

As was demonstrated in the physical analysis, the resultant mmf of the total three-phase winding is found by summing the instantaneous contributions of the alternating mmf's of each phase. Expressing it algebraically we have

$$\mathscr{F}_r = \mathscr{F}_a + \mathscr{F}_b + \mathscr{F}_c = \mathscr{F}_m[\cos \omega t \cos \alpha + \cos (\omega t - 120°) \cos (\alpha - 120°)$$
$$+ \cos (\omega t - 240°) \cos (\alpha - 240°)] \quad (4\text{-}5)$$

Upon introducing the trigonometric identity

$$\cos x \cos y = \tfrac{1}{2} \cos (x - y) + \tfrac{1}{2} \cos (x + y) \qquad (4\text{-}6)$$

there results

$$\mathscr{F}_r = \frac{\mathscr{F}_m}{2} [\cos (\omega t - \alpha) + \cos (\omega t + \alpha) + \cos (\omega t - \alpha)$$
$$+ \cos (\omega t + \alpha - 240°) + \cos (\omega t - \alpha) + \cos (\omega t + \alpha - 120°)] \quad (4\text{-}7)$$

Keeping in mind that this last expression describes a *space* field, it should be clear that the second, fourth, and sixth terms, being equal in amplitude and 120° apart, yield a net value of zero. Hence Eq. (4-7) simplifies to

$$\boxed{\mathscr{F}_r = \tfrac{3}{2}\mathscr{F}_m \cos (\omega t - \alpha)} \qquad (4\text{-}8)$$

This is the equation of a revolving field of constant amplitude. Note the correspondence with the results depicted in Fig. 4-2. There it is shown that the cooperation of the three alternating fields yields a resultant field having an amplitude equal to $\tfrac{3}{2}$ the amplitude of any phase. This is consistent with the factor $\tfrac{3}{2}\mathscr{F}_m$ appearing in Eq. (4-8). To complete the correspondence we must now show that $\cos (\omega t - \alpha)$ is the mathematical expression for a traveling wave.

In this connection let us consider an observer fixed in space at a position $\alpha = 30°$ and investigate what happens as time is allowed to progress. By direct application of Eq. (4-8) for $\alpha = 30°$ we obtain:

1. At $\omega t_1 = 0°$,

$$\mathscr{F}_r = \tfrac{3}{2}\mathscr{F}_m \cos(-30°) = \tfrac{3}{2}\mathscr{F}_m \frac{\sqrt{3}}{2}$$

2. At $\omega t_2 = 30°$,

$$\mathscr{F}_r = \tfrac{3}{2}\mathscr{F}_m$$

3. At $\omega t_3 = 120°$,

$$\mathscr{F}_r = \tfrac{3}{2}\mathscr{F}_m \cos(120° - 30°) = 0$$

Depicted in Fig. 4-4 are the various positions that the constant amplitude cosinusoid must take to yield values consistent with those just computed. Note that, as time progresses by 30°, the wave position at ωt_1 must travel 30° in the positive direction of α. Only in this way can the observer fixed at the space position $\alpha = 30°$ see the peak value of the resultant wave as called for by Eq. (4-8). For an additional elapse of time corresponding to 90° the observer sees a zero value for the resultant wave. This happens when the wave shown in position 2 moves along in the positive direction of α by 90° to position 3. Therefore, Eq. (4-8) does in fact represent the equation for a traveling wave involving a trigonometric function. The same reasoning employed in the foregoing can be used to illustrate that a traveling wave of trigonometric character traveling in the negative α direction bears the form $\cos(\omega t + \alpha)$, where as before t is a time variable and α is a space variable.

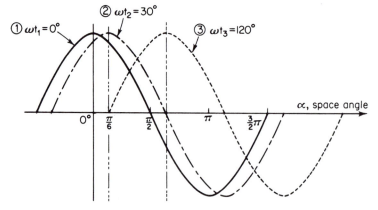

Figure 4-4   Graphical demonstration that Eq. (4-8) is the expression for a traveling wave.

4-2 INDUCTION MOTOR SLIP

The three-phase induction motor may be compared with the transformer because it is a singly energized device which involves changing flux linkages with respect to the stator and rotor windings. In this connection assume that the rotor is of the wound type and Y-connected as illustrated in Fig. 4-5. With the rotor winding open-circuited no torque can be developed. Hence the application of a three-phase voltage to the three-phase stator winding gives rise to a rotating magnetic field which cuts both the stator and rotor windings at the line frequency f_1. The rms value of the induced emf per phase of the rotor winding is given by Eq. (3-21) as

$$E_2 = 4.44 f_1 N_2 K_{w2} \Phi \tag{4-9}$$

where the subscript 2 denotes rotor winding quantities. Note that the stator frequency f_1 is used here because the rotor is at standstill. Hence E_2 is a *line*-frequency emf. Of course the flux Φ is the flux per pole, which is mutual to the stator and rotor windings.

A similar expression describes the rms value of the induced emf per phase occurring in the stator winding. Thus

$$E_1 = 4.44 f_1 N_1 K_{w1} \Phi \tag{4-10}$$

From Eqs. (4-9) and (4-10) we can formulate the ratio

$$\frac{E_1}{E_2} = \frac{N_1 K_{w1}}{N_2 K_{w2}} \tag{4-11}$$

Note the similarity of this expression to the voltage transformation ratio of the transformer. The difference lies in the inclusion of the winding factors of the motor, necessitated by the use of a distributed winding for the motor as contrasted to the concentrated coils used in the transformer. In essence, then, the induction motor at standstill exhibits the characteristics of a transformer wherein the stator winding is the primary and rotor winding is the secondary.

Next let us consider the behavior of the induction motor under running conditions—again with the intention of pointing out similarities to the transformer. To produce a starting torque (and subsequently a running torque) it is necessary to have a current flowing through the rotor winding. This is readily

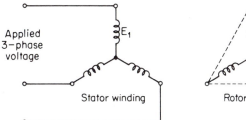

Applied
3-phase
voltage

Stator winding

Rotor winding

Figure 4-5   Schematic representation of the three-phase wound-rotor induction motor. The dashed line indicates short-circuit links for normal operation.

accomplished by short-circuiting the winding in the manner indicated by the broken line in Fig. 4-5. Initially the induced emf E_2 causes a rotor current per phase I_2 to flow through the short-circuit, producing an ampere-conductor distribution which acts with the flux field to produce the starting torque. The sense of this torque is always to cause the rotor to travel in the same direction as the rotating field. An examination of Fig. 4-6 makes this apparent. Assume that the flux field is revolving clockwise at a speed corresponding to the applied stator frequency and the number of poles of the stator winding. This speed is called the *synchronous speed* and is described by Eq. (4-1). Thus

$N_s \cdot 5a_s$

$$\boxed{n_s = \frac{120f_1}{p}} \qquad \text{rpm} \qquad (4\text{-}12)$$

By the $\bar{v} \times \bar{B}$ rule the induced emf in a typical conductor lying beneath a south-pole flux is directed into the paper as indicated in Fig. 4-6. Then for this direction of current the $\bar{I}_2 \times \bar{B}$ rule reveals the torque to be directed clockwise. Therefore the rotor moves in a direction in which it tries to catch up with the stator field.

As the rotor increases its speed, the rate at which the stator field cuts the rotor coils decreases. This reduces the resultant induced emf per phase, in turn diminishing the magnitude of the ampere-conductor distribution and yielding less torque. In fact this process continues until that rotor speed is reached which yields enough emf to produce just the current needed to develop a torque equal to the opposing torques. If there is no shaft load, the opposing torque consists chiefly of frictional losses. It is important to understand that as long as there is an opposing torque to overcome—however small or whatever its origin—*the rotor speed can never be equal to the synchronous speed.* This is characteristic of singly excited electromechanical energy-conversion devices. Since the rotor (or secondary) winding current is produced by induction, there must always be a

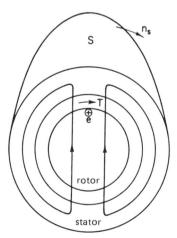

Figure 4-6 Showing the direction of induced voltage and developed torque on a typical conductor.

difference in speed between the stator field and the rotor. In other words, trans-former action must always be allowed to take place between the stator (or pri-mary) winding and the rotor (or secondary) winding.

This speed difference, or *slip*, is a very important variable for the induction motor. In terms of an equation we may write

$$\text{slip} \equiv n_s - n \qquad \text{rpm} \tag{4-13}$$

where n denotes the actual rotor speed in rpm. The term slip is used because it describes what an observer riding with the stator field sees looking at the rotor—it appears to be slipping backward. A more useful form of the slip quantity results when it is expressed on a per unit basis using synchronous speed as the refer-ence. Thus the slip in per unit is

$$\boxed{s = \frac{n_s - n}{n_s}} \tag{4-14}$$

For the conventional induction motor the values of s lie between zero and unity.

It is customary in induction-motor analysis to express rotor quantities (such as induced voltage, current, and impedance) in terms of line-frequency quantities and the slip as expressed by Eq. (4-14). For example, if the rotor is assumed to be operating at some speed $n < n_s$, then the actual emf induced in the rotor winding per phase may be expressed in terms of the line-frequency quantity E_2 as sE_2. This formulation has definite advantages, as described in the next section. In a similar fashion it is possible to express the rotor winding impedance per phase as

$$z_2 = r_2 + jsx_2 \tag{4-15}$$

where z_2 denotes the rotor phase impedance, r_2 is the rotor resistance per phase, and x_2 is the line-frequency leakage reactance per phase of the rotor winding. Of course the effective value of this reactance when the rotor operates at a speed n (or slip s) is only s times as large. Keep in mind that the frequency of the currents in the rotor is directly related to the relative speed of the stator field to the rotor winding. Accordingly, we may write

$$f_2 = \frac{p(\text{slip rpm})}{120} = \frac{p(n_s - n)}{120} \tag{4-16}$$

where f_2 is the frequency of the emf and current in the rotor winding. By means of Eq. (4-14) it is possible to rewrite Eq. (4-16) as

$$\boxed{f_2 = \frac{psn_s}{120} = s\frac{pn_s}{120} = sf_1} \tag{4-17}$$

which indicates that the rotor frequency f_2 is obtained by merely multiplying the stator line frequency by the appropriate per-unit value of the slip. For this reason f_2 is often called the *slip frequency*.

4-3 THE EQUIVALENT CIRCUIT

It is desirable to have an equivalent circuit of the three-phase induction motor in order to direct the analysis of operation and to facilitate the computation of performance. From the remarks made in the preceding section it should not come as a surprise that the equivalent circuit assumes a form identical to that of the exact equivalent circuit of the transformer. The derivation proceeds in a similar fashion with necessary modifications introduced to account for the fact that the secondary winding (the rotor) in this instance revolves and thereby develops mechanical power.

The Magnetizing Branch of the Equivalent Circuit

All the parameters of the equivalent circuit are expressed on a per-phase basis. This applies whether the stator winding is Y- or Δ-connected. In the latter case the values refer to the equivalent Y connection. Appearing in Fig. 4-7(a) is the portion of the equivalent circuit that has reference to the stator (or primary) winding. Note that it consists of a stator phase winding resistance r_1, a stator phase winding leakage reactance x_1, and a magnetizing impedance made up of the core-loss resistor r_c and the magnetizing reactance x_ϕ. There is no difference in form between this circuit and that of the transformer. The difference lies only in the magnitude of the parameters. Thus the total magnetizing current $\bar{I}_m$ is considerably larger in the case of the induction motor because the magnetic circuit necessarily includes an air gap. Whereas in the transformer this current is about 2% to 5% of the rated current, here it is approximately 25% to 40% of the rated current depending upon the size of the motor. Moreover, the primary leakage reactance for the induction motor also is larger because of the air gap as well as because the stator and rotor windings are distributed along the periphery of the air gap rather than concentrated on a core as in the transformer. The effects of the actions that take place in the rotor (or secondary) winding must reflect themselves at the proper equivalent-voltage level at terminals a-b in Fig. 4-7(a). We next investigate the manner in which this comes about.

The Actual Rotor Circuit per Phase

For any specified load condition that calls for a particular value of slip s, the rotor current per phase may be expressed as

$$\bar{I}_2 = \frac{s\bar{E}_2}{r_2 + jsx_2} \qquad (4\text{-}18)$$

where $\bar{E}_2$ and x_2 are the standstill values. The circuit interpretation of Eq. (4-18) is depicted in Fig. 4-7(b). It illustrates that $\bar{I}_2$ is a slip-frequency current produced by

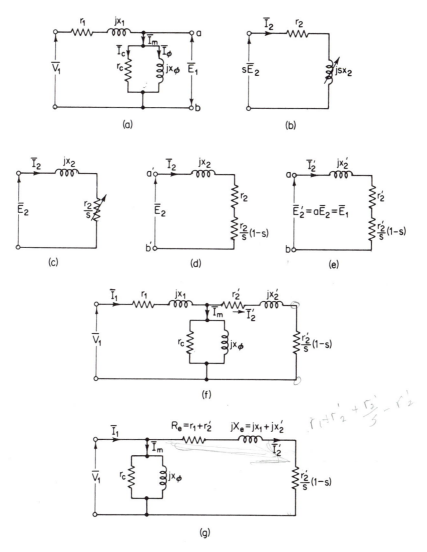

Figure 4-7 Derivation of the equivalent circuit: (a) stator winding section; (b) actual rotor circuit; (c) equivalent rotor circuit; (d) modified equivalent rotor circuit; (e) stator-referred equivalent rotor circuit; (f) exact equivalent circuit; (g) approximate equivalent circuit.

the slip-frequency induced emf sE_2 acting in a rotor circuit having an impedance per phase of $r_2 + jsx_2$. In other words, this is the current that would be seen by an observer riding with the rotor winding. Furthermore, <u>the amount of real power involved in this rotor circuit is the current squared times the real part of the rotor impedance.</u> In fact, this power represents the rotor copper loss per phase. Hence the total rotor copper loss may be expressed as

$$P_{\text{Cu}\,2} = q_2 I_2^2 r_2 \qquad (4\text{-}19)$$

where q_2 denotes the number of rotor phases.

The Equivalent Rotor Circuit

By dividing both the numerator and the denominator of Eq. (4-18) by the slip s we get

$$\bar{I}_2 = \frac{\bar{E}_2}{(r_2/s) + jx_2} \qquad (4\text{-}20)$$

The corresponding circuit interpretation of this expression appears in Fig. 4-7(c). Note that the magnitude and phase angle of $\bar{I}_2$ remain unaltered by this operation. However, there is a significant difference between Eqs. (4-18) and (4-20). In the latter case $\bar{I}_2$ is considered to be produced by a line-frequency voltage $\bar{E}_2$ acting in a rotor circuit having an impedance per phase of $r_2/s + jx_2$. Hence the $\bar{I}_2$ of Eq. (4-20) is a *line-frequency* current, whereas the $\bar{I}_2$ of Eq. (4-18) is a *slip-frequency* current. It is important that this distinction be understood.

Manipulation of Eq. (4-18) by s has enabled us to go from an actual rotor circuit characterized by constant resistance and variable leakage reactance [see Fig. 4-7(b)] to one characterized by variable resistance and constant leakage reactance [see Fig. 4-7(c)]. Moreover, the real power associated with the equivalent rotor circuit of Fig. 4-7(c) is clearly

$$P = I_2^2 \frac{r_2}{s} \qquad (4\text{-}21)$$

Hence the total power for q_2 phases is

$$P_g = q_2 I_2^2 \frac{r_2}{s} \qquad (4\text{-}22)$$

A comparison of this expression with Eq. (4-19) indicates that the power associated with the equivalent circuit of Fig. 4-7(c) is considerably greater. For example, in a large machine a typical value of s is 0.02. Hence P_g is greater than the actual rotor copper loss by a factor of 50.

What is the meaning of this power discrepancy? The answer lies in the fact that by Eq. (4-20) $\bar{I}_2$ is a *line-frequency* current. This means that the point of reference has changed from the rotor (where slip-frequency quantities exist) to the stator (where line-frequency quantities exist). In accordance with the circuit representation of Fig. 4-7(c) the observer changes his or her point of reference from the rotor to the stator. This shift is significant because now, upon looking into the rotor, the observer sees not only the rotor copper loss but the mechanical power developed as well. The latter quantity is included because with respect to a stator-based observer the rotor speed is no longer zero as it is relative to a rotor-based observer. As a matter of fact Eq. (4-22) gives the total power input to the rotor. It is the power transferred across the air gap from the stator to the rotor. We can rewrite Eq. (4-22) in a manner that stresses this fact:

$$P_g = q_2 I_2^2 \frac{r_2}{s} = q_2 I_2^2 \left[r_2 + \frac{r_2}{s}(1 - s) \right] \tag{4-23}$$

In other words, the variable resistance of Fig. 4-7(c) may be replaced by the actual rotor winding resistance r_2 and a variable resistance R_m, which represents the mechanical shaft load. That is,

$$\boxed{R_m \equiv \frac{r_2}{s}(1 - s)} \tag{4-24}$$

This expression is useful in analysis because it allows any mechanical load to be represented in the equivalent circuit by a resistor. Figure 4-7(d) depicts the modi-fied version of the rotor equivalent circuit. Finally, it should be apparent on the basis of the foregoing remarks that the rotor equivalent circuit is equivalent only insofar as the magnitude and phase angle of the rotor current per phase are concerned.

The Stator-Referred Rotor Equivalent Circuit

The voltage appearing across terminals a-b in Fig. 4-7(a) is a line-frequency quan-tity having $N_1 K_{w1}$ effective turns. The voltage appearing across terminals a'-b' in Fig. 4-7(d) is also a line-frequency quantity but has $N_2 K_{w2}$ effective turns. In general $E_1 \neq E_2$, so that a'-b' in Fig. 4-7(d) cannot be joined to a-b in Fig. 4-7(a) to yield a single-line equivalent circuit. To accomplish this it is necessary to replace the actual rotor winding with an equivalent winding having $N_1 K_{w1}$ effective turns as was done with the transformer. In other words, all the rotor quantities must be referred to the stator in the manner depicted in Fig. 4-7(e). The prime notation is used to denote stator-referred rotor quantities. The derivation of the reduction factors to be used in referring a rotor quantity to the stator appears in Appendix D. Both the wound rotor and squirrel-cage rotor machines are treated there.

The Complete Equivalent Circuit

The voltage appearing across terminals a-b in Fig. 4-7(e) is the same as that appearing across terminals a-b in Fig. 4-7(a). Hence these terminals may be joined to yield the complete equivalent circuit as it appears in Fig. 4-7(f). Note that the form is identical with that of the two-winding transformer.

The Approximate Equivalent Circuit

Considerable simplification of computation with little loss of accuracy can be achieved by moving the magnetizing branch to the machine terminals as illustrated in Fig. 4-7(g). This modification is essentially based on the assumption that $\bar{V}_1 \approx \bar{E}_1 = \bar{E}'_2$. All performance calculations will be carried out using the approximate equivalent circuit.

4-4 COMPUTATION OF PERFORMANCE

When the three-phase induction motor is running at no-load, the slip has a value very close to zero. Hence the mechanical load resistor R_m has a very large value, which in turn causes a small rotor current to flow. The corresponding electromagnetic torque, as described by Eq. (3-27), merely assumes that value which is needed to overcome the *rotational losses* consisting chiefly of friction and windage. If a mechanical load is next applied to the motor shaft, the initial reaction is for the shaft load to drop the motor speed slightly and thereby increase the slip. The increased slip subsequently causes $\bar{I}_2$ to increase to that value which, when inserted into Eq. (3-27), yields sufficient torque to provide a balance of power to the load. Thus equilibrium is established and operation proceeds at a particular value of s. In fact for each value of load horsepower requirement there is a unique value of slip. This can be inferred from the equivalent circuit, which shows that once s is specified then the power input, the rotor current, the developed torque, the power output, and the efficiency are all determined.

The use of a power-flow diagram in conjunction with the approximate equivalent circuit makes the computation of the performance of a three-phase induction motor a straightforward matter. Depicted in Fig. 4-8(a) is the power flow in statement form. Note that the loss quantities are placed on the left side of a flow point. Appearing in Fig. 4-8(b) is the same power-flow diagram but now expressed in terms of all the appropriate relationships needed to compute the performance. It should be clear that to calculate performance one must first compute the currents I_2 and I_1 from the equivalent circuit and then make use of the pertinent relationships depicted in Fig. 4-8(b).

Example 4-1

A three-phase, four-pole, 30-hp, 220-V, 60-Hz, Y-connected induction motor draws a current of 77 A from the line source at a power factor of 0.88. At this operating condition, the motor losses are known to be the following:

Stator copper losses = P_{cu1} = 1033 W

Rotor copper losses = P_{cu2} = 1299 W

Stator core losses = P_c = 485 W

Rotational losses (friction, windage, and iron losses due to rotation)

$$= P_{rot} = 540 \text{ W}$$

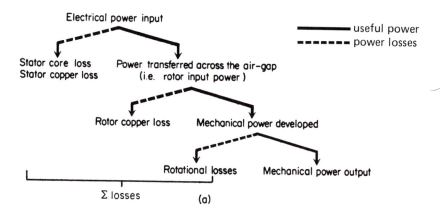

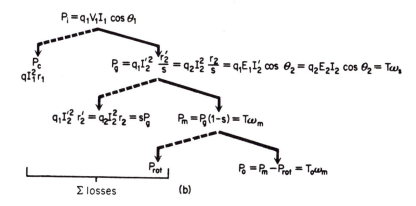

Figure 4-8 Power-flow diagram: (a) statement form; (b) equation form.

Determine: (a) the power transferred across the air gap, (b) the internally developed torque in newton-meters, (c) the slip expressed in per unit and in rpm, (d) the mechanical power developed in watts, (e) the horsepower output, (f) the motor speed in rpm and in radians per second, (g) the torque at the output shaft, (h) the torque needed to overcome the rotational losses, (i) the efficiency of operation at the stated condition.

Solution   The solution to this problem provides an exercise in the application of the power flow diagram.

(a) Sufficient information is cited to permit finding the input power. Thus

$$P_i = \sqrt{3}\; V_L I_L(\text{pf}) = \sqrt{3}(220)77(0.88) = 25{,}820 \text{ W}$$

Figure 4-8(a) then shows that the gap power is

$$P_g = P_i - P_c - P_{\text{cu1}} = 25{,}820 - 485 - 1033 = 24{,}302 \text{ W}$$

(b) Here use is made of the important relationship

$$T\omega_s = P_g$$

where in this case

$$\omega_s = \frac{\pi}{30} n_s = \frac{\pi}{30} \left(\frac{120f}{p}\right) = \frac{\pi}{30} (1800) = 188.5 \text{ rad/s}$$

Hence the internally developed torque is

$$T = \frac{P_g}{\omega_s} = \frac{24{,}302}{188.5} = 128.93 \text{ N-m}$$

(c) By Eq. (4-22), we have

$$s = \frac{P_{cu2}}{P_g} = \frac{1299}{24{,}302} = 0.0535 \text{ p-u}.$$

Then

$$\text{slip in rpm} = sn_s = (0.0535)(1800) = 96.2 \text{ rpm}$$

(d) Again from the power flow diagram the mechanical power developed can be expressed as

$$P_m = P_g - P_{cu2} = 24{,}302 - 1299 = 23{,}003 \text{ W}$$

(e) For the output power we get

$$P_o = P_m - P_{rot} = 23{,}003 - 540 = 22{,}463 \text{ W}$$

Accordingly

$$\text{hp}_o = \frac{P_o}{746} = \frac{22{,}463}{746} = 30.1$$

(f) The actual motor speed is the synchronous speed less the slip in rpm. Thus

$$n = n_s - sn_s = 1800 - 96.2 = 1703.8 \text{ rpm}$$

(g) The expression for the output torque is

$$T_o = \frac{P_o}{\omega_m}$$

where ω_m is the actual motor speed expressed in radians per second, or

$$\omega_m = \frac{\pi}{30} (n) = \frac{\pi}{30} (1703.8) = 178.42 \text{ rad/s}$$

Therefore,

$$T_o = \frac{P_o}{\omega_m} = \frac{22{,}463}{178.42} = 125.9 \text{ N-m}.$$

(h) The difference between the internally developed torque of 128.93 N-m and the output torque of 125.9 N-m is the torque required to supply the rotational losses associated with the operating speed of 1703.8 rpm. Thus

$$T_{rot} = T - T_o = 128.93 - 125.9 = 3.03 \text{ N-m}$$

It is instructive to observe that this same torque can be found from the expression that relates torque, speed, and power. Accordingly, we can also write

$$T_{rot} = \frac{P_{rot}}{\omega_m} = \frac{540}{178.42} = 3.03 \text{ N-m}$$

(i) The efficiency is

$$\eta = \frac{P_o}{P_1} = \frac{22,463}{25,820} = 0.87$$

Thus the motor operates at an efficiency of 87%.

Example 4-2

A three-phase, four-pole, 50-hp, 480-V, 60-Hz, Y-connected induction motor has the following parameters per phase:

$$r_1 = 0.10 \ \Omega, \qquad x_1 = 0.35 \ \Omega$$
$$r_2' = 0.12 \ \Omega, \qquad x_2' = 0.40 \ \Omega$$

It is known that the stator core losses amount to 1200 W and the rotational losses equal 950 W. Moreover, at no-load the motor draws a line current of 19.64 A at a power factor of 0.089 lagging.

When the motor operates at a slip of 2.5%, find: (a) the input line current and power factor, (b) the developed electromagnetic torque in newton-meters, (c) the horsepower output, (d) the efficiency.

Solution (a) The computations are carried out on a per-phase basis. Hence the phase voltage is $480/\sqrt{3}$ or 277.13 V, and the equivalent circuit is depicted in Fig. 4-9. The stator-referred rotor current then follows from

$$\bar{I}_2' = \frac{V_1}{r_1 + (r_2'/s) + j(x_1 + x_2')} = \frac{277.13}{4.9 + j0.75}$$

$$= 55.97\underline{/-8.7°} = 55.33 - j8.47$$

For all practical purposes the magnetizing current may be taken equal to the no-load current because the corresponding rotor current is negligibly small. Thus

$$\bar{I}_m = 19.64\underline{/-85°} = 1.75 - j19.58$$

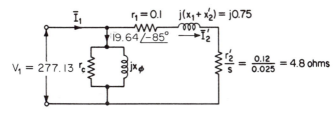

Figure 4-9 Equivalent circuit for Example 4-2.

Hence, the input line current is

$$\bar{I}_1 = \bar{I}_m + \bar{I}_2' = (55.33 + 1.75) - j(19.58 + 8.47)$$
$$= 57.08 - j28.05 = 63.6 \underline{/-26.2°} \tag{4-25}$$

and

$$\text{power factor} = \cos \theta_1 = \cos 26.2° = 0.895 \text{ lagging}$$

(b) The developed torque is found from

$$T = \frac{P_g}{\omega_s} \tag{4-26}$$

Also

$$\omega_s = \frac{2\pi n_s}{60} = \frac{2\pi(1800)}{60} = 60\pi \text{ rad/s}$$

and

$$P_g = q_1 I_2'^2 \frac{r_2'}{s} = 3(55.97)^2 4.8 = 45,110 \text{ W}$$

Therefore

$$T = \frac{45,110}{60\pi} = 239.32 \text{ N-m}$$

(c) From the power-flow diagram the power output is

$$P_o = P_m - P_{\text{rot}} = P_g(1 - s) - P_{\text{rot}} = 43,982.3 - 950 = 43,032.3 \text{ W}$$

The horsepower output is therefore

$$\text{horsepower} = \frac{43,032.3}{0,746.0} = 57.68$$

Note that this is slightly greater than the rated horsepower of 50. Rated horsepower occurs at a slip somewhat less than 2.5%.

(d) It is more accurate to find the efficiency from the relationship

$$\boxed{\eta = 1 - \frac{\Sigma \text{ losses}}{P_i}} \tag{4-27}$$

rather than from the output-to-input ratio. The tabulation of the losses is as follows:

$$P_c = \text{core loss} = 1200 \text{ W}$$

$$\text{stator copper loss} = q_1 I_1^2 r_1 = 3(63.6)^2 0.1 = 1213.5 \text{ W}$$

$$\text{rotor copper loss} = q_1 I_2'^2 r_2' = sP_g = 0.025(45,110) = 1127.8 \text{ W}$$

$$P_{\text{rot}} = \text{rotational loss} = 950 \text{ W}$$

$$\Sigma \text{ losses} = 4491.3 \text{ W}$$

Also, the input power is

$$P_i = \sqrt{3}\ 480(63.6)0.895 = 47{,}324\ \text{W}$$

Hence

$$\eta = 1 - \frac{4491.3}{47{,}324} = 1 - 0.095 = 0.905$$

The efficiency is 90.5%.

4-5 CORRELATION OF INDUCTION MOTOR OPERATION WITH THE BASIC TORQUE EQUATIONS

The variation of torque with speed (or slip) is an important characteristic of the three-phase induction motor. The general shape of this curve can be identified in terms of the basic torque equation [Eq. (3-22)] and knowledge of the performance computational procedure. When the motor operates at a very small slip, as at no-load, Eq. (4-18) indicates that the rotor current is very small—just enough to develop torque to supply the rotational losses. Moreover, the power-factor angle of the rotor current is practically zero. That is,

$$\theta_2 = \tan^{-1} \frac{sx_2}{r_2} \approx 0° \text{ for } s \text{ very small} \tag{4-28}$$

Appearing in Fig. 4-10 is a cross-sectional view of a two-pole squirrel-cage induction motor showing the rotating flux field produced by the stator winding and the corresponding induced ampere-conductor distribution of the rotor. For the instant shown and the assumed counterclockwise rotation of the field the emf induced in bar 1 is a maximum and directed out of the plane of the paper. Why? Note that the emf induced in bars 2 and 10 is less than the maximum value and that induced in bars 3 and 9 at the same time instant is less still. If the value of the bar emf's are plotted as a function of the spatial displacement along the rotor periphery, a discrete distribution of sinusoidal shape results. See Fig. 3-12(b) of Appendix to Chapter 3. Furthermore, because the power-factor angle is practically zero at low slips, the distribution of rotor currents is almost identical with the emf distribution. The current distribution is denoted in Fig. 4-10 by dots and crosses placed outside the rotor surface. The thicker the dot (or cross), the larger is the value of the current represented.

Let us now relate this picture to the quantities appearing in Eq. (3-22). Because there are two forms of the basic electromagnetic torque equation, two interpretations are possible. Taking the last form of Eq. (3-22) first, we note again that torque depends upon the presence of a flux field Φ, an ampere-conductor distribution (or current sheet) as implied by J_m, and a space displacement angle ψ. Examination of Fig. 4-10 discloses that for slips near zero the ampere-conductor distribution is in space phase with the field distribution, i.e., $\psi \approx 0°$. This

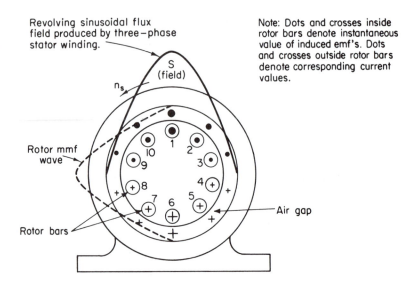

Figure 4-10 Cross-sectional view of two-pole squirrel-cage induction motor, depicting the flux field and corresponding induced rotor mmf distribution at very small slips.

means that the best field pattern for the development of torque exists—each bar produces a positive torque. However, in spite of this, very little net torque is developed at such low slips because the value of rotor ampere-conductor distribution is also near zero.

In the second interpretation the rotor is treated like a solenoid. In Fig. 4-10 note that the rotor current distribution of dots and crosses creates an mmf directed in quadrature with it (along the horizontal toward the right in this case). This mmf is represented for one pole by the dashed-line sine wave and is assumed to have a peak value of $\mathcal{F}_p$. Keep in mind that the magnitude of $\mathcal{F}_p$ is directly dependent upon the rotor current and is practically zero for very low values of slip. Since the torque angle is the angle between the axis of the flux field distribution and the axis of the armature (or rotor) mmf distribution, we see that its value here is almost 90°, again attesting to the existence of an optimum field pattern for the production of torque.

What happens to the field pattern as mechanical load is applied to the motor shaft? As already described in Sec. 4-4 the immediate reaction is for the motor speed to decrease slightly, which increases the slip. The increased slip produces an almost direct increase in rotor current because for values of slip below 10% the sx_2 term in Eq. (4-18) is negligibly small. When rated horsepower is delivered the slip lies between 3% and 5% for most induction motors and the value of the rotor power-factor angle lies in the vicinity of 10° (compare with Example 4-1). In spite of this, however, and for the convenience of pictorial representation we shall assume that at rated load the slip is such that Eq. (4-28) yields a rotor power-factor angle of 36°. Note that this is the number of degrees separating the rotor bars in the machine configuration of Fig. 4-10. If the field distribution is assumed to take the same instantaneous position as depicted in Fig. 4-10, then the emf distribution must be the same. This is shown with the appropriate dot-cross notation inside the rotor bars of Fig. 4-11(a). Note that the emf in bar 1 is a maximum because it is under the influence of the maximum value of the flux density. As a result of the counterclockwise rotation of the stator flux field we know that the emf was a maximum in bar 2 by an amount of time equal to that which the flux field takes to travel the distance between two bars, or more specifically of $\omega t = 36°$. At this point it is helpful to recall the meaning of the power-factor angle θ_2, which is the phase angle between $\bar{E}_2$ and $\bar{I}_2$. When θ_2 is a lag angle, it states that the peak value of rotor current occurs θ_2 electrical degrees in time after the peak value of the rotor induced emf has occurred. Hence for the time instant illustrated in Fig. 4-11(a) the peak outward-directed rotor current exists in bar 2. By the same reasoning outward-directed currents can be found to exist in bars 1, 10, 3, and 4, but the magnitudes will be less than the peak current by the cosine of the angle that these bars are displaced from bar 2. As before, our attention is confined just to fundamental components of all distributions. The resulting rotor ampere-conductor distribution then takes the position shown in Fig. 4-11(a) and repeated for clarity's sake in Fig. 4-11(b). Note that bars 4 and 9 produce a negative (or clockwise) torque. This is the reason that the field pattern is not optimum. Nonetheless, the developed torque is much greater here than it is in Fig. 4-10 because the rotor current is many times greater. A glance at Eq. (3-27), which is the practical form of Eq. (3-22), emphasizes this point. Of the quantities appearing there, Φ is essentially fixed by the applied terminal voltage. Hence, if I_2 increases at a greater rate than cos ψ decreases, the torque increases. In normal operation the value of cos ψ rarely dips below 0.95.

These discussions revolving about the situations depicted in Figs. 4-10 and 4-11 lead to a very useful conclusion: *In the polyphase induction motor the space-displacement angle ψ is equal to the power-factor angle of the armature winding*, i.e.,

$$\psi = \theta_2 = \tan^{-1} \frac{sx_2}{r_2} \qquad (4\text{-}28\text{a})$$

One final comment is in order concerning the development of torque in the polyphase induction motor. Keep in mind that the results of Eqs. (3-27) and (3-22) are valid provided that the flux field and the ampere-conductor distributions are stationary in relation to one another. How is this condition satisfied in the situation of Fig. 4-11? To begin with we know that the speed of the stator flux field relative to the stator structure is ω_s. It therefore remains to be shown that the rotor ampere-conductor distribution has also a speed relative to the stator structure of ω_s. Since the rotor winding carries slip-frequency currents, the speed of

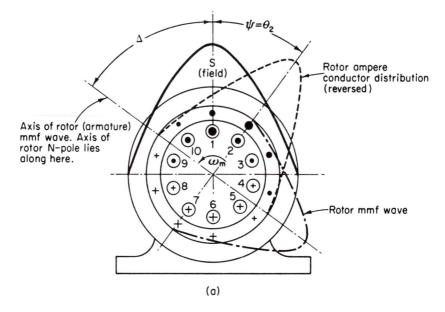

(a)

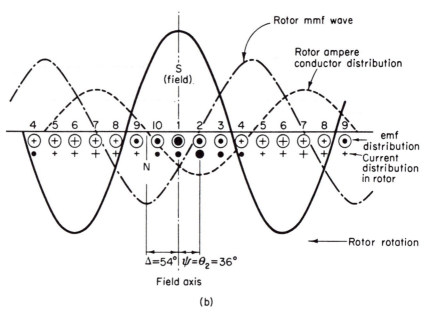

(b)

Figure 4-11 Depicting the field distribution and associated ampere-conductor and mmf waves of the rotor for $\theta_2 = \tan^{-1}(sx_2/r_2) = 36°$. Only the fundamental components are shown. The field travels from right to left. (See p. 257 for definition of Δ.)

the rotor ampere-conductor distribution relative to the rotor is at slip frequency, i.e., $s\omega_s$. However, because the rotor bars are imbedded in the rotor iron, which in turn revolves at a speed of $\omega_m = \omega_s(1 - s)$ with respect to the stator structure, it follows that the total speed of rotation of the rotor ampere-conductor distribution is the sum of $s\omega_s$ and $\omega_s(1 - s)$. Clearly this is ω_s. Hence <u>the rotor ampere-conductor distribution and the flux field distribution are stationary in relation to each other</u>. In fact this condition is fulfilled for all values of slip and is a consequence of the singly energized nature of the induction motor.

4-6 TORQUE-SPEED CHARACTERISTIC: STARTING AND MAXIMUM TORQUES

The torque-speed characteristic of the induction motor can be explained in light of Eqs. (3-27) and (3-22). As slip is allowed to increase from nearly zero to about 10%, Eq. (4-18) shows that the rotor current increases almost linearly. Moreover, for this same range of slip, ψ varies over a range of about zero to 15 degrees. This means that cos ψ remains practically invariant over the specified slip range and so torque increases almost linearly in this region.

As slip is allowed to increase still further, the rotor current continues to increase but much less rapidly than at first. The reason lies in the increasing importance of the sx_2 term of the rotor impedance. In addition, the space angle ψ now begins to increase at a rapid rate which makes the cos ψ diminish more rapidly than the current increases. Since the torque equation now involves two opposing factors, it is entirely reasonable to expect that a point is reached beyond which further increases in slip culminate in decreased developed torque. In other words, the rapidly decreasing cos ψ factor predominates over the slightly increasing I_2 factor in Eq. (3-27). As ψ increases, the field pattern for producing torque becomes less and less favorable because more and more conductors that produce negative torque are included beneath a given pole flux. A glance at Fig. 4-11 makes this statement self-evident. Accordingly, the composite torque-speed curve takes on a form similar to that shown in Fig. 4-12.

The *starting torque* is the torque developed when s is unity, i.e., the speed n is zero. Figure 4-12 indicates that for the case illustrated the starting torque is

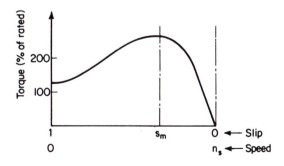

Figure 4-12 Typical torque-speed curve for a three-phase induction motor.

somewhat in excess of rated torque, which is fairly typical of such machines. The starting torque is computed in the same manner as torque is computed for any value of slip. Here it merely requires using $s = 1$. Thus the magnitude of the rotor current at standstill is

$$I'_2 = \frac{V_1}{\sqrt{(r_1 + r'_2)^2 + (x_1 + x'_2)^2}} \tag{4-29}$$

The corresponding gap power is then

$$P_g = q_1 I'^2_2 r'_2 = \frac{q_1 V^2_1 r'_2}{(r_1 + r'_2)^2 + (x_1 + x'_2)^2} \tag{4-30}$$

It is interesting to note that higher starting torques can result from increased rotor copper losses at standstill.

At unity slip the input impedance is very low so that large starting currents flow. Equation (4-29) makes this apparent. In the interest of limiting this excessive starting current, motors whose ratings exceed 3 hp are usually started at reduced voltage by means of line starters. This matter is discussed further in Sec. 4-10. Of course, starting with reduced voltage also means a reduction in the starting torque. In fact if 50% of the rated voltage is used upon starting, then clearly by Eq. (4-30) it follows that the starting torque is only one-quarter of its full-voltage value.

Another important torque quantity of the three-phase induction motor is the maximum developed torque. This quantity is so important that it is frequently the starting point in the design of the induction motor. The maximum (or breakdown) torque is a measure of the reserve capacity of the machine. It frequently has a value of 200% to 300% of rated torque. It permits the motor to operate through momentary peak loads. However, the maximum torque cannot be delivered continuously because the excessive currents that flow would destroy the insulation.

Since the developed torque is directly proportional to the gap power, it follows that the torque is a maximum when P_g is a maximum. Also, P_g is a maximum when there is a maximum transfer of power to the equivalent circuit resistor r'_2/s. Applying the maximum-power-transfer theorem to the approximate equivalent circuit leads to the result that

$$\frac{r'_2}{s_m} = \sqrt{r^2_1 + (x_1 + x'_2)^2} \tag{4-31}$$

That is, maximum power is transferred to the gap power resistor r'_2/s when this resistor is equal to the impedance looking back into the source. Accordingly, the slip s_m at which the maximum torque is developed is

$$\boxed{s_m = \frac{r'_2}{\sqrt{r^2_1 + (x_1 + x'_2)^2}}} \tag{4-32}$$

Note that the slip at which the maximum torque occurs may be increased by using a larger rotor resistance. Some induction motors are in fact designed so that the maximum torque is available as a starting torque, i.e., $s_m = 1$.

With the slip s_m known, the corresponding rotor current can be found and then inserted into the torque equation to yield the final form for the breakdown torque. Thus

$$T_m = \frac{1}{\omega_s} q_1 I_2'^2 \frac{r_2'}{s_m} = \frac{1}{\omega_s} \frac{q_1 V_1^2}{2[r_1 + \sqrt{r_1^2 + (x_1 + x_2')^2}]} \qquad (4\text{-}33)$$

An examination of Eq. (4-33) reveals the interesting information that the maximum torque is independent of the rotor winding resistance. Thus increasing the rotor winding resistance increases the slip at which the breakdown torque occurs, but it leaves the magnitude of its torque unchanged. Figure 4-13 shows the effect of increasing the rotor resistance on a typical torque-speed curve.

A word of caution is appropriate at this point. Keep in mind that the results appearing in Eqs. (4-29), (4-30), and (4-33) are predicated on the approximate equivalent circuit, which is used whenever the voltage drop across the primary leakage impedance is small compared to the stator induced emf per phase. This assumption is quite valid for normal operating conditions spanning the region from no-load to rated load. However, when interest becomes focused on maximum torque or starting torque at rated voltage, then the primary leakage voltage drop begins to exert some noticeable influence. Therefore, if more accurate results are required, we can base the analysis on the more exact equivalent circuit of Fig. 4-7(f) (see Example 4-4). Usually, errors of the order of 3% are incurred when these calculations are based on the simpler, approximate equivalent circuit. Because in practical situations the extent of these errors is often less than the effects due to saturation and skin effects, which are not accounted for either, our policy here is to stay with the approximate equivalent circuit.

Example 4-3

Refer to the induction motor of Example 4-2. Assuming that operation takes place at full-rated voltage with all saturation effects neglected, determine (a) the slip at

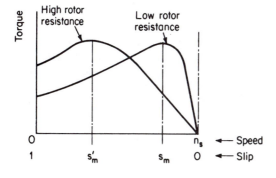

Figure 4-13 Showing the effect of increased rotor resistance on torque-speed curves.

which the maximum torque is developed, (b) the current at maximum torque, and (c) the value of the maximum torque. Compare these results with those found in Example 4-2.

Solution The solution is obtained here employing the results obtained from use of the approximate equivalent circuit.
(a) Application of Eq. (4-32) yields

$$s_m = \frac{r_2'}{\sqrt{r_1^2 + (x_1 + x_2')^2}} = \frac{0.12}{\sqrt{(0.1)^2 + 0.75^2}} = 0.159$$

(b) The rotor current referred to the stator is

$$\bar{I}_2' = \frac{277.13}{r_1 + (r_2'/s) + j(x_1 + x_2')} = \frac{277.13}{0.1 + (0.12/0.159) + j0.75} = \frac{277.13}{1.137\,\underline{/41.3°}}$$

$$= 243.74\,\underline{/-41.3°}\ \text{A}$$

This current is about 3.83 times as large as the current when the motor delivers a little in excess of 57 hp as computed in Example 4-2 at near rated conditions.
(c) The maximum torque follows directly from Eq. (4-33). Thus

$$T_m = \frac{q_1 V_1^2}{2\omega_s}\frac{1}{r_1 + \sqrt{r_1^2 + (x_1 + x_2')^2}} = \frac{3(277.13)^2}{2(188.5)}\frac{1}{r_1 + \sqrt{r_1^2 + (x_1 + x_2')^2}}$$

$$= (611.5)\frac{1}{0.1 + \sqrt{(0.1)^2 + (0.75)^2}}$$

$$= 611.15(1.181) = 721.77\ \text{N-m}$$

which is approximately three times as large as the value found near rated conditions in Example 4-2.

Example 4-4

Repeat Example 4-3 using the exact equivalent circuit in place of the approximate one.

Solution In order to use the exact equivalent circuit, the value of the magnetizing reactance, x_ϕ, is needed. This is readily found from the data appearing in Example 4-2 as

$$x_\phi = \frac{V_1}{I_\phi} = \frac{277.13}{19.58} = 14.2\ \Omega$$

The exact equivalent circuit appears in Fig. 4-14(a). Observe that the core-loss resistor is omitted for simplicity; its absence causes negligible effects.
To obtain a result analogous to Eq. (4-31) by reducing the circuit in Fig. 4-14(a) to an equivalent series circuit, we invoke Thévenin's theorem. We replace the circuit to the left of terminals ab by an equivalent voltage source of value

$$\bar{V}_i = \frac{jx_\phi}{r_1 + j(x_\phi + x_1)}\,\bar{V}_1$$

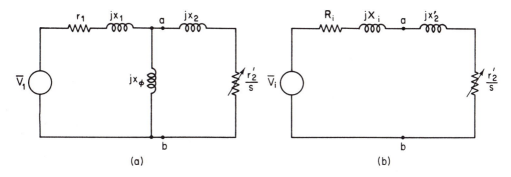

Figure 4-14 (a) Exact equivalent circuit with the core-loss resistor omitted; (b) the Thévenin equivalent of (a).

and an internal impedance which is simply the parallel combination of $(r_1 + jx_1)$ and jx_ϕ. Thus

$$\bar{Z}_i = \frac{jx_\phi(r_1 + jx_1)}{r_1 + j(x_1 + x_\phi)} = R_i + jX_i$$

The Thévenin equivalent circuit is shown in Fig. 4-14(b). It is instructive to note that this circuit is now equivalent in form to the approximate equivalent circuit which upon analysis led to Eq. (4-31). The role of V_1 is now replaced by $\bar{V}_i$, and R_i replaces r_1, and X_i replaces x_1. Accordingly, Eqs. (4-32) and (4-33) are now applicable to the exact equivalent circuit provided that the just-mentioned replacements are made. Hence, before proceeding with the evaluation of the items specified in this example, we first determine these replacement quantities.

$$\bar{V}_i = \frac{j14.2}{0.1 + j(14.2 + 0.35)} 277.1\underline{/0°} = 270.5\underline{/0.39°} \text{ V}$$

$$Z_i = \frac{j14.2(0.1 + j0.35)}{0.1 + j14.55} = 0.09 + j0.343 = R_i + jX_i$$

Observe the relatively small change that occurs in the value of R_i from r_1 and X_i from x_1. In other words, the shunting effect of the magnetizing reactance is not very great.

(a) The slip at which the maximum torque now occurs is

$$s_m = \frac{r_2'}{\sqrt{R_i^2 + (X_i + x_2')^2}} = \frac{0.12}{\sqrt{(0.09)^2 + (0.343 + 0.45)^2}} = 0.161$$

This represents a change of slightly more than 1% from the value found using the approximate equivalent circuit.

(b) The value of the current at maximum torque is

$$I_2' = \frac{V_i}{R_i + r_2' + j(X_i + x_2')} = \frac{270.5}{0.09 + 0.7453 + j(0.343 + 0.4)} = 2\cancel{X}9\underline{/41.6°} \text{ A}$$

$$242\underline{/-41.7}$$

Here again the difference is about 1%. Of course, there are motors where the differences can be of the order of 3% or 4%. In such cases the use of the more exact equivalent circuit might appear to be justified. But it is important to note here that this amounts to quibbling because there is a more important effect taking place at the maximum torque condition which we have not accounted for. With currents as large as three times the rated current flowing at the maximum torque condition, a considerable amount of saturation occurs in the teeth of the stator and rotor. A subsequent decrease in the values of the stator and rotor leakage reactances of the order of 50% and more can easily occur. *The effect of this change is much more serious than whether the approximate or the exact version of the equivalent circuit is used.*

(c) Continuing with the original premise in the statement of the problem as set forth in Example 4-2, we get for the maximum torque

$$T_m = \frac{q_1 V_i^2}{2\omega_s[R_i + \sqrt{R_i^2 + (X_i + x_2')^2}]} = \frac{3(270.5)^2}{2(188.5)[0.09 + \sqrt{(0.09)^2 + (0.343 + 0.4)^2}]}$$

$$= (582.26)(1.193) = 694.77 \text{ N-m}$$

Here the difference is nearly 4%.

4-7 EQUIVALENT CIRCUIT PARAMETERS FROM NO-LOAD TESTS

The performance computations for a three-phase induction motor presuppose knowledge of the parameters of the equivalent circuit. This information may be available either from design data or from appropriate tests. When the design data are not available information about the magnetizing branch can be obtained from a no-load test that is performed by applying a balanced three-phase voltage to the motor which is uncoupled from its load. Often the input power and line current are measured at rated voltage and rated frequency. Sometimes this no-load test is performed with a varying voltage, which leads to the plot depicted in Fig. 4-15. Point *a* corresponds to operation at rated voltage. Experimental points below *b* are not taken because the speed will no longer be close to synchronous speed.

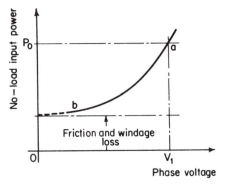

Figure 4-15 Showing the variation of no-load input power with applied voltage; *ab* is the curve obtained from experimental data.

Extrapolation of the curve to the ordinate axis gives a good indication of the friction and windage loss at normal speeds.

At no-load and rated voltage the input power is used to supply three losses: the stator copper loss, the stator core loss P_c, and the rotational losses P_{rot}. In equation form we have

$$P_0 = q_1 I_0^2 r_1 + P_c + P_{rot} \tag{4-34}$$

where q_1 denotes the number of stator phases and r_1 is the effective stator resistance per phase. The rotational losses include the friction and windage losses as well as iron losses caused by the pulsations of flux in the stator teeth as the rotor revolves. These iron losses due to rotation are larger in machines with open slots than in those with semiopen slots. When specific information about P_c or P_{rot} is not known, it is customary to assume these quantities to be equal for they are frequently found to be so in conventional machines. Based on this assumption, the current through the core-loss resistor can then be found as follows:

$$I_c = \frac{P_c}{q_1 V_1} = \frac{(P_0 - q_1 I_0^2 r_1)}{2 q_1 V_1} \tag{4-35}$$

Hence

$$\boxed{r_c = \frac{V_1}{I_c}} \tag{4-36}$$

To find the magnetizing reactance we must first obtain the reactive component of the no-load current, I_ϕ. The power-factor angle at no-load is found from

$$\theta_0 = \cos^{-1} \frac{P_0}{q_1 V_1 I_0} \tag{4-37}$$

Therefore

$$I_\phi = I_0 \sin \theta_0 \tag{4-38}$$

so that

$$\boxed{x_\phi = \frac{V_1}{I_\phi}} \tag{4-39}$$

The phasor diagram at no-load is depicted in Fig. 4-16. The angle θ_0 is usually large ($65°$ to $80°$) because of the need for a large magnetizing current to produce the required flux per pole in a magnetic circuit containing air gaps. The total in-phase component is represented by three terms to take care of the three losses appearing in Eq. (4-34). Note that the phasor sum of $\bar{I}_c$ and $\bar{I}_\phi$ yields the total magnetizing current $\bar{I}_m$ of the magnetizing branch of the equivalent circuit. The quantity $\bar{I}'_{20}$ is assumed to be equal to $\bar{I}_c$ and represents the small rotor current

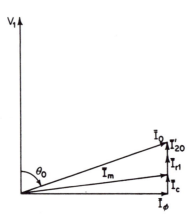

Figure 4-16 Phasor diagram of the induction motor at no-load.

that flows at no-load to supply the rotational losses. Specifically, this quantity may be written as

$$I'_{20} = \frac{P_{\text{rot}}}{q_1 V_1} = \frac{P_0 - q_1 I_0^2 r_1}{2 q_1 V_1} \tag{4-40}$$

The third part of the in-phase component of the no-load current, I_{r1}, represents the stator copper loss at no-load. This is,

$$I_{r1} = \frac{I_0^2 r_1}{V_1} \tag{4-41}$$

This too can be readily computed from the measured no-load data.

Information about the winding resistances and leakage reactances is obtained from a blocked-rotor test. This test is analogous to the short-circuit test of the transformer. It requires that the rotor shall be blocked to prevent rotation and that the rotor winding shall be short-circuited in the usual fashion. Furthermore, since the slip is unity, the mechanical load resistor R_m is zero and so the input impedance of the equivalent circuit is quite low. Hence, in order to limit the rotor current in this test to reasonable values, a reduced voltage must be used—usually about 10% to 25% of rated value. Moreover, operation at such reduced voltages renders the core loss as well as the magnetizing current negligibly small. Accordingly, the equivalent circuit in this test takes on the configuration shown in Fig. 4-17.

Assume now that the following instrument readings are taken in performing a blocked-rotor test on a Y-connected three-phase wound-rotor induction motor:

$$P_b = \text{total wattmeter reading, W}$$

$$I_b = \text{line current of Y connection}$$

$$V_b = \text{line voltage of Y connection}$$

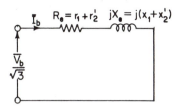

Figure 4-17 Induction motor equivalent circuit for the rotor-blocked test.

Assume too that the dc phase winding resistances of the stator and rotor windings, $r_{1\,dc}$ and $r_{2\,dc}$, are made available from a simple dc test and that the ratio of transformation, a, from stator to rotor is also available. Keep in mind that the instrument readings are consistent with operation at the rated frequency of the motor. Hence, if present, the influence of skin effect† is reflected in these readings. From the foregoing measurements it follows that the effective value of the equivalent winding resistance is given by

$$R_e \equiv r_1 + r_2' = \frac{P_b}{3I_b^2} \qquad (4\text{-}42)$$

Here both r_1 and r_2' denote effective winding resistances per phase. To separate out the effective value of r_1 use is made of the following equation:

$$r_1 = \frac{r_{1\,dc}}{r_{1\,dc} + a^2 r_{2\,dc}} R_e \qquad (4\text{-}43)$$

This last expression states that the ratio of the effective value of r_1 to the effective value of the equivalent resistance bears the same relationship as the dc value of r_1 to the dc value of the equivalent resistance. It is important at this point to understand that, in the equivalent circuit of the induction motor as it is used to compute performance at normal values of slip, the r_1 to be used is the effective value as determined by Eq. (4-43) but the r_2 to be used is the dc value. The latter quantity is used because at normal slips (3 to 5%) the frequency of the rotor current is very low (2 to 3 Hz).

The equivalent phase impedance is obtained from

$$Z_e = \frac{V_b}{\sqrt{3}\,I_b} \qquad (4\text{-}44)$$

---

†Skin effect refers to the tendency of time-varying current to crowd toward the surface of a conductor which reduces the effective cross-sectional area and thereby increases the resistance over the dc value.

Finally, the equivalent leakage reactance is determined from

$$X_e = \sqrt{Z_e^2 - R_e^2} = x_1 + x_2' \qquad (4\text{-}45)$$

Note that, as long as computations are carried out using the approximate equivalent circuit, it is sufficient to deal directly with X_e without a further breakdown into x_1 and x_2'.

In view of the importance of working with the dc value of r_2 in the equivalent circuit for low values of slip, how is this value found for the squirrel-cage induction motor if the design data are not available? The use of the effective value of r_2 leads to very serious error because the skin effect in the squirrel cage is very large. This is caused by the use of solid copper bars imbedded in iron. The ratio of effective to dc value at 60 Hz is often $3:1$. Such large ratios do not occur in the wound-rotor machine because of the use of stranded conductors. Since the phase terminals of a squirrel cage are not available, an indirect means of determining the dc value of the rotor phase resistance must be employed. If the blocked-rotor test is performed for various frequencies at reduced voltage, the effective equivalent resistance can be found and plotted to yield the curve depicted in Fig. 4-18. Note

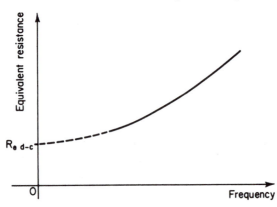

Figure 4-18 Illustrating how dc equivalent resistance of a squirrel-cage induction motor can be found from a blocked-rotor test for varying frequency.

that as the frequency is reduced the effective equivalent resistance assumes correspondingly smaller values. By extrapolating the experimental curve to the ordinate axis the dc equivalent resistance for the squirrel-cage motor is obtained. Then the stator-referred dc resistance per phase of the rotor is found as

$$r_{2\,\text{dc}}' = R_{e\,\text{dc}} - r_{1\,\text{dc}} \qquad (4\text{-}46)$$

Accordingly, in the equivalent circuit for the squirrel-cage induction motor at low slips, the effective value of r_1 and the dc value of r_2' [Eq. (4-46)] must be used.

Example 4-5

The following no-load tests are performed on a three-phase, four-pole, 60-Hz, 20-hp, 550-V, Y-connected squirrel-cage induction motor and the results are found to be:

No-load test: 550 V,   5.8 A,   754 W,   60 Hz
Blocked-rotor test at rated frequency: 123 V,   25   A,   2419 W,   60 Hz
Blocked-rotor test at low frequency: 55 V,   25   A,   2063 W,   15 Hz
Dc test on stator per phase: 15 V,   25   A

The friction and windage losses for this motor are known to be 328 W. Determine the parameters of the approximate equivalent circuit to be used at normal operating slips (i.e., 3% to 5%).

Solution   The dc stator winding resistance per phase is

$$r_{1\ dc} = \frac{15}{25} = 0.6\ \Omega$$

Although the stator winding is a wound winding using stranded wire, some small skin effect is present. Therefore, the ac value of the winding resistance is needed. This can be determined from the blocked-rotor test performed at rated frequency. Thus

$$R_{e\ ac} = \frac{2419}{3(25)^2} = 1.29\ \Omega$$

From the blocked rotor test which is performed at low frequency, the skin effect is not considered to be important and so the equivalent "dc" resistance is found to be

$$R_{e\ dc} = \frac{2063}{3(25)^2} = 1.1\ \Omega$$

Then, in accordance with Eq. (4-43), the ac stator winding resistance has the value

$$r_{1\ ac} = r_{1\ dc}\frac{R_{e\ ac}}{R_{e\ dc}} = (0.6)\left(\frac{1.29}{1.1}\right) = 0.7\ \Omega$$

This value is a bit on the pessimistic side for the reason that during the blocked-rotor test at full frequency, the skin effect that takes place in the iron-embedded bars of the rotor is greater than for the stranded wire of the stator winding. Nonetheless, this is one of the six values of the equivalent circuit. A second parameter, the "dc" value of the rotor resistance referred to the stator, is determined by applying Eq. (4-46), which here yields

$$r'_{2\ dc} = R_{e\ dc} - r_{1\ dc} = 1.1 - 0.6 = 0.5\ \Omega$$

The use of the approximate equivalent circuit means that the two parameters x_1 and x'_2 may be evaluated and used as a lumped quantity. Because these leakage reactances are the standstill values, the data of the rotor-blocked test at rated frequency are employed. Accordingly,

$$Z_e = \frac{V_b}{\sqrt{3}\ I_b} = \frac{123}{\sqrt{3}\ (25)} = 2.84\ \Omega$$

$$X_e = \sqrt{Z_e^2 - R_{e\ ac}^2} = \sqrt{2.84^2 - 1.29^2} = 2.53\ \Omega$$

The two remaining parameters are found from the no-load test data. By Eq. (4-37) the no-load power-factor angle is

$$\theta_0 = \cos^{-1} \frac{P_0}{q_1 V_1 I_0} = \cos^{-1} \frac{754}{3(550/\sqrt{3})5.8} = 82.2°$$

so that the corresponding value of the reactive component of the no-load current becomes

$$I_\phi = I_0 \sin \theta_0 = 5.8 \sin 82.2° = 5.75 \text{ A}$$

which in turn yields a magnetizing reactance of

$$x_\phi = \frac{V_1}{I_\phi} = \frac{550}{\sqrt{3}\,(5.75)} = 5.25 \ \Omega$$

Finally, the core-loss resistor is found by first calculating the value of core loss from

$$P_c = P_0 - P_{\text{fw}} - 3I_0^2 r_{1\,\text{ac}} = 754 - 328 - 3(5.8^2)(0.7)$$

$$= 355.4 \text{ W}$$

and then noting that

$$q_1 \frac{V_1^2}{r_c} = P_c$$

Hence the value of the core-loss resistor computes to be

$$r_c = \frac{qV_1^2}{P_c} = \frac{3(550)^2}{(\sqrt{3})^2 P_c} = \frac{(550)^2}{355.4} = 851.2 \ \Omega$$

4-8 SPEED CONTROL

In most industrial applications the essentially constant speed characteristic of the induction motor is a desirable one. However, there are some applications (e.g., conveyors, hoists, and elevators) where speed control capability is a transcendent factor. Therefore in this section we explore the extent to which the induction motor lends itself to adjustment of speeds under various load conditions.

The squirrel-cage induction motor is examined first because it is simpler owing to the closed nature of its rotor winding, which eliminates all possibility of control from that end. Any control that is possible must take place through manipulations at the stator. Here three possibilities present themselves, two of which are evident from Eq. (4-12). The first is to *change the line frequency*. As the line frequency is raised or lowered the synchronous speed is correspondingly raised or lowered, which provides satisfactory control. However, the serious shortcoming of this procedure is that a variable frequency source is not available generally. Moreover, if such a source is to be provided locally, a motor-generator set is needed equipped with appropriate controls and power rated comparably to that of the induction motor it is to control. Furthermore, if maximum iron-flux densities are not to be exceeded as the frequency is dropped, corresponding decreases in the applied voltage must be effected.

A second method of control is to *change the number of poles.* Recall that the number of poles is determined by the winding layout. It is possible to arrange the winding in appropriately paired sections for each phase. When the two sections, for example, are connected in a way to ensure symmetrical orientations of current flow, a four-pole machine results. However, if the current through the second section is made to flow in the reversed direction by throwing a switch, a two-pole machine results. Accordingly, the synchronous speed can be changed by a factor of 2. Such a machine is called a *multispeed motor.* The motor can operate at one of two speeds, but unfortunately once a particular speed is selected there is no further control available. Speed control in this instance occurs in a discrete fashion. For a line frequency of 60 Hz such a machine allows operation near either 3600 or 1800 rpm. If speed control is desired between 1800 and 3600 rpm, it cannot be achieved with the multispeed squirrel-cage induction motor.

The third method of speed control involves *reducing the applied line voltage.* Operation at reduced voltages means that the ordinate values of torque on the torque-speed curve assume new values in accordance with the square of the reduced voltage ratio. Thus, if the squirrel-cage motor is operated at 70.7% of rated voltage, all points on the new torque-speed curve have half the value of the original curve. The maximum torque is reduced by one-half; the starting torque is also reduced to one-half its original value; and so on. Figure 4-19 depicts the distinctions over the complete speed range. Also shown are the torque-speed curves of a constant-torque load so characteristic of conveyors and the variable load torque characteristic of fans and other propeller type loads. It is assumed that at rated voltage the induction motor delivers rated torque to each load type.

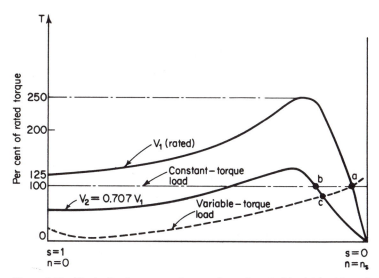

Figure 4-19   Illustrating how some degree of speed control is obtained by reducing the applied voltage.

The operating point is denoted by a, where the slip is in the vicinity of about 5%. A reduction of 30% in the applied voltage causes the operating point for the constant-torque load to move from a to b and from a to c for the variable-torque load. Note that this scheme may allow the slip to double or perhaps triple. A little thought indicates that this is not an appreciable amount of speed adjustment—and certainly not for the price paid. Speed control by voltage control has serious disadvantages which account for its infrequent use, especially for polyphase motors. First, it is expensive. Variable three-phase voltage is obtained efficiently through the employment of a motor-generator set or a three-phase variac having ratings at least as large as the motor being controlled. Second, when the motor is operating at reduced voltage, its reserve capacity is dangerously reduced. For the situation illustrated in Fig. 4-19, note that the reserve capacity for the constant-torque load is 250% of rated torque at full voltage and only 125% at 70.7% of rated voltage. Third, if an attempt were made to operate the motor over the full range at the reduced voltage, a glance at Fig. 4-19 would make it clear that only the variable-torque load can be so operated. At reduced voltage the motor's starting torque of 62.5% is insufficient to move the constant-torque load.

In conclusion therefore it can be stated that speed adjustment of the squirrel-cage induction motor is difficult to achieve and expensive besides. This motor is best suited for constant-speed loads.

Speed control is easier to achieve in the wound-rotor induction motor than in the squirrel-cage motor because of the availability of the rotor terminals. Of course the adjustment of speed by the three methods already outlined apply with equal validity to the wound-rotor machine.

Three additional methods by which the speed can be controlled include the following: increasing the total rotor reactance per phase, x_2'; increasing the total rotor resistance per phase, r_2'; and injecting suitable voltages at the rotor terminals. The use of an external three-phase reactor connected to the rotor terminals is rarely, if ever, used because of its very adverse effect on the line power factor. In addition the size of the reactor needs to be large because of the effect of the slip term. By far the most common method of adjusting the speed of a three-phase induction motor is to use an external three-phase resistor connected to the rotor terminals. The larger the value of this external resistance per phase, the lower the speed is. The effect is illustrated in Fig. 4-20. Recall from Eq. (4-32) that, as resistance is added to the rotor winding, the slip at which the maximum torque is developed increases. When this is combined with, say, the constant-torque characteristic of the load, we see that the same torque can be delivered over a considerable range of speed. It is customary with this method to obtain speed control to about 50% below synchronous speed. Beyond this point the efficiency becomes quite poor, being less than 50%. In fact this is the price paid for attaining speed control. When rated torque is delivered to a load at a slip of 50% rather than the usual 5% the power delivered to the load is less by about 45 per cent. This difference of 45% of rated power is consumed as heat in the external resistors and so is unrecoverable.

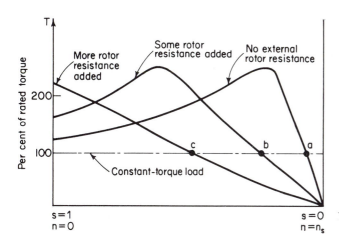

Figure 4-20 Speed control by external rotor resistance.

From the nature of the torque-speed curve we know that at a specified slip s there occurs a developed torque T. How do we determine the amount of resistance to be added to the rotor winding per phase so that this same developed torque prevails at an increased slip s'? To preserve correspondence with the discussion involving the situation illustrated in Fig. 4-20, we continue with the condition of a load that calls for constant developed torque. Because $T\omega_s = P_g$ and is a fixed quantity, it follows that constant developed torque by the motor implies constant gap power. When no external rotor resistors are used, the expression for gap power is given by Eq. (4-22), which is repeated here for convenience. Thus,

$$P_g = q_2 I_2^2 \frac{r_2}{s} \tag{4-47}$$

If we now assume that a resistor of value R_a Ω/phase is added to each of the three phase windings of the rotor, then Eq. (4-47) becomes

$$P_g = q_2 I_2^2 \frac{R_a + r_2}{s'} \tag{4-48}$$

where s' is used to indicate that the insertion of R_a changes the operating slip as demonstrated in Fig. 4-20. It is important to note that no distinction is made in the symbol for the rotor current in Eqs. (4-47) and (4-48). This is because under constant torque conditions the rotor current remains essentially fixed. Reference to Eq. (3-27) makes this statement self-evident since all terms on the right side are virtually constant, including $\cos \psi$, which moves ever so slightly closer to unity. Consequently, on equating the last expression to Eq. (4-47), the equation for R_a is found. The result is

$$\boxed{R_a = \frac{s' - s}{s} r_2} \qquad \Omega/\text{phase} \tag{4-49}$$

Equation (4-49) states that by adding R_a to each phase of the rotor winding, a given torque that is developed at a slip s with the rotor winding short-circuited can now be developed at an arbitrarily selected slip s' subject to the constraint $s \leq s' \leq 1$. Observe that by choosing $s = s_m$ = slip at maximum torque and $s' = 1$, the maximum torque can be made to be the starting torque.

To illustrate, let us find the amount of resistance to be added to the rotor winding per phase to yield the maximum torque at starting. Assume that this torque normally occurs at a 10% slip with the rotor short-circuited. In this case $s' = 1$ and $s = 0.1$; hence

$$R_a = \frac{1 - 0.1}{0.1} r_2 = 9r_2 \qquad (4\text{-}50)$$

Thus, by adding an external resistance equal to nine times the actual rotor winding resistance per phase, the starting torque is equal to the maximum torque.

The achievement of speed control by injecting a three-phase voltage into the rotor winding can be understood by studying the sequence depicted in Fig. 4-21. For simplicity it is assumed that rated torque is to be delivered to a constant-torque load at varying speeds. Moreover, the rotor leakage reactance is assumed to be negligibly small. This means that the space-displacement angle ψ of Eq. (4-28a) is zero so that the basic torque equation is $T = K\Phi I_2$. Since Φ is fixed by the constant applied voltage, it follows that to maintain constant torque at various speeds it is necessary merely to keep I_2 constant. Depicted in Fig. 4-21(a) is the situation existing in the rotor at normal operation, i.e., with the rotor short-circuited and rated motor torque delivered to the load. Note that the resultant slip voltage $s\bar{E}_2$ need only be large enough to overcome the rotor winding resistance drop. Now assume that a voltage of rms value $\bar{V}_i$ per phase is injected into the rotor in such a direction as to oppose the original $s\bar{E}_2$ voltage. The immediate effect of $\bar{V}_i$ is to cause the rotor current to drop, which in turn causes the developed torque to decrease. Then, since the motor can no longer supply rated torque to the load, the speed decreases. As the speed decreases, the slip emf increases and thereby gradually overcomes the influence of $\bar{V}_i$ on the rotor current. Equilibrium is again established when the resultant rotor voltage per phase is once more equal to $\bar{I}_2 r_2$. This happens when algebraically

$$s'E_2 = sE_2 + V_i \qquad (4\text{-}51)$$

where s' denotes the new increased slip. The condition is depicted in Fig. 4-21(b). The net effect of the injected voltage is to bring about an increase in slip in a manner not unlike that achieved with external rotor resistance. As a matter of fact the corresponding value of external rotor resistance that yields the same slip can be computed from

$$I_2 R_a = V_i \qquad (4\text{-}52)$$

The basic difference between the two methods, however, lies in the fact that, whereas with the resistance method power is wasted as heat, the injected voltage

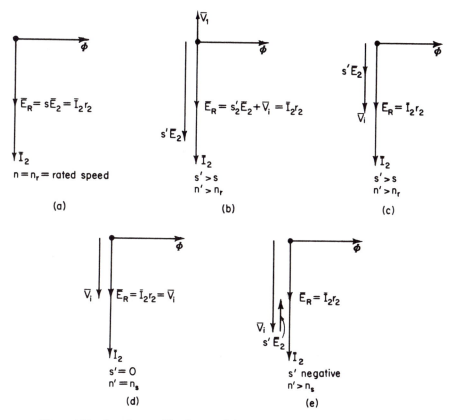

Figure 4-21 Speed control by the rotor injected-voltage method: (a) normal operation, with rated motor torque delivered to constant-torque load; (b) V_i injected
opposed to sE_2; (c) V_i injected in-phase with sE_2 of part (a) but less in magnitude;
(d) V_i injected in-phase with original sE_2 and equal to it; (e) V_i injected in-phase
with original sE_2 but greater in magnitude.

source can either usefully consume the power associated with the drop in speed or
else return it to the line source.

What happens to the induction motor speed when the injected voltage is
introduced in phase with the original slip emf but smaller in magnitude? The
condition is depicted in Fig. 4-21(c). The immediate effect of $\bar{V}_i$ is to cause $\bar{I}_2$ to
increase beyond the rated value, which causes an accelerating torque that *raises*
the motor speed beyond the normal value. Equilibrium is then established when
$s'\bar{E}_2 + \bar{V}_i = \bar{I}_2 r_2$. It is interesting to note by way of comparison that this is a
situation that cannot be realized with external rotor resistance. The use of external rotor resistance allows speed control at increased slips—never at reduced
slips.

An interesting situation occurs when the injected voltage is made just large
enough to establish the rotor current entirely by itself. Refer to Fig. 4-21(d). The

motor is thus relieved of having to play any part in producing the rotor current. Accordingly, the slip becomes zero and the motor *operates at synchronous speed*. This should not cause too much surprise because in point of fact the motor is no longer singly but, rather, doubly excited and so has every right to behave like a synchronous motor. Depicted in Fig. 4-21(e) is the situation where $\bar{V}_i$ is made larger than is needed to produce the required rated rotor current. Consequently, the motor will accelerate above synchronous speed and cause operation to take place at negative slips.

What price is paid for speed control by the injected rotor voltage method? To begin with, an auxiliary device is needed to generate the injected voltage. An important characteristic of the unit is that it must make available at the rotor terminals of the induction motor a voltage for injection that *always has the correct slip frequency;* otherwise, addition to the motor slip-frequency emf cannot be achieved. Moreover, the size and rating of the auxiliary unit must be consistent with the degree of control to be affected. Thus, if the speed is to be controlled to 50% below synchronous speed, its rating must be at least equal to one-half that of the motor being controlled. However, with the injected voltage source, speed control to the extent of 50% is possible above as well as below synchronous speed so that the total range over which control is possible is 3:1 as compared with 2:1 for the external rotor resistance method. The Schrage motor is an induction motor that employs injected armature voltages to obtain speed control. This motor has been manufactured by the General Electric Company for many years to supply up to 50 hp. GE calls it the "brush-shift" motor.

Example 4-6

A 15-hp, 60-Hz, four-pole, 480-V, three-phase, wound-rotor, Y-connected induction motor has a line frequency rotor induced emf per phase of 141 V, a rotor leakage reactance per phase of 0.27 Ω, and a rotor winding resistance of 0.16 Ω per phase. When the motor operates with the rotor winding short-circuited, it develops a torque of 62.4 N-m at a slip of 3.5% and carries a rotor current of 31.1 A. Determine the magnitude and phase of the injected voltage to be introduced at the terminals of the rotor winding (with the short-circuit removed) so that the same torque is developed at a slip which is 40% above synchronous speed. The rotor leakage reactance at this high slip is not to be neglected.

Solution We know from the basic torque equation [see Eq. (3-27)] of ac machines that the condition of constant torque requires that

$$I_2 \cos \psi_2 = I_2^* \cos \psi_2^*$$

where the * notation denotes operation at the new slip condition. The angle ψ of course denotes the power factor angle at the rotor winding at the particular operating slip. For the case at hand, we have

$$\psi_2 = \tan^{-1} \frac{sx_2}{r_2} = \tan^{-1} \frac{(0.035)(0.27)}{0.16} = 3.4°$$

and

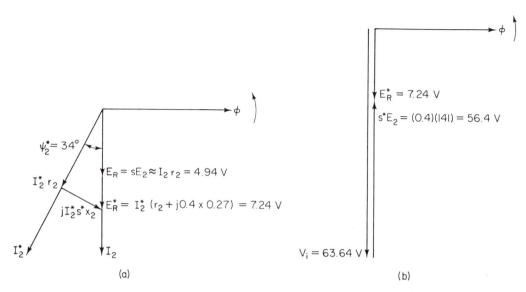

Figure 4-22   Diagram for Example 4-6: (a) showing relationships between slip-induced emf's and currents at two values of slip; (b) determination of the required injected voltage to achieve speed control above synchronous speed.

$$\psi_2^* = \tan^{-1} \frac{s^* x_2}{r_2} = \tan^{-1} \frac{(0.4)(0.27)}{0.16} = 34°$$

Accordingly the rotor current at the new slip condition is

$$I_2^* = I_2 \frac{\cos \psi_2}{\cos \psi_2^*} = 31.1 \left(\frac{\cos 3.4°}{\cos 34°} \right) = 37.5 \text{ A}$$

When the motor operates with the rotor short-circuited, the resultant emf that produces the current of 31.1 A has a magnitude of

$$E_R = I_2|r + jsx_2| = 31.1|0.16 + j(0.035)(0.27)| = 4.94 \text{ V}$$

This quantity is shown in Fig. 4-22(a) placed along a vertical line in quadrature with the flux phasor. However, when the motor operates at a high slip, the rotor winding phase angle can be appreciable as already demonstrated ($\psi_2^* = 34°$). Although the resultant rotor emf at slip s^* must still bear a quadrature relationship to the flux that produces it, the magnitude now is no longer just dependent upon the winding resistance. The leakage reactance also makes a contribution. Thus the magnitude of the resultant rotor winding induced emf now computes to be

$$E_R^* = I_2^*|r_2 + js^* x_2| = 37.5|0.16 + j(0.4)(0.27)| = 7.24 \text{ V}$$

This quantity is also depicted in Fig. 4-22(a) together with its phasor components. Observe that the projection of I_2^* on the vertical is the same for both slip conditions, thus ensuring constant torque.

Keep in mind that by design the injected voltage is introduced always in quadrature with the flux phasor for the achievement of speed control. Because in this case we want speed control above synchronous speed, the injected voltage must be introduced 90° *lagging* the flux phasor in the manner illustrated in Fig. 4-22(b). The quantities E_R^*, $s*E_2$, and V_i are all collinear and the relationship between them expressed arithmetically is simply

$$V_i - s*E_2 = E_R^*$$
$$V_i = E_R^* + s*E_2 = 7.24 + (0.4)(141) = 63.64 \text{ V}$$

4-9 ELECTRONIC METHOD OF SPEED CONTROL

Reference to Eq. (4-12) makes it obvious that an attractive means of speed control is the availability of an adjustable frequency source. It is mentioned in the preceding section that one method of obtaining such a source is to use a dc motor-ac generator set. Unfortunately, this solution projects its own shortcomings such as the need for a suitable dc supply, the need to purchase two additional machines, a more restrictive limit on the range of speed control as well as reduced efficiency. A better solution lies in the use of an *inverter* which can be made to change a dc source into a variable frequency source. Because dc power is not readily available, it becomes necessary to generate it through the use of suitable rectifier circuits. A block diagram of this *electronic scheme* of speed control of a squirrel-cage induction motor is shown in Fig. 4-23. It is significant to note that this is one of the few methods that is available to achieve speed control for the highly rugged squirrel-cage induction motor.

The three-phase rectifier in Fig. 4-23 serves the purpose of converting the commonly available three-phase voltage to a dc source. Usually, the output of the rectifier contains higher harmonics of the fundamental frequency of the ac source and these are conveniently removed by an appropriate filter. Figure 4-24(a) illustrates a typical three-phase half-wave rectifier the output voltage of which is drawn in Fig. 4-24(b). By passing this waveshape through the *LC* filter, the output voltage becomes essentially dc. It then becomes the task of the inverter to generate a new three-phase voltage source which in general exhibits the properties of variable frequency, adjustable voltage, and even adjustable phase. Simultaneous adjustment of the output voltage of the inverter with frequency is needed to

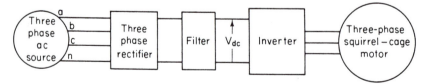

Figure 4-23 Block diagram to illustrate the principal equipment needed in the electronic speed control of a squirrel-cage three-phase motor.

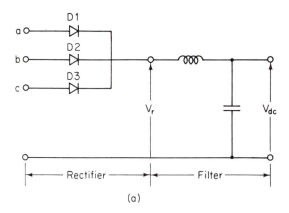

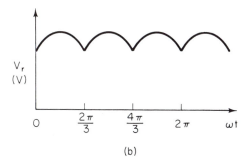

Figure 4-24 (a) A three-phase half-wave rectifier using semiconductor diodes; (b) output waveshape of the rectifier.

prevent operation at values of flux per pole which deviate sharply from the rated value of the controlled motor. This can be achieved by designing the speed control system so that it keeps the ratio of the inverter output voltage to the controlled frequency a constant. By Eq. (4-9) it then becomes clear that by this technique the flux per pole can be preserved over the operating range. If this precaution were not to be observed, then at low frequency severe saturation of the magnetic circuit could easily occur, while at high frequencies there would be a diminution of the motor's reserve capacity.

The circuitry of Fig. 4-24(a) shows the use of diodes in the rectifying circuit. It is useful to note that such diodes are available in voltage ratings up to 5000 V and current ratings of 7500 A. The output voltage can be made adjustable by replacing the diodes with silicon-controlled rectifiers (SCRs). By placing a suitable voltage pulse on the gate terminal of the SCR, the instant of firing of the diode section can be delayed, thereby reducing the value of the rectified output. Although the voltage ratings of the SCRs match those of the power diodes, the current ratings are about half as much.

A schematic diagram of a typical electronic speed control system employing the foregoing notions is depicted in Fig. 4-25. The components shown here are those called for by the block diagram of Fig. 4-23 but modified to include the control circuit for the silicon-controlled rectifiers. Moreover, the details of the

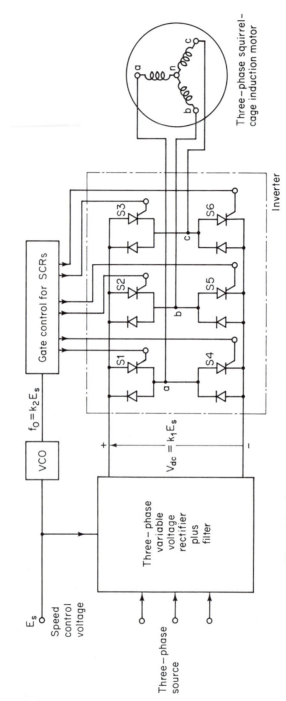

Figure 4-25 Schematic diagram for electronic speed control of a three-phase squirrel-cage induction motor. The silicon-controlled rectifiers (SCR) are marked S1 through S6.

inverter are also indicated. The switches marked S1 through S6 are the SCRs and each is appropriately gated by signals obtained from the gate control section. It is the purpose of the gate signal to sequence the ON and OFF times of each SCR so that the output voltage appearing at terminals a, b, c will exhibit the characteristics of a three-phase voltage set. The speed control voltage E_s serves two functions. First, it controls the rectifying action of the three-phase rectifier in a fashion that allows the magnitude of the dc output to be directly proportional to E_s; and second, the speed control voltage also determines the output frequency of a voltage-controlled oscillator (VCO) in a direct manner. Expressed mathematically, we can write

$$V_{dc} = k_1 E_s \qquad (4\text{-}53)$$

and

$$f_o = k_2 E_s \qquad (4\text{-}54)$$

where V_{dc} denotes the output voltage of the three-phase rectifier plus filter and f_o is the output frequency of the oscillator. The quantities k_1 and k_2 are appropriate scale factors. In fact, it is this frequency which the inverter imposes at the terminals of the squirrel-cage induction motor. The diodes that are placed in parallel with the SCRs in a reversed orientation are needed to provide a path for reactive currents when the corresponding SCR is gated off. They also are useful in allowing energy to flow back to the source whenever the induction motor is made to operate as a generator.

We now undertake a description of how the inverter takes the dc source and manipulates it to produce a three-phase ac voltage set. The gate control section of the system plays a crucial role in the process, for it provides the gating signals, which in turn establish the proper gating sequence as well as the duration of the gating (i.e., the portion of the frequency cycle each SCR is allowed to stay conductive). The switching sequence used in this explanation is shown in Fig. 4-26. One period of the frequency f_o is shown and it is divided into six equal parts of $\pi/3$ radians. In the interest of realizing a maximum dc voltage each SCR is gated to be in the ON mode for close to π radians. As a practical matter, the ON-time must be kept just under π radians in order subsequently to allow S4 to be gated ON at exactly π radians. A glance at Fig. 4-25 makes clear the importance of gating S1 to the OFF state before S4 is turned on. Failure to do so places a short-circuit across the dc supply.

The switching sequence appearing in Fig. 4-26 is determined as follows. Start with S1 and draw a line over the first three 60° periods, thus indicating that S1 is to be kept in a gated ON state for the first 180 electrical degrees. Next, because we are dealing with three-phase, start the conduction period of S2 at a position 120 electrical degrees later than the start of S1. Similarly, start the conduction period of S3 at a position 120 electrical degrees later than the start of S2. This pattern is then repeated for the second set of SCRs. Accordingly, the ON-time for S2 is made to begin at π radians and to endure for 180°. The conduction

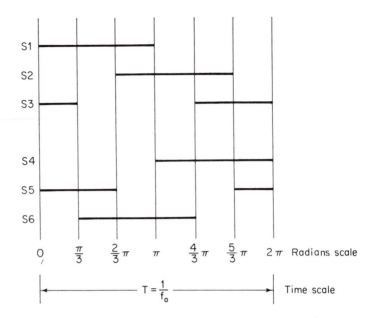

Figure 4-26   Switching sequence diagram for the SCRs in Fig. 4-25.

period for S5 is then made to begin 120° later than for S4 and to endure for 180°. Finally, conduction for S6 is programmed to commence 120° after S4 is placed into conduction and again allowed to stay conductive for almost π radians. By examining the first 60° column of Fig. 4-26, we can properly conclude that over this period the SCRs that are gated on are S1, S3, and S5. The remaining three SCRs are simultaneously and intentionally gated to the OFF state. During the period that a specific SCR is programmed to be in the ON state, it is important to maintain the gate signal so that the switch continues to stay closed whether or not current is flowing at the moment. Upon repeating the procedure for each 60° period, the following table is obtained that describes the switching sequence.

Interval	SCRs in ON state
0–60°	S1, S5, S3
60–120°	S1, S5, S6
120–180°	S1, S2, S6
180–240°	S4, S2, S6
240–300°	S4, S2, S3
300–360°	S4, S5, S3

Once the switching sequence is established, it is a routine matter to identify the line-to-line voltages as they appear at the terminals of the squirrel-cage motor, which is assumed to have a Y-connected stator winding. The inverter output appears at terminals a, b, c. During the first 60° interval, switches S1, S3, and S5 are in the ON state. Because S1 and S3 are conductive, terminals a and b are connected to the + side of the dc source and through the action of S5 terminal c is joined to the negative side. This arrangement places phases a and c of the induction motor in parallel with each other and the combination in series with phase b. Refer to the schematic diagram for the 60° interval in line (a) of Table 4-1, which illustrates the distribution of the dc input voltage to the inverter during the first three 60° intervals of the operating frequency determined by the voltage-controlled oscillator in response to the speed control voltage E_s. For balanced phase windings it follows that the distribution of the dc source voltage puts $V_{dc}/3$ across the two parallel-connected phases (i.e., coils an and cn) and $2/3\ V_{dc}$ across phase b. A simple application of Kirchhoff's voltage law then shows that the line voltages are $V_{ab} = V_{dc}$, $V_{bc} = -V_{dc}$, and $V_{ca} = 0$.

During the second 60° interval, the foregoing table shows that switches S1, S5, and S6 are in the ON state. In this case, only terminal a is joined to the positive side of the dc source while b and c are both connected to the negative side. The effect is to put phases b and c in parallel and the combination in series

TABLE 4-1 DISTRIBUTION OF THE DC INPUT VOLTAGE DURING THE FIRST THREE 60° INTERVALS OF THE OPERATING FREQUENCY

Interval	ON-gated SCRs	Voltage distribution on induction motor phases	Corresponding line voltages
(a) 0–60°	S1, S3, S5		$V_{ab} = V_{dc}$ $V_{bc} = -V_{dc}$ $V_{ca} = 0$
(b) 60–120°	S1, S5, S6		$V_{ab} = V_{dc}$ $V_{bc} = 0$ $V_{ca} = -V_{dc}$
(c) 120–180°	S1, S2, S6		$V_{ab} = 0$ $V_{bc} = V_{dc}$ $V_{ca} = -V_{dc}$

with phase a in the manner illustrated in line (b) of Table 4-1. While reading these diagrams, it is helpful to keep in mind that terminals marked with the same polarity are joined together for the specified interval. Now the distribution of the dc voltage puts two-thirds across the winding of phase a and one-third across each of the coils b and c with the polarity indicated. The result is that the line voltages during this second period are described by $V_{ab} = V_{dc}$, $V_{bc} = 0$, and $V_{ca} = -V_{dc}$.

The situation that prevails in the third 60° interval leads to the results shown in line (c) of Table 4-1. Here the coils of phase a and b are in parallel. The corresponding line voltages become $V_{ab} = 0$, $V_{bc} = V_{dc}$, and $V_{ca} = -V_{dc}$. A continuation of this procedure through to the sixth 60° interval leads to the results depicted in Fig. 4-27. An examination of these waveshapes reveals that the line voltages do exhibit the characteristics of a balanced three-phase voltage system. They are of equal magnitude and the individual line voltages are displaced in time by the required 120 electrical degrees. Of course, these waveshapes are not sinusoidal but they do contain a large fundamental component which is the one that serves to drive the induction motor.

The peak value of this fundamental component is readily found by applying

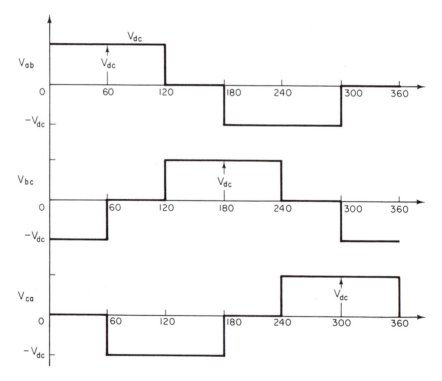

Figure 4-27 Variations of the inverter output line-to-line voltages corresponding to the switching sequence of Fig. 4-26 for one full cycle of the controlled frequency f_0.

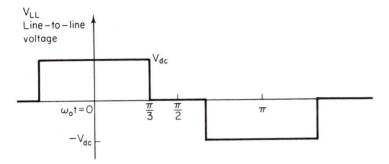

Figure 4-28 Variation of each line-to-line voltage with zero time placed at the center of the rectangular wave.

Fourier series analysis to the rectangular waveshapes that represent the line-to-line voltages. For convenience, this basic waveform is redrawn in Fig. 4-28 with the zero position relocated in the interest of taking advantage of the quarter-wave symmetry contained therein. Then from Fourier series analysis the amplitude of the fundamental component is

$$V_{1p} = \frac{1}{\pi} \int_0^{2\pi} V_{dc} \cos \omega t \; d(\omega t) = \frac{4}{\pi} \int_0 V_{dc} \cos \omega t \; d(\omega t)$$

or

$$V_{1p} = \frac{2 \sqrt{3}}{\pi} V_{dc} \qquad (4\text{-}55)$$

The corresponding rms value of this voltage is then

$$V_1 = \frac{V_{1p}}{\sqrt{2}} = \frac{\sqrt{6}}{\pi} V_{dc} = 0.78 \; V_{dc} \qquad (4\text{-}56)$$

The rectangular nature of the line voltages means, of course, that harmonics do exist in these voltages. However, the harmonic situation is not as bad as it may at first appear. For example, there are no even harmonics in these line voltages because of the mirror symmetry. Furthermore, although a third harmonic voltage exists in the line-to-neutral voltages, they do not appear in the line-to-line voltages because cancellation occurs when adding the phase voltages to obtain the line voltages. Hence the first important harmonic becomes the fifth. But Fourier analysis tells us that the magnitude of the fifth harmonic is only one-fifth that of the fundamental. When this is coupled with the fact that the frequency of the fifth harmonic is five times that of the fundamental frequency and that the induction motor presents an inductive load, which exerts a substantial filtering effect, the total effect of this harmonic is not serious. The other higher harmonics cause even less disturbance. The influence of the harmonic content of the line voltages can be further reduced by employing a switching arrangement that yields a more nearly sinusoidal waveshape for the line voltages (see Prob. 4-62). Unfortunately, this usually carries with it the penalty of reduced magnitude for the rms output voltage.

As a final point, let us examine the variation, if any, that takes place in the flux per pole of the induction motor over its operating range. Insertion of Eq. (4-53) into Eq. (4-56) gives the rms value of the voltage applied to the induction motor as

$$V_1 = 0.78V_{dc} = 0.78k_1E_s = k_0E_s \tag{4-57}$$

where k_0 denotes a new constant equal to $0.78k_1$. Except at very low frequencies, where machine inductance plays a reduced role in determining motor performance, Eq. (4-57), when divided by $\sqrt{3}$, also expresses the induced emf that occurs per phase in the induction motor as it operates at various frequencies in response to the speed control voltage E_s. Hence we may write

$$\frac{V_1}{\sqrt{3}} \approx E_1 = 4.44NK_wf_0\phi = \frac{k_0}{\sqrt{3}}E_s$$

or

$$\phi = \frac{k_0E_s}{\sqrt{3}\,(4.44)NK_wf_0} = \frac{k_0}{K}\frac{E_s}{f_0} \tag{4-58}$$

where $K = (\sqrt{3})4.44NK_W$, which is a constant. But the commanded frequency is established by the voltage-controlled oscillator in accordance with the relationship appearing in Eq. (4-54). Upon introducing this expression for f_0 into Eq. (4-58), there results

$$\phi = \frac{k_0E_s}{Kk_2E_s} = \frac{k_0}{Kk_2} = \text{a constant} \tag{4-59}$$

Therefore, the speed control offered by the electronic system depicted in Fig. 4-25 is achieved in a manner that ensures operation at essentially constant flux.

4-10 RATINGS AND APPLICATIONS OF THREE-PHASE INDUCTION MOTORS

Now that the theory of operation, the characteristics, and the performance of the three-phase induction motor are understood, we can study their standard ratings and typical applications. Of course, before it is possible to specify a particular motor for a given application, the characteristics of the load must be known. These include such items as horsepower requirement, starting torque, acceleration capability, speed variation, duty cycle, and the environment in which the motor is to operate. Once this information is available, it is often possible to select a general-purpose motor to do the job satisfactorily. Table 4-2 is a list of such motors which are readily available and standardized in accordance with generally accepted criteria established by the National Electrical Manufacturers Association (NEMA). The table is essentially self-explanatory.

A brief description of the salient features of the common classes of squirrel-cage induction motors follows. These distinctions arise from the differences in the construction details of the rotor slots which accommodate the squirrel cage.

NEMA Class A: The General-Purpose Motor

The rotor stampings for this machine usually have semiclosed slots of medium depth in order to limit the skin effect. See Fig. 4-29(a). The low-resistance cage that is placed in these slots leads to a performance characterized by high efficiency and high power factor under rated load conditions. The disadvantage of such a construction is that at rated voltage the motor draws from the line source very large starting currents in a ratio of five to eight times rated current. The limitations imposed by the electric power companies in restricting the size of the motor starting current often means that motors rated in excess of 7.5 hp cannot be started directly across the supply lines. Instead, these motors must be energized through a voltage-reducing device such as an autotransformer (also called a starting compensator). When this procedure becomes necessary, the starting torque as well as the accelerating torque available to a shaft-connected load will be reduced. Recall that the starting torque is directly proportional to the square of

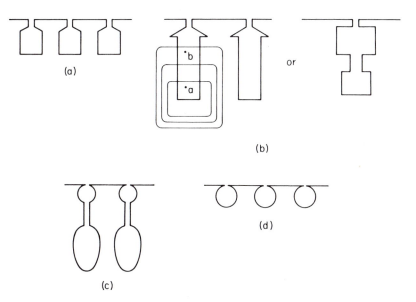

Figure 4-29 Construction details of the rotor slots of squirrel-cage induction motors: (a) class A; (b) class B; (c) class C; (d) class D.

TABLE 4-2 CHARACTERISTICS AND APPLICATIONS OF POLYPHASE 60-CYCLE AC MOTORS

Type classification	hp range	Starting torque[a] (%)			Maximum torque[a] (%)	Starting current[a] (%)
General-purpose, normal torque and starting current, NEMA class A	0.5–500	HP	Poles	Torque	Up to 225 but not less than 200	500–1000
		1.5	2	175		
		1.5	4	250		
		5.0	2	150		
		5.0	4	185		
		100	2	105		
		100	4	125		
		500	2	70		
		500	4	80		
General-purpose, normal torque, low starting current, NEMA class B	0.5–500	Same as above			About the same as class A but may be less	About 500–550, less than average of class A
High torque, low starting current, NEMA class C	3–200	200–250			190–225	About same as class B
High torque, medium and high slip, NEMA class D	3–150	Medium slip 350 High slip 275–315			Usually same as standstill torque	Medium slip 400–800, high slip 300–500
Low starting torque, either normal starting current, NEMA class E, or low starting current, NEMA class F	40–200	Low, not less than 50			Low, but not less than 150	Normal 500–1000, low 300–500
Wound-rotor	0.5–5000	Up to 300			225–275	Depends upon external rotor resistance but may be as low as 150

[a]Figures are given in percent of rated full-load values.

Source: By permission from M. Liwschitz-Garik and C. C. Whipple, *Electric Machinery,* vol. II (Princeton, N.J.: D. Van Nostrand Co., Inc.).

Slip (%)	Power factor (%)	Efficiency (%)	Typical applications
Low, 3–5	High, 87–89	High, 87–89	Constant-speed loads where excessive starting torque is not needed and where high starting current is tolerated. Fans, blowers, centrifugal pumps, most machinery tools, woodworking tools, line shafting. Lowest in cost. May require reduced voltage starter. Not to be subjected to sustained overloads, because of heating. Has high maximum torque
3–5	A little lower than class A	87–89	Same as class A—advantage over class A is lower starting current, but power factor slightly less
3–7	Less than class A	82–84	Constant-speed loads requiring fairly high starting torque and lower starting current. Conveyors, compressors, crushers, agitators, reciprocating pumps. Maximum torque at standstill
Medium 7–11; high, 12–16	Low	Low	Medium slip. Highest starting torque of all squirrel-cage motors. Used for high-inertia loads such as shears, punch presses, die stamping, bulldozers, boilers. Has very high average accelerating torque. High slip used for elevators, hoists, and so on, on intermittent loads
1 to 3½	About same as class A or class B	About same as class A or class B	Direct-connected loads of low inertia requiring low starting torque, such as fans and centrifugal pumps. Has high efficiency and low slip
3–50	High, with rotor shorted same as class A	High, with rotor shorted the same as class A, but low when used with rotor resistor for speed control	For high-starting-torque loads where very low starting current is required or where torque must be applied very gradually and where some speed control (50%) is needed. Fans, pumps, conveyors, hoists, cranes, compressors. Motor with speed control more expensive and may require more maintenance.

the applied voltage. Hence a 20% reduction in voltage applied to the motor at starting yields a 36% reduction in the starting torque.

In sizes and circumstances where line starting is tolerable, however, this motor will be found to be the least costly of the group and also the least demanding in maintenance. Moreover, the class A motor is designed to have a generous reserve capacity, as indicated by its value of maximum developed torque, which typically is close to 225% of rated torque. This puts the motor in a good position to manage successfully temporary overloads that are likely to be encountered in general-purpose applications. Figure 4-30 depicts the shape of its torque-speed curve.

NEMA Class B: Low Starting Current, Normal Starting Torque

The rotor laminations in the class B motor may be either of two types, as illustrated in Fig. 4-29(b). The slot is shaped to accommodate either a single deep bar or two bars of slightly different sizes with the upper one of larger cross-sectional area. At line frequency the leakage flux produced by a current flowing through the deep bar assumes the pattern depicted in Fig. 4-29(b). As this flux changes with the changing bar current, the associated emf induced at the deep end of the bar (point *a*) is greater than that produced at the upper end of the bar (point *b*). Accordingly, current tends to flow more in the upper portion of the bar than in the

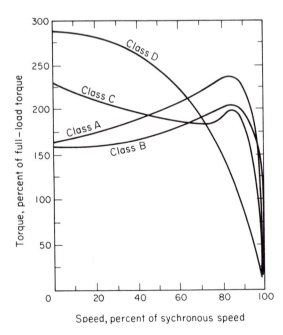

Figure 4-30 Typical torque-slip curves for various classes of squirrel-cage induction motors.

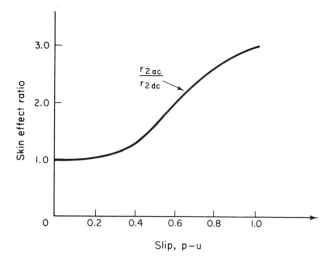

Figure 4-31   Illustrating the skin effect in a deep rotor bar.

lower portion, thus reducing the effective cross-sectional area of the conductor for current flow. Of course, for a nonvarying current flow (dc) the counter emf is zero, so the entire cross-sectional area of the bar is effective in supporting current flow. As the frequency of the bar current increases there is a greater squeezing of the current toward the upper portion of the bar. Figure 4-31 depicts the effects of this situation (the skin effect ratio) for a typical solid deep bar as a function of the slip. Observe that at standstill the ac resistance is three times greater than the dc resistance. In essence, then, the class B motor at starting exhibits a high rotor resistance, which in turn yields ample starting torque at reduced starting current. The latter is about 75% of the value of the class A motor used in application. On the debit side for the class B motor, however, it is necessary to list a lower maximum torque, slightly lower power factor, and a somewhat greater cost.

NEMA Class C: Low Starting Current, High Starting Torque

The rotor slot construction detail for the class C motor appears in Fig. 4-29(c). A double squirrel cage is always used. At standstill the impedance of the lower cage is much greater than the impedance of the upper cage. The upper cage is characterized by high resistance and low leakage reactance while the lower cage exhibits high leakage reactance and low resistance. Accordingly, at starting, the upper cage plays the dominant role; and because it has high rotor resistance, the advantages of a high starting torque at low starting current are realized. Then, as the motor reaches its normal operating speeds (where the cage current frequency is very low), control passes over to the lower cage, which displays a much lower resistance than the upper cage, thus improving efficiency during normal operation. Of course, the high leakage reactance of the lower cage is diminished in importance by operation at low slips. However, both efficiency and power factor for the class C motor will be less than for either classes A or B. A glance at the torque-speed curve for the class C motor shows that it can provide more accelerating torque than either the A or B motors. In fact, a distinguishing characteristic of this motor is that it develops a starting torque that is larger than its pull-out torque.

The availability of the class C motor has eliminated the need to specify the wound-rotor induction motor in some applications.

NEMA Class D: Highest Starting Torque, Lowest Starting Current, High Slip

A single squirrel cage is used for the class D motor. A slot of shallow depth and small cross-sectional area is used. Refer to Fig. 4-29(d). Bars of higher resistivity are also often employed. The objective is to obtain the high resistance needed to obtain very high starting and accelerating torques, as revealed by the typical torque-speed characteristic shown in Fig. 4-30. Of course, the use of a single high-resistance cage means that the efficiency will be lower and the slip higher than the preceding classes. However, the high slip characteristic is a very useful one in applications where this motor excels, such as for punch presses, hoists, cranes, and so on. When equipped with a flywheel in these applications, the drop in speed that occurs during the work portion of the duty cycle enables the flywheel to release some of its stored energy, thus permitting a decrease in the peak level of power the motor draws from the line source.

The horsepower rating of an ac motor is limited by the allowable temperature rise under operating conditions. Although copper, aluminum, and iron as individual elements can withstand very high temperature increases, this does not apply to the insulating materials that are used for the stator winding in both the wound-rotor and the squirrel-cage induction motors as well as for the rotor winding of the wound-rotor motor. Insulating materials deteriorate rapidly whenever the maximum temperatures they are designed to withstand are exceeded. Consequently, the capability of an insulating material to withstand high temperature rises plays a crucial role in establishing the physical dimensions of a motor with a specific horsepower rating. When a motor is equipped with a superior insulating material, the motor size can be made smaller for a given horsepower because it is possible to work the motor harder magnetically (i.e., operate at higher flux densities in the iron) as well as electrically (i.e., operate at higher current densities in the copper or aluminum conductors).

There are three classes of NEMA-rated insulating materials. Class B insulation is composed of such materials as mica, glass fibers, and asbestos in a manner that allows a maximum temperature rise of 80°C when operation occurs with fan cooling. Class F insulation includes similar elements in addition to some synthetic substances all designed to provide an allowable temperature rise of 100°C. The class H category refers to a silicone-type insulation which permits an even higher temperature rise of 125°C.

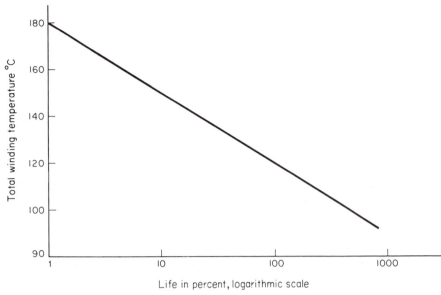

Figure 4-32   Illustrating the effect of winding temperature on motor life.

If an electric motor is operated at overload for a sustained period of time, the life of the motor can be markedly reduced. This fact is graphically demonstrated in Fig. 4-32 for an insulating material of the class B type. An examination of the plot reveals an interesting bit of information: if a motor equipped with this insulation is operated continuously at an overload condition that causes an increased temperature rise of 8 to 10°C over the rated value of the insulation, the life of the motor is approximately halved. Moreover, because of the linear character of this semilogarithmic plot, the statement is true for each additional increase in operating temperature of 10°C.

4-11 CONTROLLERS FOR THREE-PHASE INDUCTION MOTORS

After the right motor is selected for a given application, the next step is to select the appropriate controller for the motor. The primary functions of a controller are to furnish proper starting, stopping, and reversing without damage or inconvenience to the motor, to other connected loads, or to the power system. However, the controller fulfills other useful purposes as well, especially the following:

  1. It limits the starting torque. Some connected shaft loads may be damaged if excessive torque is applied upon starting. For example, fan blades can be sheared off or gears with excessive backlash can be stripped. The controller supplies reduced voltage at the start and as the speed picks up the voltage is increased in steps to its full value.

2. It limits the starting current. Most motors above 7.5 hp cannot be started directly across the three-phase line because of the excessive starting current that flows. Recall that at unity slip the current is limited only by the leakage impedance, which is usually quite a small quantity, especially in the larger motor sizes. A large starting current can be annoying because it causes light to flicker and may even cause other connected motors to stall. Reduced-voltage starting readily eliminates these annoyances.

3. It provides overload protection. All general-purpose motors are designed to deliver full-load power continuously without overheating. However, if for some reason the motor is made to deliver, say, 150% of its rated output continuously, it will proceed to accommodate the demand and burn itself up in the process. The horsepower rating of the motor is based on the allowable temperature rise that can be tolerated by the insulation used for the field and armature windings. The losses produce the heat that raises the temperature. As long as these losses do not exceed the rated values, there is no danger to the motor, but, if they are allowed to become excessive, damage will result. There is nothing inherent in the motor that will keep the temperature rise within safe limits. Accordingly, it is also the function of the controller to provide this protection. Overload protection is achieved by the use of an appropriate time-delay relay which is sensitive to the heat produced by the motor line currents.

4. It furnishes undervoltage protection. Operation at reduced voltage can be harmful to the motor, especially when the load demands rated power. If the line voltage falls below some preset limit, the motor is automatically disconnected from the three-phase line source by the controller.

Controllers for electric motors are of two types—manual and magnetic. We shall consider only the magnetic type, which has many advantages over the manual type. It is easier to operate. It provides undervoltage protection. It can be remotely operated from one or several different places. Moreover, the magnetic controller is automatic and reliable, whereas the manual controller requires a trained operator, especially when a sequence of operations is called for in a given application. The one disadvantage is the greater initial cost of the magnetic type.

Appearing in Fig. 4-33 is the schematic diagram of a magnetic full-voltage starter for a three-phase induction motor. The operation is simple. When the start button is pressed, the relay coil M is energized. This moves the relay armature to its closed position, thereby closing the main contactors M, which in turn apply full voltage to the motor. When the relay armature moves to its energized position, it also closes an auxiliary contactor M_a which serves as an electrical interlock, allowing the operator to release the start button without de-energizing the main relay. Of course contactors M are much larger in size than M_a. The former set must have a current rating that enables it to handle the starting motor current. The latter needs to accommodate just the exciting current of the relay

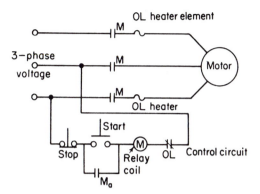

Figure 4-33 Full-voltage magnetic starter.

coil. Figure 4-33 also shows that the motor line current flows through two over-load heater elements. If the temperature rise becomes excessive, the heater element causes the overload contacts in the control circuit to open. In controller diagrams it is important to remember that all contactors are shown in their de-energized state. Thus the symbol ⊣ ⊢ means that the contactors M are open when the coil is not energized. Similarly, the symbol ⊣⊬ means that these contactors are closed in the de-energized state.

Undervoltage protection is inherent in the magnetic starter of Fig. 4-33. This comes about as a result of designing the coil M so that, if the coil voltage drops below a specified minimum, the relay armature can no longer be held in the closed position.

A full-voltage magnetic starter equipped with the control circuitry to permit reversing is illustrated in schematic form in Fig. 4-34. A three-phase induction motor is reversed by crossing two of the three line leads going to the motor terminals. In this connection note the criss-cross of two of the R contractors. Pressing the forward (FWD) button energizes coil F which in turn closes the main contactors F as well as the interlock F_a. This allows the motor to reach its forward operating speed. To reverse the motor the REV button is pushed. This

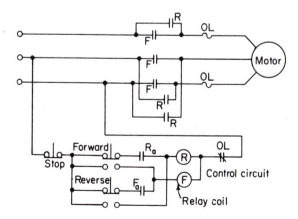

Figure 4-34 Full-voltage starter equipped with reversing control.

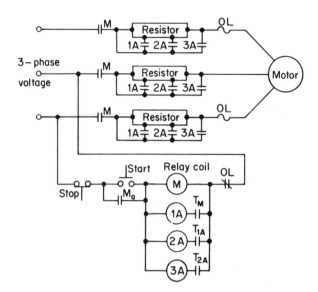

Figure 4-35 Three-step reduced-voltage starter for a three-phase induction motor.

does two things. One, it de-energizes coil F, thus opening the F contactors. Two, it energizes the R relay coil, thus closing the R contactors which apply a reversed-phase sequence to the motor causing it to attain full speed in the reverse direction. Putting the REV switch in the F_a interlock circuit is a safety measure which prevents having both the R and F contactors closed at the same time.

An illustration of a reduced-voltage magnetic controller using limiting resistors in the line circuit appears in Fig. 4-35. This unit is frequently referred to as a three-step acceleration starter because the line resistors are removed in three steps. Pushing the start button energizes coil M and closes contactors M, thus applying a three-phase voltage to the motor through the full resistors. In addition to contactors M and M_a coil M is also equipped with a time-delayed contactor T_M. This contactor is so designed that it does not close until a preset time *after* the armature of coil M is closed. The delay is usually obtained through a mechanical escapement of some sort which is actuated by the relay armature. Of course the time delay is needed to permit the motor to accelerate to a speed corresponding to the reduced applied voltage. After the elapse of the preset time, the T_M contacts close, energizing coil $1A$, which in turn closes contactors $1A$, short-circuiting the first part of the series resistor. Coil $1A$ is also equipped with a time-delay contactor T_{1A}, which is designed to allow the motor to accelerate to a higher speed before it closes. When contactors $1A$ do close, coil $2A$ is energized. This immediately closes contactors $2A$, shorting out the second section of the line resistor. Then, after still another time delay, contactor T_{2A} closes, applying an excitation voltage to coil $3A$. With the closing of contactors $3A$ the full line voltage is applied to the motor. In this manner the motor is brought up to speed in a "soft," smooth fashion without drawing excessive starting current or developing large starting torques.

PROBLEMS

4-1. A three-phase armature winding is shown in Fig. P4-1. Sinusoidal currents having an amplitude of 100 A flow in the three phases. Each coil consists of three turns. Carefully sketch to scale the actual mmf distribution along the air gap for a span of two poles for the indicated time instants t_1 and t_2.

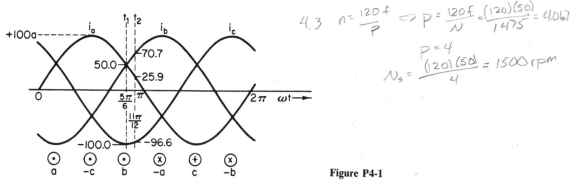

$$4.3 \quad n = \frac{120 f}{P} \Rightarrow P = \frac{120 f}{N} = \frac{(120)(50)}{1475} = 4.067$$

$$P = 4$$

$$N_s = \frac{(120)(50)}{4} = 1500 \, rpm$$

Figure P4-1

4-2. A 60-Hz polyphase induction motor runs at a speed of 873 rpm at full load. What is the synchronous speed? Find the frequency of the rotor currents.

4-3. A 50-Hz polyphase induction motor runs at a speed of 1475 rpm at full load. What is the synchronous speed of the motor? How many poles is it designed for?

4-4. State the general conditions under which it is possible for m alternating fields (with axes fixed in space) to yield a revolving field constant in amplitude and traveling at constant speed.

4-5. Determine the magnitude and direction of the resultant flux field in the machine configuration of Fig. 4-2 corresponding to time instants t_2 and t_4 in Fig. 4-1.

4-6. In the time variation of the phase currents depicted in Fig. 4-1 assume that the amplitude of phase b current is one-half that of phases a and c but that each is displaced by 120° from the other. Find the magnitude and direction of the resultant flux for the time instants t_1, t_3, and t_5, and compare with the results shown in Fig. 4-2.

4-7. An unbalanced three-phase mmf flows through three coils that are space displaced by 120° as shown in Fig. 4-2. The nature of the unbalance is such that phase b lags behind phase a by 90° instead of 120°. However, each phase has the same amplitude. Investigate the character of the resultant flux field corresponding to various instants of time.

4-8. Demonstrate that a resultant mmf that is described by an equation of the form

$$\mathcal{F}_r = \mathcal{F} \cos (\omega t + \alpha)$$

represents a wave that travels in the negative α direction. Keep in mind that t is a time variable and α is a space variable.

4-9. A balanced two-phase voltage is applied to identical coils, the axes of which are 90° apart. Derive the expression for the resultant mmf in terms of the mmf amplitude of either coil ($\mathcal{F}_m$), the angular frequency ω of the exciting currents, and the space variable α. Assume that each coil produces a sinusoidal field distribution.

4-10. A balanced, three-phase, 60-Hz voltage is applied to a three-phase, four-pole induction motor. When the motor delivers rated output horsepower, the slip is found to be 0.05. Determine the following:

    (a) The speed of the revolving field relative to the stator structure, which accommodates the exciting winding.

    (b) The frequency of the rotor currents.

    (c) The speed of the rotor mmf relative to the rotor structure.

    (d) The speed of the rotor mmf relative to the stator structure.

    (e) The speed of the rotor mmf relative to the stator field distribution.

    (f) Are the conditions right for the development of a net unidirectional torque? Explain.

4-11. Repeat Prob. 4-10 for the case where the rotor structure is blocked, thus preventing rotation in spite of the application of a balanced three-phase voltage to the stator.

4-12. Repeat Prob. 4-10 for the case where the slip has a value of 0.03.

4-13. Determine the no-load speed of a six-pole, wound-rotor, three-phase induction motor, the stator of which is connected to a 60-Hz line and the rotor of which is connected to a 25-Hz line, when:

    (a) The stator field and the rotor field revolve in the same direction.

    (b) The stator field and the rotor field revolve in opposite directions.

4-14. Determine the speed of operation of a four-pole, wound-rotor, three-phase induction motor the stator of which is connected to a 60-Hz line and the rotor of which is connected to a 50-Hz line when:

    (a) The stator field and the rotor field revolve in the same direction.

    (b) The stator field and the rotor field revolve in opposite directions.

4-15. Answer whether or not the following statements are true or false. When the statement is false, provide the correct answer.

    (a) In a three-phase induction motor the rotor ampere-conductor distribution is stationary with respect to the stator magnetic field distribution at *all* speeds of the rotor.

    (b) The load on the shaft of a three-phase induction motor is increased. This causes the speed of the rotor mmf relative to the stator mmf to change, thus permitting an increased input current to provide power balance.

4-16. The shaft output of a three-phase, 60-Hz induction motor is 75 kW. The friction and windage losses are 900 W, the stator core loss is 4200 W, and the stator copper loss is 2700 W. If the slip is 3.75%, what is the percent efficiency at this output?

4-17. Refer to Prob. 4-16. The rotor winding resistance referred to the stator is known to be 0.1 Ω. Determine the value of the stator-referred rotor current for the specified operating conditions.

4-18. A 15-hp, 220-V, three-phase, 60-Hz, six-pole, Y-connected induction motor has the following parameters per phase: $r_1 = 0.128$ Ω, $r_2' = 0.0935$ Ω, $x_1 + x_2' = 0.496$ Ω, $r_c = 183$ Ω, $x_\phi = 8$ Ω. The rotational losses are equal to the stator hysteresis and eddy-current losses. For a slip of 3%, find:

    (a) The line current and power factor.

    (b) The horsepower output.

    (c) The starting torque.

4-19. A 10-hp, 220-V, Y-connected, four-pole, 60-Hz, three-phase induction motor draws a line current of 26.2 A at a pf of 0.78 lag when it operates at a slip of 5%. Rotational losses amount to 250 W. The motor is known to have the following parameters expressed in ohms per phase:

$$r_1 = 0.3 \qquad x_1 = x_2' = 1.25 \qquad r_c = 150 \qquad x_\phi = 18$$

(a) Calculate the rotor current per phase referred to the stator.
(b) Find the value of output horsepower.
(c) Determine the efficiency.
(d) Find the developed torque in N-m.
(e) Calculate the starting line current at rated line voltage.

4-20. A three-phase, 60-Hz, six-pole, wound-rotor induction motor draws 10 kW when driving its normal load. It draws 700 W when the load is disconnected. The rotor and stator copper losses under normal load are 295 W and 310 W, respectively. Calculate the efficiency, the speed, and the shaft torque of this motor under normal load conditions. Assume equal rotational and core losses and negligible no-load copper losses.

4-21. A three-phase induction motor has a Y-connected rotor winding. At standstill the rotor induced emf per phase is 100 V rms. The resistance per phase is 0.3 Ω, and the leakage reactance is 1.0 Ω per phase.
(a) With the rotor blocked, what is the rms value of the rotor current? What is the power factor of the rotor circuit?
(b) When the motor is running at a slip of 0.06, what is the rms value of the rotor current? What is the power factor of the rotor circuit?
(c) Compute the value of the developed power in part (b).

4-22. A three-phase, 12-pole, 60-Hz, 2200-V induction motor runs at no-load with rated voltage and frequency impressed and draws a line current of 20 A and an input power of 14 kW. The stator is Y-connected, and its resistance per phase is 0.4 Ω. The rotor resistance r_2' is 0.2 Ω per phase. Also, $x_1 + x_2' = 2.0$ Ω per phase. The motor runs at a slip of 2% when it s delivering power to a load. For this condition compute:
(a) The developed torque.
(b) The input line current and power factor.

4-23. A three-phase, 440-V, 60-Hz, Y-connected, eight-pole, 100-hp induction motor has the following parameters expressed per phase:

$$r_1 = 0.06 \ \Omega, \qquad x_1 = x_2' = 0.26 \ \Omega$$

$$r_2' = 0.048 \ \Omega \qquad r_c = 107.5 \ \Omega$$

$$x_\phi = 8.47 \ \Omega$$

The rotational losses are 1600 W. Using the approximate equivalent circuit determine for $s = 0.03$:
(a) The input line current and power factor.
(b) The efficiency.

4-24. A three-phase, 335-hp, 2000-V, six-pole, 60-Hz, Y-connected squirrel-cage induction motor has the following parameters per phase that are applicable at normal slips:

$$r_1 = 0.2 \ \Omega \qquad x_1 = x_2' = 0.707 \ \Omega$$

$$r_2' = 0.203 \ \Omega \qquad r_c = 450 \ \Omega$$

$$x_\phi = 77 \ \Omega$$

The rotational losses are 4100 W. Using the approximate equivalent circuit, compute for a slip of 1.5%:
   (a) Line power factor and current.
   (b) Developed torque.
   (c) Efficiency.

4-25. A six-pole, three-phase, 40-hp, 60-Hz, induction motor has an input when loaded of 35 kW, 51 A, 440 V, and a speed of 1152 rpm. When uncoupled from the load, the readings are found to be: 440 V, 21.3 A, 2.3 kW, and 1199 rpm. The resistance measured between terminals for the stator winding is 0.25 Ω for a Y-connection. The stator core losses and the rotational losses are known to be equal. Determine:
   (a) The power factor of the motor when loaded.
   (b) The motor efficiency when loaded.
   (c) The horsepower rating of the load.

4-26. Answer each part briefly:
   (a) What is the effect of doubling the air gap of an induction motor on the magnitude of the magnetizing current and of the maximum value of the flux per pole? Neglect the effect of the leakage impedance.
   (b) Describe the effect of a reduced leakage reactance on the maximum developed torque, the power factor at full load, and the starting current of a three-phase induction motor.

4-27. The manufacturers' specifications concerning the ratings of induction motors invariably list the maximum (or breakdown) torque as a percentage of rated torque. Its value is usually 200%. Explain the significance of choosing so high a value.

4-28. In a given situation only 220-V, 30-Hz, three-phase service is available. A plant manager has an opportunity to buy cheaply a 10-hp, 440-V, three-phase, 60-Hz, four-pole squirrel-cage motor, which is to be used to supply power to a constant torque load. The load torque corresponds to that of the 10-hp motor.
   (a) Can the motor be used? Justify your answer.
   (b) If so, what would be its rating?
   (c) How would the efficiency of operation be affected?
   (d) Find the approximate change in starting torque.

4-29. Would you recommend use of a 115-V, 60-Hz, three-phase induction motor on a 460-V, 400-Hz system to deliver power to a constant torque load? Explain.

4-30. When a 10-hp, 220-V, 60-Hz, three-phase induction motor operates at 220 V, 50 Hz delivering the same constant rated torque, what change occurs in the rotor current?

4-31. In what way, if any, does a reduced applied voltage affect the breakdown torque of a three-phase induction motor? In what way, if any, does an increased frequency affect the breakdown torque?

4-32. A 115-V, three-phase, 60-Hz induction motor is used on a 115-V, 50-Hz source. Assuming that the delivered torque is to be the same when operating at either frequency, determine:

(a) The change in operating flux.

(b) The change in rotor current.

(c) The change in synchronous speed.

(d) Would you permit this motor to be used continuously on the 50-Hz source? Explain.

4-33. Refer to the motor of Prob. 4-19.

(a) Compute the slip at which maximum torque occurs. $.096$

(b) Determine the value of the maximum torque.

4-34. Refer to the motor of Prob. 4-19.

(a) For the operating condition described determine the missing parameter of the exact equivalent circuit: namely, the stator-referred rotor resistance per phase.

(b) Using the exact equivalent circuit obtain the speed at which maximum torque occurs.

(c) Find the stator-referred rotor current at maximum torque. How does this differ from the input stator current?

(d) Compute the value of the maximum torque.

4-35. A three-phase, Y-connected, 440-V, 200-hp induction motor has the following blocked-rotor data: $P_b = 10$ kW, $I_b = 250$ A, $V_b = 65$ V, and $r_1 = 0.02$ Ω. Find the value of the rotor resistance referred to the stator. $-.033 .\text{r}$

4-36. A six-pole, 60-Hz, three-phase, Y-connected wound-rotor (three phases also) induction motor has a standstill induced rotor voltage of 130 V per phase. At short-circuit with the rotor blocked, this voltage produces a current of 80 A at a power factor of 0.3 lagging. At full-load the motor runs at a slip of 9%. Find the full-load developed torque.

4-37. A 500-hp, three-phase, 2200-V, 25-Hz, 12-pole, Y-connected wound-rotor induction motor has the following parameters: $r_1 = 0.225$ Ω, $r_2' = 0.235$ Ω, $x_1 + x_2' = 1.43$ Ω, $x_\phi = 31.8$ Ω, $r_c = 780$ Ω. A no-load and a blocked-rotor test are performed on this machine.

(a) With rated voltage applied in the no-load test, compute the readings of the line ammeters as well as the total wattmeter reading.

(b) In the blocked-rotor test the applied voltage is adjusted so that 228-A line current is made to flow in each phase. Calculate the reading of the line voltmeter and the total wattmeter reading.

4-38. A three-phase, 2000-V, Y-connected wound-rotor induction motor has the following no-load and blocked-rotor test data:

No-load:	2000 V,	15.3 A,	10.1 kW
Blocked-rotor:	440 V,	170.0 A,	36.4 kW

The resistance of the stator winding is 0.22 Ω per phase. The rotational losses are equal to 2 kW. Calculate all the necessary data for the approximate equivalent circuit at a slip of 2%, and draw the circuit showing all parameter values.

4-39. For the machine of Prob. 4-37 compute the following:

    (a) the slip at which maximum torque occurs.

    (b) The input line current and power factor at the condition of maximum torque.

    (c) The value of the maximum torque.

4-40. Refer to the motor of Prob. 4-37, and find the value of resistance that must be externally connected per phase to the rotor winding in order that the maximum torque be developed at starting. What is the value of this torque?

4-41. A three-phase, 440-V, 50-hp, 60-Hz, Y-connected induction motor yields the following no-load test data:

Dc test on stator per phase:	5 V,	50 A
Blocked-rotor test at rated frequency:	106 V,	65 A, 3423 W, 60 Hz
Blocked-rotor test at low frequency:	42 V,	65 A, 2910 W, 15 Hz
No-load test:	440 V,	16.3 A, 2157 W, 60 Hz

    The friction and windage losses for this motor are 834 W. Compute the parameters of the approximate equivalent circuit to be used at normal operating slips.

4-42. From the no-load test on a 10-hp, four-pole, 230-V, 60-Hz, three-phase, Y-connected induction motor with rated voltage applied, the no-load current was found to be 9.2 A and the corresponding input power 670 W. Also, with 57 volts applied in the blocked-rotor test, it was found that the motor took 30 A and 950 W from the line. The stator winding resistance was measured to be 0.15 Ω per phase. When this motor is coupled to its mechanical load, the input to the motor is found to be 9150 W at 28 A and a power factor of 0.82. The stator core losses are equal to the rotational losses.

    (a) Compute the rotor current referred to the stator.

    (b) Find the developed torque.

    (c) What is the value of the slip?

    (d) At what efficiency is the motor operating?

4-43. A Y-connected, wound-rotor induction motor is found to have the following parameters per phase:

$$r_1 = 0.26\ \Omega, \qquad x_1 = 0.6\ \Omega, \qquad r_c = 143\ \Omega$$

$$r_2' = 0.40\ \Omega, \qquad x_2' = 1.4\ \Omega, \qquad x_\phi = 22.2\ \Omega$$

    (a) If rated voltage of 220 V is applied to the motor at no-load, compute the line current and power factor. Assume the stator core losses are equal to the rotational losses.

    (b) If the rotor is short-circuited and blocked and a line voltage of 60 V is applied, find the power input in watts.

4-44. A three-phase induction motor has the following data: 20 hp rated, 220 V, Y-connected, four-poles, 60 Hz. Stator effective resistance is 0.118 Ω per phase; rotor effective resistance referred to the stator is 0.102 Ω per phase. Equivalent stator referred reactance is 0.208 Ω per phase. Total core loss is 408 W, of which one-half is for the rotor. Friction and windage is 400 W.

    (a) What is the starting torque at full voltage?

    (b) If the motor is driven at 1800 rpm against the revolving field, what is the brake effect obtained in watts at the shaft? Assume that rotor resistance and core loss remain unchanged.

4-45. A three-phase wound-rotor induction motor delivers fixed torque to a load. The rotor leakage reactance is negligible. For this load condition the rotor resistance drop is found to be 20 V. Determine the rotor injected voltage needed to produce twice this torque at 10 times the slip.

4-46. A 625-V (line voltage), three-phase, Y-connected wound-rotor induction motor has a ratio of transformation from stator to rotor of 3.0. At a specified load condition with the rotor short-circuited, the motor operates with a slip emf (i.e., sE_2) of 6 V per phase.

  Find the value of the injected voltage needed to increase the slip to a value of 40%. Assume that the torque varies in a manner proportional to the square root of the slip. Neglect rotor leakage reactance and rotational losses.

4-47. A four-pole, three-phase, wound-rotor induction motor is running at 1700 rpm with the rotor short-circuited and rated voltage at rated frequency applied to the stator winding. The rotor slip emf (sE_2) is 10 V per phase and the rotor current is 40 A per phase. An injected voltage of slip frequency and constant rms value of 80 V per phase is then introduced. This voltage is in phase opposition to the normal slip emf. Rotor leakage reactance is negligible.

  (a) What is the new motor speed if the torque required by the load is constant as the speed varies?

  (b) Repeat part (a), assuming that the torque varies directly with the speed.

4-48. A three-phase wound-rotor induction motor delivers rated torque and draws a line current of 100 A at a pf of 0.707. The effective turns ratio from stator to rotor is 1.41 and the rotor resistance per phase is 0.1 Ω. The rotor is wound for three phases. Assume that the stator core loss, stator winding resistance, and stator and rotor leakage reactances are all negligible.

  (a) Calculate the value of rotor injected voltage needed to increase the line pf to 90% lagging.

  (b) Find the power supplied by the source of injected voltage.

4-49. A 60-Hz, six-pole, three-phase induction motor has its speed adjusted by a rotor injected voltage. When it develops rated torque, the motor slip is 5%, and the normal slip frequency voltage is 20 V. The rotor resistance per phase is 0.1 ohm and the rotor leakage reactance is negligible. An injected voltage of 100 V per phase is inserted into the rotor in phase with the original sE_2 voltage. If the torque demand remains unchanged, determine:

  (a) The new value of slip.

  (b) The new operating speed in rpm.

  (c) The power supplied by the injected voltage source.

4-50. A three-phase, 1000-hp, 2200-V, 25-Hz, 12-pole, wound-rotor induction motor has a ratio of transformation from stator to rotor of 22 to 15. Both stator and rotor windings are Y-connected. At a slip of 2% the motor develops its rated torque and draws from the line a current of 180 A at a pf of 0.866. Consider the quantities r_1, x_1, x_2, and r_c negligible. The motor is to develop the same torque at a slip of 10%, while simultaneously operating at unity pf, by means of a rotor injected voltage. Determine the magnitude and phase of the injected voltage relative to the original slip emf.

4-51. A three-phase, Y-connected, wound-rotor induction motor delivers rated torque at a slip of 4%, and draws a line current of 100 A at a pf of 80%. The effective turns ratio from stator to rotor is unity, and the rotor is wound for three phases. The rotor resistance per phase is 0.1 Ω, and the rotor standstill emf is 200 V per phase. For simplicity assume that $r_1 = x_1 = x_2' = 0$.

(a) Find the rotor injected voltage needed to give the same torque at a slip of 20% below synchronous speed.

(b) Calculate the rotor injected voltage required to make the line pf unity.

4-52. A three-phase induction motor with slip rings delivers power to a constant torque load. The rotor current is 40 A per phase; the rotor resistance is 0.4 Ω per phase. To bring the pf to unity, a voltage is injected into the rotor whose magnitude in this case is equal to the normal slip emf induced in the rotor at this particular slip. The speed of the rotor is now to be decreased so that the slip is five times greater than before. The torque remains fixed. What must now be the magnitude and phase of the injected voltage to produce unity pf?

4-53. A three-phase, wound-rotor induction motor delivers rated torque at a slip of 2%. At this slip the value of the rotor induced emf is 10 V. The rotor resistance per phase is 0.2 Ω per phase. If it is desirable to produce *twice* rated torque at the same slip of 2%, determine:

(a) The magnitude, phase, and frequency of the voltage that must be injected into the rotor.

(b) The amount of power that the auxiliary source must supply to the induction motor.

Note: At the low slip of 2% the rotor leakage reactance may be neglected.

4-54. A 1000-hp, three-phase, 2200-V, 25-Hz, 12-pole, wound-rotor induction motor has the following parameters per phase:

$$r_1 = 0.093 \ \Omega, \qquad x_1 = 0.295 \ \Omega$$

$$r_2 = 0.072 \ \Omega, \qquad x_2 = 0.15 \ \Omega$$

stator to rotor effective turns ratio: 22/15.

When operating at a slip of 1.72% this machine develops a torque of 15,000 lb-ft, corresponding to a rotor current per phase of 205 A. The friction and windage losses total 6 kW.

(a) Determine the magnitude and phase of the minimum value of injected rotor voltage needed to change the slip to 50% above synchronous speed at constant torque. Do not neglect rotor leakage reactance.

(b) Determine the mechanical power output of the rotor in part (a).

(c) Determine the rotor electrical power "output" in part (a).

(d) Find the total rotor copper loss. How is this loss supplied?

4-55. In the configuration of Fig. P4-55 balanced three-phase currents of 20 Hz variation are fed to the brushes of the dc winding. The rotor rotates at a frequency of 35 Hz. What is the frequency of the emf induced at the brushes? Why?

4-56. The resistance measured between each pair of slip rings of a three-phase, 60-Hz, 300-hp, 16-pole induction motor is 0.04 Ω. With the rings short-circuited, the full-load slip is 0.025, and it may be assumed that the slip-torque curve is a straight line from no-load to full-load. The motor drives a fan which requires 300 hp at the full-load speed of the motor. The torque required to drive the fan varies as the square of the speed. What value of resistance should be connected in series with each slip ring so that the fan will run at 300 rpm?

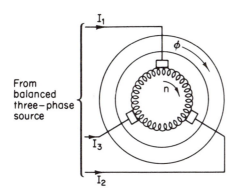

From
balanced
three-phase
source

Figure P4-55

4-57. A three-phase, 60-Hz, four-pole, wound-rotor induction motor has a rated speed of 1746 rpm. The actual rotor resistance per phase is 0.8 Ω. What starting resistance must be inserted in the rotor circuit so that rated torque will be developed at starting? Neglect any change of rotor resistance with frequency.

4-58. For the polyphase induction motor, sketch the variation of torque as a function of horsepower output over the speed range from zero to synchronous speed. *Hint:* Apply the circle diagram.

4-59. Continue the analysis displayed in Table 4-1 by completing the tabulation for the remaining three 60° intervals of the cycle.

4-60. Draw the line-to-neutral voltage variations for each phase over one complete cycle for the speed control system of Fig. 4-25 corresponding to the switching sequence displayed in Fig. 4-26.

4-61. In the interest of generating a more nearly sinusoidal waveform for the line voltages in the speed control system of Fig. 4-25, the gated ON interval for each SCR is specified as 120°. Draw the switching sequence diagram for the inverter.

4-62. **(a)** Use the results of Prob. 4-61 and sketch the variation of each line-to-neutral voltage over one cycle of the control frequency.

   (b) Sketch the variations of the line-to-line voltages over one complete cycle of the control frequency.

   (c) Determine the expression for the rms value of the line-to-line voltages and compare it with the result that appears in Eq. (4-56).

5

Three-Phase
Synchronous Generators

The synchronous generator is universally used by the electric power industry for supplying three-phase as well as single-phase power to its customers. The single-phase power that is brought to homes, shops, and offices originates from one phase of the three-phase system. Moreover, the assignment of commercial load circuits to each phase is made in an effort to keep the phases balanced.

The basic construction features of these machines are described in Sec. 3-3 and illustrated in Fig. 3-17. It is worth noting here that synchronous generators are classified in two types. The first is the *low-speed* (engine- or water-driven) type, which is characterized physically by having salient poles, a large diameter, and small axial length. The second is the *turbogenerator*, which uses the steam turbine as the prime mover. The usual salient-pole rotor construction is abandoned in these high-speed generators in favor of the cylindrical (or smooth) rotor because the protruding-pole construction gives rise to dangerously high mechanical stresses. For 60-Hz three-phase power the two-pole generator must operate at 3600 rpm. Moreover, since turbogenerators are invariably designed for two poles, it follows that these machines are further characterized by a small diameter and long axial length.

5-1 GENERATION OF A THREE-PHASE VOLTAGE

How is a three-phase voltage generated? This can be readily understood from a study of Fig. 5-1(a). Depicted here is a two-pole rotor, the field winding of which is assumed to be energized from a dc source to create the pole flux. It is further assumed that the pole pieces are shaped to produce a sinusoidal flux field. Appearing in the stator is a balanced three-phase winding with the axis of each phase displaced by 120°. The complete winding of each phase is represented in Fig. 5-1 by a single coil. Now consider that the rotor is driven by the prime mover in a counterclockwise direction at synchronous speed. Applying the $\bar{v} \times \bar{B}$ rule for the field direction shown reveals that the instantaneous voltage induced in coil sides a, b', c' is directed out of the paper while in coil sides a', b, c it is directed into the paper. Furthermore, since coil sides a and a' are located beneath the maximum value of the flux-density wave, the induced emf in phase a is at its maximum value. The corresponding induced emf in phase b, as identified by coil side b, is seen to be of opposite sign (it is under a south-pole flux) and of smaller magnitude. In fact, since coil sides b and b' are both displaced from the maximum flux-density position by 60°, it follows that this instantaneous voltage is at one-half (cos 60°) its maximum value. A similar reasoning yields the same result for phase c.

Next consider that the rotor has advanced by 60° in the counterclockwise direction. This puts the amplitude of the north-pole flux directly beneath b', indicating that the induced emf is now a negative maximum in phase b. Also, coil c-c' now finds itself under the influence of flux of reversed polarity as indicated in Fig. 5-1(b). Hence the direction of the emf in c is now out of the paper and into the paper for c'. The instantaneous values of the three-phase voltages for the first and second time instants are depicted in Fig. 5-2 and are identified as t_1 and t_2, respectively.

By repeating the foregoing procedure many times and plotting the results for each instant of time we obtain the complete curves of Fig. 5-2. Note that the induced-emf variation for each phase is identical with the other two except for time displacements of 120° and 240°, respectively. As a matter of fact this variation is a direct consequence of having the beginning of each phase space-displaced

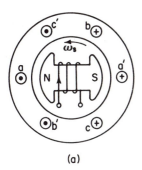

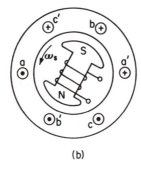

(a)   (b)

Figure 5-1 Induced-voltage distribution in the stator winding of a synchronous generator: (a) phase a at a positive maximum (time t_1 in Fig. 5-2); (b) phase b at a negative maximum (time t_2).

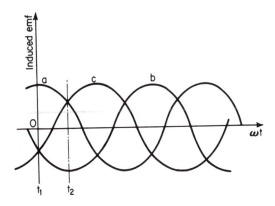

Figure 5-2 Time history of three-phase induced emf's for the generator of Fig. 5-1.

by 120 electrical degrees. Physical connection of the ends of each phase (points a', b', c') gives a Y-connected stator winding.

5-2 LINEAR ANALYSIS BY THE GENERAL METHOD

The basic principles and characteristics of the cylindrical-pole synchronous generator can be readily described by developing the phasor diagram as was done for the transformer. However, there exists an important distinction that must be understood right at the start. The transformer (as also the three-phase induction motor) is essentially a constant-flux device, i.e., the amplitude of the resultant operating air-gap flux is fixed by the applied ac voltage. In contrast the resultant air-gap flux in the synchronous machine depends upon the magnitude of the exciting current in the field winding (located on the rotor structure) as well as on the value of the armature (or load) current.† Recall that for the synchronous machine the armature winding is the stator winding. Consequently, when analyzing the synchronous generator, it becomes necessary to introduce the magnetization curve that applies for its magnetic circuit. If the design data are available, the magnetization curve can be computed by following the procedure outlined in Chapter 1. Otherwise, if the machine is available in the laboratory, the magnetization curve can be found by performing a simple *open-circuit test*. The wiring diagram for such a test is shown in Fig. 5-3. The rotor is driven at synchronous speed throughout the test by the prime mover. The dc field current is varied by means of the rheostat and at each setting the corresponding line-to-neutral voltage is measured. A plot of these data points is depicted as curve (a) in Fig. 5-4 and is often called the *no-load characteristic*. The straight-line portion of this curve is called the *air-gap characteristic*. In this section our interest is confined to *linear*

---

†Here we are treating the synchronous machine as an isolated unit. If it were assumed to be connected to an infinite bus system (i.e., in parallel with other synchronous generators many times its size), the air-gap flux would be invariant.

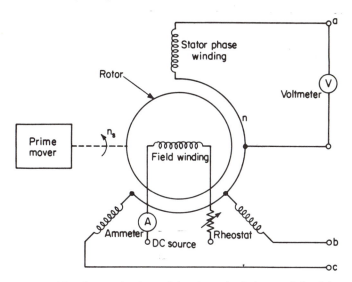

Figure 5-3   Wiring diagram for determining the no-load characteristic of the synchronous generator.

analysis so that for our immediate purposes the magnetization curve to be used will be curve (b) of Fig. 5-4.

Another useful characteristic in connection with synchronous generator analysis is the *short-circuit characteristic*. This is found by short-circuiting terminals *a, b,* and *c* in the diagram of Fig. 5-3, driving the rotor at synchronous speed, and measuring the short-circuit line currents for various values of field current. A plot of the short-circuit line current versus field current leads to curve (c) of Fig. 5-4. This characteristic is linear because the amount of air-gap flux needed to produce even as much as twice rated armature current is so small as to cause little or no saturation of the magnetic-circuit iron. Only enough flux is needed to produce an emf which overcomes the leakage impedance drop.

Before proceeding with a description of the phasor diagram, a few remarks are in order at this point concerning the inclusion in one phasor diagram of both the time-varying quantities (such as voltage and current) and the mmf space sinusoids. Figure 5-5 depicts a two-pole turbogenerator. The field winding is located on the rotor and is distributed in suitably placed slots. When a direct current is made to flow through this winding, an ampere-conductor distribution is produced. The illustration shows the current moving into the conductors on the right side, and flowing out of the conductors on the left side. The corresponding mmf produced by this field winding has a sinusoidal distribution in space with its axis directed along the vertical. Because the maximum value of the flux density lies along this vertical too, it follows that the voltage induced in coil e_f-e_f' is a maximum value. Moreover, application of the $\bar{v} \times \bar{B}$ rule indicates that this emf is directed out of the paper for the top conductor and into the paper for the bottom

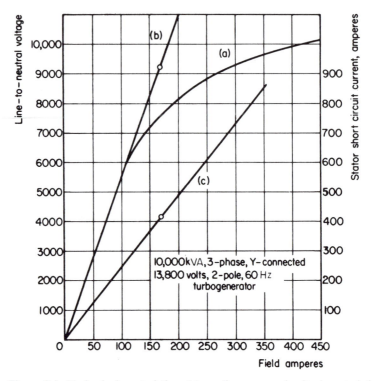

Figure 5-4 No-load characteristics: (a) nonlinear open-circuit characteristic; (b) linear air-gap characteristic; (c) short-circuit characteristic.

conductor. Although it is assumed that the stator is equipped with a three-phase winding, essentially coil e_f-e_f' can be considered to represent the emf induced in this entire winding for the instant shown. Accordingly, if one were to look at the emf induced in each coil side of the entire three-phase winding, a sinusoidal emf distribution would result, with its peak lying along the vertical. As a matter of convenience, therefore, it can be said that insofar as the armature induced emf is concerned, the entire three-phase winding can be represented by the single coil e_f-e_f'. Note that the axis of this coil lies along the horizontal. In general, the axis of the coil that represents the emf distribution in the three-phase winding is always 90 electrical degrees displaced from the axis of the field mmf $\mathscr{F}_F$. As a result of this quadrature relationship it will be customary for us in the phasor diagrams which follow to locate the field mmf $\mathscr{F}_F$ at a position 90 degrees leading the direction of the excitation emf phasor. The excitation emf is the voltage induced by the field winding flux alone. It is helpful to note in Fig. 5-5 that when the emf in coil e_f-e_f' is a maximum, the flux linkage is zero.

A similar convention is adopted concerning the phasor representation of the armature current and the corresponding mmf associated with this current. For example, in Fig. 5-5, coil i_a-i_a' is used to denote the sinusoidal ampere-conductor

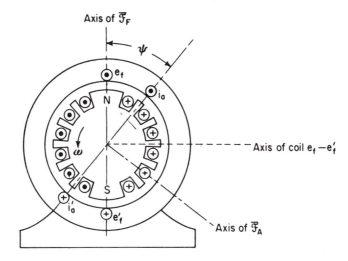

Figure 5-5 Diagram illustrating the relationship between mmf space sinusoids and the associated time-varying quantities of excitation voltage and armature current. The rotor is assumed to be driven counterclockwise at ω radians per second.

distribution of the entire three-phase armature winding for a phase lag relative to the excitation emf of ψ degrees. Note that the axis of this coil coincides with the location of the peak value of the corresponding sinusoidal armature mmf $\mathcal{F}_A$ and is space displaced from the field mmf axis $\mathcal{F}_F$ by $(90° + \psi)$. Another point worthy of note is that as the phase angle between armature current and excitation emf changes, a corresponding change occurs in the axis of $\mathcal{F}_A$. This fact is noted in subsequent phasor diagrams by placing the armature mmf coincident with the phasor for armature current.

In synchronous machine analysis it is important to note that the two space sinusoids $\bar{\mathcal{F}}_F$ and $\bar{\mathcal{F}}_A$, having the same frequency of rotation and the same number of poles, can be added to yield a resultant space sinusoid denoted by $\bar{\mathcal{F}}_R$, which in turn is responsible for generating the resultant induced emf in the armature winding $\bar{E}_r$. As a convenience, the result of this addition is also represented in the phasor diagram in spite of the fact that these mmf's are space sinusoids and not time-varying sinusoids. This procedure is permissible because space sinusoids can also be summed by phasor methods.

One final point. Application of the force rule $(\bar{I} \times \bar{B})$ discloses that an average torque is produced on the armature (stator) winding to cause counterclockwise rotation. However, because the stator is immobile, there occurs instead a clockwise torque on the rotor. The latter represents a counter-torque to the prime mover of the turbogenerator.

The development of the phasor diagram is done on a per-phase basis. Assume that the armature winding resistance r_a and leakage reactance x_l per phase are known along with the air-gap characteristic. Assume too that the synchronous generator is delivering rated armature current to a lagging power-factor load at rated terminal voltage. Call the rated terminal voltage per phase V_t and the load power-factor angle θ. Let it be desired that the value of the field current corresponding to this load condition should be found. Before proceeding, it is helpful to keep in mind the factors that must be reckoned with in establishing the rated terminal voltage when rated current flows. It is shown in Sec. 4-1 that the existence of three-phase currents in a three-phase winding leads to a revolving mmf $\bar{\mathscr{F}}_A$. It follows then that the resultant mmf $\bar{\mathscr{F}}_R$ is made up of the vector sum of the field mmf $\bar{\mathscr{F}}_F$ and the armature mmf $\bar{\mathscr{F}}_A$. In equation form we have

$$\bar{\mathscr{F}}_F + \bar{\mathscr{F}}_A = \bar{\mathscr{F}}_R \qquad (5\text{-}1)$$

In turn, this resultant mmf is responsible for the resultant air-gap flux Φ_r that induces the actual armature winding voltage $\bar{E}_r$. The terminal voltge is then obtained by subtracting the armature leakage impedance drop per phase from $\bar{E}_r$.

We shall now develop the phasor diagram by starting with the known terminal conditions: rated armature current I_a lagging the rated terminal voltage V_t by θ degrees. Refer to Fig. 5-6. The induced armature winding voltage per phase is obtained as the phasor sum of $\bar{V}_t$ and the armature leakage impedance drop. Thus, employing the horizontal axis as the zero reference line for angle measurements, we can compute

$$\bar{E}_r = \bar{V}_t + \bar{I}_a(r_a + jx_l)$$

or

$$E_r\underline{/\gamma} = V_t\underline{/0°} + I_a\underline{/-\theta}\,(r_a + jx_l) \qquad (5\text{-}2)$$

The quantity R that produces E_r is found by entering the air-gap characteristic. Note that R is expressed in field amperes rather than field ampere-turns $\bar{\mathscr{F}}_R$. This is irrelevant here in view of our objective, which is to find the required field current for the specified load condition.

The complete expression for the field current representing the resultant flux is then

$$\bar{R} = R\underline{/90° + \gamma} \qquad (5\text{-}3)$$

where γ is determined from the right side of Eq. (5-2). It is convenient now to write the counterpart of Eq. (5-1) in field amperes, which after all is the quantity that usually appears as the abscissa quantity in the experimentally established curves of Fig. 5-4. Thus

$$\bar{F} + \bar{A} = \bar{R} \qquad (5\text{-}4)$$

where $\bar{A}$ denotes the armature mmf expressed in *equivalent field amperes*.

A glance at Eq. (5-4) shows that with $\bar{R}$ available $\bar{F}$ can be determined once $\bar{A}$ is known. There are two ways to determine $\bar{A}$, either from the design data or from appropriate no-load tests. The latter approach is discussed in Sec. 5-4.

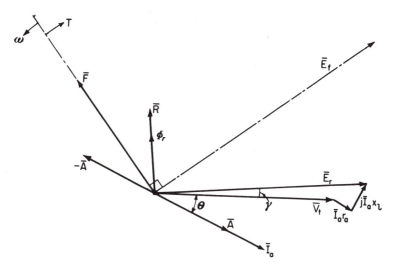

Figure 5-6 Phasor diagram of a synchronous generator for a lagging power-factor load. The general method of analysis is indicated.

When the design data are available, Eq. (3-24) may be used to represent the armature mmf per pole. Keep in mind that this equation is derived (see Appendix C) as the amplitude of the fundamental component of the mmf distribution of the armature winding. Also, the distribution of the field winding on the cylindrical-rotor surface is such that it leads practically to a sinusoidal mmf wave having a peak mmf per pole of $N_f I_f$, where N_f is the number of turns per pole. Accordingly, since the armature mmf and the field mmf waves are both of the same shape, it follows that the actual armature mmf as described by Eq. (3-24) can be replaced by an equivalent field winding current A. Thus

$$A = \frac{1}{N_f} 0.9 \, q \, \frac{N_2 K_{w2}}{p} I_a \qquad \text{equivalent field amperes} \qquad (5\text{-}5)$$

where q = number of phases
$\quad p$ = number of poles
$N_2 K_{w2}$ = effective armature winding turns per phase
$\quad I_a$ = armature current per phase.

Equation (5-5) indicates that A and I_a differ by a fixed relationship as already described. Hence the phasor expression for the equivalent field current representing the armature mmf for lagging power-factor load is

$$\bar{A} = A \underline{/-\theta} \qquad (5\text{-}6)$$

where A is described by Eq. (5-5). In the phasor diagram of Fig. 5-6 note that $\bar{A}$ and $\bar{I}_a$ are shown on the same line as a matter of convenience.

With $\bar{R}$ and $\bar{A}$ determined, the required field current is found to be

$$\bar{F} = \bar{R} - \bar{A} = R\underline{/90° + \gamma} - A\underline{/\theta} \qquad (5\text{-}7)$$

The quantity $\bar{E}_f$ appearing in Fig. 5-6 and shown $90°$ behind $\bar{F}$ is the voltage produced by $\bar{F}$ at the armature terminals with the load removed (i.e., $I_a = 0$) and no saturation of the iron. It is usually called the excitation voltage. The foregoing procedure for finding the field current is referred to as the *general method*.

The positive direction of rotation for the phasor quantities and the space sinusoids depicted in Fig. 5-6 is counterclockwise and is indicated by the direction of the arrow marked ω. It was also previously pointed out that for generator action a counter torque exists on the rotor. This is indicated by the arrow marked T in Fig. 5-6. It is useful to note here that in generator action the torque on the field structure is always such as to bring the field axis in alignment with the resultant flux.

Example 5-1

A 10,000-kVA, three-phase, Y-connected, two-pole, 60-Hz, 13,800-V, line-to-line turbogenerator has the air-gap characteristic depicted in Fig. 5-4. The armature winding resistance is 0.07 Ω per phase and the armature winding leakage reactance per phase is 1.9 Ω. Also, the armature mmf at rated current is known to be equal to 155 equivalent field amperes. Find the field current that produces rated terminal voltage when rated armature current is delivered to a 0.8 lagging power-factor three-phase balanced load.

Solution   The rated phase voltage is

$$V_t = \frac{13,800}{\sqrt{3}} = 7967.4 \text{ V} \qquad (5\text{-}8)$$

and the rated armature current is

$$I_a = \frac{10,000}{\sqrt{3}\ 13.8} = 418.4 \text{ A} \qquad (5\text{-}9)$$

The actual armature induced emf is found from Eq. (5-2) to be, $\cos\theta = .8$ lagging

$$\bar{E}_r = E_r\underline{/\gamma} = 7967.4\underline{/0°} + 418.4\underline{/-36.8°}\ (0.07 + j1.9) \qquad (5\text{-}10)$$

Note that r_a is small enough to be neglected. Thus,

$$\bar{E}_r = 7967.4 + 418.4(0.8 - j0.6)j1.9$$

$$= 7967.4 + 477 + j636$$

$$= 8444.4\underline{/4.3°} \qquad (5\text{-}11)$$

The corresponding value of R from the air-gap characteristic is 157 A. Hence by Eq. (5-3)

$$\bar{R} = 157\underline{/90° + 4.3°} = 157\underline{/94.3°} \qquad (5\text{-}12)$$

Equation (5-7) then yields the desired result:

$$\bar{F} = \bar{R} - \bar{A} = 157 \underline{/94.3°} - 155 \underline{/-36.8°}$$

$$= -11.7 + j156.6 - 124 + j92.8$$

$$= -135.79 + j249.1 = 284 \underline{/119°} \text{ A} \qquad (5\text{-}13)$$

The excitation voltage produced by this field current is found on the air-gap line to be

$$\bar{E}_f = 15,300 \underline{/29°} \text{ V} \qquad (5\text{-}14)$$

This quantity is almost twice the rated terminal voltage. It is a highly exaggerated result because of the assumption of no saturation. Practical and useful results are obtained when this assumption is dropped as described in Sec. 5-4.

5-3 LINEAR ANALYSIS BY THE SYNCHRONOUS REACTANCE METHOD. THE EQUIVALENT CIRCUIT

In any treatment of the behavior of synchronous machines it is necessary to take due account of the effects produced by the current that flows in the armature winding—frequently referred to as *armature reaction*. In the preceding section this was accomplished by working directly with the armature winding mmf expressed in equivalent field amperes. A second approach is possible by treating the armature reaction in terms of a suitable flux and its associated induced voltage. Since the assumption of no saturation has been made, it is to be expected that either viewpoint will lead to the same result. As a matter of fact this is precisely the reason for imposing the linearity condition. In this way the introduction of new concepts can readily be correlated with preceding ideas with consistency.

By means of superposition the effect of the rotating armature mmf can be treated individually. Accordingly, the set of revolving mmf poles created by the three-phase armature current in the three-phase winding can be considered to produce an armature reaction flux Φ_{ar}. Of course this flux bears a fixed relationship to its mmf $\bar{A}$. Moreover, it revolves with respect to the stationary armature winding at a speed corresponding with 60 Hz. Hence Φ_{ar} can be said to induce a voltage (rise) of value E_{ar} which can be placed 90° behind Φ_{ar} as depicted in Fig. 5-7(a). In situations of this kind it is customary to replace such an armature flux voltage by a reactance drop that involves the associated current (I_a in this case) directly. Recall that this same technique was used to replace the effect of leakage flux in the transformer as well as armature leakage flux occurring around the slots of the stator windings. Expressed mathematically this is

$$\boxed{\bar{E}_{ar} \equiv -j\bar{I}_a x_\phi} \qquad (5\text{-}15)$$

where x_ϕ is the *armature reaction reactance*. The $-j$ factor accounts for the 90° lag existing between $\bar{I}_a$ and $\bar{E}_{ar}$. Furthermore, the linearity condition enables a constant value of x_ϕ to replace the effect of armature flux. Thus a 50% increase in armature reaction flux brought about by a 50% increase in armature current causes E_{ar} to increase to 1.5 times its previous value. But the presence of I_a on the right side of Eq. (5-15) automatically accounts for this. If I_a and Φ_{ar} were not assumed to be linearly related, this correspondence would not prevail. Finally, it should be apparent that, when the synchronous generator operates at a field current $\bar{F}$, the associated excitation voltage must contain a component equal and opposite to that specified by Eq. (5-15).

Based on this background the complete phasor diagram of the effect of armature mmf treated in terms of its equivalent reactance drop can be developed. We begin by assuming rated armature current flows to a balanced three-phase lagging power-factor load at rated terminal voltage. As with the general method, the armature leakage impedance drop can be added to $\bar{V}_t$ to yield $\bar{E}_r$. Refer to Fig. 5-7(b). However, at this point, instead of dealing with mmf's, the analysis continues in voltages. Thus to $\bar{E}_r$ is added an armature reaction reactance drop that is equal and opposite to $\bar{E}_{ar}$. The resultant is the excitation voltage. An important

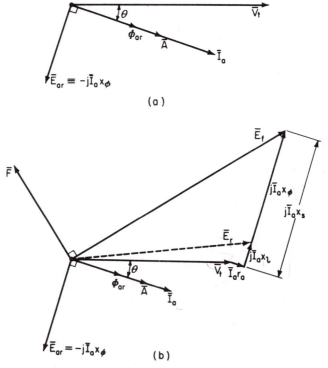

(a)

(b)

Figure 5-7 Illustrating the replacement of the effect of armature mmf by an equivalent induced emf E_{ar}.

observation to make in Fig. 5-7(b) is that the leakage reactance drop and the armature reaction reactance drop add along the same line, both being perpendicular to $\bar{I}_a$. This allows the two quantities to be added algebraically, which leads to a total reactance called the *synchronous reactance* and denoted by x_s. In equation form

$$x_s \equiv x_l + x_\phi$$

(5-16)

where x_l is the armature winding leakage reactance per phase and x_ϕ is defined by Eq. (5-15).

A study of Fig. 5-7(b) discloses that with the use of the synchronous reactance the required field excitation for any specified load condition can be found in two steps. The first step involves finding $\bar{E}_f$ directly from the equation

$$\bar{E}_f = \bar{V}_t + \bar{I}_a(r_a + jx_s) \approx V_t\underline{/0^\circ} + I_a\underline{/-\theta}\,jx_s$$

(5-17)

In step two the air-gap characteristic is entered at the level of E_f and the corresponding value of F is recorded. This procedure is called the *synchronous reactance method*. It is useful to keep in mind that the synchronous reactance is a fictitious quantity that replaces the effect of the armature winding leakage flux and the armature winding rotating mmf.

A circuit interpretation of Eq. (5-17) leads directly to the equivalent circuit of the synchronous generator and is depicted in Fig. 5-8. This is an immediate consequence of introducing the concept of synchronous reactance.

How is the synchronous reactance determined? If the design data are available so that the effective turns of the stator as well as the reluctance of the magnetic circuit can be computed, then an expression like Eq. (3-84) of Appendix to Chapter 3 can be used. Otherwise x_s can be found from the air-gap characteristic and the short-circuit characteristic, i.e., through the use of curves (b) and (c) of Fig. 5-4. To understand how this comes about let us examine the phasor diagram of the synchronous generator under short-circuit conditions with the field current adjusted to the value that causes rated armature current to flow. The armature winding mmf remains the same, but the terminal voltage is zero. Accordingly, the phasor diagram of Fig. 5-7(b) reduces to that shown in Fig. 5-9(a). Note that the field current is used almost entirely to neutralize the effect of the armature winding mmf. The remainder, R_{sc}, is only large enough to overcome the leakage impedance drop, which for all practical purposes is equal to the leakage reactance drop. Note too that the armature current in short-circuit conditions lags

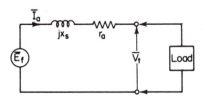

Figure 5-8 The per-phase equivalent circuit of the synchronous generator lies to the left of the load terminal.

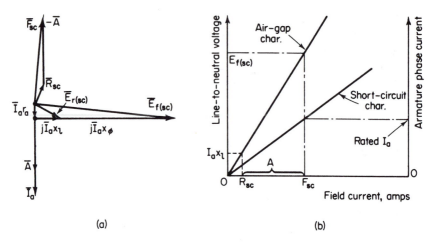

Figure 5-9   Determining the synchronous reactance from no-load tests: (a) phasor diagram at short-circuit; (b) no-load characteristics.

the excitation voltage by nearly 90°. For obvious reasons at such low power factors the quantities $\bar{R}$, $\bar{A}$, and $\bar{F}$ are essentially collinear. However, the important thing to note in Fig. 5-9(a) is that the excitation voltage at short-circuit is practically equal to the synchronous impedance drop. This observation provides the key for the determination of x_s. If F_{sc} is maintained and the short-circuit is removed, the open-circuit voltage produced by this field current is obviously E_{fsc}. Therefore, with F_{sc} known, the air-gap characteristic can be used in the manner illustrated in Fig. 5-9(b) to obtain E_{fsc}. Similarly, by entering the short-circuit characteristic the corresponding current produced by F_{sc} can be found. The synchronous reactance is then the ratio of these two quantities. Thus

$$x_s = \frac{E_{fsc}}{I_a} \tag{5-18}$$

Keep in mind that this is the *unsaturated* value of synchronous reactance because we are still working with the air-gap line.

Example 5-2

Determine the following information for the machine of Example 5-1: (a) the value of the unsaturated synchronous reactance and (b) the field current needed to establish rated terminal voltage when rated current is delivered to a balanced three-phase load of 0.8 power factor lagging, using the synchronous reactance method.

Solution   (a) A glance at Fig. 5-4 indicates that a field current of $F_{sc} = 170$ A is needed to force rated current of 418.4 A through the armature winding at short-circuit. Furthermore, the excitation voltage corresponding to 170 field amperes is $E_{fsc} = 9200$ V. Hence

$$x_s = \frac{E_{fsc}}{I_{a\,(\text{rated})}} = \frac{9200}{418.4} = 22\ \Omega \tag{5-19}$$

(b) By Eq. (5-17) we have

$$\bar{E}_f = V_t\,\underline{/0°} + j\bar{I}_a x_s = 7967.4\,\underline{/0°} + 418.4(0.8 - j0.6)j22$$

$$= 7967.4 + 5523 + j7364 = 13{,}490 = j7364$$

$$= 15{,}369\,\underline{/28.6°} \tag{5-20}$$

Then from the air-gap characteristic

$$F = 285\ \text{A} \tag{5-21}$$

A comparison of these results with Eqs. (5-13) and (5-14), which are obtained by the general method, shows them to be almost identical. Such correspondence is entirely in keeping with the linearity assumption.

5-4 NONLINEAR MACHINE ANALYSIS

The power output capability of electric machines is directly dependent upon the degree to which the material is worked magnetically and electrically. The higher the operating flux density in the iron parts of the machine and the current density in the copper parts are, the greater is the output. Consequently, in most electric machines the level of flux density chosen puts operation in the saturated or nonlinear region. If analytical procedures are to be used to predict machine performance, then the effect of the nonlinearity must be appropriately accounted for. Otherwise realistic results will be almost impossible to attain. This is a point to keep well in mind when sophisticated procedures are employed to predict machine behavior, formulated in terms of inductance parameters based on linearity.

The obvious way to account for operation of the machine in its nonlinear region is to work directly with the nonlinear characteristic usually represented in a graphic form such as curve (a) in Fig. 5-4. This curve is the actual *open-circuit characteristic* and may be either computed from the design data by employing the procedures discussed in Chapter 1 or found experimentally as described in Sec. 5-2. It should not be inferred that the mere use of the actual nonlinear characteristic guarantees realistic results. That this need not be so will be illustrated presently. Attention is now turned to the application of the general method and the synchronous reactance method to find the necessary excitation currents for synchronous generators by utilizing the actual, nonlinear open-circuit characteristic instead of the linear air-gap characteristic. It is worthwhile to note that information about the range of the excitation currents is necessary for the proper design of a voltage regulator which will act to maintain constant terminal voltage in the presence of changing load demands.

The General Method

Again, let us find the excitation needed to produce rated terminal voltage when rated current flows to a balanced lagging power-factor load. The phasor diagram of Fig. 5-6 applies with undiminished validity. The actual armature winding emf induced by the resultant flux is computed as in the linear case. Thus

$$\bar{E}_r = V_t \underline{/0°} + \bar{I}_a(r_a + jx_l) = 7966.4 + 418.4\underline{/-36.8°}\,(0.07 + j1.9)$$

$$= 8444.4\underline{/4.3°} \text{ V} \qquad (5\text{-}22)$$

The value of the resultant mmf expressed in field amperes that produces this result is found by entering the nonlinear open-circuit characteristic (Fig. 5-4) at 8444.4 volts and reading the associated abscissa value. This yields

$$R = 222 \text{ A}$$

so that the complete phasor expression becomes

$$\bar{R} = 222\underline{/94.3°} \text{ A} \qquad (5\text{-}23)$$

A comparison with the result obtained through linear analysis indicates a wide discrepancy (222 versus 157 A), which attests to the fact that the machine is operating at a fairly high level of saturation.

The armature reaction mmf is again treated in terms of an equivalent field current of 155 A. Application of Eq. (5-7) then provides the required value of the field current:

$$\bar{F} = \bar{R} - \bar{A} = 222\underline{/94.3°} - 155\underline{/-36.8°}$$

$$= -140.6 + j314$$

$$= 344\underline{/114°} \text{ A} \qquad (5\text{-}24)$$

Eq. (5-13) shows that the result obtained by linear analysis is considerably lower and, if used for the design of a voltage regulator, would obviously lead to an inadequate design.

The field excitation voltage that corresponds to the field current of 344 amperes is found on the curve of the open-circuit characteristic to be

$$E_f = 9560 \text{ V} \qquad (5\text{-}25)$$

This value stands in sharp contrast to the figure of 15,300 found in the linear case. The latter quantity is highly unrealistic, and it is easy to appreciate why by noting in Fig. 5-4 the wide divergence existing between the open-circuit and the air-gap characteristics for field current in the vicinity of 300 A.

Equation (5-25) identifies the value of the terminal voltage that occurs if the excitation current is maintained at 344 A and the load is disconnected. In other words the no-load terminal voltage is 9560 V. Accordingly, the voltage regulation of this machine in percent is found to be

$$\text{voltage regulation} = \frac{E_f - V_t}{V_t} 100 = \frac{9560 - 7960}{7960} 100 = 20.1\% \quad (5\text{-}26)$$

where E_f is the excitation voltage found on the magnetization curve. This figure is not only realistic but typical of such machines. Calculation of the voltage regulation for the linear case yields a meaningless result.

The Saturated Synchronous Reactance Method

It is shown in Sec. 5-3 that the use of the unsaturated synchronous reactance method in finding excitation currents leads to unrealistic solutions for reasons therein described. However, it is possible to alter the method of computation in a way that is consistent with the state of saturation that the machine is in and thereby obtain results that are comparable with those of the general method. Interested readers are referred to Appendix E for details of a method that is based on the flux density conditions associated with the resultant mmf R.

The generally accepted way of accounting for a saturated value of synchronous reactance is to deal with the flux density condition that is associated with the mmf needed to produce rated voltage at no load. To illustrate, let the field current that is needed to generate rated terminal voltage (V_R) be represented by Oa in Fig. 5-10. Now, if this same field current is used in a short-circuit test, there results a short-circuit armature current (I_{sc}) that is often in the vicinity of the rated value (I_R). In Fig. 5-10 this current is shown to be slightly less than the rated current. The conditions that prevail in the short-circuit case are such that there exists a synchronous reactance drop, $I_{sc}x_s$, that is virtually equal in value to the rated terminal voltage because the field current Oa is common. Thus, we can write

$$I_{sc}x_s = V_R \quad (5\text{-}27)$$

or

$$x_s = \frac{V_R}{I_{sc}} = \text{saturated synchronous reactance} \quad (5\text{-}28)$$

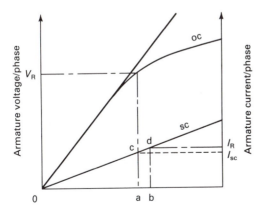

Figure 5-10 Graphical data used to find the saturated value of synchronous reactance; also the short-circuit ratio (SCR).

By way of comparison it is instructive to note that the unsaturated value of synchronous reactance is, of course, a larger quantity than that displayed in Eq. (5-28) since in the calculation use is made of the voltage on the air-gap line rather than on the open-circuit characteristic for excitation Oa.

Short-Circuit Ratio.   A useful synchronous generator parameter results when Eq. (5-27) is expressed in terms of per-unit values. Accordingly, we have

$$(\text{p-u } I_{sc})(\text{p-u } x_s) = \text{p-u } V_R$$

Since in machine analysis, the rated voltage is taken as the base quantity, the corresponding per-unit value is unity. Therefore, it follows from the last expression that the per-unit value of the saturated synchronous reactance can be given simply by

$$\text{p-u } x_s = \frac{1}{\text{p-u } I_{sc}} \tag{5-29}$$

A modified form of this last expression is obtained by recalling the definition of the per-unit value of the armature current, namely,

$$\text{p-u } I_{sc} = \frac{I_{sc} \text{ (in amperes)}}{I_R \text{ (in amperes)}} \equiv \text{SCR} \tag{5-30}$$

which is also defined as the *short-circuit ratio* (SCR). Hence,

$$\text{p-u } x_s = \frac{1}{\text{p-u } I_{sc}} = \frac{1}{\text{SCR}} \tag{5-31}$$

Another view of the short-circuit ratio is possible by noting from the similar triangles found in Fig. 5-10 that

$$\frac{Oa}{Ob} = \frac{ac}{bd} = \frac{I_{sc}}{I_R} \tag{5-32}$$

Accordingly, the short-circuit ratio can also be identified as the ratio of field current (Oa) that generates rated voltage at no load to the field current (Ob) that produces rated current at short-circuit. The short-circuit ratio is useful because its reciprocal denotes the per-unit value of the saturated synchronous reactance, one of the important parameters of the synchronous machine.

Example 5-3

For the machine of Example 5-1 find the excitation emf corresponding to rated terminal voltage when rated armature current is delivered to a 0.8 lagging pf three-phase balanced load. Use the saturated synchronous reactance method. Also find the short-circuit ratio and the per-unit value of the saturated synchronous reactance.

Solution   The field current needed to generate the no-load rated phase voltage of 7967.4 V is found from Fig. 5-4 to be

$$Oa = 188 \text{ A}$$

The armature current on the short-circuit characteristic produced by this same field current is

$$I_{sc} = 462 \text{ A}$$

Hence, the value of the saturated synchronous reactance becomes by Eq. (5-28)

$$x_s = \frac{7967.4}{462} = 17.25 \ \Omega/\text{phase}$$

The per-unit value of the saturated synchronous reactance can be found by two procedures. The first uses the short-circuit ratio which by Eq. (5-30) is found in this case to have a value of

$$\text{SCR} = \text{p-u } I_{sc} = \frac{I_{sc}}{I_R} = \frac{462}{418.4} = 1.105$$

Then invoking Eq. (5-31) the per-unit saturated synchronous reactance is

$$\text{p-u } x_s = \frac{1}{\text{SCR}} = \frac{1}{1.105} = 0.905$$

The second procedure makes use of the base impedance which for the example at hand is given by

$$Z_{\text{base}} = \frac{V_R}{I_R} = \frac{7967.4}{418.4} = 19.04 \ \Omega$$

Accordingly,

$$\text{p-u } x_s = \frac{x_s}{Z_{\text{base}}} = \frac{17.25}{19.04} = 0.906$$

The excitation voltage is found by applying Kirchhoff's voltage law to the equivalent circuit of Fig. 5-8. Thus,

$$\bar{E}_f = \bar{V}_t + \bar{I}_a(r_a + jx_s) \approx \bar{V}_t + \bar{I}_a \underline{/-36.8^\circ} \ (jx_s)$$

$$= 7967.4 \underline{/0^\circ} + (418.4)(0.8 - j0.6)j17.25$$

$$= 13{,}585.84 \underline{/25.15^\circ} \text{ V}$$

Comparing this result with that found for the unsaturated case displayed in Eq. (5-14) indicates a marked improvement. Still more improvement can be realized by employing the modified saturated synchronous reactance method described in Appendix E.

The field excitation current needed to maintain constant terminal voltage at fixed lagging load power factor rises with increasing kVA demand at the load terminals. This is obvious from the results of the preceding examples. In those circumstances where the load draws a leading current, the need for field excitation actually diminishes with increasing load kVA because of the magnetizing effect caused by the armature mmf. These results are graphically portrayed in Fig. 5-11.

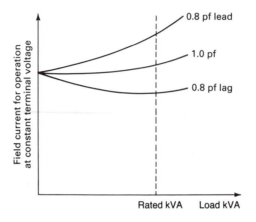

Figure 5-11 Change in field current needed to maintain constant terminal voltage as load kVA varies for different load power factors.

5-5 EXPERIMENTAL DETERMINATION OF LEAKAGE REACTANCE x_l AND ARMATURE REACTION MMF A

Nonlinear machine analysis by the general method presupposes knowledge of the armature winding leakage reactance as well as the armature winding mmf expressed in equivalent field amperes. When the complete design data of a machine are available, these quantities can be readily calculated. However, in the absence of such data other means must be sought to furnish the information.

As a review, let us return to Fig. 5-9, which depicts the situation in the synchronous generator at short-circuit with rated current flowing. Recall that the armature current flows at almost zero power factor and so the quantities $\bar{F}_{sc}$, $\bar{R}_{sc}$, and $\bar{A}$ are collinear. In Fig. 5-9(b), F_{sc} is a measured quantity and is accordingly available. If the winding data are known, the quantity A can be computed from Eq. (5-5) and then algebraically subtracted from F_{sc} to yield R_{sc}. Then on the air-gap line corresponding to R_{sc}, the leakage reactance drop is found. Strictly speaking, this ordinate quantity is $E_{r(sc)}$, but because $x_l > r_a$ little error is made by taking $E_{r(sc)} = I_a x_l$. Since $E_{r(sc)}$ and I_a are then both known, x_l can be found. Unfortunately, this determination of x_l is dependent upon knowledge of A which is generally not available and must also be found.

On the basis of the foregoing it appears worthwhile to explore further the operation of the synchronous generator at zero power factor. The wiring diagram for such a test is shown in Fig. 5-12(a). The zero power-factor load consists of purely inductive coils arranged in a Y-connection. Initially the coils are set for zero reactance and the field current is adjusted to yield rated current. This corresponds with point F_{sc} in Fig. 5-9(b). The coil reactance is then increased, which causes a drop in current. The field current is then increased to reestablish the rated value of the armature current. Now the terminal voltage is no longer zero but, rather, equal to $I_a X_L$, where X_L is the load reactance. By continuing this procedure, rated terminal voltage can be made to appear across the zero power-

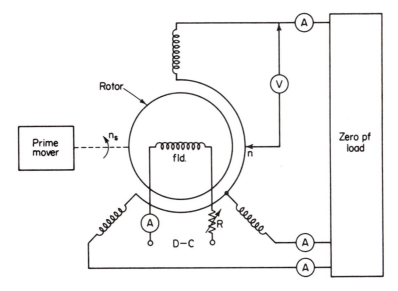

Figure 5-12(a) Wiring diagram for obtaining the zero power-factor characteristic.

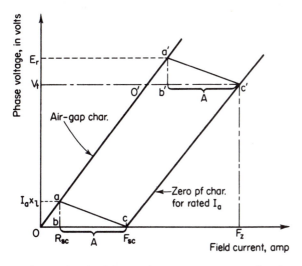

Figure 5-12(b) Illustrating how the zero power-factor characteristic can be derived from the short-circuit data for conditions of no saturation.

factor load with rated armature current flowing. A plot of terminal voltage versus field current yields the *zero power-factor characteristic*.

If the machine were entirely linear (i.e., with no saturation occurring), the zero power-factor characteristic could be readily drawn from a knowledge of F_{sc} alone. The construction should be obvious from an inspection of Fig. 5-12(b). The zero power-factor characteristic is generated merely by displacing line Oc to higher and higher voltage levels. Note that to produce rated terminal voltage at zero power factor, the total field current must be F_z. By subtracting A from F_z, R

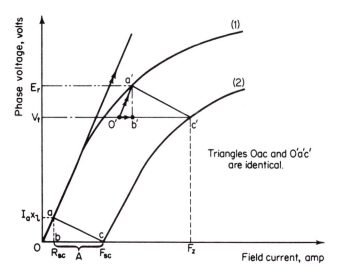

Triangles *Oac* and *O'a'c'*
are identical.

Figure 5-13 Graphical determination
of leakage reactance x_e and armature
mmf expressed in field amperes.
Triangle *a'b'c'* is known as the *Potier
triangle*.

is obtained corresponding to which is the armature induced voltage, E_r on the air-gap line. The vertical distance between E_r and V_t (line *a'b'* in Fig. 5-12(b)) continues to be the leakage reactance drop because at zero power factor E_r and V_t are also essentially collinear in the phasor diagram.

Let us next consider what happens to the situation depicted in Fig. 5-12(b) when the actual nonlinear characteristics are used. Refer to Fig. 5-13. Curve 1 denotes the open-circuit characteristic and curve 2 denotes the zero power-factor characteristic for rated armature current. A little thought should make it clear that the zero power-factor curve cannot be derived in the manner just described for the linear case because of saturation. Hence this characteristic must be found experimentally (or at least points *c* and *c'*). A useful thing happens if we attempt to construct triangle *O'a'c'* of Fig. 5-12(b) on the characteristics of Fig. 5-13. To accomplish this, it is necessary to mark off line *O'c'* equal to *Oc*. Then at *O'* a line is drawn parallel to the air-gap line until it intersects the open-circuit characteristic at point *a'*. Thus the triangle *Oac* is duplicated and, what is more, as a result of this construction the two items being sought are immediately available. Line *a'b'* represents the leakage reactance drop and can easily be measured in volts. Then

$$x_l = \frac{a'b' \text{ (volts)}}{I_a \text{ (rated)}} \qquad (5\text{-}33)$$

This quantity is called the *Potier reactance*. Moreover, the armature winding mmf expressed in equivalent field amperes is immediately available upon measuring line *b'c'* against the abscissa axis. Thus

$$A = b'c' \quad \text{in field amperes} \qquad (5\text{-}34)$$

The rationale for drawing a line at O' that is parallel to the air-gap line lies in the fact that triangle Oac involves quantities in which saturation bears no influence, namely, stator leakage reactance, x_l (which involves chiefly air paths) and the armature winding mmf A (which is dependent upon solely winding data and current). This is not to say that saturation is not present; it is and a comparison of F_z corresponding to c' in Fig. 5-12b with F_z in Fig. 5-13 makes this obvious. However, by using c' in Fig. 5-13 as a starting point and then constructing the Potier triangle, $O'a'c'$, in a manner that ignores saturation, valid information about x_l and A thereby results.

5-6 THEORY FOR SALIENT-POLE MACHINES

In contrast to the cylindrical-pole synchronous machine, which has a uniform air gap, the salient-pole machine has a highly nonuniform air gap because of the use of a protruding-pole structure. A glance at Fig. 3-7 makes this obvious. A flux path that includes two protruding poles has two small gaps to cross. This path is indicated in Fig. 5-14 by ϕ_d and is called the *direct axis*. It is the path of minimum reluctance. On the other hand the path denoted by ϕ_q in Fig. 5-14 involves two large air gaps and is the path of maximum reluctance. This is called the *quadrature axis*. In normal operation the armature (or stator) winding mmf is distributed with its peak value located somewhere between the direct and quadrature axes. Accordingly it produces a significant effect in each axis—but in a different way because of the considerable difference in the reluctance variation of each axis. Consequently, the cylindrical-rotor theory developed in the preceding sections cannot be used for the salient-pole machine with satisfactory results. A modified theory is required that takes due account of this difference. In this section such a theory is developed for the salient-pole machine in both a two-reactance method (analogous to the synchronous reactance method for smooth-rotor machines) and a method based on mmf's.

Two-Reactance Method for Salient-Pole Machines

To provide background, let us consider first the manner of locating the position of the armature winding mmf relative to the rotor field distribution when current flows. Call the angle between the excitation voltage and the armature current ψ. Again refer to Fig. 5-14(a). For the indicated direction of the rotor flux and speed the induced emf in those conductors under the influence of the north-pole flux is directed outward; and in those inductors under the influence of the south-pole flux, it is directed into the plane of the paper. Coil e_f-e_f' is a single coil that corresponds to the location of the amplitude of the fundamental sinusoidal emf distribution that takes place in the entire three-phase winding for the instant shown. The resultant distribution of the currents in the same three-phase winding for a lagging power-factor load is represented by coil i_a-i_a', which is located behind

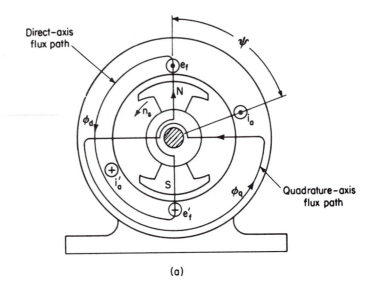

(a)

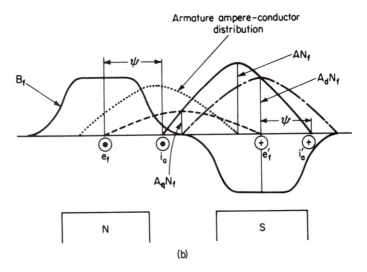

(b)

Figure 5-14 (a) Identification of direct- and quadrature-axis flux paths; (b) illustrating how the armature mmf, AN_f, is resolved into two components: one acting in the direct axis, A_dN_f; the other acting in the quadrature axis, A_qN_f.

the emf distribution by the specified phase angle ψ. Appearing in Fig. 5-14(b) is the developed representation of Fig. 5-14(a). The armature ampere-conductor distribution is shown dotted and lagging behind the field flux distribution by ψ. On the other hand the armature mmf, which in synchronous machines is being denoted in terms of equivalent field mmf, is represented by AN_f, where A is the equivalent field current as described by Eq. (5-5). It is always desirable to work in equivalent field current so that the abscissa axis of the open-circuit characteristic may be used directly in accounting for armature reaction mmf. Again note that the armature mmf is 90 degrees behind the ampere-conductor distribution. Accordingly, the armature mmf curve AN_f takes the position shown in Fig. 5-14(b). Clearly, the location of this wave is dependent upon the phase angle between E_f and I_a, which in turn is dependent upon the power factor of the load and the amount of power delivered (see Sec. 5-7).

It is helpful to consider for a moment the general effects caused by various locations of the armature mmf wave. If the load conditions are such that ψ becomes equal to 90°, the armature mmf wave lies entirely in the direct axis and directly opposes the field flux. In other words, the armature mmf has a *demagnetizing* effect on the field flux. When ψ assumes a value of 0 degrees (as nearly happens for a unity power-factor load), then clearly the armature wave lies entirely in the cross axis and its effect is to cause a distortion of the field flux. That is, the armature mmf wave now has a *cross-magnetizing* effect on the field flux. When the armature current leads the excitation voltage (as happens with leading power-factor loads) by, say, 90°, the effect of the armature mmf is to produce *magnetization* of the field flux. Because of the nature of industrial and domestic loads, the armature current usually lags behind the excitation voltage so that the situation depicted in Fig. 5-14 is the customary one. This means that generally the armature mmf wave is so located that it produces simultaneously an effect on both the direct and the cross axes.

In view of the fact that the direct axis is influenced by iron saturation as mmf is added or subtracted in this axis whereas iron saturation is of little consequence in the cross axis, it is imperative that for any specified ψ the armature mmf wave be composed of two components—one that acts in the direct axis and the other that acts in the quadrature axis. The amplitude of the direct-axis component is denoted by A_dN_f in Fig. 5-14. A study of Fig. 5-14(b) reveals that

$$\bar{A}_d = \bar{A} \sin \psi \qquad \text{A} \qquad (5\text{-}35)$$

where $\bar{A}_d$ is the direct-axis armature mmf expressed in equivalent field amperes. Similarly, the component of the quadrature-axis mmf can be expressed as

$$\bar{A}_q = \bar{A} \cos \psi \qquad \text{A} \qquad (5\text{-}36)$$

Appearing in Fig. 5-15 are the respective flux-density wave forms produced by the direct-axis and quadrature-axis armature mmf components. The direct-axis mmf produces a quasi-truncated sinusoid, whereas the quadrature-axis mmf produces a saddle-shape flux-density wave because of the high reluctance of the interpolar space. The effects are indeed quite different and accordingly must be treated separately.

The key to the reactance method of treating the salient-pole synchronous machine lies in Eqs. (5-35) and (5-36). In accordance with these equations the

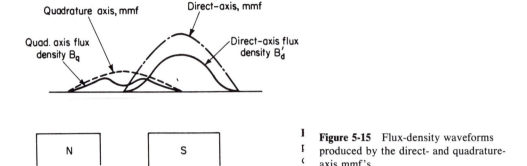

Figure 5-15 Flux-density waveforms produced by the direct- and quadrature-axis mmf's.

armature current can be considered composed of a direct-axis and a quadrature-axis current. That is,

$$\bar{I}_d = \bar{I}_a \sin \psi \tag{5-37}$$

$$\bar{I}_q = \bar{I}_a \cos \psi \tag{5-38}$$

Hence, if we assume for the moment that $\bar{E}_f$, $\bar{I}_a$, and ψ are known, then $\bar{I}_d$ and $\bar{I}_q$, as well as their associated mmfs $\bar{A}_d$ and $\bar{A}_q$, will be represented as depicted in Fig. 5-16. Note that the direct axis is that along which the field mmf $\bar{F}$ acts. By superposition each mmf is then considered to produce its own induced emf in the armature winding. Thus by following the same procedure outlined in the synchronous reactance method for the cylindrical rotor machine, through superposition we can associate with $\bar{A}_d$ an emf which is replaced by the reactance drop in the direct axis, i.e., $j\bar{I}_d x_{ad}$. The j term provides the necessary 90-degree lead that must exist between a current and its related reactance drop. Here x_{ad} is *the armature reaction reactance in the direct axis.* In a similar fashion the effect of $\bar{A}_q$ is replaced by a reactance drop $+j\bar{I}_q x_{aq}$ leading $\bar{A}_q$ (or $\bar{I}_q$) by 90°. The quantity x_{aq}

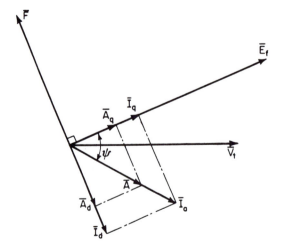

Figure 5-16 Illustrating the composition of the armature current and its associated mmf's into direct- and quadrature-axis components.

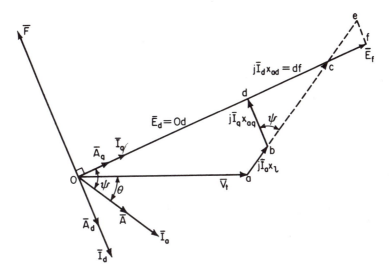

Figure 5-17 Development of a phasor diagram for the salient-pole synchronous machine, by the armature-reaction reactances.

is called the armature reaction reactance in the quadrature axis. A single arma-ture reaction cannot be used here because of the wide difference in the flux-density curves associated with the two axes. The value of x_{aq} is always less than x_{ad} since the emf induced by a given mmf acting on each axis is always smaller for the quadrature axis owing to its higher reluctance.

We next describe the development of the phasor diagram for the salient-pole synchronous machine, assuming the following data to be available: rated terminal voltage per phase, V_t; rated armature current per phase, I_a; load power-factor angle, θ; armature winding leakage reactance per phase, x_l; and the armature reaction reactance in each axis expressed per phase, x_{ad} and x_{aq}. For simplicity the armature winding resistance is assumed negligible. Refer to Fig. 5-17. Add $j\bar{I}_a x_l$ to $\bar{V}_t$ to get the induced emf in the armature winding caused by the resultant gap flux. This is quantity Ob. At b mark off the quantity $+j\bar{I}_q x_{aq}$ leading $\bar{I}_q$ by 90° and terminating at point d. The excitation voltage $\bar{E}_f$ lies along line Od. Finally, the magnitude of $\bar{E}_f$ is found by adding the armature reaction reactance drop in the direct axis, i.e., $j\bar{I}_d x_{ad}$, along line Od starting at point d. The quantity Of mea-sured on a suitable voltage scale yields the value of the excitation voltage.

Generally, unless the design data are furnished, the quantities x_{aq} and x_{ad} are not available. Even an appropriate no-load test does not provide information about the armature reactances directly. Rather, it is information about the com-bined effects of the leakage reactance with the armature reaction reactances that becomes directly available. In this connection then let us study the phasor dia-gram of Fig. 5-17 in more detail. First note that

$$\sphericalangle \, cbd = \psi$$

It then follows that

$$bc = \frac{bd}{\cos \psi} = \frac{I_q x_{aq}}{\cos \psi} = \frac{I_a \cos \psi \, x_{aq}}{\cos \psi} = I_a x_{aq} \tag{5-39}$$

In addition,

$$ac = ab + bc = I_a(x_l + x_{aq}) \tag{5-40}$$

Therefore we can define

$$\boxed{x_q \equiv x_l + x_{aq}} \tag{5-41}$$

where x_q is the *quadrature-axis synchronous reactance*. Note the correspondence with Eq. (5-16) for the cylindrical rotor machine. Equation (5-40) furnishes an alternative way of locating the direction along which the $\bar{E}_f$ phasor lies. It merely requires that the quantity $j\bar{I}_a x_q$ be added to $\bar{V}_t$ to locate point c in Fig. 5-17.

By proceeding in a similar fashion a synchronous reactance associated with the direct axis can be defined. From the geometry of Fig. 5-17 we can write that

$$be = \frac{df}{\sin \psi} = \frac{I_d x_{ad}}{\sin \psi} = \frac{I_a \sin \psi \, x_{ad}}{\sin \psi} = I_a x_{ad}$$

Also,

$$ae = ab + be = I_a(x_l + x_{ad})$$

Therefore

$$\boxed{x_d \equiv x_l + x_{ad}} \tag{5-42}$$

where x_d denotes the *direct-axis synchronous reactance*. This quantity is comparable to x_s for the smooth-rotor machine. Fairly accurate values for x_q and x_d can be obtained from a no-load test called the *slip test*, which is described in Appendix F. Our discussions hereinafter proceed on the assumption that information about x_q and x_d is available either from the design data or from an appropriate no-load test such as the slip test.

The determination of $\bar{E}_f$ (and thence the corresponding field current $\bar{F}$) by the use of x_d and x_q has caused the method to be called the *two-reactance method*. Depicted in Fig. 5-18 is the phasor diagram expressed solely in terms of synchronous reactance drops. The solution procedure then involves just three steps as outlined next:

1. Locate the direction of $\bar{E}_f$ from

$$\psi = \tan^{-1} \frac{V_t \sin \theta + I_a x_q}{V_t \cos \theta} \tag{5-43}$$

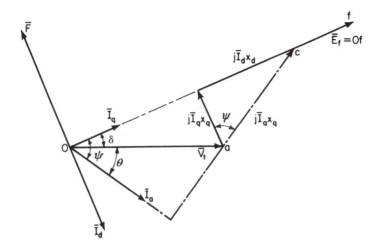

Figure 5-18   Phasor diagram of the salient-pole machine in terms of the synchronous reactances in the direct and quadrature axes.

2. Compute the phasor expressions for direct and quadrature currents.

$$\bar{I}_q = (I_a \cos \psi) \underline{/\delta} \qquad (5\text{-}44)$$

$$\bar{I}_d = (I_a \sin \psi) \underline{/\delta - 90°} \qquad (5\text{-}45)$$

where

$$\delta \equiv \psi - \theta \qquad (5\text{-}46)$$

and the angles are all positive for lagging power factor.

3. Finally, as can be seen from Fig. 5-18,

$$\bar{E}_f = \bar{V}_t + j\bar{I}_q x_q + j\bar{I}_d x_d \qquad (5\text{-}47)$$

The field current that produces this excitation voltage is then found from the open-circuit characteristic.

The same limitations that apply to the synchronous reactance method apply to the two-reactance method. To obtain reliable results the direct-axis synchronous reactance must be found corresponding to the proper flux conditions in the machine. Of course, x_q presents no such problem because it is associated with large air paths where saturation is of little or no consequence. Surely, whenever there is a choice, the preferred way of dealing with the effect of armature reaction in the direct axis is to work directly with the open-circuit characteristic. This is possible by employing the general method for salient-pole machines which is described next.

General Method for Salient-Pole Machines

The phasor diagram to be used is the one depicted in Fig. 5-17. The procedure is as follows:

1. Find ψ by means of Eq. (5-43).
2. Compute the emf induced by the resultant flux in the direct axis. This is the quantity Od and shall now be called E_d. Its magnitude is given by

$$E_d = V_t \cos(\psi - \theta) + I_a x_l \sin \psi \qquad (5\text{-}48)$$

3. Corresponding to E_d on the open-circuit characteristic, find the resultant mmf expressed in field amperes. Call this R_d.
4. Find A by means of Eq. (5-34) and compute

$$A_d = A \sin \psi$$

5. Then for ψ lagging the required field current is found from the algebraic sum of steps 3 and 4. Thus

$$F = R_d + A \sin \psi \qquad (5\text{-}49)$$

It is interesting to note that by computing E_d in the manner described by Eq. (5-48), there is no need to have knowledge of x_{aq}. Of course the magnitude of $\bar{E}_d$ is the same as that which would be obtained from the expression

$$\bar{E}_d = \bar{V}_t + jI_q x_l + jI_q x_{aq} \qquad (5\text{-}50)$$

Finally, note that steps 4 and 5 of the foregoing procedure are the mmf equivalent of adding the reactance drop $I_d x_{ad}$ to E_d to obtain the excitation voltage. The advantage of the mmf version is that one automatically accounts for the correct level of saturation by working directly with the nonlinear open-circuit characteristic. This method yields highly realistic results.

Example 5-4

A three-phase, 15-kVA, 220-V, Y-connected, 60-Hz, six-pole salient-pole synchronous generator has the open-circuit, short-circuit, and zero power factor characteristics depicted in Fig. P5-10. The armature winding resistance is negligible. It is also known that for this type machine the synchronous reactance in the quadrature axis is 60% of the value associated with the direct axis. Moreover, the leakage reactance is known to have a value of 0.9 Ω per phase.

(a) Identify the base quantities to be used in calculating the performance employing the per-unit notation.

(b) Determine the per-unit values of the leakage reactance and the direct-axis synchronous reactance per phase as derived from the graphical data.

(c) Determine the excitation emf when this machine delivers rated kVA to a 0.8 pf lagging load using the two reactance method. Make all calculations in terms of per-unit quantities.

(d) Find the excitation emf for the conditions of part (c) using the general method for the salient-pole machine.

Solution   (a) It is customary to use the rated phase voltage as the base voltage and the rated phase current as the base current. Thus in this case

$$V_{base} = V_{phase} = \frac{V_L}{\sqrt{3}} = \frac{220}{\sqrt{3}} = 127 \text{ V/phase}$$

$$I_{base} = I_{phase} = \frac{15,000}{\sqrt{3}\,(220)} = 39.4 \text{ A/phase}$$

Correspondingly, the base impedance becomes

$$Z_{base} = \frac{V_{base}}{I_{base}} = \frac{127}{39.4} = 3.22 \text{ }\Omega/\text{phase}$$

A useful, alternative form for the calculation of the base impedance can be obtained by writing

$$Z_{base} = \frac{V_{phase}}{I_{phase}} = \frac{V_L/\sqrt{3}}{\text{VA}/\sqrt{3}\,V_L} = \frac{V_L^2}{\text{VA}}$$

where V_L is the line-to-line voltage (irrespective of whether or not the stator winding is Y- or Δ-connected) and VA denotes the volt-ampere rating of the generator. Observe that with this formulation the $\sqrt{3}$ is not a factor in the evaluation. Applied to the case at hand, we get

$$Z_{base} = \frac{(220)^2}{15,000} = 3.23 \text{ }\Omega/\text{phase}$$

(b) The per-unit value of the leakage reactance per phase readily follows from the definition of a per-unit value. Thus

$$\text{p-u } x_l = \frac{x_l(\Omega)}{Z_B(\Omega)} = \frac{0.9}{3.22} = 0.28$$

A convenient way to find the per-unit value of the direct-axis synchronous reactance when graphical data is available is to find the *short-circuit ratio* (SCR). By definition the SCR is the armature current, expressed in per unit, which flows in a short-circuit test corresponding to that value of field excitation that produces rated voltage in an open-circuit test. Algebraically stated, we have

$$\text{SCR} = \text{p-u } I_{sc}$$

where the right side is the short-circuit armature current caused by that field current that on open circuit yields rated voltage. Applying this definition to the graphical data, we see that to produce a rated voltage of 127 V on the open-circuit characteristic a field current of 4.15 A is needed. However, this same field current causes a short-circuit current of 41 A, which expressed as a per-unit quantity is p-u $I_{sc} = 41/39.4 = 1.041 = $ SCR. Consequently, the per-unit value of x_d becomes simply

$$\text{p-u } x_d = \frac{\text{p-u } V_R}{\text{p-u } I_{sc}} = \frac{1}{\text{p-u } I_{sc}} = \frac{1}{\text{SCR}} = \frac{1}{1.041} = 0.961$$

(c) Delivering rated kVA means that both the voltage and current are at their rated values which on a per-unit basis means values of unity. The solution procedure for this part is outlined by Eqs. (5-43) through (5-47). For a 0.8 pf the power factor angle is

$$\theta = \cos^{-1} 0.8 = 36.9°$$

Also, the p-u value of the synchronous reactance in the quadrature axis is

$$\text{p-u } x_q = (0.6)(\text{p-u } x_d) = 0.577$$

Hence

$$\psi = \tan^{-1} \frac{(1)\sin 36.9° + (1)(0.577)}{(1)\cos 36.9°} = \tan^{-1} \frac{0.6 + 0.577}{0.8} = 55.8°$$

$$\text{p-u } \bar{I}_q = (\text{p-u } \bar{I}_a \cos \psi)\underline{/\delta} = \cos 55.8°\underline{/\delta} = 0.562\underline{/\delta}$$

where

$$\delta = \psi - \theta = 55.8° - 36.9° = 18.9°$$

$$\text{p-u } \bar{I}_d = (\text{p-u } I_a \sin \psi)\underline{/-90° + \delta} = 0.827\underline{/-71.1°}$$

Therefore,

$$\text{p-u } \bar{E}_f = \bar{V}_t + j\bar{I}_q x_q + j\bar{I}_d x_d = 1\underline{/0°} + (0.562)\underline{/18.9°}(0.577)\underline{/90°} + (0.827)\underline{/-71.1°}(0.961)\underline{/90°}$$

$$= 1\underline{/0°} + (0.562)\underline{/18.9°}(0.577)\underline{/90°} + (0.827)\underline{/-71.1°}(0.961)\underline{/90°}$$

$$= 1\underline{/0°} - 0.105 + j0.307 + 0.752 + j0.257 = 1.647 + j0.564$$

$$= 1.74\underline{/18.9°}$$

In accordance with this result it follows that the voltage regulation of this machine is a high 74%. This value is highly pessimistic, which we already know is characteristic of the synchronous reactance method. A more realistic result can be obtained by using the general method, which is done in part (d).

(d) The general method also begins with the evaluation of ψ. The emf induced by the resultant flux in the air gap of the direct axis is then found with Eq. (5-48). Thus

$$\text{p-u } E_d = V_t \cos (\psi - \theta) + I_a x_l \sin \psi$$

$$= (1) \cos (18.9°) + (1)(0.28) \sin 18.9° = 0.946 + 0.091 = 1.037$$

The corresponding voltage value is

$$E_d = 1.037(127) = 131.7 \text{ V}$$

and the resultant field current, R_d, associated with this voltage as obtained from the open-circuit characteristic is found to be

$$R_d = 4.6 \text{ A}$$

Also, by construction of the Potier triangle the armature reaction mmf expressed in terms of equivalent field amperes is found to be $A = 2.96$ A. Accordingly, the resultant field current that produces the excitation emf is determined by Eq. (5-49) to be

$$F = R_d + A \sin \psi = 4.6 + 2.96 \sin 55.8° = 7.05 \text{ A}$$

Entering the open-circuit characteristic for this value of field current yields an excitation voltage of

$$E_f = 166 \text{ V/phase}$$

The per-unit value is

$$\text{p-u } E_f = \frac{166}{127} = 1.31$$

Thus a much more realistic value of 31% voltage regulation is now obtained.

5-7 POWER

The expression for power in an electromechanical energy-conversion device is always a matter of primary importance. It was shown in Chapter 3 that, once the developed torque is computed, the developed power can be found from

$$P_m = T\omega_m \tag{5-51}$$

where ω_m denotes the mechanical speed of rotation expressed in radians per second. However, it is frequently useful to have an equation for power that is expressed in terms of appropriate machine parameters, voltages, and phase angles because of the insight it gives regarding machine performance. Such an equation can be derived from the phasor diagram for the cylindrical and also the salient-pole machine.

Cylindrical-Rotor Machine

Use is made of the phasor diagram of Fig. 5-7(b), which is repeated in Fig. 5-19. Since the effect of armature winding resistance is very small, it is neglected here. Consequently, the expression for the developed power can be written in one of two ways, i.e.,

$$P_m = qV_tI_a \cos \theta = qE_fI_a \cos \psi \tag{5-52}$$

This statement follows from Fig. 5-19 upon noting that the projection of $\bar{E}_f$ on the $\bar{I}_a$ phasor is identical with the projection of $\bar{V}_t$ on $\bar{I}_a$. Moreover, from the geometry of Fig. 5-19 we see that the quantity ab may also be expressed in two ways. Thus

$$ab = E_f \sin \delta = I_a x_s \cos \theta \tag{5-53}$$

from which it follows that

$$I_a \cos \theta = \frac{E_f}{x_s} \sin \delta \tag{5-54}$$

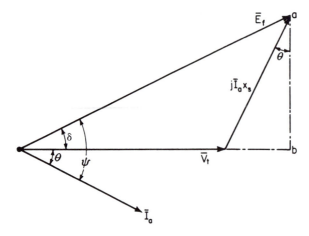

Figure 5-19   Phasor diagram of the cylindrical-rotor machine for lagging power-factor.

Insertion of Eq. (5-54) into Eq. (5-52) yields the desired result. Hence the power developed in the synchronous generator where armature winding resistance is negligible is given by

$$P_m = q \, \frac{V_t E_f}{x_s} \sin \delta \qquad\qquad (5\text{-}55)$$

where q denotes the number of phases.

Because of the dependence of the power upon the angle δ, this angle has come to be called the *power angle*. Equation (5-55) indicates that, if the power angle is zero, the synchronous machine cannot develop useful power. It further reveals that the developed power is a sinusoidal function of the power angle and that it has its maximum value at δ equal to 90°. Figure 5-20 depicts the variation P_m as a function of positive and negative values of δ. As described in Chapter 6, negative values of power angle refer to the operation of the synchronous machine

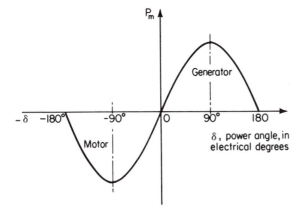

Figure 5-20   Graphical representation of developed power as a function of the power angle for a cylindrical-rotor synchronous machine.

as a motor. It is also worthwhile to note the correspondence of Eq. (5-55) with Eq. (3-22) and Eq. (3-97) of Appendix to Chapter 3. Both the developed electromagnetic torque and the developed power are sinusoidal functions of an appropriate power angle.†

Salient-Pole Machine

The derivation of the expression for the developed power of the salient-pole machine proceeds in a similar fashion. However, in this case use is made of the phasor diagram based on the two-reactance theory. The diagram is repeated for convenience in Fig. 5-21. Again, the phasor diagram as it applies to the synchronous generator (i.e., positive power angle) is used, but it should be understood that the results apply equally for operation as a synchronous motor.

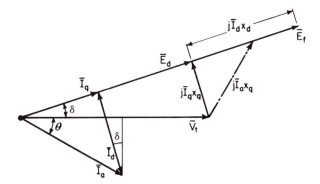

Figure 5-21 Synchronous-generator phasor diagram based on the two-reactance theory.

Because of the assumption of negligible armature resistance, the power output and the power developed are equal. Accordingly we may write

$$P_m = qV_tI_a \cos \theta \qquad (5\text{-}56)$$

A study of Fig. 5-21 discloses that the component of armature current that is in phase with the terminal voltage is related to I_d and I_q by

$$I_a \cos \theta = I_q \cos \delta + I_d \sin \delta \qquad (5\text{-}57)$$

Moreover, an expression for I_q can be found by noting that by the geometry of Fig. 5-21

$$I_q x_q = V_t \sin \delta \qquad (5\text{-}58)$$

---

†Take care not to confuse the torque angle Δ of Eq. (3-97) with the power angle δ of Eq. (5-55). In the former, Δ is defined as the angle between the axis of the flux-density wave produced by the field winding and the axis of the armature winding mmf $\mathscr{F}$ (which is 90° displaced from the armature ampere-conductor distribution). In the latter case, δ is defined as the time angle between the excitation voltage (produced by the field flux) and the terminal voltage (produced by the resultant flux minus the leakage flux). See Fig. 6-18 for further clarification.

from which

$$I_q = \frac{V_t}{x_q} \sin \delta \qquad (5\text{-}59)$$

In a similar fashion an expression for I_d can be found expressed in terms of voltage and the machine parameters. Thus,

$$E_f = V_t \cos \delta + I_d x_d \qquad (5\text{-}60)$$

Therefore,

$$I_d = \frac{E_f - V_t \cos \delta}{x_d} \qquad (5\text{-}61)$$

Upon substituting Eqs. (5-59) and (5-61) in Eq. (5-57), there results

$$I_a \cos \theta = \frac{V_t}{x_q} \sin \delta \cos \delta + \frac{E_f}{x_d} \sin \delta - \frac{V_t}{x_d} \sin \delta \cos \delta \qquad (5\text{-}62)$$

The expression for the developed power then becomes

$$P_m = q \frac{V_t E_f}{x_d} \sin \delta + q \frac{V_t^2}{x_d x_q} (x_d - x_q) \sin \delta \cos \delta \qquad (5\text{-}63)$$

Introducing the trigonometric identity

$$\sin 2\delta = 2 \sin \delta \cos \delta \qquad (5\text{-}64)$$

into the last equation yields the desired result. Hence,

$$\boxed{P_m = q \frac{V_t E_f}{x_d} \sin \delta + q \frac{V_t^2}{2 x_d x_q} (x_d - x_q) \sin 2\delta} \qquad (5\text{-}65)$$

A plot of this equation appears in Fig. 5-22 for motor as well as generator operation. Curve (a) is a plot of the first term of Eq. (5-65) and is identical with the result obtained for the cylindrical-rotor machine. It represents the power associated with the electromagnetic torque produced by the mutual coupling between the rotor field winding (indicated by E_f) and the stator armature winding (indicated by V_t). Curve (b) is a graphical representation of the second term of Eq. (5-65). It is independent of excitation voltage which discloses that this term exists even when the field current is zero. Of course it is assumed that the machine is connected to an infinite bus system, otherwise the second term would be meaningless. When connected to the infinite bus, an armature current can flow without field excitation. Accordingly, a revolving mmf exists and it attempts to align itself with the minimum reluctance path as explained in Chapter 3. Since it arises because of the difference in reluctance between the direct and quadrature axes, it is called the *reluctance power* and has a value usually in the vicinity of 20% to 25%

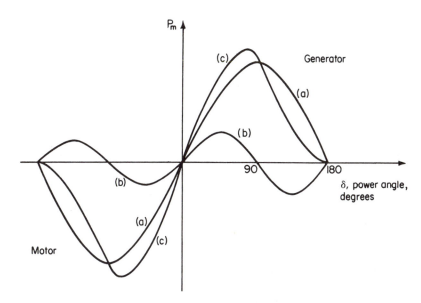

Figure 5-22 Developed power as a function of the power angle for the synchronous salient-pole machine.

of the rating of the machine. Moreover, it involves double the power angle and
thereby causes a peaking of the resultant developed power curve as indicated by
curve (c) in Fig. 5-22. The effect of the reluctance power term in the expression
for developed power is not only to increase the maximum value of developed
power but also to cause it to occur at a value of δ less than 90°. Finally, it is
interesting to note that, when Eq. (5-65) is applied to the cylindrical-rotor machine, the expression reduces to that of Eq. (5-55) because x_d and x_q are equal.

Example 5-5

(a) Calculate the electrical power delivered to the specified load by the three-phase
turbogenerator of Example 5-1.

(b) Using the results found in the solution to Example 5-2, find the value of the
mechanical power developed by the turbogenerator assuming negligible armature
winding losses.

(c) Repeat part (b) for Example 5-3.

Solution   (a) The power output using phase quantities is given by

$$P_o = 3V_t I_a \cos \theta = 3(7.9674)10^3(0.4184)10^3(0.8) = 8 \times 10^6 \text{ W} = 8 \text{ MW}$$

(b) Invoking Eq. (5-55) yields

$$P_m = 3\frac{V_t E_f}{x_s} \sin \delta = 3\frac{(7.9674)10^3(15.369)10^3}{22} \sin \underline{/28.6°}$$

$$= 7.993 \times 10^6 \text{ W} = 7.993 \text{ MW}$$

(c) Again by Eq. (5-55) the result is

$$P_m = 3 \frac{V_t E_f}{x_s} \sin \delta = 3 \frac{(7.9674)10^3(13.586)10^3}{17.25} \sin \underline{/25.15^\circ} = 8 \times 10^6 \text{ W} = 8 \text{ MW}$$

It is instructive to observe that the result for power is the same whether the calculation is based on the results of the excitation emf and the associated power angle found using the unsaturated or the saturated synchronous reactances.

5-8 PARALLEL OPERATION OF SYNCHRONOUS GENERATORS

The electrical needs of industry, commercial establishments, and individual consumers are supplied almost exclusively by synchronous generators. In many large regions of the country, there are virtually hundreds of such generators operating in parallel to meet these needs. It is the purpose of this section to describe some of the principles that govern the steady-state behavior of alternators that operate in parallel.

The Need to Synchronize

Initially we focus on the parallel operation of two isolated synchronous generators of equal ratings. Assume that the first generator G_1 is already connected to a three-phase resistive load as indicated in Fig. 5-23 by the closed load switch and that it is operating at rated frequency and rated voltage. It is now desirable to place the second alternator in parallel with the first in order that it might help share the current load as well as to prepare for the future increased power requirements of this load.

If severe shocks to both generators are to be avoided, a definite procedure, called the *synchronizing procedure,* must be followed before the switch (or circuit breaker) that parallels the two machines can be closed. The process is started by adjusting the speed of the prime mover of generator G_2 to correspond exactly to

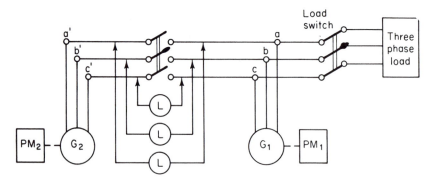

Figure 5-23 Synchronizing two synchronous generators (alternators) for parallel operation.

that of generator G_1. The field current of G_2 is then adjusted to yield the same nominal voltage between lines a', b', and c'. A suitable bank of lamps† is placed in series with lines a'-a, b'-b, c'-c, as shown in Fig. 5-23. These lamps will light up if either the voltage or the frequency of G_2 differs slightly from those of G_1. When the two frequencies are identical but the magnitudes of the line voltages are slightly different, the lamps will be on steadily. A subsequent adjustment of the field current of G_2 can then be used to cause the lamps to darken, at which time the synchronizing switch can be closed, thereby placing both machines in parallel without incident. On the other hand, if the two sets of three-phase voltages are equal in magnitude but a bit different in frequency, the lamps will again register light but now they will flicker at a rate equal to the difference frequency. At one point in the difference cycle, the voltage across the lamps will reach twice the line voltage; a half-cycle later it will be zero. An adjustment in the prime mover speed can serve to reduce this difference frequency to a very small value. Then, as a dark period is approached, the line switch can be closed with little or no disturbance.

Effect of Field Excitation (Control of Reactive AC Power)

Once G_2 is placed in parallel with G_1, in the manner described, and no further changes are made, the second machine assumes a state of simply idling on the line. If the assumed fixed load demand is to be equally shared by G_1 and G_2, it is necessary to make appropriate adjustments on both prime movers in a fashion to be described presently. For the moment, however, let us assume that each generator is delivering half the load power at rated frequency and rated load. Now we investigate the effect of varying the field excitation of each machine that is operating in parallel. This initial state of affairs is depicted in Fig. 5-24, corresponding to the unprimed quantities. This diagram is drawn for the case of negligible armature (stator) winding resistance. Moreover, the excitation current is initially assumed to be set at that value which yields unity power factor for each machine. This is illustrated by placing I_1 and I_2, the respective G_1 and G_2 stator currents, in phase with the voltage phasor. It is instructive to note here that synchronous generators operating in parallel have a unique field excitation that yields unity power factor at a fixed load share.

To better understand the sequence of events that unfolds with variation of excitation, we begin by first increasing the field current of G_1. In turn, this causes an increase in the field excitation phasor E_{f1} to a larger value E'_{f1}. The new position of phasor E'_{f1} in Fig. 5-24 must be such that its tip lies on the horizontal locus line to ensure that the vertical distance between this line and the line along which V lies remains a constant. This condition is demanded by Eq. (5-55), which for constant power requires that $E_{f1} \sin \delta_1 = E'_{f1} \sin \delta'_1$. In addition, the operating

---

†The equivalent commercial version of these lamps is called a *synchroscope*.

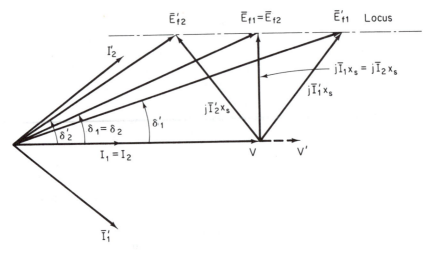

Figure 5-24 Phasor diagram drawn to illustrate the effect of field excitation on two synchronous generators operating in parallel. Real power remains fixed.

power angle of G_1 must assume a reduced value in the presence of the augmented E'_{f1}.

What happens to this power system if no subsequent adjustment is made on the second generator? The phasor diagram for the new condition in G_1 immediately reveals that G_1 can no longer operate at unity power factor. A reaction is, therefore, set in motion to reduce the effect of the increased excitation. In a generator overexcitation calls for a lagging reactive current component because it produces a demagnetizing effect. But because the load is assumed to be resistive, this generator lagging current component cannot be accommodated by the load. Therefore, it must exist as a circulating current between the two stator windings and as such must necessarily take on the character of a leading current in G_2. Since a leading current in a generator produces a magnetizing effect on the air-gap flux, it follows that the net effect of increasing only one field current is to cause the terminal voltage of the parallel set to increase to V' located midway between the horizontal projections of E_{f1} and E'_{f1} onto the V line.

On the basis of the foregoing analysis, it should be apparent that to maintain operation at the original value of terminal voltage (after the field current of G_1 is increased) it becomes necessary to *decrease* the field current of G_2 by a corresponding amount. The required condition is depicted in Fig. 5-24 by the prime notation on the quantities of G_2. Here again note that the phasor voltage differences between the new excitation voltages and the terminal voltage call for a reactive current flow—lagging in the overexcited machine G_1 and leading in the underexcited machine G_2. Accordingly, we are led to the conclusion that the effect of changing the field excitation of generators operating in parallel is merely to change the amount of reactive volt-amperes associated with each machine. There is no change in the amount of real power delivered.

Effect of Prime Mover Power (Control of Real Power)

Attention is next directed to a study of altering the power delivered to the shaft of each of two synchronous generators operating in parallel with fixed field excitations. The starting point is again taken to be that corresponding to which each generator delivers the same electrical power to the load at that excitation that puts the power factor of each machine at unity. It is important to understand at this juncture that the electrical energy developed by each generator is mechanically derived from the individual prime movers. These prime movers may be steam turbines, gasoline engines, waterwheels, or any practical primary energy source. The speed-power characteristics of these prime movers are decisive in establishing the manner in which generators operating in parallel will share a common load. Moreover, these characteristics assume a drooping bias in the fashion illustrated in Fig. 5-25. Because G_1 and G_2 are assumed to be identical machines, their speed-torque characteristics are also assumed to be the same. They are depicted as mirror images in Fig. 5-25 for reasons of clarity. At the initial operating point, the prime movers are revolving at a speed that yields the frequency required at the electrical load. At this frequency the prime mover of G_1 delivers power P_1 while the prime mover of G_2 delivers power $P_2 = P_1$.

    Suppose now that by suitable adjustment of the prime mover of G_1 the power delivered to G_1 is reduced. Simultaneously, in the interest of maintaining a constant frequency, let the input power to G_2 from its prime mover be increased in order to continue to operate at constant frequency. These prime mover adjustments are represented in Fig. 5-25 by a displacement of the speed-power characteristics as illustrated. Observe that to reduce the share of load supplied by generator G_1 its speed-power characteristic is displaced downward from position G_1 to G_1'. Similarly, the characteristic of the prime mover of machine two is raised in reference to its initial position. The result is that, at the maintained frequency, G_1 is now assuming a smaller share of the load $P_1' < P_1$, while G_2 takes on a greater share $P_2' > P_2$. The sum of the primed and unprimed powers is constant, however.

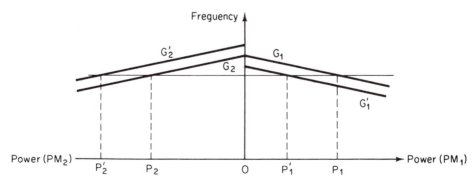

Figure 5-25 Speed-power characteristics of the prime movers of generators G_1 and G_2 of Fig. 5-23.

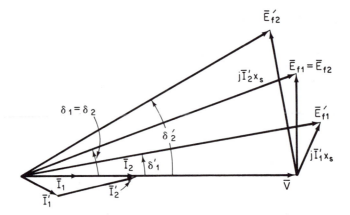

Figure 5-26 Phasor diagrams illustrating the effect of changing the prime mover power at constant excitation for two generators operating in parallel.

The effects of these prime mover adjustments on the electrical quantities of the individual generators are shown in Fig. 5-26. The reduced share of the electrical load supplied by G_1 is accompanied by a decrease in its power angle from δ_1 to δ_1' as demanded by Eq. (5-55). The phasor voltage difference between E_{f1} and V is correspondingly diminished and this in turn produces a smaller stator current $\bar{I}_1'$ for G_1. Two points are worth noting about this new value of G_1 current. First, it has a smaller in-phase component, which is consistent with the fact that it is delivering a smaller share of the electrical power taken by the load. Second, the power factor of this current is no longer unity. In fact, it is a lagging power factor. Corresponding changes take place in generator G_2. Here the power angle increases from δ_2 to δ_2', thus producing a larger phasor voltage difference which acts to establish an increased value of stator current in G_2 as represented by $\bar{I}_2'$. This current exhibits a greater in-phase component because it delivers a greater load share. In addition, it too now possesses a reactive current which is a leading component. Since the load is resistive, the reactive current of both generators constitute a common circulating current. Hence, if it is of a lagging nature in one generator, it must be leading in the other. Once again, we have an illustration of the fact that at each load condition there is a unique excitation for operation at unity power factor.

Connection to an Infinite Bus

For the sake of argument, let G_1 in Fig. 5-23 represent many large synchronous generators connected to a common power grid. This interconnection of generators frequently involves the use of solid copper bars and has come to be called a *bus*. Appearing at such a bus are characteristics of the electric power system that exhibit virtually constant voltage and frequency. It has come to be called an *infinite bus* because no single generator on the interconnected system can by itself in normal circumstances influence either of these quantities to any noticeable extent.

The synchronizing procedure needed to connect G_2 to the infinite bus is the same as for the interconnection of two isolated generators as described previously. Assume that this has been done and that G_2 is now idling on the power grid line. Any increase in the excitation of G_2 produces a lagging current with a power factor very close to zero. This reactive current serves to neutralize the effect of the increased field mmf in order that the terminal voltage stay invariant as demanded by the infinite bus. Actually, this reactive generator current serves to relieve the other generators on the power grid of some of their reactive volt-ampere burden. Correspondingly, a decrease in excitation causes G_2 to draw a leading current from the bus system. This current serves to produce a magnetizing effect of sufficient quantity to ensure an air-gap flux that is consistent with the bus voltage. A return of the excitation to a value of 100% (i.e., $E_f = V$), puts the machine once again in the idling state.

Let the prime mover of G_2 now be adjusted to raise its speed-power characteristic. This has the effect of increasing the input power to G_2, which in turn responds by delivering a corresponding amount to the power grid. As the speed of the prime mover of G_2 is momentarily increased to raise its speed-power characteristic, it does make an effort to raise the frequency of the entire power system. Of course, it fails to do so because it is no match for the enormous total capacity already on the grid to which G_2 would need to supply synchronizing power in order to bring about a frequency change. Instead, G_2 remains content with simply relieving the system of a tiny fraction of the delivered electrical power.

PROBLEMS

5-1. A synchronous generator has a short-circuit characteristic such that rated armature current is obtained by 0.75 per-unit excitation. Rated voltage on the air-gap line is obtained by 1.25 per-unit excitation. What per-unit excitation is required to supply a load at rated voltage, rated kVA, and 80% pf lagging?

5-2. Repeat Prob. 5-1 for the case where the load power factor is 80% leading.

5-3. A synchronous generator has the following rating:

   180-kVA, 440-V, 300-rpm, 60-Hz, three-phase, Y-connection,

   $r_i \approx 0,$ and $x_l = 0.296 \ \Omega$

The air-gap line is described by the equation $E = 17 I_f$, expressed per phase. The short-circuit characteristic is described by $I_{sc} = 10.75 I_f$. Compute the field current needed to provide rated voltage when rated current is delivered to a load at a pf of 0.8 *leading*. Use the general method linear analysis. Assume that $I_A = 17.9$ equivalent field amperes.

5-4. Repeat Prob. 5-3 for the case where the pf of the load is 0.8 lagging. $|\bar{A}|$

5-5. When the machine of Prob. 5-3 delivers rated armature current to a unity power factor load, the field winding current is measured to be 30 A. Moreover, the armature current is found to be lagging the excitation voltage by 50 electrical degrees.

   (a) Compute the resultant voltage induced in the armature winding by the air-gap flux.

   (b) What is the value of the terminal voltage per phase?

   (c) Find the power delivered to the load.

5-6. Repeat Prob. 5-5 for the case where the pf of the load is 0.8 lagging.

5-7. A 5-kVA, 220-V, 60-Hz, six-pole, Y-connected synchronous generator has a three-phase winding with 135 effective turns per phase. Moreover, its field winding is equipped with 200 turns. The leakage reactance per phase is 0.78 Ω and the armature winding resistance is negligible.

   (a) Compute the field current required to establish rated voltage across the terminals of a unity power factor load that draws rated generator armature current.

   (b) Determine the field current needed to provide rated terminal voltage to a load that draws 125% of rated current at 0.8 lagging pf.

5-8. A 5-kVA, 220-V, six-pole, 60-Hz, Y-connected synchronous generator has a three-phase winding with 135 effective turns per phase and an armature winding leakage reactance of 0.8 Ω per phase. The equation for the air-gap line is $E = 25I_f$, expressed in induced volts per phase. The generator has 200 field turns. When this machine delivers rated armature current to its connected load, like poles of the armature mmf are known to lag behind its field poles by 150 electrical degrees. Assume negligible saturation.

   (a) Find the value of the terminal voltage per phase.

   (b) What is the pf of the load?

   (c) How much power does the load take?

5-9. The machine of Prob. 5-3 has a cylindrical rotor.

   (a) Find the value of the unsaturated synchronous reactance.

   (b) What is the value of the armature reaction reactance?

   (c) By means of the synchronous reactance method find the field current needed to yield rated terminal voltage at rated current for a 0.8 lagging pf load.

   (d) Prepare a carefully drawn phasor diagram.

5-10. A three-phase, 15-kVA, 220-V, Y-connected, six-pole synchronous generator has a cylindrical rotor and negligible armature resistance. The open-circuit, short-circuit, and full-load zero pf characteristics are shown in Fig. P5-10.

   (a) Determine the armature leakage reactance per phase.

   (b) Compute the voltage regulation at unity pf by the general method.

   (c) Find the voltage regulation at pf = 1 by the use of the saturated synchronous reactance.

   (d) Compare the results of parts (b) and (c) and comment.

5-11. The number of field turns per pole for the machine of Prob. 5-10 is 400. Determine the number of effective armature turns per pole.

5-12. At no-load, the field current of the machine of Prob. 5-10 is adjusted to 5.8 A and kept at this value. A resistive load is then applied to the generator until rated current flows through the armature winding. At this condition the angle between the no-load excitation voltage E_f and the full-load terminal voltage is measured to be 46 electrical degrees. Find the value of the terminal voltage per phase by the general method.

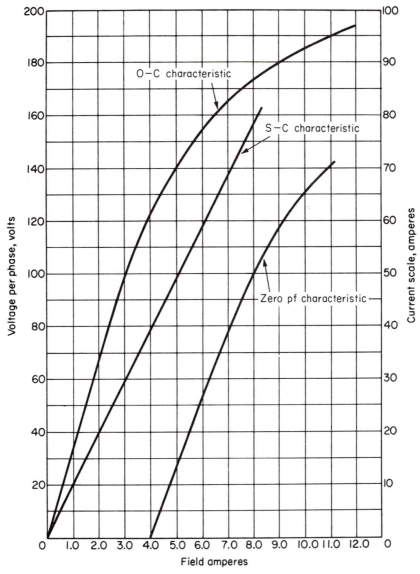

Figure P5-10

5-13. A cylindrical pole synchronous generator has the following rating:

$$200\text{-kVA, } 440\text{-V, } 300\text{-rpm, } 60\text{-Hz, three-phase, Y-connection, } r_a \approx 0$$

The open-circuit and rated current zero pf characteristics are shown in Fig. P5-13. Compute the field current needed to provide rated voltage when rated current is delivered to a load at a pf of 0.8 lagging. Use the general method.

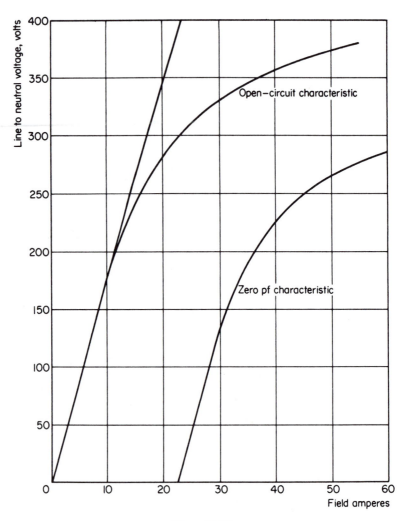

Figure P5-13

5-14. A 40,000-kVA, 14,000-V, three-phase, Y-connected alternator has negligible armature resistance and a leakage reactance of 1 Ω per phase. Other pertinent data are as
follows:

Short-circuit characteristic: $I_a = 7I_f$

Air-gap line, volts per phase: $E = 33I_f$

Open-circuit characteristic volts per phase: $E = \dfrac{21,300I_f}{430 + I_f}$

The equation for the open-circuit characteristic is not valid for values of I_f about the
origin. For a constant terminal voltage of 14,000 V (line) compute the variation of

field current from no-load to full-load kVA at 80% lagging pf. Use the general method for nonlinear analysis.

5-15. A 2100-kVA, 6600-V, Y-connected, 25-Hz, three-phase alternator is operated at synchronous speed with the armature short-circuited. The excitation is adjusted to produce full-load current in the armature. Without changing the field current, the armature circuit is opened and the terminal voltage is found to rise to 2900 V. Armature resistance is negligible.
  (a) Determine the voltage regulation for a 0.8 lagging pf by the synchronous reactance method.
  (b) Repeat for a pf of 0.8 leading.

5-16. A three-phase, Y-connected synchronous generator is rated at 200 kVA and 2300 V. It is supplying rated current at rated terminal voltage to a 0.8 lagging pf load. When the load is completely disconnected (without any change of field excitation), the terminal voltage rises to 3800 V.
  (a) Draw a phasor diagram, illustrating the foregoing load condition. Mark each phasor by its symbol and its numerical value.
  (b) Calculate the synchronous reactance of this generator. Assume r_a negligible.

5-17. A 50-kVA, 550-V, 25-Hz, single-phase alternator has an effective resistance of 0.25 Ω and a synchronous reactance of 3.2 Ω. Determine the percent regulation for rated load and 0.8 lagging pf.

5-18. A cylindrical-pole synchronous generator supplies rated kVA at a leading pf of 80%. Its per-unit synchronous reactance is 1.0, and the armature resistance may be assumed negligible. The base impedance for such a machine is defined as the ratio of the rated phase voltage to rated phase current. Compute the per-unit value of the excitation voltage and the associated power angle.

5-19. A synchronous generator operates at rated voltage and rated current supplying power to a 0.866 leading pf load. The per-unit value of excitation voltage is 1.732. The base voltage is the rated value of the phase voltage. Armature resistance is negligible. Find the per-unit value of the synchronous reactance.

5-20. A 1800-rpm, 60-Hz, 50,000-kVA, 13,800-V, synchronous generator has the following test data available:

Open-circuit test:

I_f (A):	175	200	225	250	275
V (line to line) kV:	12	13.0	13.8	14.5	15.1

Short-circuit test:
  $I_f = 193$ A, $I_a = 2090$ A

Zero pf characteristic:
  $I_f = 468$ A, $I_a = 2090$ A, $V = 13.8$ kV

Compute the field current needed to maintain rated terminal voltage for rated kVA at 80% pf.

5-21. A 5-kVA, 220-V, Y-connected, three-phase, salient-pole synchronous generator is used to supply power to a unity pf load. By means of a slip test, the direct-axis synchronous reactance is found to be 12 Ω and the quadrature-axis synchronous reactance 7 Ω. Assume that rated current is delivered to the load at rated voltage and that armature resistance is negligible.

     (a) Compute the corresponding value of the excitation voltage.
     (b) What is the value of the internal phase displacement angle ψ?
     (c) What is the value of the power angle?
     (d) Draw a carefully labeled phasor diagram.

5-22. Repeat Prob. 5-21 for the case where the load pf is 0.8 lagging.

5-23. Repeat Prob. 5-21 for the case where the load pf is 0.8 leading.

5-24. Assume that the characteristics depicted in Fig. P5-10 apply to the machine of Prob. 5-21. Moreover, assume that rated kVA is delivered to a unity pf load at rated voltage.
     (a) By the general method for salient-pole machines compute the value of excitation voltage E_f and compare with the result of Prob. 5-21(a).
     (b) What is the field current?
     (c) Draw a carefully labeled phasor diagram for this load condition.

5-25. Repeat Prob. 5-24 for the case where the load pf is 0.8 leading.

5-26. Repeat Prob. 5-24 for the case where the load pf is 0.8 lagging.

5-27. A six-pole, 60-Hz, Y-connected synchronous generator is rated at 15-kVA, 220-V, three-phase. A slip test is performed at a fixed reduced voltage of 96 V (line to line). The maximum and minimum readings of the ammeter used in the test are 20 A and 12 A, respectively. Compute the value of the excitation voltage when this machine delivers rated current to a 0.8 lagging pf load. Neglect armature resistance.

5-28. What physical significance can be attached to the synchronous reactances x_d and x_q as they relate to the salient-pole synchronous machine? Which reactance is larger and why?

5-29. A three-phase, four-pole, 60-Hz cylindrical-rotor synchronous machine is rated at 10,000 kW, 5000 V, and 0.8 pf lead. The synchronous reactance is 0.8 p-u. The normal open-circuit field current is 200 A. Neglect saturation and armature resistance. Assume a constant field excitation. Calculate each of the following when the machine is taking rated power from the line at rated current and voltage: (a) reactive volt-amperes, (b) field current, and (c) power angle.

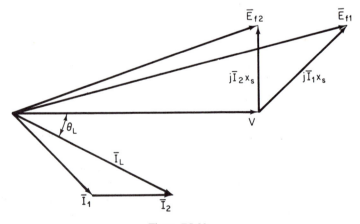

Figure P5-32

5-30. Refer to the schematic diagram of Fig. 5-23. Describe the reactions that occur during the synchronizing procedure when the speeds of the two generators are identical and the switch is closed at an instant when the voltages of the two generators are not exactly equal.

5-31. Repeat Prob. 5-30 for the case where the voltages are exactly the same but the frequencies of the two generators are slightly different.

5-32. Two generators of equal rating are operating in parallel supplying power to a lagging pf load for a condition as depicted in Fig. P5-32. Describe the adjustments to be made so that each generator is made to operate at the same power factor of the load. Is there any advantage to the new operating condition? Explain.

5-33. For the situation depicted in Fig. P5-32, describe the adjustments that must be made in order to permit both generators to operate at unity power factor when the load is unequally shared. What happens to the power angles of the individual generators?

5-2 $I_a = 1$ $(.8 \text{ leading})$ $\phi = 36.87°$

$E_{fsc} = \left(\frac{1}{1.25}\right)(.75) = 0.6 pu$ $E_f = V_t + I_a(r_a + j X_s)$

$X_s = \frac{.6 pu}{1}$ $= 1 + (.8 + j.6)(j.6) = 0.8 \angle 36.9°$

$\boxed{F = .8 \angle 126.9°}$

5-6a) $\bar{E} = 17 I_f$ $\bar{R} = \bar{F} + \bar{A} = \bar{F} + 17.9 \angle 0°$

$\bar{I}_f = \bar{R}$ $\boxed{\bar{E} = (\bar{F} + 17.9 \angle 0°)17}$

b) $\bar{V}_t = \bar{E} - j I_a X_R = (\bar{F} + 17.9 \angle 0°)17 - j I_a(.296)$

$= \boxed{(\bar{F} + 17.9)17' - j 69.9}$

$I_a = \frac{S}{3 \frac{V_\phi}{\sqrt 3}} = \frac{180K}{3\frac{440}{\sqrt3}} = 236.2$

5-16   200 KVA   a) $E_f = 2192$ $I_a X_s = 1121.7$
2300 V
.8 lag $\Rightarrow \phi = 36.8°$ 1327 / 36.8°

b) $I_a = \frac{200k}{3(2300)/\sqrt3} = 50.2A$ 50.2

$V_t = \frac{2300}{\sqrt3} = 1327.9 v$ $c^2 = a^2+b^2 - 2ab \cos\theta$

$E_f = \frac{3800}{\sqrt3} = 2192.8$ $1121.7^2 = 1327^2 + 2192^2 - 2(1327)(2192)\cos\theta$ $\theta = 24.2°$

$|E_f| = V_t + I_a(r_a + j X_s)$

$E_f^2 = (V_t \cdot .8 + I_a r_a)^2 + (V_t .6 + I_a X_s)^2$

$I_a X_s = \sqrt{2192^2 - ((1327.8)(.8))^2} - (1327.9)(.6) = 1121.7$

$X_s = \frac{1121.7}{50.2} = 22.3 \Omega$

6

Three-Phase Synchronous Motors

The significant and distinguishing feature of synchronous motors in contrast to induction motors is that they are doubly excited. Electrical energy is supplied to both the field and the armature windings. When this is done torque can be developed at only one† speed—the *synchronous speed*. At any other speed the average torque is zero. The synchronous speed refers to that rotor speed at which the rotor flux field and the armature ampere-conductor distribution (or the armature flux field) are stationary with respect to each other. To illustrate this point consider the conventional synchronous motor equipped with two poles. It has a dc voltage applied to the rotor winding and a three-phase ac voltage applied to the stator winding which is usually 60 Hz. The three-phase voltage produces in the stator an ampere-conductor distribution which revolves at a rate of 60 Hz. The dc (or zero-frequency) currents in the rotor set up a two-pole flux field which is stationary so long as the rotor is not turning. Accordingly, we have a situation in which there exists a pair of revolving armature poles and a pair of stationary rotor poles. A little thought should make it clear that for a period of half a cycle (or $\frac{1}{120}$ s) a positive torque is developed by the cooperation of the field flux and the revolving ampere-conductor distribution. However, in the next half-cycle this torque is reversed so that the average torque is zero. This is the reason that a synchronous motor per se has no starting torque.

---

†There is a second speed, too, but it is only of academic interest because of the excessive currents that flow.

To develop a continuous torque, then, it is necessary not only to have a flux field and an appropriately displaced ampere-conductor distribution, *but the two quantities must be stationary with respect to each other*. In singly fed machines such as the induction motor this comes about automatically. The relative speed between the stator field and the rotor winding induces rotor currents at those slip frequencies which yield a revolving rotor mmf. Superposing this upon the rotor speed gives a revolving rotor mmf, which in turn is stationary relative to the stator field. In the synchronous motor this condition cannot take place automatically because of the separate excitation used for the field and armature windings. In light of these comments it is reasonable to expect that, if an induction motor is doubly energized, it too will behave as a synchronous motor. Refer to Sec. 4-8.

On the basis of the foregoing comments it follows that to produce a continuous nonzero torque it is first necessary to bring the dc excited rotor to synchronous speed by means of an auxiliary device. Sometimes the auxiliary device takes the form of a small dc motor mounted on the rotor shaft. Most often, however, use is made of a squirrel-cage winding similar to that used in induction motors and imbedded in the pole faces. By means of this winding (also called the *amortisseur winding*) the synchronous motor is brought up to almost synchronous speed. Then, if the field winding is energized at the right moment, a positive torque will be developed for a sufficiently long period to allow the armature poles to pull the rotor poles into synchronism.

A general expression may be written to identify that frequency of rotation required of the rotor so that the stator and rotor fields are stationary with respect to each other. Let

f_1 = frequency of the currents through the stator winding

f_2 = frequency of the currents through the rotor winding

f_r = frequency of rotation of the rotor structure

Then the defining equation is

$$f_1 = f_2 + f_r \tag{6-1}$$

In the conventional synchronous motor the frequency of the rotor current f_2 is zero, since it is direct current. Hence for an f_1 of 60 Hz the rotor must revolve at 60 Hz in order for nonzero torque to be developed. On the other hand, note that, if the rotor winding were energized with 20-Hz current in place of direct current, then for an f_1 of 60 Hz the synchronous speed would be 40 Hz.

6-1 PHASOR DIAGRAM AND EQUIVALENT CIRCUIT

When the synchronous machine makes the transition from generator to motor action, there occurs a reversal of power flow. Thus, instead of current flowing *out of* the armature winding to a suitable electrical load, the current flows *into* the armature winding originating from a suitable electrical source. Since this transition occurs with the direction of rotation and the direction of field current flow unaltered from the generator case, it follows that the relationship between F and its associated excitation voltage E_f remains invariant. An observer looking only at this relationship certainly could not distinguish whether operation is as a motor or generator. One way to make this distinction, however, is to determine whether the armature current flows in a direction that is essentially opposed to that of the excitation voltage. If it is, the machine is behaving as a motor. In this condition the phasor diagram of the synchronous motor can then be shown to differ from that of the generator in two respects. The first involves the voltage equation for the stator circuit. In the generator, $\bar{E}_f$ (the excitation voltage) played the role of a source voltage and the terminal voltage $\bar{V}_t$ was dependent upon the magnitude of $\bar{E}_f$ and the synchronous impedance drop. In the motor the roles are interchanged. Now $\bar{V}_t$ is the source voltage applied to the synchronous motor armature winding and $\bar{E}_f$ is a reaction or counter emf which is internally generated. It is assumed that the terminal voltage originates from an infinite bus system† and so remains invariant. Applying Kirchhoff's voltage law to the synchronous motor for negligible armature resistance leads to

$$\boxed{V_t\underline{/0^\circ} = \bar{E}_f + j\bar{I}_a x_s}\qquad(6\text{-}2)$$

Clearly, this equation states that the applied stator voltage is equal to the sum of the drops. The excitation voltage is treated as a reaction voltage drop in much the same way as $\bar{E}_1$ was treated in the case of the transformer.

The circuit interpretation of Eq. (6-2) leads to the equivalent circuit as it applies to the synchronous motor. This appears in Fig. 6-1.

The second point of view in which the phasor diagram of the motor differs from that of the generator involves the angle δ. Physically, for the generator, the phase of $\bar{E}_f$ is ahead of $\bar{V}_t$ in time because of the driving action of the prime mover. Keep in mind that $\bar{E}_f$ may be considered associated with the rotor field axis. A similar line of reasoning pertains for the motor. At no-load, δ is zero and so the axis of a field pole and the resultant flux axis associated with the terminal voltage are coincident. As shaft load is applied, however, the rotor falls slightly behind its no-load position and in this way causes δ to increase but in a sense opposite to that for the generator.

On the basis of these modifications the phasor diagram of the synchronous motor becomes that shown in Fig. 6-2. Note that $\bar{E}_f$ now lags $\bar{V}_t$ and that it is the phasor sum of $\bar{E}_f$ and $j\bar{I}_a x_s$ which equals $\bar{V}_t$ as called for by Eq. (6-2). Another interesting point about this diagram is revealed upon comparison with Fig. 5-19. In both cases the magnitude of the excitation voltage $\bar{E}_f$ exceeds that of the terminal voltage $\bar{V}_t$. This is referred to as a condition of *overexcitation* (i.e.,

---

†A power system of tremendous capacity compared with the rating of the synchronous motor.

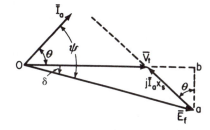

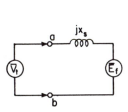

Figure 6-1 The equivalent circuit of the synchronous motor appears to the right of terminals *ab*.

Figure 6-2 Phasor diagram of the overexcited synchronous motor.

$|E_f| > |V_t|$). Note, however, that overexcitation in the synchronous generator is accompanied by a current of *lagging* power factor, whereas in the synchronous motor overexcitation is accompanied by a current of *leading* power factor. When the magnitude of $\bar{E}_f$ is equal to $\bar{V}_t$ the condition is referred to as *100% excitation*. A situation where $|E_f| < |V_t|$ is described as *underexcitation*.

6-2 PERFORMANCE COMPUTATION

Cylindrical-Pole Motor

The analysis of the performance of a synchronous motor is readily determined by the phasor diagram, the power-flow diagram, and the expression for the mechanical power developed. The usefulness of the power-flow diagram has already been demonstrated in connection with the induction motor. Hence it is presented in Fig. 6-3 without much additional comment. There is one fewer step in this flow graph because the ac source does not have to supply the rotor winding copper loss

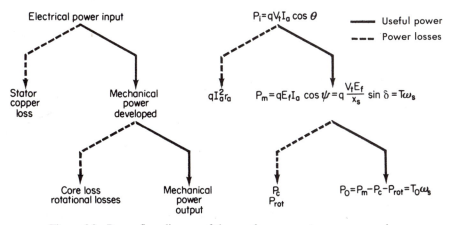

Figure 6-3 Power-flow diagram of the synchronous motor; ac power only.

since the rotor is traveling at synchronous speed. This is not to say that the rotor winding copper losses do not exist. They do, of course, but the power is supplied from the separate dc source. Figure 6-3 represents the flow of power from the ac source to the shaft of the synchronous motor.

What is the mechanism by which the synchronous motor becomes aware of the presence of a shaft load, and how is the electric energy source made aware of this so that it proceeds to provide energy balance? To answer this question we start first with a study of the conditions prevailing at no-load. As the power-flow diagram indicates, the only mechanical power needed is that which is required to supply the rotational and core losses. This calls for a very small value of δ, as depicted in Fig. 6-4(a). The machine is assumed to be overexcited since $\bar{E}_f > \bar{V}_t$. Note that δ_0 is just large enough so that the in-phase component of $\bar{I}_a$ is sufficient to supply the losses. Consider next that a large mechanical load is suddenly applied to the motor shaft. The first reaction is to cause a momentary drop in speed. In turn this appreciably increases the power angle and thereby causes a phasor voltage difference to exist between $\bar{V}_t$ and the excitation voltage $\bar{E}_f$. The result is the flow of an increased armature current at a very much improved power factor compared with no-load. As a matter of fact the speed changes momentarily by a sufficient amount to allow the power angle to assume that value which enables the armature current and input power factor to assume those values which permit the input power to balance the power required at the load plus the losses. Figure 6-4(b) is typical of what the final phasor diagram looks like. Note that the power component of the current is considerably increased over the no-load case. The increase in power angle from no-load to the load case can be measured by means of any stroboscopic instrument. The stroboscope furnishes a convincing demonstration of the physical character of the power angle.

The elements of the phasor diagram of the synchronous machine are the same whether it is used as a motor or generator. The geometry of the diagrams

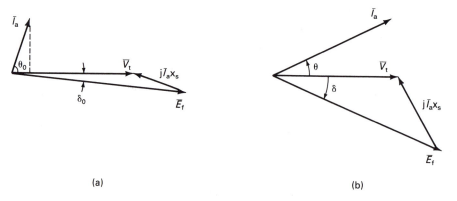

(a)   (b)

Figure 6-4   (a) Phasor diagram of synchronous motor at no-load with overexcitation; (b) under load at fixed excitation.

differs in two details as already indicated: The signs of the power angles are reversed and Eq. (6-2) is used for the motor, whereas Eq. (5-22) applies for the generator. In either case the expression for the mechanical power developed is given by Eq. (5-55) and is repeated here for convenience.

$$P_m = q \frac{V_t E_f}{x_s} \sin \delta \qquad (6\text{-}3)$$

Power Angle and Static Stability

When the synchronous motor is connected to a line source that behaves as an infinite bus, it follows that the maximum power developed corresponds to a power angle of 90° so that Eq. (6-3) becomes

$$(P_m)_{max} = q \frac{V_t E_f}{x_s} \qquad W \qquad (6\text{-}4)$$

It is useful to keep in mind, however, that during normal operation the power angle is considerably less than 90° even at rated conditions. The differential serves as a reserve capacity to help preserve static stability (i.e., maintain synchronism) during load changes.

Unfortunately, this stability limit is seriously compromised when the synchronous motor is supplied by a local generator which is of comparable rating. To illustrate the situation, refer to Fig. 6-5. For simplicity, armature resistances and other losses are assumed negligible. The generator is assumed electrically coupled to the synchronous motor through the synchronous reactances of both machines as depicted in Fig. 6-5(a). Moreover, the motor field excitation is adjusted

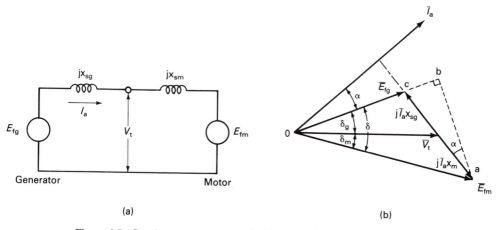

(a)   (b)

Figure 6-5 Synchronous motor supplied by a synchronous generator of comparable size. (a) Line diagram using per-phase quantities; (b) phasor diagram for an overexcited synchronous motor and underexcited synchronous generator.

for overexcitation and, correspondingly, the excitation in the generator is set for underexcitation. The power developed by the generator can be expressed as

$$P = qE_{fg}I_a \cos \alpha \qquad (6\text{-}5)$$

where E_{fg} is the generator excitation voltage and α is the phase angle between E_{fg} and I_a. Also, in triangle Oab of Fig. 6-5(b), the distance ab can be identified as

$$ab = E_{fm} \sin (\delta_m + \delta_g) = E_{fm} \sin \delta \qquad (6\text{-}6)$$

In triangle abc this quantity is alternatively given by

$$ab = I_a(x_{sg} + x_{sm}) \cos \alpha \qquad (6\text{-}7)$$

From the last two equations it then follows that

$$I_a \cos \alpha = \frac{E_{fm}}{x_{sg} + x_{sm}} \sin \delta \qquad (6\text{-}8)$$

Inserting Eq. (6-8) into Eq. (6-5) yields the result

$$P = q \frac{E_{fg}E_{fm}}{x_{sg} + x_{sm}} \sin \delta \qquad (6\text{-}9)$$

where

$$\delta = \delta_g + \delta_m \qquad (6\text{-}10)$$

Here δ_g is the power angle of the synchronous generator and δ_m is the power angle of the synchronous motor.

The maximum power represented by this motor-generator combination is obtained from Eq. (6-9) by again setting δ equal to 90°. But a comparison of this result with the case where the synchronous motor is supplied from an infinite bus reveals two important differences. (1) In the motor-generator case the angle δ includes both the generator as well as the motor power angles. Thus, in the case where the generator has the same nominal rating as the motor the individual power angles are each equal to 45°. Hence the maximum allowable motor power angle needed to maintain synchronism has been reduced to one-half the value allowed when the motor is supplied from an infinite bus. (2) The coefficient of sin δ in Eq. (6-9) is smaller than the corresponding coefficient of Eq. (6-3) because the denominator reactance term is now greater by a factor of two and, furthermore, because E_{fg} is less than V_t. The conclusion then follows that the static stability of the synchronous motor is much less when it is supplied by a comparably rated synchronous generator than when it is supplied from an infinite bus.

Example 6-1

A 2300-V, three-phase, 60-Hz, Y-connected, cylindrical-rotor synchronous motor has a synchronous reactance of 11 Ω per phase. When it delivers 200 hp the efficiency is found to be 90% exclusive of field loss, and the power angle is 15 electrical degrees as measured by a stroboscope. Neglect ohmic resistance and determine (a) the induced excitation voltage per phase, E_f, (b) the line current I_a, and (c) the power factor. Assume the motor is supplied from an infinite bus.

Solution   (a) The power input and the mechanical power developed are the same. Hence

$$3 \frac{V_t E_f}{x_s} \sin \delta = \frac{200(746)}{0.9}$$

Inserting $V_t = 2300/\sqrt{3}$, $\delta = 15°$, and $x_s = 11$ yields

$$E_f = 1793 \text{ V/phase}$$

(b) The armature current follows directly from Eq. (6-2). Thus

$$\bar{I}_a = \frac{1327\,\underline{/0°} - 1793\,\underline{/-15°}}{11\,\underline{/90°}} = 55.9\,\underline{/41°}$$

Thus since the angle of $\bar{I}_a$ is positive, the power factor is leading.
(c) The line power factor is merely

$$\text{power factor} = \cos 41° = 0.755 \text{ leading}$$

Salient-Pole Motor

The phasor diagram for an overexcited salient-pole synchronous motor is depicted in Fig. 6-6. The motor is assumed to be connected to an infinite bus and operating at a leading power factor angle θ. A comparison of this diagram with Fig. 5-18, which illustrates the phasor diagram of an overexcited synchronous generator, reveals a considerable similarity. There are differences, however. One is that overexcitation produces opposite effects regarding the power factor; it is leading for the motor and lagging for the generator. Another is that for the generator the terminal voltage plus the reactance drops yields the excitation voltage, whereas

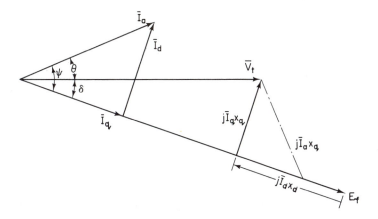

Figure 6-6 Phasor diagram for an overexcited salient-pole synchronous motor.

for the motor it is the excitation voltage summed with the reactance drops that equates to the terminal voltage.

Because the geometry of the phasor diagram is the same for Fig. 6-6 as for Fig. 5-18, the expression for the determination of the angle ψ, which locates the direction along which the excitation phasor lies, is identical to that of Eq. (5-43), which is repeated here for convenience. Thus

$$\psi = \tan^{-1} \frac{V_t \sin \theta + I_a x_q}{V_t \cos \theta} \tag{6-11}$$

The angle θ is to be used as an absolute quantity because the expression derives from geometrical considerations. Also, although the expressions for the magnitudes of the direct-axis and quadrature-axis currents are the same in both cases, the angles of these complex quantities are reversed in sign. Accordingly, we have

$$\bar{I}_q = (I_a \cos \psi)\underline{/-\delta}$$
$$\bar{I}_d = (I_a \sin \psi)\underline{/90° - \delta} \tag{6-12}$$

where in absolute terms the power angle takes on the value

$$\delta = \psi - \theta$$

Finally, with $\bar{V}_t$ taking on the role of the applied line voltage, Kirchhoff's voltage law applied on a per phase basis to the stator winding yields

$$\bar{V}_t = \bar{E}_f + j\bar{I}_d x_d + j\bar{I}_q x_q \tag{6-13}$$

from which the expression for the excitation voltage readily follows. Hence

$$\bar{E}_f = \bar{V}_t - j\bar{I}_d x_d - j\bar{I}_q x_q \tag{6-14}$$

The expression for the mechanical power developed by a salient-pole synchronous machine is derived in Sec. 5-7 specifically for the synchronous generator. The result is displayed in Eq. (5-65) and graphically portrayed in Fig. 5-22. Because the result applies with equal validity to the salient-pole synchronous motor, it is repeated here for ease of reference. Thus,

$$P_m = q \frac{E_f V_t}{x_d} \sin \delta + q \frac{V_t^2}{2 x_d x_q} (x_d - x_q) \sin 2\delta \tag{6-15}$$

Several points are worth noting in connection with Eq. (6-15). First, it is helpful to keep in mind that the derivation is based on the assumption that the energy source for the motor comes from an infinite-bus system. Also, the second term on the right side is called the *reluctance power* of the salient-pole synchronous motor because it owes its presence to a difference in reluctance between the direct axis and the quadrature axis of the machine. This term is independent of the excitation voltage and so does not require a field current for it to exist. As long as an armature winding current is permitted to flow, there exists a rotating armature mmf which produces a revolving air-gap flux that produces a force that acts to keep the rotor locked in a minimum reluctance position as the rotor revolves in synchronism with the revolving armature flux field. This reluctance power can be as large as 25% of the motor's rating. Finally, because the reluctance power is a function of double the power angle associated with the excitation component of developed power, the superposition of the two components plotted versus power angle causes the maximum power developed to occur at a power angle that is less than 90°. See Fig. 5-22.

Example 6-2

Consider that the synchronous machine of Example 5-4 is now operated as an over-excited synchronous motor drawing rated current at rated voltage from the three-phase line source at a leading pf of 0.8.

(a) Determine the excitation voltage of the synchronous motor at this operating point.

(b) Obtain the value of the mechanical power developed in watts as it derives from the presence of the field excitation.

(c) Find the reluctance power in watts.

(d) Compute the total mechanical power developed and compare it with the input power drawn from the line.

(e) Calculate the developed motor torque.

Solution   (a) Angle ψ is found to be the same as the value found for the overexcited generator. Thus

$$\psi = \tan^{-1} \frac{V_t \sin \theta + I_a x_q}{V_t \cos \theta} = \tan^{-1} \frac{\sin 36.9° + 0.577}{0.8} = 55.8°$$

Also,

$$\bar{I}_d = (I_a \sin \psi)\underline{/90° - \delta} = 0.827\underline{/71.1°}$$

$$\bar{I}_q = (I_a \cos \psi)\underline{/-\delta} = 0.562\underline{/-18.9°}$$

where δ is found from

$$\delta = \psi - \theta$$

Finally, from Eq. (6-14) we get

$$\text{p-u } \bar{E}_f = \text{p-u } \bar{V}_t - j(\text{p-u } \bar{I}_d)x_d - j(\text{p-u } \bar{I}_q)x_q$$

$$= 1\underline{/0°} - (0.827\underline{/71.1°})(0.961\underline{/90°}) - (0.562\underline{/-18.9°})(0.577\underline{/90°})$$

$$= 1 - (-0.752 + j0.257) - (0.105 + j0.307)$$

$$= 1.74\underline{/-18.9°}$$

It should not be surprising that this result is identical to that found in Example 5-4 for the overexcited synchronous generator except for the sign of the power angle.

(b) The developed mechanical power that originates from the excitation of the field winding is described by the first term on the right side of Eq. (6-15). Thus, evaluated in terms of per-unit quantities, we get

$$\text{p-u } P_{mf} = \frac{(\text{p-u } V_t)(\text{p-u } E_f)}{\text{p-u } x_d} \sin \delta = \frac{1(1.74)}{0.961} = \sin 18.9° = 0.5865$$

To obtain this result in watts it is simply necessary to multiply by the base kVA which is the rating of the motor, namely, 15 kVA. Thus,

$$P_{mf} = (0.5865)(15,000) = 8797 \text{ W}$$

(c) The expression for the reluctance power is the second term of Eq. (6-15). Calling this part P_{mr}, we evaluate

$$\text{p-u } P_{mr} = \frac{(\text{p-u } V_t)^2}{2(\text{p-u } x_d)(\text{p-u } x_q)} (\text{p-u } x_d - \text{p-u } x_q) \sin 2\delta$$

$$= \frac{1^2}{2(0.961)(0.577)} = (0.961 - 0.577) \sin 37.8°$$

$$= 0.2122$$

and expressed in watts this becomes

$$P_{mr} = (0.2122)(15,000) = 3183 \text{ W}$$

(d) The total mechanical power developed by this motor is the sum of the excitation power and the reluctance power. On a per-unit basis this value is

$$\text{p-u } P_m = \text{p-u } P_{mf} + \text{p-u } P_{mr} = 0.5865 + 0.2122 = 0.7987$$

This compares favorably with the per-unit input power, which is

$$\text{p-u } P_i = (\text{p-u kVA}) \cos \theta = 1(\cos 36.9) = 0.8$$

Expressed in watts this input power is

$$P_i = (0.8)(15,000) = 12,000 \text{ W}$$

The corresponding expression derived from the total mechanical power developed is

$$P_m = (0.7987)(15,000) = 11,985 \text{ W}$$

(e) As a six-pole machine operating on a 60-Hz power supply, the synchronous motor shaft speed is

$$n_s = \frac{120(60)}{6} = 1200 \text{ rpm}$$

Or, expressed in radians per second,

$$\omega_s = 1200 \left(\frac{\pi}{30}\right) = 125.7 \text{ rad/s}$$

Hence the torque expressed in newton-meters is

$$T = \frac{P_m}{\omega_s} = \frac{11,985}{125.7} = 95.4 \text{ N-m}$$

which in English units is

$$T = (0.7376)95.4 = 70.35 \text{ lb-ft}$$

6-3 POWER-FACTOR CONTROL

For a fixed mechanical power developed (or load) it is possible to adjust the reactive component of the current drawn from the line by varying the dc field current. This feature is achievable in the synchronous motor precisely because it is a doubly excited machine. Thus, although operation at a constant applied voltage demands a fixed resultant flux, both the dc source and the ac source may cooperate in establishing this resultant flux. If the field winding current is made excessively large, then clearly the resultant air-gap voltage in the motor tends to be larger than that demanded by the applied voltage. Accordingly, a reaction occurs which causes the armature current to assume such a power-factor angle that the armature mmf exerts that amount of demagnetizing effect which is needed to restore the required resultant flux. Similarly, if the field is underexcited, then the resultant gap flux tends to be too small. This also creates a phasor voltage difference between the line voltage and the motor excitation voltage which acts to cause the armature current to flow at that power-factor angle which enables it to magnetize the air gap to the extent needed to provide the necessary resultant flux.

Whether the reactive component of current must be leading or lagging readily follows from an investigation of the phasor diagrams depicted in Fig. 6-7 for various values of excitation voltage. The diagram applies to an assumed constant mechanical power developed. Hence as $\bar{E}_f$ is changed, the $\sin \delta$ must change correspondingly to keep $E_f \sin \delta$ invariant. In Fig. 6-7 this means that, as the excitation is varied, the locus of the tip of the $\bar{E}_f$ phasor is the broken line. Furthermore, the in-phase component of the current in each case must be the same. In the first case, where the excitation voltage is $\bar{E}_{f1}$, the field current is producing too much flux. This creates a reaction between the motor and the source which calls for a leading current of such a magnitude that it provides an amount and direction of synchronous reactance drop that when added to $\bar{E}_{f1}$ yields the fixed terminal voltage $\bar{V}_t$. Physically, what is happening is that a leading reactive current is made to flow, which acts to demagnetize the flux field to the extent needed. When the excitation is reduced to $\bar{E}_{f2}$, there is no excess flux produced by the field winding. Consequently, the ac line current contains no reactive component. It merely has the value of the in-phase component needed to

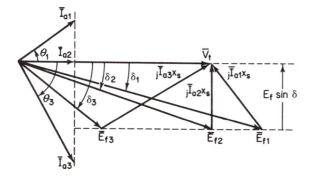

Figure 6-7 Showing the effect of varying excitation on power-factor.

supply power to the load. At the excitation corresponding to $\bar{E}_{f3}$ the machine is greatly underexcited. To compensate for this a reaction occurs which allows the line to deliver a large lagging reactive current, which helps to establish the value of air-gap flux demanded by the terminal voltage. Note, too, that as the excitation is decreased the power angle must increase for a fixed mechanical power developed.

The foregoing description can be neatly summarized by observing that at a given load condition, δ, the synchronous motor has no need to draw reactive current whenever

$$E_f \cos \delta = V_t$$

However, it draws a *leading* reactive current whenever

$$E_f \cos \delta > V_t$$

and a *lagging* reactive current whenever

$$E_f \cos \delta < V_t$$

If a plot is made of the armature current as the excitation is varied for fixed mechanical power developed, the current is observed to be large for underexcitation and overexcitation and passing through a minimum at some intermediate point. The plot in fact resembles a V shape as demonstrated in Fig. 6-8.

Synchronous Motor as a Synchronous Capacitor

The ability of the synchronous motor to draw leading current when overexcited can be used to improve the power factor at the input lines to an industrial establishment that makes heavy use of induction motors and other equipment drawing power at a lagging power factor. Many electric power companies charge increased power rates when power is bought at a poor lagging power factor. Over the years such increased power rates can result in an appreciable expenditure of money. In such instances the installation of a synchronous motor operated at overexcitation can more than pay for itself by improving the overall input power factor to the point that the penalty clause no longer applies.

When the synchronous motor is operated solely for the purpose of furnishing power factor correction, it is called a *synchronous capacitor*. In such a situation

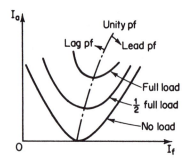

Figure 6-8 Synchronous-motor V curves.

the synchronous motor serves the function of supplying the reactive volt-amperes required by the singly excited induction motors that are operating on the same power lines and thereby relieves the power source of the need to do so. In addition to the favorable power rates that can result, the overall efficiency of the plant is improved because of the reduced line losses that accompany the improved line power factor.

Example 6-3

An industrial plant is supplied from a 2300-V, three-phase infinite-bus transmission line. The plant uses several large three-phase induction motors drawing a total of 2600 kW at a lagging power factor of 0.87 when fully loaded.

(a) Find the line current drawn from the transmission line.

(b) Determine the kVA rating of a synchronous motor used as a synchronous capacitor that is needed to improve the power factor of the plant input to unity.

(c) What is the new value of the input current to the plant?

(d) If the selected synchronous capacitor is found to have a synchronous reactance of 1.6 Ω, find the required excitation voltage per phase needed for part (c).

Solution   (a) From the expression for power in a three-phase system

$$I_L = \frac{2600 \times 10}{\sqrt{3}(2.3)10^3\ (0.87)} = 750.2 \text{ A}$$

(b) The current rating of the synchronous capacitor is the total reactive component of the input current found in part (a). Thus,

$$I_{sm} = I_L \sin \theta = 750.2 \sin (\cos^{-1} 0.87) = 750.2 \sin 29.54° = 369.9 \text{ A}$$

$$\therefore \ \ \text{kVA rating} = \sqrt{3}(2.3)(369.9) = 1473$$

(c) The new value of the input current from the transmission line is simply the in-phase component of the condition described in part (a). Hence,

$$\text{new } I_L = 750.2 \cos 29.54° = 750.2(0.87) = 652.7 \text{ A}$$

(d) From Eq. (6-2)

$$\bar{E}_f = \bar{V}_t - j\bar{I}_{sm}x_s = \frac{2300}{\sqrt{3}} - j369.9 \underline{/90°}\ (1.6)$$

where $I_{sm} = 369.9 \underline{/90°}$ because the synchronous capacitor draws leading current. Accordingly,

$$E_f = \frac{2300}{\sqrt{3}} \underline{/0°} - j1.6(j369.9) = 1327.9 + 591.8 = 1919.7 \text{ V}$$

6-4 SYNCHRONOUS MOTOR APPLICATIONS

Synchronous motors are rarely used below 50 hp in the medium-speed range because of their much higher initial cost compared with induction motors. In addition they require a dc excitation source, and the starting and control devices are usually more expensive—especially where automatic operation is required. However, synchronous motors do offer some very definite advantages. These include constant-speed operation, power-factor control, and high operating efficiency. Furthermore, there is a horsepower and speed range at which the disadvantage of higher initial cost vanishes, even putting the synchronous motor at advantage. This is demonstrated in Fig. 6-9. Where low speeds and high horsepower are involved, the induction motor is no longer cheaper because it must use large amounts of iron in order not to exceed a gap flux density of 0.7 T. In the synchronous machine, on the other hand, a value twice this figure is permissible because of the separate excitation.

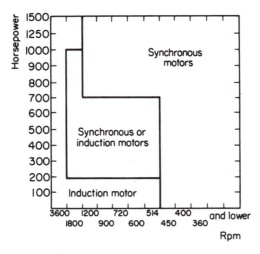

Figure 6-9 The general areas of application of synchronous and induction motors.

Appearing in Table 6-1 are some of the more important characteristics of the synchronous motor along with some typical applications. In addition to the area of application listed there, these motors are also quite prominently found in the following applications, which are characterized by operation at low speeds and high horsepower: large, low-head pumps; flour-mill line shafts; rubber mills and mixers; crushers; shippers; pulp grinders and jordans and refiners used in the papermaking industry.

6-5 FIELD EXCITATION BY RECTIFIED SOURCES

A distinguishing feature of the conventional synchronous motor is its need for dc as well as an ac electrical supply. Because the ac source is the one that is frequently available, special provision must be made for a dc source. In years past, one solution was to equip the synchronous motor with an auxiliary self-excited dc generator attached to the motor shaft. Once the synchronous motor reached nearly synchronous speed through the induction-motor action of its squirrel-cage winding, the dc generator field rheostat was adjusted to furnish the proper excita-

TABLE 6-1 SYNCHRONOUS MOTOR CHARACTERISTICS AND APPLICATIONS

Type designation	Synchronous, high speed, above 500 rpm	Synchronous, low speed, below 500 rpm
Starting torque (% of normal)	Up to 120	Low 40
Pull-in torque (% of normal)	100–125	30
Pull-out torque (%)	Up to 200	Up to 180
Starting current (%)	500–700	200–350
Slip	Zero	Zero
Power factor	High, but varies with load and with excitation	High, but varies with excitation
Efficiency, %	Highest of all motors, 92–96	Highest of all motors, 92–96
Typical applications	Fans, blowers, dc generators, line shafts, centrifugal pumps and compressors, reciprocating pumps and compressors. Useful for power-factor correction. Constant speed. Frequency changers	Lower-speed direct-connected loads such as reciprocating compressors when started unloaded, dc generators, rolling mills, band mills, ball mills, pumps. Useful for power-factor control. Constant speed. Flywheel used for pulsating loads.

Source: By permission from M. Liwschitz-Garik and C. C. Whipple, *Electric Machinery,* vol. II (Princeton, N.J.: D. Van Nostrand Co., Inc., 1946).

tion for the synchronous motor, which in turn pulled the synchronous motor into synchronism with the line voltage. In applications of large synchronous motors today, the exciter generator is conveniently replaced by a dc source obtained by electronic methods employing diodes and silicon-controlled rectifiers (SCRs), which are fully capable of handling a large range of currents over a wide range of voltages.

A typical controlled electronic rectifier circuit is shown in Fig. 6-10. This circuit is a single-phase, half-wave controlled rectifier which can be used to adjust the field current in the field winding of a synchronous motor. The field winding is represented by the resistance R and the inductance L and is called the load. Two electronic elements are used: the diode D_{fw} that serves to shunt the load and the silicon-controlled rectifier S1 which behaves as a controllable switch. Of course the source voltage is the frequently available ac supply. The manner in which this circuit performs the conversion from a fixed ac supply to an adjustable dc source can be understood by referring to Fig. 6-11. It is helpful to recall that the SCR conducts when two conditions are satisfied at its three terminals. The first is that the voltage from plate (P) to cathode (K) be positive. The second is that the gate-cathode voltage be set to allow conduction to take place during the time the plate-cathode voltage is positive. By proper adjustment of this gate voltage it is possible to permit conduction to occur over the entire time the plate-cathode voltage is

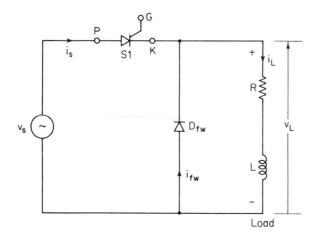

Figure 6-10 Single-phase, half-wave controlled rectifier circuit.

positive or during no part of this time. Figure 6-11(a) shows two cycles of the ac source voltage. In Fig. 6-11(b) the rectified version of the ac source is shown. Observe that a positive potential appears across S1 (the SCR) only during the positive portion of the ac input voltage variation. During the negative portion of the cycle, a reversed (negative) voltage appears across the SCR and so conduction is not possible. This accounts for the zero level of load voltage from $\omega t = \pi$ to $\omega t = 2\pi$. In the first half-cycle, however, S1 is made to conduct at the angle θ, which is determined by the voltage on the gate terminal. This angle is often referred to as the *firing angle* of the SCR and is adjustable between 0 and π radians in this case. When firing of the thyristor (SCR) is made to occur at $\theta = 0°$, the voltage developed across the load is a maximum. The associated average value of this maximum voltage can easily be shown to be V_m/π where V_m denotes the peak value of the ac supply voltage.

The variation of the load current produced by the rectified voltage of Fig. 6-11(b) is depicted in Fig. 6-11(c). It is particularly interesting here to note that action of the inductance causes the current to exist beyond the time when the thyristor ceases to conduct. Since the current at $\omega t = \pi$ is not zero and there exists an amount of energy stored in the magnetic field of the inductor consistent with this nonzero current, a path must be provided to permit this current to continue to flow. As soon as the SCR cuts off at $\omega t = \pi$, the load current tries to change instantaneously. Then, by Lenz's law, an internally generated emf is induced in the inductor of a polarity that acts to sustain the original direction of the current. This action puts a positive potential across the shunt diode, D_{fw}, which makes it conductive, thus providing a path for the load current. Current decay subsequently proceeds at a rate determined by the field winding time constant. The shunt diode is called a *free-wheeling* diode for the reason that it freely responds to conditions that call for an auxiliary current path whenever inductive elements are involved.

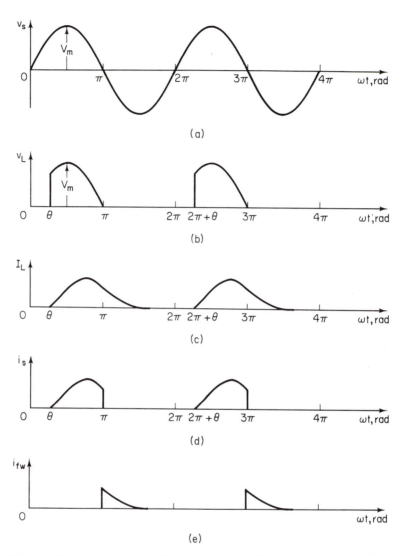

Figure 6-11   Time variations of the electrical quantities appearing in Fig. 6-10: (a) source voltage; (b) rectified voltage across the R-L load; (c) the R-L load current variation; (d) source current; (e) current in the free-wheeling shunt diode. The firing angle of the SCR (S1) is θ.

The variations with time of the source current and the current in the free-wheeling diode are depicted in Figs. 6-11(d) and (e), respectively. The abrupt change in the source current at $\omega t = \pi$ is allowed on the assumption that the ac source has negligible internal inductance, which is reasonable. The magnitude of the step current that appears in i_{fw} of course matches the cutoff value that is found in the variation of i_s.

The half-wave controlled rectifier has the disadvantage of relatively low currents because of the absence of any contributions from the source when its potential is undergoing a negative excursion. This situation, however, is nicely resolved by employing the circuitry illustrated in Fig. 6-12, which is called a full-wave controlled rectifier. By adding an additional SCR and diode along with appropriate modifications in the gate control circuitry, conduction to the load can be achieved in a controlled fashion during each successive half-cycle of the source frequency. For an explanation of the operation of this circuit refer to Fig. 6-13. During the first positive half-cycle of the source voltage, a positive potential appears across the plate-cathode terminals of thyristor S1. In Fig. 6-12 the left side of the source is placed positive while the cathode of D2 is placed negative. When the gate terminal is pulsed at θ, S1 conducts and thereby applies the source voltage to the R-L load by completing the circuit through diode D2. In other words, when the source voltage is positive, there exists a series path connecting the source to the load through thyristor S1 and diode D2. As soon as the source goes negative just beyond π radians in Fig. 6-13(a), S1 goes off as does D2. But now the right side of the source in Fig. 6-12 is positive and the left side is negative, which means that a positive potential appears across both S2 and D1. Again a closed path becomes established between the source and the load upon the arrival of the gate pulse at $\omega t = \pi + \theta$, but now it is thyristor S2 and diode D1 that are effective in providing the connection. Observe too that the sequencing of the switching operation is such as to maintain the polarity of the voltage across the load the same as it was in the preceding half-cycle. See Fig. 6-13(b). This ensures a unidirectional current flow. Because a contribution occurs for each half-cycle of the source voltage, the average voltage that is available to the load doubles over that of the half-wave rectifier to a maximum value of $(2/\pi)V_m$. Moreover, control

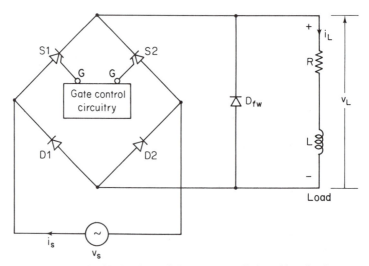

Figure 6-12 Single-phase, full-wave controlled rectifier circuit.

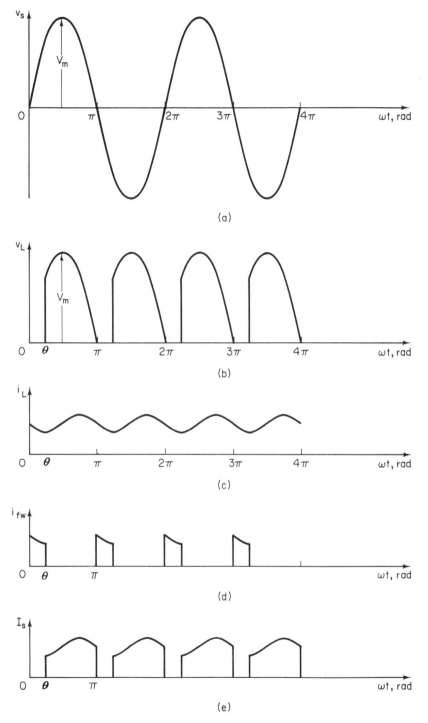

Figure 6-13 Time variations of the electrical quantities found in Fig. 6-12: (a) source voltage; (b) rectified load voltage; (c) resultant load current variation in steady state; (d) current in the free-wheeling diode; (e) the source current. The SCR firing angle is set to θ.

is available from zero to this maximum through the action of the gate control circuit, which varies θ from π radians to zero.

The corresponding variation of the current in the load is depicted in Fig. 6-13(c). This graph shows the variation after steady state is reached following a suitable transient response. Note the appreciable improvement over the half-wave circuit. A glance at Fig. 6-13(d) reveals the essential contribution of the free-wheeling diodes which is to provide a current path over the time interval between the start of any half-cycle of the source voltage and the point of firing of the thyristors. Figure 6-13(e) illustrates the variations of the source current during the time that the SCRs are in the ON state. These current pulses are supplied alternately through thyristors S1 and S2 to the load from the source.

The foregoing analysis illustrates the improvement that occurs when a full-wave rectifier circuit is used in place of a half-wave rectifier circuit. However, a further improvement is possible when a *three-phase* full-wave arrangement is used in place of the *single-phase* full-wave circuit. Such a circuit is depicted in Fig. 6-14. It is useful to note that a total of six thyristors (SCRs) is needed with this arrangement. The availability of a rectifier circuit that uses a three-phase source is attractive for the reason that the three-phase source is already needed to drive the three-phase synchronous motor and, accordingly, is assumed to be available. By the proper sequential pulsing of the gate terminals of the thyristors in the circuitry of Fig. 6-14 it is possible to vary the average voltage across the field winding over a range that varies from zero to $(3/\pi)V_m$, where V_m denotes the maximum value of the ac voltage between lines.

The success of this circuit depends upon a biasing scheme that permits only two thyristors to be in a conducting state at any time while simultaneously ensuring that the remaining SCRs are in a state of reversed bias. A typical switching sequence for this circuit is illustrated in Fig. 6-15 as it applies to a line voltage sequence of a, b, c. Each thyristor is scheduled for a maximum ON period of 120

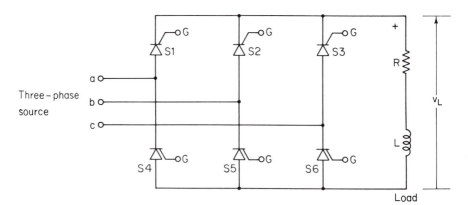

Figure 6-14   Three-phase, full-wave controlled rectifier circuit.

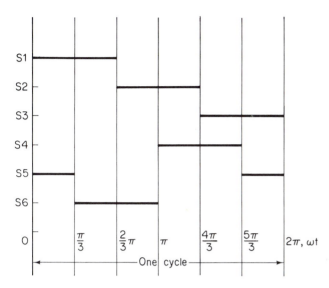

Figure 6-15 Switching sequence diagram for the thyristors of Fig. 6-14. The heavy lines indicate the specific portions of a cycle when each thyristor is forward biased for conduction to occur subject to the action of the gate signal.

electrical degrees for each cycle of the source voltage. The gating pulse is designed to allow a full variation over this interval. Starting with the beginning of a cycle, S1 is forward-biased to be in the ON state potentially over the first 120 electrical degrees while S2 and S3 are scheduled potentially to be in the ON state for each of the next 120° intervals respectively. Then, beginning with the start of the next half-cycle at π radians, a similar schedule is established for thyristors S4, S5, and S6. Accordingly, S4 is arranged to have a total possible ON period of 120° extending from π to $\frac{4}{3}\pi$ radians. Similarly, S5 is forward-biased for a possible ON period extending from $\frac{5}{3}\pi$ radians right on into the first $\pi/3$ radians of the next cycle. As a matter of convenience this second 60° portion is shown reflected into the first cycle in Fig. 6-15. Thyristor S6 is subsequently scheduled for a potential ON period covering the next 120 electrical degrees. Reflected back into the first cycle, S6 is scheduled to occupy the period from $\pi/3$ to π radians.

As a result of the switching sequence schedule depicted in Fig. 6-15 for the line voltage sequence a, b, c, it follows that both thyristors S1 and S2 are forward-biased for conduction during the first $\pi/3$ radians of a cycle that begins with v_{ab} at $(\sqrt{3}/2)V_m$ and moving to its peak while v_{ac} is simultaneously passing through zero and moving positively. The specific point in this interval at which conduction actually occurs is determined by the gate pulsing signals. Similarly, in the next 60° interval, it is thyristors S1 and S6 that are involved in the conduction process. In the third 60° interval the responsibility for conduction passes over to thyristors S2 and S6. This process continues for the remainder of the cycle and then repeats itself for each cycle. The result is an effective control of the field winding dc current at entirely acceptable levels.

Brushless Excitation for Synchronous Generators

Synchronous generators employed by the public utility companies can have ratings of 1000 MVA and more (where MVA denotes megavolt-amperes). To equip such machines with dc power to be conducted to the rotating field winding through

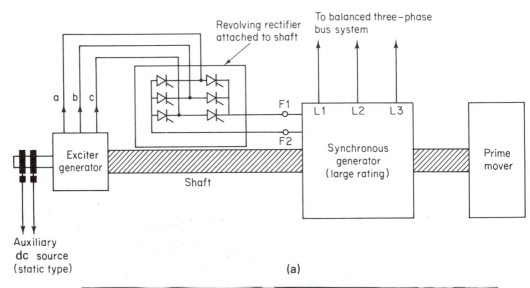

(a)

(b)

Figure 6-16 (a) Schematic diagram illustrating how brushless excitation of a large synchronous generator is obtained. (b) Rotor of synchronous generator showing the exciter generator as well as the revolving rectifier.

slip rings presents a significant maintenance problem as well as the loss of a substantial amount of power consumed as a heat loss in the carbon brushes used to conduct the large dc currents to the slip rings. An innovative application of the use of the circuitry of Fig. 6-14 can be used to eliminate these problems. The arrangement is illustrated in Fig. 6-16(a) and a photograph of an actual machine appears in Fig. 6-16(b).

The object of this scheme is to make available a dc source which is revolving at the same speed as the field winding of the main generator. A direct connection to the terminals of the field winding is then possible thereby eliminating the need for brushes. Attached to the shaft of the main generator is an exciter generator the size of which is sufficient to provide the power required by the field winding to maintain the level of dc current needed by the synchronous generator. Also attached to the shaft is the rectifier of Fig. 6-14. A small dc current (in comparison to that required by the synchronous generator) is supplied to the field winding of the exciter generator. An auxiliary dc source derived from a static rectifier can be used because the currents involved here are manageable. When the prime mover of the synchronous generator reaches synchronous speed, the ac exciter generator develops a balanced three-phase voltage. Application of this voltage set to the three-phase full-wave rectifier attached to the generator shaft produces a dc voltage source at terminals F1–F2 which revolves at the same velocity as the generator field winding itself. Hence, a firm, terminal connection is possible and slip rings and brushes for the main generator are no longer necessary.

6-6 CORRELATION OF SYNCHRONOUS MOTOR OPERATION WITH THE BASIC TORQUE EQUATION

It is the purpose of this section to outline in some detail the manner in which the factors appearing in the basic torque equation [Eq. (3-97) of Appendix to Chapter 3] are manifested in synchronous motor operation for several power factor conditions. The doubly excited feature of the synchronous motor complicates the situation somewhat because field winding conditions of overexcitation or underexcitation influence the reactive component of the armature current to become leading or lagging, respectively. See Sec. 6-3. Attention is directed first to the case where ψ equals $0°$, i.e., the armature current is in phase with the excitation voltage (and treated as a voltage drop). Refer to Fig. 6-17(a). Our concern is with the cylindrical-rotor machine in spite of the use of a salient-pole structure; the latter is employed merely to facilitate the graphical representation. For the indicated polarity of the rotor field flux and direction of rotation, the location of the amplitude of the emf distribution of the entire three-phase winding is represented by coil e_f-e_f' having the instantaneous directions shown. When the machine is operating as a motor, the armature current flows *opposed* to the polarity of the e_f shown because e_f is represented here as a *voltage rise* with respect to the field flux that

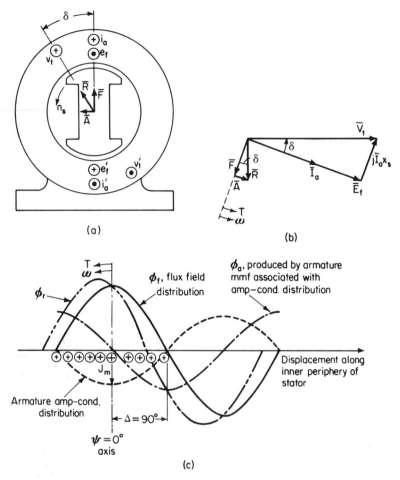

Figure 6-17 Space and time relationships in the synchronous motor for $\psi = 0°$: (a) space diagram depicting the axis of ampere-conductor distribution; (b) corresponding phasor diagram; (c) detailed representation of (a). Field motion is toward the left.

produces it. However, when viewed with respect to the current,† the $\bar{E}_f$ phasor is treated as a *voltage drop* and is so represented in Fig. 6-17(b). The coil i_a-i_a' in Fig. 6-17(a) represents the location of the amplitude of the assumed sinusoidal ampere-conductor distribution. Note that the armature current flows inward for the upper conductor and outward for the lower conductor because i_a is specified as in-phase with the excitation voltage expressed as a voltage drop and hence must be opposite to the direction of e_f treated as a voltage rise. The armature mmf (represented by $\bar{A}$) is therefore directed from right to left and cooperates with $\bar{F}$ to

†See Sec. 2-1.

produce the resultant $\bar{R}$. Note that the $\bar{R}$ in Fig. 6-17(a) is consistent with that shown in Fig. 6-17(b). For simplicity the armature leakage reactance is assumed to be negligible so that the armature induced emf $\bar{E}_r$ may be taken equal to the terminal voltage $\bar{V}_t$. It is for this reason that $\bar{R}$ is drawn perpendicular to $\bar{V}_t$ in Fig. 6-17(b). Then to be consistent with this the amplitude of the distribution of the terminal voltage in the armature winding can be represented in the space picture of Fig. 6-17(a) by coil v_t-v_t'. It is worthwhile to note at this point that by means of such a representation a physical interpretation of the power angle δ is obtained.

Also worthy of note is the fact that application of the $\bar{I} \times \bar{B}$ rule indicates a clockwise torque produced on the armature (stator) winding, while the torque on the rotor is counterclockwise. For the motor, therefore, the direction of torque and the direction of rotation are the same. In Fig. 6-17(b) note again that the direction of the torque on the rotor is such that it tends to align the field axis with the axis of the resultant flux, as was the case for generator action.

The information contained in Figs. 6-17(a) and (b) can also be represented in more detail by employing the developed form depicted in Fig. 6-17(c). The fundamental component of the field flux produced by $\bar{F}$ is denoted by ϕ_f. The ampere-conductor distribution is denoted by the dashed line. The armature flux produced by the armature mmf associated with the ampere-conductor distribution is represented by ϕ_a; this ϕ_a wave is properly displaced 90° behind the ampere-conductor distribution and in-phase with the armature mmf wave. Of course the resultant flux is produced by summing ϕ_f and ϕ_a and is permitted by the linearity assumption. This is denoted by ϕ_r. Keep in mind that for a constant terminal voltage the amplitude of the ϕ_r wave remains invariant. A study of these distributions makes the correspondence with the basic torque equation self-evident. Because $\psi = 0°$, each conductor beneath a given field pole carries current in the same direction and thereby makes a contribution to the developed torque. This viewpoint is consistent with that version of the basic torque equation described by

$$T = \frac{\pi}{8} p^2 \Phi J_m \cos \psi = \frac{\pi}{8} p^2 \Phi J_m \qquad (6\text{-}16)$$

where Φ is the flux produced by the field winding and J_m denotes the amplitude of the ampere-conductor distribution.

The second viewpoint involves the use of the *torque angle* Δ. This angle is defined in Fig. 3-18 (p 574) as the angle between the axis of the field pole and the axis of the armature mmf pole. Moreover, Δ is always the complement of angle ψ. A glance at Fig. 6-17(c) discloses that Δ is 90° for this case so that the second form of the basic torque equation becomes

$$T = \frac{\pi}{8} p^2 \Phi \mathscr{F}_p \sin \Delta = \frac{\pi}{8} p^2 \Phi \mathscr{F}_p \qquad (6\text{-}17)$$

where $\mathscr{F}_p$ denotes the armature mmf per pole.

Figure 6-18 depicts the situation that exists in the synchronous motor when $\psi = 60°$ leading. Note that the field pattern for producing torque is not optimum

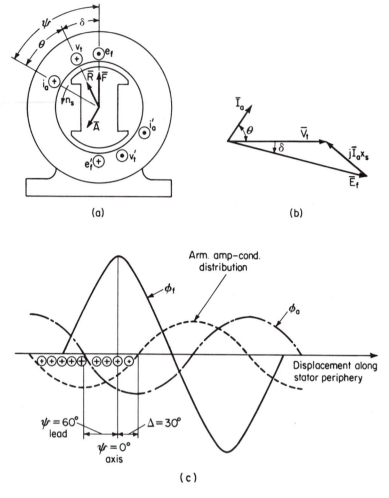

Figure 6-18   Space and time relationships in the synchronous motor for $\psi = 60°$
leading: (a) space diagram; (b) phasor diagram; (c) detailed form of (a).

because now some of the armature conductors find themselves under a pole flux of
the wrong polarity. This is accounted for in Eq. (6-16) by the need to use cos $\psi =$
cos 60° rather than unity as in the preceding case. When the situation is described
in terms of the formulation of Eq. (6-17), the required value of Δ is obviously 30°.
It is worthwhile to note in Fig. 6-18(c) that the orientation of the armature mmf (or
its associated flux) with the field flux is such as to produce a net demagnetization.
This is entirely consistent with the vector relationship shown between $\bar{F}$ and $\bar{A}$ in
Fig. 6-18(a).

PROBLEMS

6-1. A three-phase, 1732-V (line to line), Y-connected, cylindrical-pole synchronous motor has $r_a = 0$ and $x_s = 10 \, \Omega$ per phase. The friction and windage plus the core losses amount to 9 kW. The motor delivers an output of 390 hp. The greatest excitation voltage that may be obtained is 2500 V per phase.
   (a) Calculate the magnitude and pf of the armature current for maximum excitation at the specified load. *.605 leading*
   (b) Compute the smallest excitation for which the motor will remain in synchronism for the given power output. *1000 V/phase*

6-2. A 2300-V, three-phase, 60-Hz, Y-connected, cylindrical-rotor synchronous motor has a synchronous reactance of 11 Ω per phase. When it delivers 200 hp, the efficiency is 90% and the power angle is 15 electrical degrees. Neglect r_a and determine: **(a)** the excitation emf per phase; **(b)** the line current; **(c)** the power factor.

6-3. A synchronous motor is operated at half-load. An increase in its field current causes a decrease in armature current. Does the armature current lead or lag the terminal voltage? Explain.

6-4. A three-phase, Y-connected synchronous motor is operating at 80% leading pf. The synchronous reactance is 2.9 Ω per phase and the armature winding resistance is negligible. The armature current is 20 A per phase. The applied line voltage is 440 V.
   (a) Find the excitation voltage and the power angle.
   (b) It is desired to increase the armature current to 40 A per phase and maintain the pf at 80% leading. Show clearly in a phasor diagram how the field excitation must be changed. Can this be accomplished if the shaft load remains fixed? Explain.

6-5. A Y-connected, three-phase, 60-Hz, 13,500-V synchronous motor has an armature resistance of 1.52 Ω per phase and a synchronous reactance of 37.4 Ω per phase. When the motor delivers 2000 hp, the efficiency is 96% and the field current is so adjusted that the motor takes a leading current of 85 A.
   (a) At what pf is the motor operating?
   (b) Calculate the excitation emf.
   (c) Find the mechanical power developed.
   (d) If the load is removed, describe (do not calculate) how the magnitude and pf of the resulting armature current compare with the original.

6-6. A 1250-hp, three-phase, cylindrical-pole, synchronous motor receives constant power of 800 kW at 11,000 V. Armature winding resistance is negligible and the synchronous reactance per phase is 50 Ω. The armature is Y-connected. The motor has a rated full-load current of 52 A. If the armature current shall not exceed 135% of this value, determine the range over which the excitation emf can be varied through adjustment of the field current.

6-7. A synchronous machine is supplied from a constant-voltage source. At no-load the motor armature current is found to be negligible when the excitation is 1.0 per unit. The per-unit motor constants are $x_d = 1.0$ and $x_q = 0.6$.
   (a) If the machine loses synchronism when the angle between the quadrature axis

and the terminal voltage phasor direction is 60 electrical degrees, what is the per-unit excitation at pullout?

(b) What is the load on the machine at pullout? Assume the same excitations as part (a).

6-8. A synchronous machine operating at rated voltage and rated current shows a leading pf of 86.6%. The per-unit excitation voltage is 1.732. The per-unit synchronous reactance is 1.00.

(a) Does this condition represent motor or generator operation? Explain.

(b) Compute the power angle.

6-9. As a synchronous motor is loaded from zero to full-load, can the pf ever become leading if the field excitation is maintained at 75%? Explain.

6-10. A synchronous motor delivers rated power to a load. Is there a lower limit on the excitation current beyond which it may not be reduced? Explain.

6-11. A 440-V, three-phase, Y-connected synchronous motor has a synchronous reactance of 6.06 Ω per phase. The armature resistance is negligible, and the induced excitation emf per phase is 200 V. Moreover, the power angle between $\bar{V}_t$ and $\bar{E}_f$ is 36.4 electrical degrees.

(a) Calculate the line current and pf.

(b) What values of excitation emf and power angle are necessary to make the pf unity for the same input?

6-12. A synchronous motor at no-load is connected to an infinite bus system. The field circuit is accidentally opened. Explain what happens to the armature current.

6-13. An eight-pole synchronous motor draws 45 kW from the 208-V, 60-Hz, three-phase power system at a pf of 0.8 lagging. The motor is Y-connected and has a synchronous reactance of 0.6 Ω per phase. Armature resistance is negligible. Without any further manipulations on the motor, what is the highest possible value of its steady-state torque?

6-14. Answer whether the following statements are true or false. When the statement is false, describe why.

(a) Increasing the air-gap of a machine reduces the synchronous reactance and raises the steady-state power limit of the machine.

(b) For a synchronous motor, the developed electromagnetic torque is in a direction opposite to the direction of rotation.

(c) In a synchronous machine, the space-phase relation between the field mmf wave and armature-reaction-mmf wave is determined by the pf of the load.

(d) The sum of the armature mmfs in the three phases of the stator winding is zero at any given instant.

(e) The sum of the main field flux ϕ_f and the armature reaction flux Φ_{ar} always add to give the resultant flux Φ_r.

(f) A synchronous motor operating at leading pf is underexcited.

(g) The high currents flowing during a short-circuit cause a synchronous machine to operate under saturated conditions.

(h) The load on a synchronous motor is increased. This causes the main field mmf axis to drop further behind the air-gap mmf axis, so as to increase the torque angle.

(i) An overexcited synchronous motor is a generator of lagging kvars. This means that it supplies lagging kvars to the system to which it is connected.

(j) For large machines the armature resistance and leakage reactance are about 10% on the machine rating and the synchronous reactance is about 100%.

(k) The equivalent circuit developed for the synchronous machine assumes that saturation is negligible and that the air gap of the machine is uniform.

(l) The electromagnetic torque developed in a machine is proportional to the magnitudes of the main-field and air-gap-mmf waves and the sine of the angle between them.

(m) By loss of synchronism we mean that the field and air-gap mmf waves are no longer stationary with respect to each other.

6-15. A synchronous motor is fed from an infinite bus with 1.00 per-unit voltage, through a line containing inductance and no resistance. With a per-unit excitation of 0.9, the motor current is negligible at no-load. However, when the motor is loaded without changing excitation, it falls out of step at a load of 0.81 per unit of rated output. Full-load current is produced by 0.75 per-unit excitation during a short-circuit test.

(a) What is the per-unit inductive reactance of the line?

(b) What is the current under the preceding conditions?

6-16. A six-pole *synchronous motor* (cylindrical rotor) has a per-unit synchronous reactance of 0.75, the resistance being negligibly small. Its excitation emf is adjusted to a per-unit value of 1.50. As the motor is loaded, the rotor is observed to drop back 10 degrees from its initial space position corresponding to zero load. Losses are negligible. For rated voltage impressed on the motor, compute:

(a) Line current in per-unit values.

(b) Power factor, lagging or leading.

6-17. Answer whether or not the following statements are true or false. When the statement is false, provide the correct answer. Assume a Y-connection and 0.8 pf lead.

(a) In a synchronous machine the relative position of the field mmf axis and the armature mmf axis is solely dependent upon the power factor of the load.

(b) The load on a synchronous motor is increased. This causes the speed of the armature mmf to drop to a lower value at steady state so that more input current can flow to supply the need for increased shaft power.

6-18. A synchronous motor with salient poles is operating at no-load with the per-unit rated terminal voltage $V_t = 1$ and per-unit excitation voltage of $E_f = 0.65$. The motor is found to draw an armature current of 0.5 per unit. Assuming that the ratio $x_d/x_q = 1.75$, determine the per-unit values of the direct-axis and quadrature-axis synchronous reactances.

6-19. The direct- and quadrature-axis synchronous reactances of a salient-pole synchronous generator are $x_d = 1.00$ per unit, and $x_q = 0.50$ per unit. The stator resistance and the rotational losses are both negligible. The generator is connected to an infinite bus of rated voltage and frequency. The generator field excitation is adjusted so that at no-load the line current is zero. The power input to the generator from its prime mover is then slowly increased until the generator is pulled out of synchronism. What is the per-unit line current and the pf when the machine is on the verge of pulling out? Prepare a neat, clear phasor diagram illustrating your calculations.

6-20. For the machine described in Example 6-2 determine the following.

(a) The power angle at which maximum power is developed.

(b) The maximum value of the developed power expressed in per unit.

6-21. The synchronous motor of Example 6-2 is operated with a change in field current that causes the power factor to change to 0.8 lagging from 0.8 leading. Determine the new value of the excitation voltage.

6-22. A cylindrical-pole synchronous motor is to be used to improve the power factor of a load of 1500 kW at a power factor of 0.85 lagging. The synchronous motor is to operate at a 0.75 leading power factor. What must be the kVA rating of the synchronous motor if it is required to raise the total power factor to 0.95 lagging?

6-23. An industrial plant uses two 3-phase induction motors rated at 460 V. At full load one draws 232 A at 0.85 pf lagging; the other draws 169 A at 0.82 pf lagging.

  Find the kVA rating of a synchronous capacitor that can serve to bring the plant power factor to 0.9 lagging.

6-24. The net input to an industrial plant supplied from a 2300 V infinite bus system is measured to be 765 A at a 0.92 pf lagging. Although most of the loads are inductive, the input pf was improved by the installation of a synchronous capacitor operating at its rating of 1000 kVA.

  Determine the original power factor of the plant.

6-25. The source voltage appearing in Fig. 6-12 is assumed to be 120 V rms and 60 Hz.

  (a) Determine the value in volts of the average load voltage when the firing angle of the SCR is set for $\pi/4$ radian.

  (b) Let the resistor in this circuit be 10 Ω and the inductor 5 H. Compute the average value of the load current for the firing angle of part (a).

6-9 $\bar{E}_f = \bar{V}_t - j \bar{I}_a X_s$ $E_f = 0.75\%$

  i) for loading p.f. $|E_f| > |V_t|$

  ii) but $|E_f| < |V_t|$ $\therefore$ NEVER

6-10 $P_m = q \left(\dfrac{V_t E_f}{X_s} \right) \sin \delta$ delivering radiated power

  $E_f = \dfrac{P_m X_s}{q V_t \sin \delta}$ $\Rightarrow$ smallest E_f when $\sin \delta = 1$ $\therefore$ $E_{f\,min}$

  Can't be reduced beyond the current value at which

  $E_{f\,min}$ is obtained

6-16 $\bar{E}_f = \bar{V}_t - j \bar{I}_a X_s$

  $I_a = \dfrac{\bar{E}_f - \bar{V}_t}{j X_s} = \dfrac{1 - 1.5 \angle{-10°}}{j .75} = 0.73 \angle{61.4°}$

  p.f. $= \cos (61.6°) = 0.48$

7

Direct-Current Generators

The dc machine is a highly versatile machine. It can provide high starting torques as well as high accelerating and decelerating torques. It is capable of quick reversals, and speed control over a range of 4 : 1 is achieved with relative ease in comparison with all other electromechanical energy-conversion devices. These are features that are responsible for its use in the really tough jobs in industry, such as are found in steel mills. Unfortunately, the need for a mechanical rectifier (in the form of a commutator) to convert the ac emf that is induced in each armature coil to a unidirectional voltage makes it one of the least rugged of electric machines as well as more expensive.

The principles underlying basic torque production and induced voltages in the dc machine are outlined in detail in Chapter 3. Here we investigate the operational characteristics of the various types of dc machines, beginning with a description of how the direct voltage is obtained. The influence of the armature winding mmf on machine behavior is explored with particular emphasis on its effect on commutation and its external characteristics. The subject of commutation occupies a position of preeminence in the study of dc machines because without good commutation the machine is rendered almost useless. The motor speed-torque curves are analyzed and also the various methods of speed control. Finally, the procedure for finding machine performance through the use of the governing equations, the equivalent circuit, the power-flow diagram, and the magnetization curve is outlined and illustrated.

7-1 GENERATION OF UNIDIRECTIONAL VOLTAGES

It is helpful at the start of our study of dc machines to understand how a direct voltage can be produced at the armature output terminals. Appearing in Fig. 7-1(a) is a plot of the flux density produced by the field winding as a function of

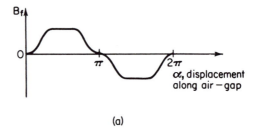

(a)

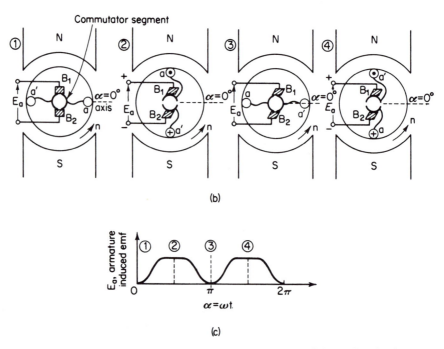

(b)

(c)

Figure 7-1 Generation of unidirectional voltage: (a) shape of air-gap flux density produced by field winding; (b) emf generated in a single-coil armature for four instants of time; (c) plot of E_a as a function of time, with specific instants of part (b) shown.

displacement along the periphery of the rotor. The flat-topped portion of the plot is attributable to the constant air gap between the rotor surface and the faces of the pole pieces. A sharp falloff occurs in the interpolar space because of the effect of the large air gaps there. Figure 7-1(b) depicts a cross-sectional view of a two-pole machine having an armature winding consisting of a simple coil, a–a'. Coil side a is joined to coil side a' by a back connection that is not shown. The front ends of each coil side are joined permanently to copper segments as indicated. Note that both coil sides are effective in voltage production with this arrangement, unlike the Gramme-ring winding. The copper segments are part of the rotor structure, thereby rotating with it. Any emf that is induced in the coil a–a' appears at these copper segments. Placed in contact with the copper segments and fixed in space are two carbon brushes B_1 and B_2. These are used to collect the voltage induced in the armature winding and to make it available to the external circuit. The coil in Fig. 7-1(b) is shown in four different positions relative to the flux field. The rotor is assumed revolving at speed n in the counterclockwise direction. At time instant 1, the emf induced in the coil is zero because each coil side finds itself at a point of zero flux density. At instant 2, there is an emf induced in each coil side and its magnitude is proportional to the value of flux density, as obtained from Fig. 7-1(a), as well as to the velocity and length of the coil sides. The direction of the induced voltage in coil side a is such that it makes the polarity of brush B_1 *positive*. As the rotor revolves an additional 90 degrees, the situation illustrated in 3 of Fig. 7-1(b) is found to prevail. Again note the zero voltage value. At instant 4, coil side a is under the influence of south-pole flux and so has an emf of reversed polarity induced. However, note that its attached commutator segment is now in contact with brush B_2, which keeps this brush at a negative polarity. A complete plot of the voltage appearing at the armature terminals is depicted in Fig. 7-1(c) with the four time instants of Fig. 7-1(b) specifically indicated. It is important to note that, although the voltage in each coil side alternates for each revolution, the voltage appearing at the brushes is unidirectional because of the effect of the segmented commutator.†

Although the use of a single coil in conjunction with the commutator furnishes a unidirectional voltage, the resulting wave shape of Fig. 7-1(c) is unsatisfactory because the magnitude is not constant over the full period. A considerable improvement can be achieved by increasing and distributing the number of armature coils. Figure 7-2(a) depicts in developed form a two-pole machine having eight slots on the rotor equipped with two coil sides per slot, yielding a total of eight coils. This representation is typical of the situation found in practical machines. The lower part of the figure illustrates the manner in which the various coil sides are joined to give a summed quantity. The winding layout shown is

---

†For better understanding the distinction with the ac generator is worth making. The arrangement of Fig. 7-1(b) becomes an ac generator when each coil side is connected to a closed copper ring and each brush is made to ride on one ring. Then the positive and negative variations of induced emfs are made to appear at the brushes and therefore the armature terminals.

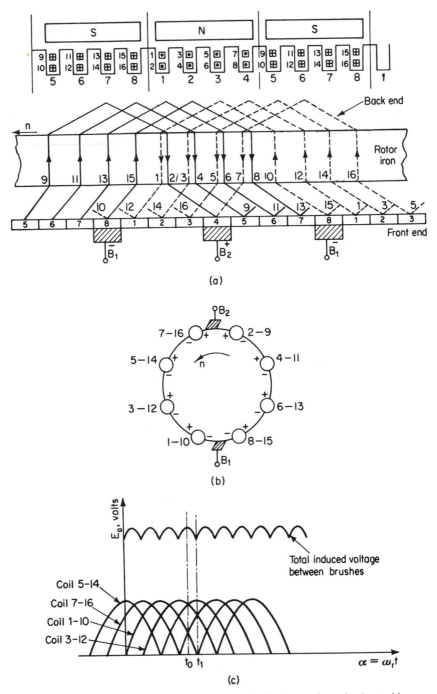

Figure 7-2 Generation of a dc voltage: (a) winding layout for a simple machine; (b) schematic representation of two armature paths existing between the brushes at the time t_0 illustrated in (a); (c) total induced armature voltage as a function of time; complete cycle is shown for each of four coils.

referred to as a *wave winding* for obvious reasons.† As a general rule the sides of a given coil are made to span one pole pitch (i.e., 180 electrical degrees). Moreover, in the double layer winding, such as the one illustrated here, each coil is so placed that one side occupies a lower position and the other side an upper position. For example, coil sides 1 and 10 are joined by the back-end connection to constitute one coil. Note that the span is 180° and that coil side 10 occupies the lower position in slot 5, while coil side 1 occupies the upper position in slot 1.

A study of Fig. 7-2(a) discloses that the armature winding completely closes on itself and that in particular the wave winding has two armature paths with respect to the brushes. The arrowheads attached to each coil side indicate the direction in which the induced emf causes current to flow when a load is placed across the brushes. In the plan view of Fig. 7-2(a) the direction of the emf for the direction of rotation shown is downward beneath a north pole and upward beneath a south pole. For clarity's sake the south pole is repeated once; on the left side the upper coil sides are specifically illustrated, while on the right side the coil sides located in the lower portions of the slots are shown. Also brush B_2 is assumed to be resting on commutator segment 4, and simultaneously brush B_1 is assumed to be in contact with commutator segment 8. Examination of the circuitry at B_2 discloses that current converges at this brush from two directions—from coil side 2 and coil side 7. Further investigation reveals that coil side 2 is associated with a path that consists of four coils: 8–15, 6–13, 4–11, and 2–9. These coils are drawn with solid lines in Fig. 7-2(a) and are denoted by small series-connected dc sources in Fig. 7-2(b). Similarly coil side 7 is associated with the remaining four coils: 7–16, 5–14, 3–12, and 1–10. These coils are represented in Fig. 7-2(a) with broken lines and in Fig. 7-2(b) by series-connected dc sources.

If attention is directed solely at the fundamental component of the flux-density curve of Fig. 7-1(a), it should be clear that for the time instant being considered the values of the coil voltages are not the same. For example, the instantaneous values of coils 3–12, 4–11, 5–14, and 6–13 are displaced from their positions of maximum value by 22.5° while the remaining coils are away by 67.5°. A partial time history of these coil voltages for various positions of the armature relative to the field distribution is depicted in Fig. 7-2(c). The instant represented in Fig. 7-2(a) is identified as t_0. Note that each coil voltage over any full cycle appears as a rectified wave because of the action of the commutator. Furthermore, only the four coils on one side of the armature winding are depicted. The total induced voltage appearing between the brushes B_1 and B_2 at time t_0 is the sum of the instantaneous coil voltages (E_{a0}), i.e., $E_{a0} = 2E_m \cos 22.5° + 2E_m \cos 67.5°$, where E_m is the peak voltage induced in the coil. If the conditions leading to Fig. 7-2(b) are analyzed for a time instant t_1—22.5° later—then coils 5–14 and 6–13 will experience their peak values, while the emf induced in coils 2–9 and 1–10 will be zero. The instantaneous values of the induced emfs in the remaining coils will be $E_m \cos 45°$. The total contribution at time t_1 is therefore $E_{a1} = E_m +$

†Another frequently used arrangement is the lap winding.

$2E_m \cos 45° + 0$. This total is identified in Fig. 7-2(c) and is less than the value occurring at time t_0.

A glance at Fig. 7-2(c) makes it obvious that an almost constant voltage now appears between the brushes in contrast to the situation of Fig. 7-1(c). The small ripple is attributable to the small number of slots per pole. In practical machines this number is very large so that the ripple is hardly detectable even with sensitive instruments.

7-2 DIRECT-CURRENT GENERATOR TYPES

The dc machine functions as a generator when mechanical energy is supplied to the rotor and an electrical load is connected across the armature terminals. In order to supply electrical energy to the load, however, a magnetic field must first be established in the air gap. The field is necessary because it serves as the coupling device permitting the transfer of energy from the mechanical to the electrical system. There are two ways in which the field winding may be energized to produce the magnetic field. One method is to excite the field separately from an auxiliary source as depicted in the schematic diagram of Fig. 7-3. But clearly this scheme is disadvantageous because of the need of another dc source. After all, the purpose of the dc generator is to make available such a source. Therefore, invariably dc generators are excited by the second method which involves a process of self-excitation. The wiring diagram appears in Fig. 7-4, and the arrangement is called the *self-excited shunt generator*. The word shunt is used because the field winding appears in parallel with the armature winding. That is, the two windings form a shunt connection.

To understand how the self-excitation process takes place we must start with the *magnetization curve* of the machine. Sometimes this is called the *saturation curve*. Strictly speaking, the magnetization curve represents a plot of air-gap flux versus field winding mmf. However, in the dc generator where the winding

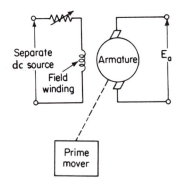

Figure 7-3 Schematic diagram of a separately excited dc generator.

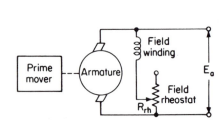

Figure 7-4 Schematic diagram of a self-excited shunt generator.

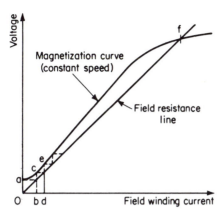

Figure 7-5 Illustrating the build-up procedure of a self-excited shunt generator.

constant K_E is known and the speed n is fixed, the magnetization curve has come to represent a plot of the open-circuit induced armature voltage as a function of the field winding current. With K_E and n fixed, Eq. (3-33) shows that ϕ and E_a differ only by a constant factor. Figure 7-5 depicts a typical magnetization curve, valid for a constant speed of rotation of the armature. It is especially important to note in this plot that even with zero field current an emf is induced in the armature of value Oa. This voltage is due entirely to residual magnetism, which is present because of the previous excitation history of the magnetic-circuit iron. The linear curve appearing on the same set of axes is the *field-resistance line*. It is a plot of the current caused by the voltage applied to the series combination of the field winding and the active portion of the field rheostat. Clearly, then, the slope of the linear curve is equal to the sum of the field-winding resistance R_f and the active rheostat resistance R_{rh}. The voltage Oa due to residual magnetism appears across the field circuit and causes a field current Ob to flow. But in accordance with the magnetization curve this field current aids the residual flux and thereby produces a larger induced emf of value bc. In turn this increased emf causes an even larger field current, which creates more flux for a larger emf, and so on. This process of voltage build-up continues until the induced emf produces just enough field current to sustain it. This corresponds to point f in Fig. 7-5. Note that in order for the build-up process to take place three conditions must be satisfied: (1) There must be a residual flux.† (2) The field winding mmf must act to aid this residual flux. (3) The total field-circuit resistance must be less than the critical value. The *critical field resistance* is that value which makes the resistance line coincide with the linear portion of the saturation curve.

In addition to the shunt generator there are the other generator types—the compound generator and the series generator. A *compound generator* is a shunt generator equipped with a *series winding*. The series winding is a coil of compara-

---

†When the iron is in its original virgin state, the residual flux is negligible. In such cases the field winding must receive an initial excitation to give the machine a usable level of residual flux. This process is called *flashing the field*.

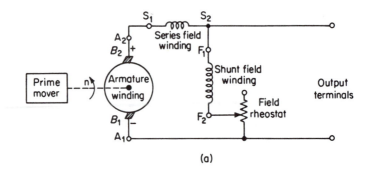

(a)

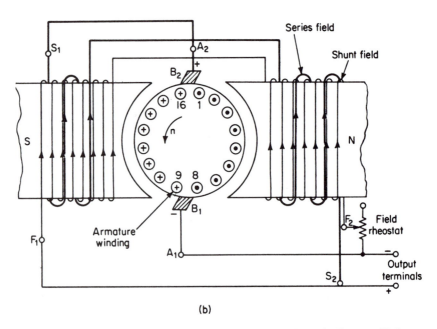

(b)

Figure 7-6 Cumulatively compounded generator: (a) schematic diagram; (b) detailed diagram depicting the location of series and shunt windings on the pole structure.

tively few turns wound on the same magnetic axis as the field winding and connected in series with the armature winding. Consult Fig. 7-6. Because the series field winding must be capable of carrying the full armature current, its cross-sectional area is much greater than that used in the shunt field winding. The purpose of the series field is to provide additional air-gap flux as increased armature current flows, in order to neutralize the armature winding resistance drop as well as the voltage drops occurring in the feeder wires leading to the load. In such cases the generator is usually referred to as a *cumulatively* compounded generator because the series field *aids* the shunt field flux. If the series field connection were reversed, its flux would oppose the shunt field flux in which case the configuration is referred to as a *differentially* compounded generator.

By imposing the appropriate constraint on the connection diagram depicted in Fig. 7-6(a), we can identify any one of the three modes of operation of the dc generator. Thus in addition to the armature winding we have the following:

Compound generator: shunt and series field windings

Shunt generator: shunt field winding

Series generator: series field winding

Since the series generator is rarely used except for special applications, all further treatment of generators is confined to the shunt and compound modes.

7-3 DEMAGNETIZING EFFECT OF THE ARMATURE WINDING MMF

The armature winding mmf produces two adverse effects: it causes a net reduction in the field flux, and it makes it more difficult for the armature current in the coils to commutate. We turn our attention to the first of these here; the problem of commutation is treated in Sec. 7-6.

A glance at Fig. 7-6(b) shows that the flow of current in the armature winding produces an ampere-conductor distribution that makes the armature behave like a solenoid. The associated mmf is directed downward along the brush axis and in a position of quadrature to the field axis. If there were no saturation of the iron present, the effect of this cross-armature mmf would be merely to cause a distortion of the flux. However, in most practical machines operation occurs around the knee of the magnetization curve and as a result a net demagnetization takes place. To understand why this happens, consider the situation illustrated in Fig. 7-7 which is that of a shunt generator delivering current to a load (the latter not shown). The armature mmf wave is assumed to be triangular. This is a valid assumption whenever the number of surface armature conductors is large; otherwise it would be a stepped trapezoid. The flux-density wave produced by this armature mmf is saddle-shaped because of the high reluctance of the interpolar space. The resultant air-gap flux density is the sum of the B_f and the B_a curves. It should be noted that the armature mmf acts to increase the flux in the trailing half of the pole, while it causes a diminution of flux in the leading half. If the increase of flux in the trailing half is equal to the decrease in the leading half, no net reduction in field flux takes place. If the increase in the trailing half is less than the decrease in the leading half, demagnetization occurs. In Fig. 7-7 the cross-hatched area represents the amount of demagnetization that occurs in the leading half of the pole, the double cross-hatched area denotes the amount of magnetization caused by the cross-magnetizing armature mmf. When a net demagnetization of the field flux results, the former area exceeds the latter. Because machines are

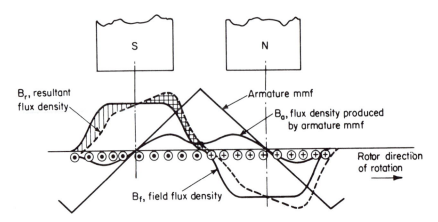

Figure 7-7 Effect of armature mmf on the air-gap flux density. Demagnetization occurs if the cross-hatched area to the left of the pole axis is greater than the double-cross-hatched area to the right of the pole axis.

operated in a partially saturated state, there often exists some degree of magnetization.

A measure of the amount of this demagnetization expressed in equivalent field amperes can be obtained from a knowledge of the magnetization curve and appropriate design data. The procedure is graphically illustrated in Fig. 7-8. The air-gap flux density per pole produced by the field winding acting alone is denoted by B_f. Let ρ be defined as the ratio of the pole face width, b_p, to the width of a pole pitch, τ_p. That is,

$$\rho \equiv \frac{b_p}{\tau_p}$$

Then the amount of armature mmf acting at either pole tip is

$$\text{armature mmf at pole tip} = \rho \,\frac{Z}{2p}\frac{I_a}{a} \qquad (7\text{-}1)$$

where Z denotes the total armature conductors, p is the number of poles and a denotes the number of armature paths. The factor 2 converts conductors to turns. To express Eq. (7-1) in equivalent field amperes, it is necessary to divide by the number of turns per pole of the field winding, N_f. This quantity is shown on the abscissa axis of Fig. 7-8 with some exaggerations for the sake of clarity. The value of the resultant air-gap flux density at the center of the pole is B_f. However, at the trailing pole tip the magnetizing action of the armature mmf raises this quantity to B_b (as is also indicated in Fig. 7-7), whereas at the leading pole tip it drops to B_c. Area acd is a measure of the amount of demagnetization that occurs in the leading half of a pole; area abe is a measure of the amount of magnetization that occurs in the trailing half of a pole. For machines that exhibit a measurable degree of saturation area acd is greater than area abe. Hence demag-

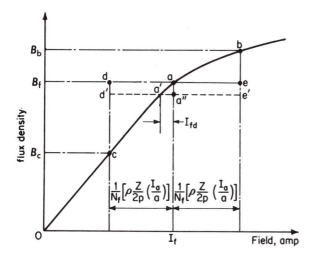

Figure 7-8 Illustrating the demagnetizing mmf by means of the magnetizing curve.

netization occurs. The value of this demagnetization is found by displacing line *dae* to that position *d'a'e'* such that area *a'd'c* equals area *a'be'*. The quantity *aa'* projected on the field current axis yields the desired result. Thus

$$I_{fd} \equiv a'a'' \quad \text{measured in field amperes} \tag{7-2}$$

where I_{fd} is used to denote the demagnetizing effect of the armature mmf in equivalent field amperes.

7-4 EXTERNAL GENERATOR CHARACTERISTICS

The external characteristic of a generator is described by a plot of terminal voltage versus load current with the speed held fixed. A generator that is equipped with an armature winding and just a single field winding can be operated either in the separately excited mode or the shunt excited mode. The external characteristic for each mode is depicted in Fig. 7-9 with solid lines. For the purpose of comparison, each mode is assumed to be adjusted so that the terminal voltage at no-load is the same. As the load current is then allowed to increase to its rated value, there occurs a fall-off of terminal voltage with the situation being more pronounced for the shunt connection than for the separately excited mode. Of course, in both cases there is a drop in voltage that is associated with the flow of the increasing armature current through the armature circuit resistance. This is indicated in Fig. 7-9 by the broken-line curve. Moreover, for the separately excited mode of operation there is an additional drop in voltage from the broken-line characteristic that comes about because of the demagnetizing effect of the armature current as described in the preceding section. These two effects—the armature circuit resistance voltage drop and armature reaction—define the resultant external characteristic of the separately excited generator.

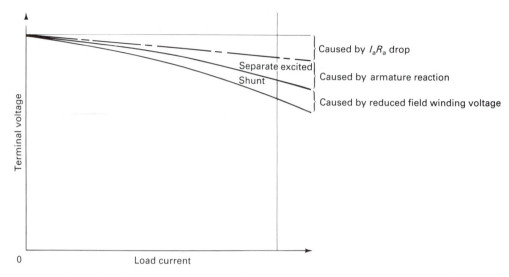

Figure 7-9 External characteristics of dc generator operated in shunt and separately excited modes. Speed is constant.

When the dc generator is operated as a self-excited shunt generator, then in addition to the two factors already identified, there is a third factor that is responsible for a further reduction of terminal voltage at a given load current. Keep in mind that with the shunt connection the supply voltage for the field current is in fact the very terminal voltage of the generator which is undergoing a decrease with increasing load demand. Consequently, this reduced voltage causes a corresponding reduction in field current and thereby in the air-gap flux which in turn causes a further reduction in terminal voltage. Accordingly, the external characteristic of the shunt generator assumes a position that is below the one obtained for the separately excited mode of operation.

A dc generator that is equipped with a series field winding in addition to the shunt field can be designed to counter the effects that cause the drooping characteristics of Fig. 7-9. By supplying the series field winding with the right number of turns, it is possible to exactly neutralize all three effects that cause the terminal voltage of the shunt motor to decrease with load current. When this happens the compound generator is said to be *flat compounded*. This condition is illustrated in Fig. 7-10. If the machine is designed to have a number of turns fewer than the number required for flat compounding, the generator is said to be *undercompounded*. In such a case only a partial compensation of the voltage drop of the shunt mode is achieved. However, more commonly, the number of series turns designed into the series winding exceeds that required for flat compounding. This gives rise to an external characteristic that is called *overcompounded*. It is helpful to note that an overcompounded generator offers the additional feature of permitting the neutralization of the effect of the voltage drop incurred in the feeder lines

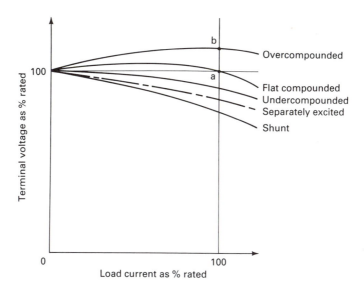

Figure 7-10   External characteristics of various dc generator types at constant speed.

that are used to connect the generator to the load. This is an especially desirable feature when long feeder lines are necessary. The voltage available for the feeder line drop at full load in Fig. 7-10 is represented by the distance *ab*.

In the interest of providing some flexibility in dealing with situations where different feeder lengths are encountered, a machine designed with a large number of series field turns can be used in conjunction with a resistor connected in parallel with the series winding. By proper adjustment of this resistor the contribution of the series field mmf to the air-gap flux can be controlled by diverting more or less of the armature current through the resistor. This resistor is known as the *di-verter*.

An alternative way of representing the external characteristics of the various types of dc generators is depicted in Fig. 7-11. Here adjustments are made in each mode of operation to establish rated terminal voltage when rated load current is delivered. As the load current is allowed to diminish to zero at constant speed, the external characteristics take on the positions shown. The value of the no-load terminal voltage of the overcompounded generator is the smallest while that of the shunt generator is the largest (why?).

Voltage Control.   In typical applications it is desirable to keep the terminal voltage at the load terminals fixed at the rated value regardless of the load condition. There are two ways to achieve this result: (1) adjust the driving speed of the generator prime mover; (2) adjust the rheostat that is placed in series with the field winding (see Fig. 7-6). Of course, for a flat-compounded generator and

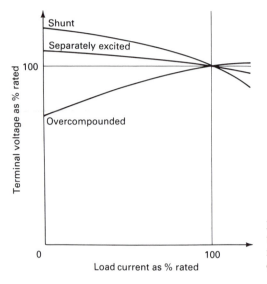

Figure 7-11 External generator characteristics drawn with full-load current at rated terminal voltage as the common operating point. Speed is constant.

negligible feeder line voltage drop, no adjustment is needed if operation is confined to no-load or to full load. However, for operation at half-load from no-load, it is obvious from the external characteristic appearing in Fig. 7-10 that either a drop in speed or a reduced field current adjustment is required to keep the terminal at the rated value. On the other hand, if the operating generator mode is that of undercompounding, separate excitation, or shunt, maintenance of constant terminal voltage calls for an increase in speed or an increase in field current. A combination of the two can also be employed. Of course, the biggest adjustments must be made for the shunt mode of operation.

Example 7-1

A 50-Kw, 500-V cumulatively compounded generator equipped with a diverter delivers rated load current to a load at rated voltage. This condition is achieved by proper adjustment of the diverter resistance and field rheostat. Refer to Fig. 7-12 which shows the shunt field placed directly across the load terminals (called the long-shunt

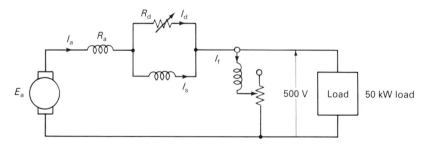

Figure 7-12 Diagram for Example 7-1.

connection).† The measured currents through the field winding and diverter are 4 A and 21 A, respectively. The armature and series field winding resistances are 0.1 and 0.02, respectively.

(a) What is the series field current?
(b) Find the diverter resistance.
(c) Compute the generated armature voltage.

Solution   (a)   The armature current is

$$I_a = I_L + I_f = \frac{50}{0.5} + 4 = 104 \text{ A}$$

$$I_s = I_a - I_d = 104 - 21 = 83 \text{ A}$$

where I_d denotes the diverter current.

(b) The voltage drop across the diverter is the same as that across the series field winding. Hence,

$$I_d R_d = 83(0.02) = 1.66 \text{ V}$$

$$R_d = \frac{1.66}{I_d} = \frac{1.66}{21} = 0.079 \ \Omega$$

(c) By Kirchhoff's voltage law in the circuit of Fig. 7-12

$$E_a = V_t + I_a R_a + I_d R_d = 500 + 104(0.1) + 1.66 = 512 \text{ V}$$

Derivation of External Characteristic of Self-Excited Shunt Generator

The precise way in which the factors shown in Fig. 7-9 combine to yield a stable operating point can be understood in the graphical representation appearing on the left side of Fig. 7-13. Assume that the following information is available: armature circuit resistance, field circuit resistance, magnetization curve, and the equivalent field current that represents the demagnetization caused by the armature mmf (i.e., I_{fd}). The last item is available either from the design data or an appropriate test. The intersection of the field resistance line with the nonlinear magnetization curve shows that at no-load the equilibrium field current is I_{f0} corresponding to the no-load terminal voltage V_0. When the load condition is such that the armature current is I_{a1}, the external characteristic shows that the terminal voltage has the value V_1. Extension of a horizontal line at the V_1 level results in a point of intersection with the field resistance line identified as O'. The projection of O' onto the field axis discloses that the current flowing in the field circuit is now I_{f1}. A comparison with the value at no-load makes it clear that a *reduction* occurs equal to

$$\Delta I_f = I_{f0} - I_{f1} \tag{7-3}$$

†There is also a short-shunt connection where F1 is joined to S1 in Fig. 7-6 instead of S2.

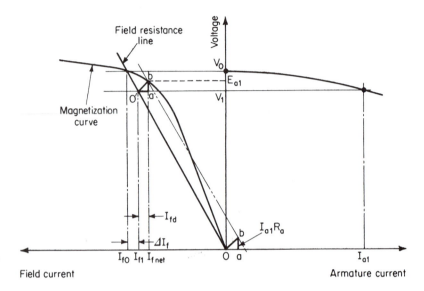

Figure 7-13   Derivation of the external characteristic of a self-excited shunt generator.

To find the net field current, which in turn is responsible for producing the air-gap flux that provides the induced armature voltage E_a, it is necessary to subtract from I_{f1} the demagnetization effect I_{fd} of the armature mmf. Thus

$$I_{f\,net} = I_{f1} - I_{fd} \tag{7-4}$$

Or expressing this more generally in terms of the field current existing at no-load, we can write

$$I_{f\,net} = I_{f0} - \Delta I_f - I_{fd} \tag{7-5}$$

The induced armature voltage for the specified load condition E_{a1} is then found on the nonlinear open-circuit characteristic corresponding to $I_{f\,net}$. This is represented in Fig. 7-13 by point b. Upon subtracting from E_{a1} the armature circuit resistance drop $I_{a1}R_a$, the terminal voltage V_1 results. The $I_{a1}R_a$ drop is denoted by line ab in Fig. 7-13. It should be noted that there is only one position in the vicinity of the upper portion of the magnetization curve where triangle $O'ab$ will probably fit between the field resistance line and the magnetization curve. In fact it is this position that defines the equilibrium point that is being described.

On the basis of the foregoing discussion it can be seen that the external characteristic can be derived from the magnetization curve, the field resistance line, and knowledge of I_{fd} and R_a. To find the terminal voltage corresponding to any specified value of I_a, first construct the triangle Oab at the origin as shown in Fig. 7-13. Side ab is equal to I_aR_a and side Oa is equal to the amount of demagnetization associated with the specified I_a. Then draw a line through point b parallel to the field resistance line and intersecting the magnetization curve in two places. Triangle Oab may then be translated to position $O'ab$ taking care to keep O' on the field resistance line and b on the magnetization curve. The terminal voltage is the ordinate value corresponding to point a. By repeating this procedure for various values of armature current the complete characteristic is obtained.

Derivation of External Characteristic of Cumulatively Compounded Generator

A similar procedure may be used to describe the external characteristic of the self-excited cumulatively compounded generator. Very often the contribution of the series field is responsible for a rising external characteristic such as the one depicted on the right side of Fig. 7-14. This characteristic can also be explained in terms of the same information that is used for the shunt generator, but in addition information about the series field contribution is required, usually expressed in terms of an equivalent field current. The last quantity is readily written as

$$I_{fs} \equiv \frac{N_s}{N_f} I_s \tag{7-6}$$

where I_{fs} is the field current equivalent of the series field mmf, N_s denotes the number of turns of the series field per pole, N_f denotes the number of shunt field turns per pole, and I_s is the current flowing through the series field winding. Often I_s is equal to the armature current, but sometimes a portion of the armature

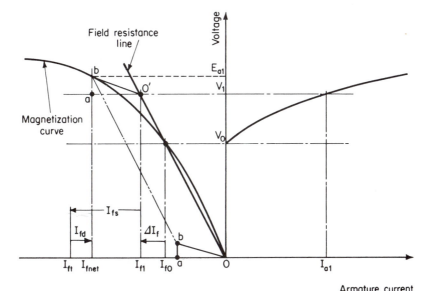

Figure 7-14 Derivation of the external characteristic of a cumulatively compounded generator.

current is diverted through a low resistance shunt (called a *diverter*) placed across the series field winding in order to limit its contribution.

At no-load the operating point is on the magnetization curve. The no-load terminal voltage is identified as V_0 and its corresponding field current as I_{f0}. It is helpful to think of this point as identical with the no-load point of the self-excited shunt generator. Next consider that load current is allowed to flow so that the armature current becomes I_{a1}, corresponding to which the terminal voltage becomes V_1 as shown in Fig. 7-14. With the assumption that (V_1, I_{a1}) is the new equilibrium point, it follows that the field current is *greater* by the amount ΔI_f which yields a total current in the field winding of value $I_{f1} = I_{f0} + \Delta I_f$. The plus sign for ΔI_f always appears whenever the external characteristic is a rising one. The current I_{f1} is associated with point O' on the field resistance line.

To the shunt field current must now be added the effect of the series field as expressed by Eq. (7-6). This yields a total main axis excitation expressed by

$$I_{ft} = I_{f1} + \frac{N_s}{N_f} I_s = I_{f1} + I_{fs} \tag{7-7}$$

The net field current, which is responsible for the armature induced emf, is found by subtracting the demagnetizing effect of the armature mmf. Thus

$$I_{f\,\text{net}} = I_{ft} - I_{fd} \tag{7-8}$$

Corresponding to $I_{f\,\text{net}}$ is the induced armature voltage E_{a1}. The terminal voltage is then found by removing the armature circuit resistance drop from E_{a1}, which yields V_1. The $I_{a1}(R_a + R_s)$ drop is represented by line ab in Fig. 7-14.

The external characteristic can be derived by following a procedure similar to that described for the shunt generator. Construct triangle Oab at the origin. Line ab is equal to the armature circuit resistance drop for the specified armature current. Line Oa is equal to the resultant of the magnetizing effect of the series field and the demagnetizing effect of the armature mmf, i.e.,

$$Oa = I_{fs} - I_{fd} \tag{7-9}$$

Then through b draw a line parallel to the field resistance line intersecting the magnetization curve at point b. Dropping down from b by the $I_{a1}(R_a + R_s)$ drop gives the desired terminal voltage V_1. A comparison with the procedure employed for the shunt generator discloses that the triangle Oab is placed on the left side of the origin rather than the right side. This is dictated by the fact that the base of the triangle for the cumulatively compounded generator represents a net magnetization rather than a net demagnetization.

7-5 COMPUTATION OF GENERATOR PERFORMANCE

Appearing in Fig. 7-15 is the equivalent circuit of the compound generator. The armature winding is replaced by a source voltage having the induced emf E_a and a resistance R_a, which represents the armature circuit resistance. The armature

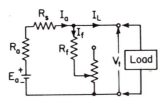

Figure 7-15 Equivalent circuit of the compound dc generator.

circuit resistance includes the armature winding resistance, interpole winding resistance, compensating winding resistance, plus the effect of the voltage drop in the carbon brushes. The series field is replaced by its resistance R_s, and the same for the shunt field.

The governing equations for determining the performance are the following:

$$E_a = \frac{pZ}{60a} \Phi n = K_E \Phi n \tag{7-10}$$

$$T = \frac{pZ}{2\pi a} \Phi I_a = K_T \Phi I_a \tag{7-11}$$

$$V_t = E_a - I_a(R_a + R_s) \tag{7-12}$$

$$I_a = I_L + I_f \tag{7-13}$$

These include the two basic relationships for induced voltage and electromagnetic torque as developed in Sec. 3-4. Equations (7-12) and (7-13) are merely statements of Kirchhoff's voltage and current laws as they apply to the equivalent circuit of Fig. 7-15. For the compound generator the expression for the air-gap flux must include the effect of the series field as well as the shunt field. Thus on the assumption of negligible saturation,

$$\Phi = \Phi_{sh} + \Phi_s \tag{7-14}$$

where Φ_{sh} denotes the flux produced by the shunt field winding and Φ_s denotes the flux caused by the series field winding. For the shunt generator, of course, Φ_s is zero in Eq. (7-14) and so too is R_s in Eq. (7-12).

The power-flow diagram for the dc generator is depicted in Fig. 7-16. Note the similarity it bears to that of the synchronous generator. The field winding loss is included in the power flow directly because it is assumed that the generator is self-excited. Of course, if the field winding is separately excited, the field losses are not supplied from the prime mover and so must be handled separately. Moreover, note that the field losses are represented in terms of the product of the field terminal voltage and the field current. This ensures that the field winding rheostat losses are included as well.

The quantity P_{rot} that appears in the power flow diagram represents the iron losses that arise from rotation of the armature (rotor) structure as well as friction and windage losses. The iron losses associated with rotation include hysteresis

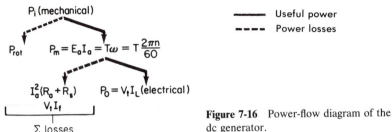

Figure 7-16   Power-flow diagram of the dc generator.

and eddy current losses. Although the slotted rotor structure, which accommodates the armature winding, is composed of laminations, eddy currents are caused to flow in the individual laminations as the rotor iron is driven alternately beneath poles of N polarity and S polarity. This very same alternation of the flux in the iron is responsible for the hysteresis losses represented by the generated hysteresis loops. Both losses are functions of the maximum flux density and the speed. Moreover, for machines with open slots, there is an additional component of P_{rot} that accounts for the losses incurred in the pole faces because of the ripple that is made to appear in the field distribution curve by the slot effect. The major part of the frictional losses is contributed by the friction of the brushes on the commutator segments. Also included but to a lesser degree are the machine bearing losses. Typically, the total of these losses represents 3% to 5% of the machine rating. However, it is useful to keep in mind that the per-unit value of all machine losses diminishes as the machine rating increases.

The armature winding copper losses are essentially losses that occur under load conditions. As such these losses account for all armature winding copper losses, including the skin effect that takes place in the individual coils of the armature winding, the electrical loss in the carbon brushes, (which is frequently taken to be $2I_a$ on the assumption that the voltage in the brushes is about 2 V), and the short-circuit loss in those coils that undergo commutation. For convenience all of these are assumed to be represented in the *armature winding circuit resistance*. Thus the parameter R_a is not to be treated as merely the dc armature winding resistance.

The per-unit value of the armature circuit resistance indicates the ratio of the armature circuit losses expressed in units of power to the rated power output of the generator. If we denote the rated armature current as the base current, I_B, the rated armature voltage as the base voltage V_B, and the rated output power as the base power P_B, then

$$\text{p-u } R_a = \frac{I_B^2 R_a}{V_B I_B} = \frac{I_B R_a}{V_B} = \frac{R_a}{V_B/I_B} = \frac{R_a}{R_B}$$

where R_B is the base resistance and is determined as the ratio of the base voltage to the base current. Thus a value of p-u $R_a = 0.05$ means that 5% of the rated output of the generator is consumed as losses in the armature circuit when rated

current is being delivered. This per-unit value takes on larger values for generators of small ratings. Similarly, for large generators the p-u R_a diminishes.

Example 7-2

The magnetization curve of a 10-kW, 250-V, dc self-excited shunt generator driven at
1000 rpm is shown in Fig. 7-17. Each vertical division represents 20 V and each
horizontal unit represents 0.2 A. The armature circuit resistance is 0.15 Ω and the
field current is 1.64 A when the terminal voltage is 250 V at rated load. Also, the
rotational losses are known to be equal to 540 W. Find at rated load (a) the armature
induced emf, (b) the developed torque, (c) the efficiency. Assume constant speed
operation.

Solution   (a) With the generator delivering rated load current it follows that

$$I_L = \frac{10,000}{250} \text{ W} = 40 \text{ A}$$

Hence from Eq. (7-13) the armature current is

$$I_a = I_L + I_f = 40 + 1.64 = 41.64 \text{ A}$$

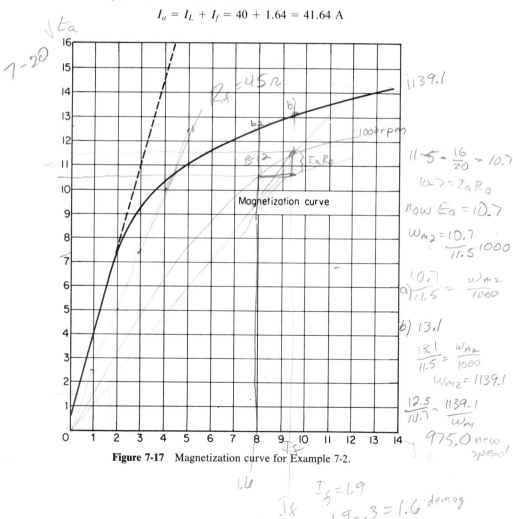

Figure 7-17   Magnetization curve for Example 7-2.

The induced armature voltage then follows from Eq. (7-12):

$$E_a = V_t + I_a R_a = 250 + 41.64(0.15) = 256.25 \text{ V} \qquad (7\text{-}15)$$

(b) To determine the developed torque it is first necessary to compute the electromagnetic power $E_a I_a$. Thus

$$E_a I_a = 256.25(41.64) = 10,670 \text{ W}$$

The power-flow diagram reveals that this power may also be found from

$$E_a I_a = P_0 + I_a^2 R_a + V_t I_f = 10 \text{ kW} + (41.64)^2 0.15 + 250(1.64)$$

$$= 10 \text{ kW} + 260 + 410 = 10,670 \text{ W} \qquad (7\text{-}16)$$

which checks with the preceding calculation. Hence the developed torque is

$$T = \frac{E_a I_a}{2\pi n/60} = \frac{10,670}{2\pi(1000)/60} = 101.9 \text{ N-m} \qquad (7\text{-}17)$$

(c) The efficiency is found from Eq. (4-27), which for the dc generator takes the form

$$\eta = 1 - \frac{\Sigma \text{ losses}}{E_a I_a + P_{\text{rot}}} \qquad (7\text{-}18)$$

In this case

$$\Sigma \text{ losses} = P_{\text{rot}} + I_a^2 R_a + V_t I_f$$

$$= 540 + 260 + 410 = 1210 \text{ W}$$

Hence

$$\eta = 1 - \frac{1210}{10,670 + 540} = 1 - \frac{1210}{11,210} = 1 - 0.108 = 0.892$$

The efficiency is therefore 89.2%.

Voltage Regulation

The voltage regulation of a dc generator is defined as the change in terminal voltage from full load to no-load expressed as a percentage of the full-load voltage. In mathematical form we can write

$$\text{VR} = \frac{\text{no-load voltage} - \text{full-load voltage}}{\text{full-load voltage}} = \frac{V_{nl} - V_{fl}}{V_{fl}} 100 = \frac{V_o - V_t}{V_t} 100$$

$$(7\text{-}19)$$

Here the full-load voltage, V_{fl}, is the rated terminal voltage (V_t) and V_{nl} is the voltage measured when the load is disconnected from the generator. The latter quantity is identified as V_o in Figs. 7-13 and 7-14.

Example 7-3

Find the voltage regulation for the load condition described in Example 7-2. Assume that the equivalent field current that represents the demagnetizing effect of the armature mmf at rated load is 0.2 A.

Solution   We resort to a graphical solution employing Fig. 7-17. On the magnetization curve of Fig. 7-17 locate point b (see Fig. 7-13) at a level of $E_a = 256.25$ V. Along the ordinate line from b mark off the armature circuit resistance drop of 6.25 V. Next, along the horizontal from a mark off the value of the equivalent field current of the demagnetizing effect of the armature mmf of value 0.2 A. This locates point o' (as identified in Fig. 7-13). Finally, draw the field resistance line that passes through o' from the origin. The intersection of this line with the magnetization curve locates the no-load terminal voltage of $V_o = 268$ V. Hence,

$$VR = \frac{268 - 250}{250} \, 100 = 7.2\%$$

7-6 COMMUTATION

Good commutation is indispensable for the satisfactory operation of dc machines. This matter is so important that it takes precedence in design considerations over the problem of heat dissipation. Without proper commutation, deterioration of the carbon brushes and the copper commutator segments can rapidly set in. This causes severe sparking and thereby renders the machine useless. Even such an important characteristic as the maximum torque is limited by commutation rather than by heating.

Commutation is concerned with providing a suitable transition of the armature current in a coil from a value $+I_a/a$ to a value of $-I_a/a$ as it passes beneath a brush from one pole to an adjacent pole. A glance at Fig. 7-18(a) makes it evident that when coil 1–10 passes beneath the brushes and occupies the position now being occupied by coil 15–8, the current in the coil will have reversed completely as indicated by the cross-dot notation. Commutation of the current from coil 1–10 begins to take place when brush B_1 makes contact with commutator segment 1 as shown in Fig. 7-18(b). Note that when this happens coil 1–10 is actually short-circuited by brush B_1 through segments 8 and 1. All commutator segments are insulated from one another by a suitable material. A similar situation occurs at brush B_2 where coil 2–9 is being short-circuited. Because of this and in the interest of preventing large circulating currents in these coils, which in turn can cause severe sparking as the circulating currents are interrupted upon leaving the brush, the coils undergoing commutation should be free of voltage sources. The time instant depicted in Fig. 7-18(b) is 22.5° later than that shown in Fig. 7-18(a). Hence coils 1–10 and 2–9 find themselves in the interpolar space where the flux density produced by the field mmf is zero. Accordingly, no Blv voltage is induced

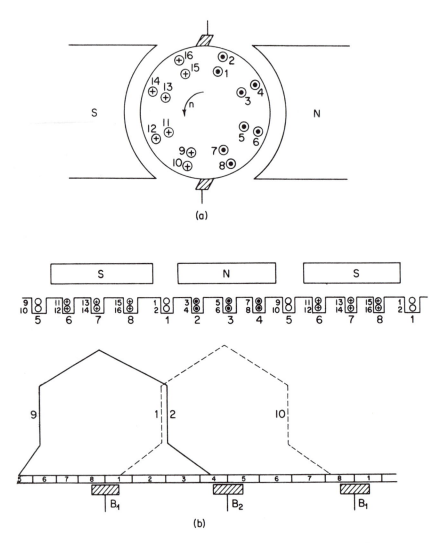

Figure 7-18 Illustrating the commutation problem: (a) coil 1-10 must undergo a complete current reversal as it advances counterclockwise to the left side of the brush axis; (b) for the position shown, coil 1-10 is short-circuited through segments 8 and 1 by B_1, and coil 2-9 is short-circuited through segments 4 and 5 by B_2.

in these coils from this source. However, the presence of the armature mmf causes the resultant flux density to have a finite value in the interpolar space as is evident from Fig. 7-7. Since the polarity of this flux is the same as that of the main field pole which the short-circuited coils are leaving, an emf is induced that acts to maintain current flow in the same direction as exists before commutation begins. Therefore this induced emf due to the armature mmf acts to hinder commutation;

it becomes more difficult for the coil current to reverse by the time it leaves the brush.

Types of Commutation

There are two types of commutation possible: resistance commutation and voltage commutation. Moreover, the latter may be divided into the two classes of undercommutation and overcommutation. The distinction between these types is depicted by the various commutation curves of Fig. 7-19.

     Curve *a* represents *linear* commutation, and it occurs only when the brush contact resistance is the exclusive factor influencing commutation. All induced emf's from whatever sources must add to zero and the coil resistance must be negligible compared with the brush contact resistance. Under these conditions the current in the coil changes uniformly during the commutation period T_c. The mechanism of commutation by this method is illustrated in Fig. 7-20. Figure 7-20(a) depicts the situation just before commutation of the current in coil 2–9 is about to occur. Brush B_2 is resting on commutator segment 4, and of the total armature current I_a flowing through this brush half comes from one armature path involving coil 16–7 and the remaining half comes from the second path involving coil 2–9. The direction of rotation of the armature with respect to the brush is towards the left. Depicted in Fig. 7-20(b) is the situation where 25% of the brush is in contact with commutator segment 5. The contact resistance of that part of the brush surface in contact with segment 5 is three times as large as the resistance of the brush in contact with segment 4. Hence division of the current coming from the two armature paths takes place in accordance with this ratio. Thus the current $I_a/2$ coming from the right side divides at junction a in the ratio of $3:1$ so that the current passing to B_2 from segment 5 from this source is $\frac{1}{4}(I_a/2)$ or $I_a/8$. The portion passing to B_2 through segment 4 is $\frac{3}{4}(I_a/2)$ or $\frac{3}{8}I_a$. Similarly, the current $I_a/2$ originating from the armature path on the left side splits, so that $\frac{3}{4}(I_a/2)$ or $\frac{3}{8}I_a$ flows to B_2 through segment 4 and $\frac{1}{4}(I_a/2)$ or $\frac{1}{8}I_a$ flows from b to a to segment 4 and

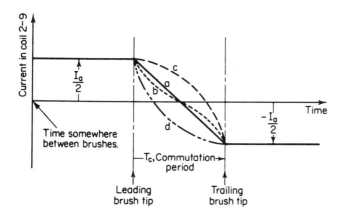

Figure 7-19 Commutation curve of coil 2-9 in Fig. 7-18. During the commutation period the curve is a plot of the circulating current in the short-circuited coil undergoing commutation.

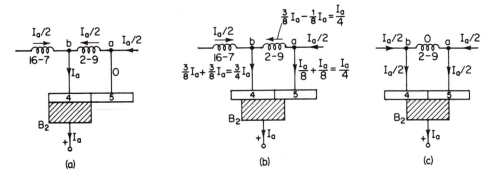

Figure 7-20 Illustrating linear commutation in coil 2-9: (a) conditions just before commutation begins; (b) conditions at a point one-quarter through the commutation period; (c) conditions at a point halfway through the commutation period.

thence to B_2. The net current that flows from segment 4 to B_2 is therefore $\frac{3}{8}I_a + \frac{3}{8}I_a$ or $\frac{3}{4}I_a$, whereas the net current to B_2 from segment 5 is $(I_a/8) + (I_a/8)$ or $I_a/4$. The resultant current in coil 2–9 flows from a to b and is equal to $\frac{3}{8}I_a - \frac{1}{8}I_a$ or $I_a/4$. Consequently, one-quarter through the period of commutation the current in coil 2–9 has changed from $I_a/2$ to $I_a/4$. Keep in mind that the commutation curve is a plot of the current in the short-circuited coil as a function of time. Appearing in Fig. 7-20(c) is the distribution of currents that prevail at a time midway through the commutation period. Note that now the current in the short-circuited coil is zero.

Curve b in Fig. 7-19 represents *resistance* commutation. It is the curve that results when the resistance of the short-circuited coil is not negligible compared with the brush contact resistance. Note that at the leading brush tip the slope of curve b is steeper than that of curve a, which indicates that the coil resistance enhances the current reversal. This is apparent from an examination of Fig. 7-20(b) where it is seen that at junction a the presence of coil resistance means that more current flows to segment 5. However, at the trailing brush tip (or during the second half of the commutation period) the coil resistance serves to retard the current reversal. The current density, which is represented by the slope of the commutation curve, is therefore higher at both the leading and trailing brush edges for resistance commutation.

The commutation curve denoted as c in Fig. 7-19 is described as *under commutation*. It is a form of voltage commutation in which current reversal has been delayed by the action of the emf of self-induction in the short-circuited coil or of the Blv voltage induced in the short-circuited coil by the flux produced by the armature mmf or of a combination of both. A decelerated commutation of this kind is characterized by high current density at the trailing brush tip, which can promote a deterioration of the brush material and also bring about undesirable chemical changes in the copper segments of the commutator.

A scheme that is frequently used to neutralize this emf of self-induction and the effect of the armature mmf is the inclusion of small poles located on the brush axis and energized by a coil carrying the armature current. Because of their location and function, these poles are called *interpoles* or *commutating poles*. The use of armature-current excitation is dictated by the fact that both the emf of self-induction in the short-circuited coil and the flux in the brush axis produced by the armature mmf are proportional to armature current. For generator action the polarity of the interpole must be the same as that of the field pole into which the coil is moving. Figure 7-21 depicts the situation for a two-pole machine. It is helpful to compare the resultant flux-density curve here with that appearing in Fig. 7-7. Note that the value of the flux density on the interpolar axis in Fig. 7-21 is opposite to that occurring in Fig. 7-7. If the magnitude of the flux density produced by the interpole is adjusted properly, it can be made to neutralize the emf of self-induction completely. Of course the effect of the armature mmf has already been neutralized in the representation of Fig. 7-21 because the resultant flux density is shown to be of opposite polarity. With such an adjustment the type of commutation that prevails is linear commutation, provided that the coil resistance compared to the brush contact resistance is inconsequential. Otherwise resistance commutation takes place.

    A considerable improvement in the commutation process can be achieved by increasing the interpole flux to the point at which, in addition to canceling the effects of the emf of self-induction and the armature mmf, there is induced a voltage equal to twice the brush voltage drop, $2I_aR_b$. Under these conditions that special condition of *overcommutation* results that leads to a value of zero current density at the trailing edge of the brush. This situation is denoted by curve *d* in Fig. 7-19. Note that the current in the coil undergoing commutation has already reached its value of $-I_a/2$ before the coil leaves the brush. *This results in no sparking at the trailing edge of the brush and represents the most favorable condition for good commutation.*

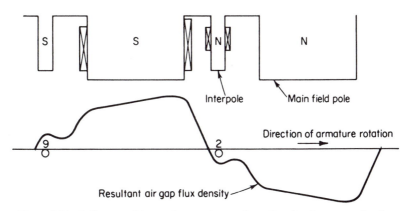

Figure 7-21 Influence of interpoles on the resultant flux-density wave of a dc machine.

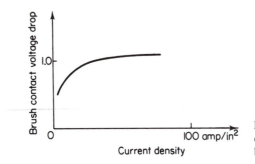

Figure 7-22 Illustrating the arc-type conduction characteristics of carbon brushes.

The relationship between the contact voltage drop at a brush and the current density is shown in Fig. 7-22. For average carbon brushes the contact drop tends to remain essentially constant at about 1 volt. This is consistent with the fact that the conduction in the space between the copper segments and the carbon brushes occurs in accordance with arc phenomenon.

When dc machines are used in applications that involve severe duty cycles (e.g., steel mills), swiftly changing loads and high overloads can cause breakdown of the air space between commutator segments. This breakdown can spread rapidly over the entire commutator, thus causing damage to the commutator as well as a short-circuit on the dc supply lines. Unfortunately, this situation can occur in spite of the presence of interpoles.

The most effective way to prevent this flashover condition is to neutralize the armature reaction that causes the severe peaking of the air-gap flux density that leads to breakdown. This is best achieved by the use of a *compensating winding* embedded in slots distributed along the pole faces. By allowing armature current to flow through this pole-face winding with a polarity opposite to that of

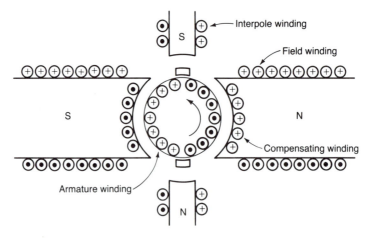

Figure 7-23 Schematic diagram of a two-pole dc machine equipped with interpole and compensating windings.

the armature winding and a suitable number of ampere-turns, there can result complete cancellation of the armature reaction. The chief disadvantage of this method of eliminating or reducing the commutation problem is the high expense of pole-face windings.

Figure 7-23 shows the schematic diagram of a two-pole dc machine equipped with a compensating winding located in the pole faces. Beneath each pole face the current carried by the compensating winding flows in the opposite direction to that which flows in the armature winding.

PROBLEMS

7-1. The rotor of a two-pole dc machine has 12 slots. The wave-wound, full-pitch armature winding has two coil sides per slot. There are 12 commutator segments. Show a plan view of the winding layout. Clearly identify the position of the brushes on the commutator and relative to the poles. Also show the direction of the induced emf in the coil sides for a counterclockwise rotation of the rotor.

7-2. The rotor of a two-pole dc machine carries a wave-wound armature winding having two coil sides per slot in its 12 slots. The winding is full-pitch.
   (a) Draw the schematic diagram that shows the two armature paths existing between the two brushes. Represent each suitably marked coil voltage by small dc sources with properly identified polarities. Consider that instant when slots 1 and 7 are located exactly between the poles.
   (b) Directing attention solely to the fundamental component of the flux density in the air gap, compute the voltage value appearing at the brushes.

7-3. Repeat Prob. 7-2 for the case where slot 1 lies to the left of the interpolar axis by 15°.

7-4. For the machine of Prob. 7-2 draw the variation of the armature induced voltage measured at the brushes as a function of $\omega_r t$ where ω_r is the angular rotor velocity. Dealing only with the fundamental component of the flux density curve, show the variation of each coil voltage as well as that of the total. Compute the maximum and minimum values of the coil and total voltages.

7-5. Repeat Prob. 7-1 for a dc machine with the following specifications: two poles, six slots, full-pitch, wave-wound armature winding having two coil sides per slot.

7-6. For the machine of Prob. 7-1 sketch the variation of the emf induced in each coil voltage as well as between brushes as a function of $\omega_r t$, where ω_r is the rotor angular velocity. Direct attention solely to the fundamental component of the air-gap flux density. Indicate the maximum and minimum values of coil and brush voltages.

7-7. Explain your answer to each part:
   (a) Can a separately excited dc generator operate below the knee of its magnetization curve?
   (b) Can a dc shunt generator operate below the knee of its magnetization curve?

7-8. The magnetization curve for a dc shunt generator driven at a constant speed of 1000 rpm is shown in Fig. 7-17. Each vertical division represents 20 V; each horizontal division represents 0.2 A.
   (a) Compute the critical field resistance.

(b) What voltage is induced by the residual flux of this machine?

(c) What must be the resistance of the field circuit in order that the no-load terminal voltage be 240 V at a speed of 1000 rpm?

(d) Determine the field current produced by the residual flux voltage when the field circuit resistance has the value found in part (c).

(e) At what speed must the generator be driven in order that it would fail to build up when operating with the field circuit resistance of part (c)?

7-9. A dc shunt generator has a magnetization curve given by Fig. 7-17, where each ordinate unit is made equal to 20 V and each abscissa unit is set equal to 0.2 A and the speed of rotation is 1000 rpm. The field circuit resistance is 156 Ω. Determine the voltage induced between brushes when the generator is operated at a reduced speed of 800 rpm.

7-10. It is desired to reverse the terminal voltage polarity of a generator that has been operating properly as a cumulative compound generator. The machine is stopped. Residual magnetism is reversed by temporarily disconnecting the shunt field and separately exciting it with reversed current. The connections are then restored exactly as they were before.

(a) Does the terminal voltage build up? Explain.

(b) If answer to part (a) is yes, will the compounding be cumulative or differential? Explain.

7-11. A two-pole dc machine is equipped with a wave winding having 120 armature conductors. The pole faces span 70% of a pole pitch and the field winding has 1050 turns. This machine operates with a fixed field current of 1.6 A. The magnetization curve is depicted in Fig. 7-17 when each ordinate unit represents 0.1 Wb and each abscissa unit represents 0.2 A. At rated load the armature current is 40 A.

(a) Determine the amount of demagnetization of the field flux caused by the armature mmf.

(b) What is the value of the air-gap flux under load?

(c) What is the value of flux at the leading pole tip?

7-12. In a lap-wound armature winding there are as many brushes as there are poles. Hence, for a four-pole dc machine it is customary to use four brushes, with each pair of brushes being responsible for conducting one-half of the total armature current. Show the distribution of current in the coils undergoing commutation for two instances: just before undergoing commutation and five-eighths through the commutation period. Assume resistance commutation throughout.

7-13. The no-load characteristic of a 10-kW, 250-V, dc self-excited shunt generator driven at 1000 rpm is shown in Fig. 7-17. Each vertical unit denotes 20 V, and each horizontal unit represents 0.2 A. This armature circuit resistance is 0.3 Ω, and the field circuit resistance is set at 157 Ω. Determine:

(a) The critical resistance of the shunt field circuit.

(b) The change in voltage from no-load to full-load. Neglect armature demagnetization.

(c) The shunt field current under rated load conditions.

7-14. Repeat Prob. 7-13(b) and (c) for the case where the armature demagnetizing effect I_{fd} is known to be 0.2 A expressed in terms of equivalent field current.

7-15. The machine of Prob. 7-13 operates with a field winding resistance of 143 Ω.

(a) What is the no-load voltage?

(b) When the generator delivers power to a specified load, the terminal voltage is found to drop to 220 V and the corresponding demagnetization effect of the armature mmf is known to be 0.25 equivalent field ampere. Determine the armature winding current.

(c) What is the value of the field current at the load condition of part (b)?

(d) Find the armature winding emf under load.

7-16. The no-load characteristic of a 5-kW, 125-V shunt generator driven at 1000 rpm is shown in Fig. 7-17. Each ordinate unit represents 10 V, and each abscissa unit represents 0.1 A. The armature circuit resistance is 0.2 Ω, and the field circuit resistance is 157 Ω. When the generator delivers rated armature current, the armature induced emf is 116 V.

(a) Find the terminal voltage.

(b) What is the value of the demagnetization of the armature mmf expressed in equivalent field amperes?

(c) What is the field current?

(d) How much power is delivered to the load?

7-17. The no-load characteristic of a dc shunt generator driven at a speed of 800 rpm is approximated in its useful range by the equation

$$E = \frac{300I_f}{2 + I_f}$$

where E is armature induced emf and I_f is the field current. The armature circuit resistance is 0.1 Ω. The field winding resistance is 20 Ω. The demagnetizing effect of armature reaction may be neglected.

(a) The terminal voltage is to be 225 V when the armature current is 150 A. The generator is driven at 800 rpm. Find the resistance of the field rheostat for this condition.

(b) Change the field rheostat setting to 10 Ω. Let the load be disconnected and the speed reduced to 720 rpm. Find the terminal voltage.

(c) Return to the conditions of part (a). Assume the rotational losses to be 2 kW. Find the output power, the efficiency, and the shaft input torque from the prime mover of the generator.

7-18. The no-load characteristic of a dc machine in its useful operating range and driven at 1500 rpm may be approximated by

$$E_a = \frac{400I_f}{3 + I_f} \qquad \text{or} \qquad I_f = \frac{3E_a}{400 - E_a}$$

This machine is operated as a dc shunt generator whose field circuit resistance is 50 Ω.

(a) Compute the no-load terminal voltage when the generator is driven at 1500 rpm, indicating any necessary assumptions.

(b) What will be the terminal voltage at no-load when the speed is reduced to 750 rpm? Account for the large difference.

7-19. The magnetization curve of a dc shunt generator is represented in Fig. 7-17, provided that each ordinate unit is multiplied by 40 V and each abscissa unit by 0.5 A. The armature circuit resistance is 0.4 Ω, and the field winding resistance is 80 Ω.

(a) This machine is made to deliver an armature current of 60 A. Find the resistance of the field rheostat (in series with the field winding) so that the terminal voltage will be 520 V at a speed of 1000 rpm.

(b) With the field rheostat set as in part (a) find the terminal voltage at no-load and 1000 rpm.

7-20. A cumulative-compound generator driven at 1000 rpm has the magnetization curve shown in Fig. 7-17, provided that each unit of the ordinate axis is multiplied by 20 V and each abscissa unit is multiplied by 1 A. The armature circuit resistance is 0.06 Ω. In addition:

Series turns per pole (N_s) = 20

Shunt turns per pole (N_f) = 1000

Armature-reaction demagnetizing effect at I_a of 100 A = 600 A-t per pole

The field circuit resistance is 45 Ω.

(a) Determine the no-load terminal voltage.

(b) Find the terminal voltage when the load is such that it causes 100 A to flow through the armature winding.

(c) Find the new value of the field current.

(d) What is the value of the armature induced emf at the specified load.

7-21. Repeat Prob. 7-20 for the case where the field circuit resistance is reduced to 31.5 Ω and the number of series field turns per pole is increased to 40. The armature current remains at 100 A.

7-22. Refer to the external characteristics depicted in Fig. 7-11.

(a) The shunt generator is operating at rated voltage, rated load current, and rated speed. Assume the load demand is reduced to one-half the rated value. Describe the adjustments you would make to maintain constant terminal voltage.

(b) Repeat part (a) for the overcompounded operating mode.

7-23. A 10-kW, 250-V generator is equipped with a ratio of shunt-to-series turns of 40:1. When operated as a self-excited generator, it takes an increase in field current of 0.3 A to maintain the terminal voltage at rated value at full-load current. The total field current for this field condition is 1.6 A. Determine whether or not a diverter is needed when this generator is operated cumulatively compounded. Assume a long-shunt connection.

7-24. A cumulative-compound generator has the magnetization curve shown in Fig. 7-17, where each ordinate unit is 20 V and each abscissa unit is 1 A. Determine the required field circuit resistance that allows the generator to operate at a terminal voltage of 240 V when the armature current is 100 A. In addition:

Armature circuit resistance drop = 10 V

Series field turns per pole = 25

Shunt field turns per pole = 1000

Armature demagnetizing A-t per pole = 800

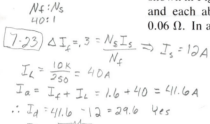

$N_f : N_s$
40:1
(7-23) $\Delta I_f = .3 = \dfrac{N_s I_s}{N_f} \Rightarrow I_s = 12A$
$I_L = \dfrac{10K}{250} = 40A$
$I_a = I_f + I_L = 1.6 + 40 = 41.6A$
∴ $I_d = 41.6 - 12 = 29.6$ Yes

$N_f : N_s$

(7-20) 1000 shunt turns/pole = N_f
a) $E_0 = 20 \cdot 11 = 220$
b) $I_{fs} = \dfrac{20 \cdot 1000}{1000} = 20$
$I_{fd} = \dfrac{600}{1000} = 0.6$
$0a = I_{fs} - I_{fd} = 2 - 0.06 = 1.4$
$ab = I_a(R_a + R_s) = 100 \times .06 = 6$
$V_t = 11.6 \times 20 = 232$
c) $I_f = (5.16)(1) = 5.16$
d) $E_a = (11.9)(20) = 238$

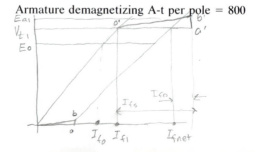

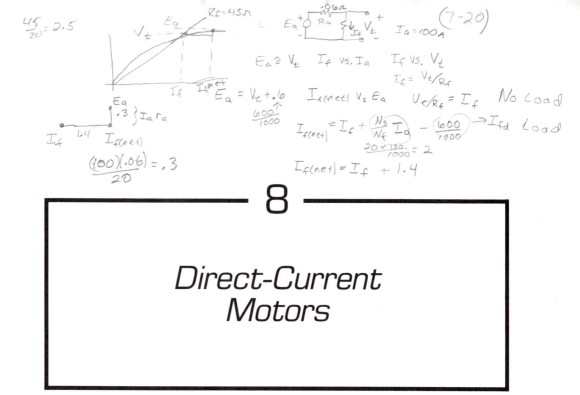

8

Direct-Current Motors

A dc motor is a dc generator with the power flow reversed. In the dc motor electrical energy is converted to mechanical form. Also, as is the case with the generator, there are three types of dc motors: the *shunt* motor, the *cumulatively compounded* motor, and the *series* motor. The compound motor is prefixed with the word cumulative in order to stress that the connections to the series field winding are such as to ensure that the series field flux *aids* the shunt field flux. The series motor, unlike the series generator, finds wide application, especially for traction-type loads. Hence due attention is given to this machine in the treatment that follows.

8-1 MOTOR SPEED-TORQUE CHARACTERISTICS

How does the dc motor react to the application of a shaft load? What is the mechanism by which the dc motor adapts itself to supply to the load the power it demands? The answers to these questions can be obtained by reasoning in terms of a pair of simple equations. Initially our remarks are confined to the shunt motor, but a similar line of reasoning applies for the others. For our purposes the two pertinent equations are those for torque and current. Thus

and

$$T = K_T \Phi I_a \tag{8-1}$$

$$I_a = \frac{V_t - K_E \Phi n}{R_a} \tag{8-2}$$

Note that the last expression results from Kirchhoff's voltage law and replacing E_a with Eq. (3-34). With no shaft load applied, the only torque needed is that which overcomes the rotational losses. Since the shunt motor operates at essentially constant flux, Eq. (8-1) indicates that only a small armature current is required compared with its rated value to furnish these losses. Equation (8-2) reveals the manner in which the armature current is made to assume just the right value. In this expression V_t, R_a, K_E, and Φ are fixed in value. Therefore the speed is the critical variable. If, for the moment, it is assumed that the speed has too low a value, then the numerator of Eq. (8-2) takes on an excessive value and in turn makes I_a larger than required. At this point the motor reacts to correct the situation. The excessive armature current produces a developed torque which exceeds the opposing torques of friction and windage. In fact this excess serves as an accelerating torque, which then proceeds to increase the speed to that level which corresponds to the equilibrium value of armature current. In other words, the acceleration torque becomes zero only when the speed is at that value which by Eq. (8-2) yields just the right I_a needed to overcome the rotational losses.

Consider next that a load demanding rated torque is suddenly applied to the motor shaft. Because the developed torque at this instant is only sufficient to overcome friction and windage and not the load torque, the first reaction is for the motor to lose speed. In this way, as Eq. (8-2) reveals, the armature current can be increased so that in turn the electromagnetic torque can increase. As a matter of fact the applied load torque causes the motor to assume that value of speed which yields a current sufficient to produce a developed torque to overcome the applied shaft torque and the frictional torque. Power balance is thereby achieved, because an equilibrium condition is reached where the electromagnetic power $E_a I_a$ is equal to the mechanical power developed, $T\omega_m$.

A comparison of the dc motor with the three-phase induction motor indicates that both are *speed-sensitive* devices in response to applied shaft loads. An essential difference, however, is that for the three-phase induction motor developed torque is adversely influenced by the power-factor angle of the armature current. Of course no analogous situation prevails in the case of the dc motor.

Shunt Motor

From the foregoing discussion it should be apparent that the speed-torque relationship of the dc motor is an important characteristic. Accordingly, it is useful to obtain an expression for speed as a function of the developed torque. We begin with Eq. (8-2) by solving for the speed. Thus,

$$n = \frac{E_a}{K_E\Phi_{sh}} = \frac{V_t - I_aR_a}{K_E\Phi_{sh}} \qquad (8\text{-}3)$$

The subscript on ϕ emphasizes that consideration for the moment is confined to the shunt motor. The torque variable can next be introduced by Eq. (8-1). Hence,

$$n = \frac{V_t}{K_E\Phi_{sh}} - \frac{R_a}{K_EK_T\Phi_{sh}^2}\,T \qquad (8\text{-}4)$$

If it is now assumed that the shunt motor is equipped with a compensating winding of negligible resistance so that the demagnetizing effect of the armature winding is essentially neutralized (i.e., $I_{fd} = 0$), Φ_{sh} stays constant and so Eq. (8-4) may be more conveniently written as

$$n = k_1V_t - k_2T = n_0 - k_2T \qquad (8\text{-}5)$$

where

$$k_1 = \frac{1}{K_E\Phi_{sh}}\ \text{rpm/V}; \qquad k_2 = \frac{R_a}{K_EK_T\Phi_{sh}^2}\ \text{rpm/N-m} \qquad (8\text{-}6)$$

and

$$n_o = k_1V_t = \text{no-load speed} \qquad (8\text{-}7)$$

Clearly, Eq. (8-5) is a straight line having a negative slope of k_2 rpm/N-m and an intercept of n_o. The solid line of Fig. 8-1 shows the plot. The broken-line curve that also appears is the speed-torque characteristic with the demagnetizing effect of the armature mmf present; in other words, when the machine is not equipped with a compensating winding. The reduction in Φ_{sh} with armature current causes some compensation for the drop in the numerator term of Eq. (8-3).

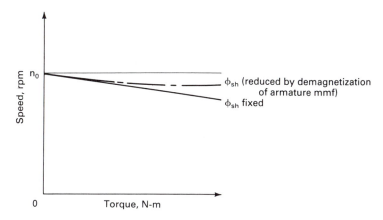

Figure 8-1 Speed-torque characteristics of a shunt motor with and without a compensating winding.

Example 8-1

A 20-hp, 230-V, 1150-rpm shunt motor has design parameters which lead to values for k_1 and k_2 in Eqs. (8-6) and (8-7) which are respectively 5.32 rpm/V and 0.4 rpm/N-m.

(a) Find the no-load speed.

(b) Determine the speed when full-load torque of 124 N-m is delivered. Assume the motor is equipped with a compensating winding.

Solution (a) From Eq. (8-7) the no-load speed is

$$n_o = 5.32(230) = 1223.6 \text{ rpm}$$

(b) From Eq. (8-5)

$$n = 1223.6 - 0.4(124) = 1150.4 \text{ rpm}$$

Speed determination at various load conditions. Speed and load conditions are related through the induced armature voltage equation, $E_a = K_E \Phi n$. Because E_a is determinable from the terminal voltage and the armature-circuit resistance drop, it follows that at each load condition E_a is known. Letting subscript 1 denote conditions at some original operating point where speed, input line current, field current, and so on are known and subscript 2 denote conditions at a new operating point, the speed at the changed load condition can be found from the relationship

$$\frac{E_{a2}}{E_{a1}} = \frac{K_E \Phi_2 n_2}{K_E \Phi_1 n_1} \tag{8-8}$$

This relationship reduces to a strikingly simple one for motors where the flux per pole remains essentially invariant with load changes. Such is surely the case with a shunt motor equipped with a compensating winding or one in which the demagnetizing effect of the armature mmf is negligible ($I_{fd} = 0$). Then the new speed is related to the original speed simply by

$$n_2 = \frac{E_{a2}}{E_{a1}} n_1 \tag{8-9}$$

The change in motor speed that occurs as the load is changed from rated value to no-load is an important performance indicator and is called the *speed regulation* (SR). Expressed in percent of the full-load value it is written as

$$\boxed{\text{SR} = \frac{\text{no-load speed} - \text{full-load speed}}{\text{full-load speed}} 100} \tag{8-10}$$

Example 8-2

A 20-hp, 230-V, 1150-rpm shunt motor equipped with a compensating winding has a total armature-circuit resistance of 0.188 Ω. At rated output the motor draws a line current of 74.6 A and a field current of 1.6 A.

(a) Find the speed when the input line current is 38.1 A.
(b) What is the no-load speed if $I_L = 1.9$ A?
(c) Determine the speed regulation.

Solution   (a) The armature induced emf at rated load is

$$E_{a1} = V_t - I_a R_a = 230 - (74.6 - 1.6)0.188 = 216.3 \text{ V}$$

$$n_1 = 1150 \text{ rpm at rated output}$$

At $I_L = 38.1$ A

$$E_{a2} = 230 - (38.1 - 1.6)0.188 = 223.14 \text{ V}$$

Hence by Eq. (8-10)

$$n_2 = \frac{E_{a2}}{E_{a1}} n_1 = \frac{223.14}{216.3} 1150 = 1186.4 \text{ rpm}$$

(b) Now

$$E_{ao} = 230 - 0.3(0.188) = 229.9 \text{ V}$$

$$\therefore \quad n_o = \frac{229.3}{216.3} 1150 = 1223.3 \text{ rpm}$$

(c) From Eq. (8-10)

$$\text{SR} = \frac{1223.3 - 1150}{1150} 100 = 6.29\%$$

Effect of demagnetization on speed.   When the shunt motor is not provided with a compensating winding, the demagnetizing effect of the armature mmf
causes a change in the flux per pole so that Eq. (8-9) is not directly applicable.
One way to account for this effect is to find I_{fd} in the manner described in Sec. 7-3
corresponding to specific armature currents. This requires working with the magnetization curve of the motor. The following example illustrates the procedure.

Example 8-3

The machine of Example 7-3 is used as a shunt motor. The motor's magnetization
curve, which is obtained at 1000 rpm, is displayed in Fig. 7-17. At a specific load
condition this motor draws a field current of 1.6 A and an armature current of 40 A.
The demagnetizing effect of this armature current on the pole flux is known in terms
of an equivalent field current to be 0.3 A. The armature-circuit resistance is 0.15 Ω.
Find the motor speed at the given load condition.

Solution   The solution strategy in this case is to use the magnetization curve to find
the state of the pole flux at the given armature current. Since the magnetization
curve is found at 1000 rpm, we have

$$E_a = K_E \Phi(1000)$$

If there were no demagnetization effect, the quantity, $K_E \Phi$, could be found by entering the magnetization curve at a field current of 1.6 A. But to account for the

demagnetizing effect of the 40-A armature current we must use a value of E_a that corresponds to a net field current of $1.6 - 0.3 = 1.3$ A. From Fig. 7-17 this yields

$$E_a = 11.8(20) = 236 = K_E\Phi(1000)$$

Hence, the state of pole flux at 40 A armature current is represented by

$$K_E\Phi = \frac{236}{1000} = 0.236 \qquad (8\text{-}11)$$

Then at the specified operating point the actual induced armature voltage is

$$E_a = 250 - I_a R_a = 250 - 40(0.15) = 244 \text{ V}$$

This value of E_a derives from a *circuit* evaluation. A *field* evaluation, which must yield the same result, is based on the result in Eq. (8-11). Hence

$$E_a = K_E\Phi n$$

$$244 = (0.236)n$$

$$\therefore \ n = 244/0.236 = 1033.9 \text{ rpm}$$

Compound Motor

The expression for the speed of a cumulatively compounded motor differs from Eq. (8-3) by the fact that the net air-gap flux contains the contribution of the series field. Moreover, the armature circuit also contains the resistance of the series field for a long-shunt connection. Thus, the analogous expression for the speed at a given armature and field excitation is

$$\boxed{n = \frac{V_t - I_a(R_a + R_s)}{K(\Phi_{sh} + \Phi_s)}} \qquad (8\text{-}12)$$

The summation of fluxes is used on the assumption of linearity for the magnetic circuit. Although the total resistance drop is larger for the compound motor than it is for the shunt mode, the strong contribution usually associated with Φ_s is the dominant factor. Consequently, for the same armature current as for the shunt mode of operation, the speed is lower and the speed-torque curve displays a more pronounced droop as illustrated in Fig. 8-2. A better appreciation of the interplay of these factors can be gained from a study of the following example.

Example 8-4

The motor of Example 8-2 is also supplied with a series field winding having a resistance of 0.06 Ω. When operating as a shunt motor the rated output is obtained with an armature current of 73 A. The same constant torque is to be supplied when this machine is arranged to operate as a cumulatively compounded motor. At this load condition the series field winding contributes a 25% increase in pole flux.

(a) Find the armature current of the compound mode of operation.

(b) What is the speed of the compound motor?

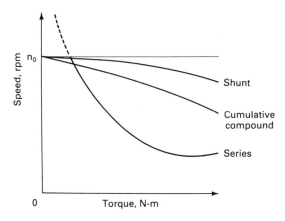

Figure 8-2   Speed-torque characteristics of various dc motor types using common no-load speed (except for series motor).

(c) Find the speed regulation (SR) assuming the series field effect is negligible at no-load.

Solution   (a) At constant torque the increased flux field calls for a smaller armature current. Thus,

$$I'_a = \frac{\Phi_{sh}}{\Phi'} I = \frac{1}{1.25} 73 = 58.4 \text{ A}$$

(b) As a shunt motor at rated load and speed the induced armature emf is

$$E_a = 230 - 73(0.188) = 216.3 \text{ V}$$

As a compound motor

$$E'_a = 230 - 58.4(0.188 + 0.06) = 215.5 \text{ V}$$

Employing Eq. (8-8) then yields

$$n' = \frac{E'_a}{E_a} \frac{\Phi_{sh}}{\Phi'} n = \frac{215.4}{216.3} \frac{1}{1.25} 1150 = 916.7 \text{ rpm}$$

(c) The no-load speed from Example 8-2 is 1222.3 rpm. Hence,

$$SR = \frac{1222.3 - 916.7}{916.7} 100 = 33.3\%$$

Note how much larger this quantity is compared to operation as a shunt motor.

Series Motor

The speed-torque characteristic of the series motor differs sharply from those of the shunt and compound motors. This is not a surprising expectation in view of the absence of a substantial and virtually constant flux supplied by a shunt field winding. In the series motor all of the working flux must be produced by the series field winding. It is instructive now to derive the expression that relates speed and torque.

For simplicity we assume no saturation in the iron paths of the motor. This permits the series winding flux to be related to the armature winding by

$$\Phi_s = K_s I_a \tag{8-13}$$

where K_s is an appropriate proportionality factor expressed in Wb/A. Inserting Eq. (8-13) into the basic torque equation of Eq. (8-1) gives

$$T = K_T \Phi_s I_a = K_T K_s I_a^2 \tag{8-14}$$

Furthermore, in the armature circuit of the series motor, Kirchhoff's voltage law states that

$$V_t = E_a - I_a(R_a + R_s) = K_T \Phi_s \omega + I_a R \tag{8-15}†$$

where for convenience $R = R_a + R_s$. Then by employing Eq. (8-13) in the last expression we get more simply

$$V_t = I_a(K_T K_s \omega + R)$$

or

$$I_a = \frac{V_t}{K_T K_s \omega + R} \tag{8-16}$$

The desired speed-torque relation follows on inserting Eq. (8-16) into Eq. (8-14). Hence,

$$T = K_T K_s \frac{V_t^2}{(K_T K_s \omega + R)^2}$$

or

$$\frac{V_t}{K_T K_s \omega + R} = \sqrt{\frac{T}{K_T K_s}} \tag{8-17}$$

In Eq. (8-17) ω is the motor speed and T is the developed electromagnetic torque; all other quantities are known. A plot of this equation takes the form depicted in Fig. 8-2.

A more manageable expression results from Eq. (8-17) by assuming R small compared to $K_T K_s n$. (See Prob. 8-7.) This step offers the advantage of making it easier to see the manner in which these two variables are related. Accordingly, we get

$$\omega = \frac{V_t}{K_T} \sqrt{\frac{K_T}{K_s}} \frac{1}{\sqrt{T}} \tag{8-18}$$

Next, upon introducing

$$K = \frac{V_t}{K_T} \sqrt{\frac{K_T}{K_s}} = \frac{V_t}{\sqrt{K_T K_s}} \tag{8-19}$$

†See Eq. (8-30) and the footnote on p 350.

we get finally,

$$\boxed{\omega = \frac{K}{\sqrt{T}}} \tag{8-20}$$

Equation (8-20) shows that an increase in torque by a factor of 4 causes the speed to halve. Similarly, a torque output that is nine times larger causes the motor to reduce its speed to one-third its former value. A complete plot of Eq. (8-20) takes the form depicted in Fig. 8-2.

A matter that is especially noteworthy about the series motor is the serious consequence that can result if it is ever allowed to operate at no-load (i.e., $T \rightarrow 0$). Equation (8-20) clearly indicates that the series motor speed tends to infinity. Of course such a speed will never be attained. What will happen instead is that the motor armature coils will fly out of their slots and be ripped apart unless appropriate circuit-breaker protection is provided soon enough. A cardinal rule that applies to the safe operation of the series motor is that it must *never* be uncoupled from its load. Belt couplings must not be used in order to avoid the possibility of their slipping and thereby causing damage.

Example 8-5

A 250-V, 25-hp, 600-rpm series motor draws an armature current of 85 A at rated load torque of 314 N-m. The armature circuit resistance is 0.12 Ω and the series field winding resistance is 0.09 Ω. Rotational losses are negligible.

Find the motor speed when the torque requirement at the motor shaft is reduced to 20 N-m.

Solution   The induced armature voltage at rated load is

$$E_a = 250 - 85(0.21) = 232.15 \text{ V}$$

To find the value of E_a at the reduced load condition, we first need to find the new armature current. This is readily determined by invoking Eq. (8-14). Thus

$$\frac{I_a'}{I_a} = \sqrt{\frac{T'}{T}} = \sqrt{\frac{20}{314}} = 0.2524$$

$$\therefore \quad I_a' = 0.2524(85) = 21.45 \text{ A}$$

where primes denote the reduced load condition. Therefore,

$$E_a' = 250 - 21.45(0.21) = 245.5 \text{ V}$$

Finally, forming the ratio of induced voltages at the specified loads, we get

$$\frac{E_a'}{E_a} = \frac{K_E K_s I_a' n'}{K_E K_s I_a n} = (0.2524) \frac{n'}{n}$$

$$\therefore \quad n' = \frac{245.5}{232.15} \frac{1}{0.2524} 600 = 2513.9 \text{ rpm}$$

Observe that this speed is more than four times the rated value and can mean trouble.

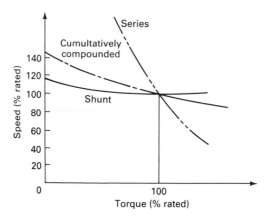

Figure 8-3 Typical speed-torque curves of dc motors.

A more useful representation of the graphs of Fig. 8-2 is shown in Fig. 8-3. These curves are drawn with a common point which is rated speed at rated torque. In this situation the series motor can share with the others the point of commonality because the flow of rated armature current no longer poses a threat to the series motor.

Torque versus Armature Current

The characteristic curves that describe the variation of developed torque as a function of the motor's drawn armature current for the various motor types readily follows from the basic torque equation, Eq. (8-1). A shunt motor equipped with a compensating winding provides the simplest relationship which is that of a straight line through the origin. That is,

$$T = K_T\Phi_{sh}I_a = K_{sh}I_a$$

where

$$K_{sh} = K_T\Phi_{sh}$$

When the shunt motor operates under the influence of a demagnetizing effect, the decreasing pole flux with increasing armature current causes the characteristic to fall below that of the linear relationship. This distinction is illustrated in Fig. 8-4.

For the cumulatively compounded motor the pertinent relationship is

$$T = K_T(\Phi_{sh} + \Phi_s)I_a = K_T\Phi_{sh}I_a + K_T\Phi_sI_a = K_T\Phi_{sh}I_a + K_tI_a^2 \qquad (8\text{-}21)$$

where

$$\Phi_s = K_sI_a \qquad \text{and} \qquad K_t = K_TK_s$$

Clearly, Eq. (8-21) emphasizes the increasing contributions to the developed torque as the armature current gets larger. Observe the increasing slope of the characteristic as I_a increases.

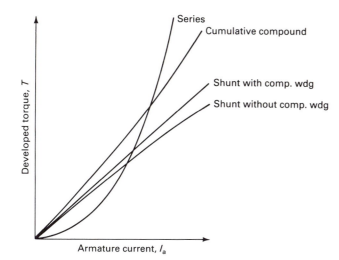

Figure 8-4 Torque versus armature current characteristic for various dc motor types.

For the series motor the expression for the developed torque is more simply described by Eq. (8-14), which is parabolic as indicated in Fig. 8-4. It is instructive to note that as the series motor reacts to develop greater torques in response to load requirements, the speed drops correspondingly. This feature is responsible for making the series motor very well suited to applications involving traction-type loads.

8-2 EQUIVALENT CIRCUITS AND THE POWER FLOW DIAGRAM

The performance of the dc motor operating in any one of its three modes can conveniently be described in terms of an equivalent circuit, a set of performance equations, a power-flow diagram, and the magnetization curve. The equivalent circuit is depicted in Fig. 8-5. It is worthwhile to note that now the armature induced voltage is treated as a reaction or counter emf. Observe that the defined direction of I_a is reversed from the case for a dc generator. The generated voltage is caused by the relative motion of the rotor with respect to the magnetic field, and its polarity is independent of the direction of armature current. By imposing constraints similar to those applied to the dc generator, we obtain the correct

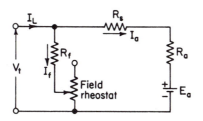

Figure 8-5 Equivalent circuit of the dc motor.

equivalent circuit for the desired mode of operation. For example, for a series motor the appropriate equivalent circuit results upon removing R_f from the circuitry of Fig. 8-5.

The set of equations needed to compute the performance of dc motors consists of the induced armature emf equation, the developed torque equation, and Kirchhoff's voltage and current laws as they apply to the equivalent circuit displayed in Fig. 8-5. These are listed here as as matter of convenience.

$$E_a = K_E \Phi n \tag{8-22}$$

$$T = K_T \Phi I_a \tag{8-23}$$

$$V_t = E_a + I_a(R_a + R_s) \tag{8-24}$$

$$I_L = I_f + I_a \tag{8-25}$$

The first two equations are identical with those used in generator analysis. Note, however, that the next two equations are modified to account for the fact that for the motor V_t is the applied or source voltage and as such must be equal to the sum of the voltage drops. Similarly the line current is equal to the *sum* rather than the *difference* of the armature current and field currents.

The power-flow diagram, depicting the reversed flow from that which occurs in the generator, is illustrated in Fig. 8-6. The electrical power input $V_t L_L$ originating from the line supplies the field power needed to establish the flux field as well as the armature circuit copper loss needed to maintain the flow of I_a. This current flowing through the armature conductors imbedded in the flux field causes torque to be developed. The law of conservation of energy then demands that the electromagnetic power $E_a I_a$ be equal to $T\omega_m$, where ω_m is the steady-state operat-

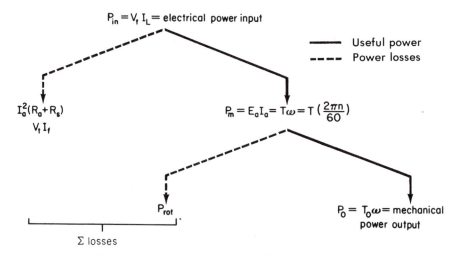

Figure 8-6   Power-flow diagram of the dc motor.

ing speed. Removal of the rotational losses from the developed mechanical power yields the mechanical output power.

The strategy associated with the particular form of the power-flow diagram illustrated in Fig. 8-6 is to start with the input power and then follow this power through the machine to the place where the useful output appears. In the process appropriate steps are included to underscore any transformations or conversions that occur. Moreover, the flow of the useful power is represented by solid lines, whereas the losses accompanied with each conversion or transformation are indicated by broken lines. A particular advantage of this representation is the ease with which the pertinent relationships that are used to provide a complete description of the behavior of the motor are displayed. In essence the power-flow diagram so formulated furnishes an excellent model of the mathematical depiction of the machine. The following examples illustrate these advantages.

Example 8-6

A 20-hp 230-V 1150-rpm shunt motor has four poles, four parallel armature paths, and 882 armature conductors. The armature-circuit resistance is 0.188 Ω. At rated speed and rated output the armature current is 73 A and the field current is 1.6 A. Calculate: (a) the electromagnetic torque, (b) the flux per pole, (c) the rotational losses, (d) the efficiency, (e) the shaft load.

Solution   (a) The developed electromagnetic torque can be found by employing the relationship that appears on the second level of the power-flow diagram which is simply a statement of the law of conservation of energy. It is at this point that the power associated with armature winding, namely $E_a I_a$, is converted to mechanical power, $T\omega_m$. Accordingly, we proceed as follows:

$$E_a = V_t - I_a R_a = 230 - 73(0.188) = 230 - 13.7 = 216.3 \text{ V}$$

Also

$$\omega_m = \frac{2\pi n}{60} = \frac{2\pi(1150)}{60} = 120 \text{ rad/s}$$

Therefore

$$T = \frac{E_a I_a}{\omega_m} = \frac{216.3(73)}{120} = 132 \text{ N-m}$$

(b) $$E_a = K_E \Phi n = \frac{p}{60} \frac{Z}{a} \Phi n = \frac{4}{60} \frac{882}{4} \Phi(1150)$$

$$\therefore \quad \Phi = \frac{216.3(60)}{882(1150)} = 0.0128 \text{ Wb}$$

(c) From the power-flow diagram

$$P_{\text{rot}} = P_m - P_0 = E_a I_a - 20(746) = 15{,}790 - 14{,}920 = 870 \text{ W}$$

(d) First find the sum of the losses. Thus

$$\Sigma \text{ losses} = P_{\text{rot}} + I_a^2 R_a + V_t I_f = 870 + (73)^2(0.188) + 230(1.6)$$

$$= 870 + 1002 + 368 = 2240 \text{ W}$$

Hence

$$\eta = 1 - \frac{\Sigma \text{ losses}}{E_a I_a + I_a^2 R_a + V_t I_f} = 1 - \frac{2240}{17,160} = 0.869$$

The efficiency is 86.9%.

(e) $$T_0 = \frac{1}{\omega_m} P_0 = \frac{1}{120} (20 \times 746) = 124 \text{ N-m}$$

The difference between this torque and the electromagnetic torque is the amount needed to overcome the rotational losses.

Example 8-7

The shaft load on the motor of Example 8-6 remains fixed, but the field flux is reduced to 80% of its value by means of the field rheostat. Determine the new operating speed.

Solution Information about the speed is available from Eq. (8-22). However, in turn, knowledge of flux and armature induced emf E_a is needed. By the statement of the problem the new flux Φ' is related to the original flux by

$$\Phi' = 0.8\Phi$$

To obtain information about E_a, we must determine the change in I_a, if any. By the constant-torque condition we have

$$K_T \Phi I_a = K_T \Phi' I_a'$$

Hence

$$I_a' = \frac{\Phi}{\Phi'} I_a = \frac{1}{0.8} 73 = 91.3 \text{ A}$$

Consequently

$$E_a' = V_t - I_a' R_a = 230 - 91.3(0.188) = 212.8 \text{ V}$$

Returning to Eq. (8-22) we can now formulate the ratio

$$\frac{E_a'}{E_a} = \frac{K_E \Phi' n'}{K_E \Phi n}$$

from which the expression for the new operating speed becomes

$$n' = \frac{E_a'}{E_a} \frac{\Phi}{\Phi'} n = \frac{212.8}{216.3} \frac{1}{0.8} 1150 = 1414 \text{ rpm}$$

8-3 SPEED CONTROL

One of the attractive features the dc motor offers over all other types is the relative ease with which speed control over a substantial range can be achieved. The various schemes available for speed control can be deduced from the speed equation which is readily obtained from the formula for the induced armature voltage and Kirchhoff's voltage law for the armature circuit. Accordingly,

$$E_a = K_E \Phi n$$

and

$$E_a = V_t - I_a(R_a + R_e)$$

Hence

$$n = \frac{E_a}{K_E \Phi} = \frac{V_t - I_a(R_a + R_e)}{K_E \Phi} \qquad (8\text{-}26)$$

Observe that the armature circuit resistance is augmented by the inclusion of an adjustable external resistor, R_e. Inspection of Eq. (8-26) reveals that the speed of the motor can be controlled by adjusting any one of three factors appearing on the right side: Φ, R_e, and V_t. Our goal next is to examine and illustrate these methods of control for each motor type.

By Adjustment of Pole Flux (Φ)

Pole flux is the simplest to adjust and involves the least expense. We consider first the shunt motor and for simplicity the analysis proceeds on the assumption of negligible demagnetizing effect from the armature winding mmf throughout.

Shunt motor.   In the case of flux control, R_e is zero in Eq. (8-26) and the flux field Φ is assumed entirely due to current that flows to the field winding. Adjustment of the pole flux is conveniently accomplished by means of the rheostat which is placed in series with the field winding in the manner depicted in Fig. 8-7. If the field rheostat resistance is increased, the pole flux diminishes and the speed increases. General-purpose shunt motors are designed to provide a 200% increase in base (rated) speed by this method of control.

We now undertake to derive the manner in which the presence of R_{rh} influences the speed-torque curve of the shunt motor. The analysis begins by getting the relationship between speed and torque when $R_{rh} = 0$. Moreover, no saturation is assumed so that the pole flux can be directly related to the field current by

$$\Phi_{sh} = K_{sh} I_f = K_{sh} \frac{V_t}{R_f} \qquad (8\text{-}27)$$

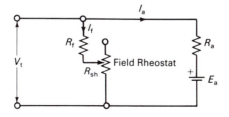

Figure 8-7 Speed control of a shunt motor equipped with a rheostat.

where K_{sh} is expressed in units of Wb/A and the field current is replaced by its equivalent expression using Ohm's law. Then by inserting Eq. (8-27) into the expression for the basic torque we get

$$T = K_T \Phi I_a = K_T K_{sh} \frac{V_t}{R_f} I_a \qquad (8\text{-}28)$$

Also,

$$I_a = \frac{V_t - E_a}{R_a} \qquad (8\text{-}29)$$

and

$$E_a = K_E \Phi_{sh} n = K_T \Phi_{sh} \omega \qquad (8\text{-}30)$$

The last equation indicates that the armature induced voltage may be represented either with speed expressed in rpm or in rad/s.† Because Eq. (8-28) already involves the K_T parameter, we choose the second alternative for use in Eq. (8-29). Accordingly,

$$I_a = \frac{V_t - K_T \Phi_{sh} \omega}{R_a}$$

and so Eq. (8-28) becomes

$$T = \frac{K_T K_{sh}}{R_a R_f} V_t^2 - \frac{(K_T K_{sh} V_t)^2}{R_a R_f^2} \qquad (8\text{-}31)$$

Rearranging yields the desired result,

$$\omega = \frac{R_f}{K_T K_{sh}} - \frac{R_a R_f^2}{K_T^2 K_{sh}^2 V_t^2} T = \omega_0 - aT \qquad (8\text{-}32)$$

where

$$\omega_0 = \frac{R_f}{K_T K_{sh}} = \text{speed intercept} \qquad (8\text{-}33)$$

and

$$a = \frac{R_a R_f^2}{K_T^2 K_{sh}^2 V_t^2} = \text{slope of speed-torque curve} \qquad (8\text{-}34)$$

Clearly, Eq. (8-32) is that of a straight line with no-load intercept speed of ω_0 and a negative slope described by Eq. (8-34). The plot of this curve is shown in Fig. 8-8 for $R_{sh} = 0$. (It is also plotted as the solid line in Fig. 8-1.)

How does Eq. (8-32) change when the field resistance is increased by the contribution of the rheostat? In such a case the total field resistance changes from

---

†A comparison of Eqs. (3-36) and (3-44) shows that $K_E = \dfrac{2\pi}{60} K_T$ so that replacing K_E by its equivalent yields the angular velocity $\omega = \dfrac{2\pi}{60} n$.

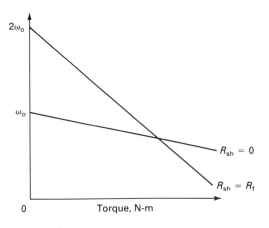

Figure 8-8 The effect of rheostat field resistance on the speed-torque curve of a dc shunt motor neglecting saturation and armature reaction.

R_f to $(R_f + R_{sh})$. The modified expression for the speed-torque curve can be found by determining the effects of the increased field resistance on both the intercept and the slope. The strategy to employ is the simple replacement of R_f by $(R_f + R_{sh})$. Accordingly, the expression for the new intercept becomes

$$\omega_0' = \frac{R_f + R_{sh}}{K_T K_{sh}} = \frac{R_f}{K_T K_{sh}}\left(1 + \frac{R_{sh}}{R_f}\right) = \omega_0\left(1 + \frac{R_{sh}}{R_f}\right) \qquad (8\text{-}35)$$

By appropriate manipulation the new intercept is expressed in terms of the original value. Clearly, for the linear case, making $R_{sh} = R_f$ has the effect of doubling the no-load speed. Of course, Eq. (8-26) also makes this self-evident because at no-load ($I_a = 0$) halving ϕ must cause n to double.

By proceeding in a similar fashion for the slope we get

$$\frac{R_a(R_f + R_{sh})^2}{K_T^2 K_{sh}^2 V_t^2} = \frac{R_a R_f^2}{K_T^2 K_{sh}^2 V_t^2}\left(1 + \frac{R_{sh}}{R_f}\right)^2 = a\left(1 + \frac{R_{sh}}{R_f}\right)^2 \qquad (8\text{-}36)$$

Again, note that by setting $R_{sh} = 0$, Eq. (8-34) results. If $R_{sh} = R_f$, Eq. (8-36) discloses that the negative slope increases by a factor of 4. Figure 8-8 illustrates this matter graphically.

The modified expression of the relationship of speed to the developed torque of a shunt motor when rheostat resistance is present is

$$\omega = \omega_0\left(1 + \frac{R_{sh}}{R_f}\right) - a\left(1 + \frac{R_{sh}}{R_f}\right)^2 T \qquad (8\text{-}37)$$

where ω_0 and a are given by Eqs. (8-33) and (8-34).

Example 8-8

A shunt motor has a field winding resistance of 144 Ω. Moreover, the values of ω_0 and a in Eq. (8-37) are 129 rad/s and 0.07 rad/s/N-m respectively. Find the value of field rheostat resistance so that a torque of 70 N-m is supplied at a speed of 1650 rpm.

Solution The new speed in rad/s is

$$\omega = \frac{2\pi}{60}\ 1650 = 86.4 \text{ rad/s}$$

Let $x = 1 + \dfrac{R_{sh}}{R_f}$. Inserting the foregoing data into Eq. (8-37) there results

$$86.4 = 129x - 0.07(70)x^2$$

$$4.9x^2 - 129x + 86.4 = 0$$

$$x_{1,2} = 13.163 \pm 9.32 = 3.84, 22.48$$

Choosing the smaller value then yields

$$1 + \frac{R_{sh}}{R_f} = 3.84$$

$$R_{sh} = 2.84(R) = 409\ \Omega$$

Compound motor. An analysis similar to that just made for the shunt motor can be employed to get an expression that relates ω to T for the compound motor. However, because the result depends on the relative strength of the series field contribution vis-a-vis the shunt field and owing to the complexity of the resulting expression, let it suffice to say that the plots for the compound motor with and without a field rheostat resistance will generally be similar to those displayed for the shunt motor in Fig. 8-8. In visualizing these characteristics it is helpful to keep in mind that as the shunt field winding contribution to the pole flux is diminished with the insertion of R_{sh}, the importance of the series field contribution increases.

Series motor. Control of the pole flux for the series motor is achieved through the use of a diverter resistance R_d in the manner shown in Fig. 8-9. The developed torque can be expressed as follows:

$$T = K_T\Phi_sI_a = K_TK_sI_sI_a = K_TK_s\frac{R_d}{R_s + R_d}I_a^2 = k\rho I_a^2 \qquad (8\text{-}38)$$

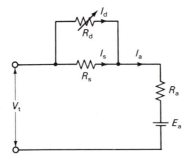

Figure 8-9 Field control of a series motor by means of a diverter resistor.

where

$$k = K_T K_s \quad \text{and,} \quad \rho = \frac{R_d}{R_s + R_d} \tag{8-39}$$

The expression for the armature current comes from Kirchhoff's voltage law for the armature circuit. Thus

$$V_t = E_a + I_a \frac{R_s R_d}{R_s + R_d} + I_a R_a = K_T K_s I_s \omega + \rho I_a R_s + I_a R_a$$

$$= k\rho\omega I_a + I_a\rho R_s + I_a R_a$$

$$= I_a(k\rho\omega + \rho R_s + R_a) \tag{8-40}$$

Accordingly,

$$I_a = \frac{V_t}{\rho R_s + R_a + k\rho\omega} \tag{8-41}$$

which upon insertion into Eq. (8-38) yields

$$T = \frac{k\rho V_t^2}{(R_a + \rho R_s + k\rho\omega)^2} \tag{8-42}$$

Equation (8-42) is the general expression that relates speed and torque for the configuration of Fig. 8-9. By Eq. (8-39) $0 \le \rho \le 1$. When the series motor is used without the diverter resistance, i.e., $R_d = \infty$, $\rho = 1$ and so Eq. (8-42) becomes more simply

$$T = \frac{k V_t^2}{(R_a + R_s + k\omega)^2} \tag{8-43}$$

where T is in N-m and ω is in rad/s.

To examine how the speed changes with the presence of a diverter resistance, introduce a diverter resistance ($\rho < 1$) and find the new speed at which the same torque of Eq. (8-43) is supplied. This requires setting Eq. (8-42) equal to Eq. (8-43). The result is

$$\omega' = \frac{1}{\sqrt{\rho}}\,\omega + \frac{R_a}{k}\left(\frac{1}{\sqrt{\rho}} - \frac{1}{\rho}\right) + \frac{R_s}{k}\left(\frac{1}{\sqrt{\rho}} - 1\right) \tag{8-44}$$

Because the first term on the right side of Eq. (8-44) is the predominant one, it follows that at constant load torque the presence of the diverter resistance causes an increase in speed.

Example 8-9

A 250-V, 25-hp series motor produces a torque of 150 N-m at a speed of 868 rpm. The armature winding and the field winding resistances are 0.12 Ω and 0.09 Ω, respectively. For a constant torque load find the new speed when a diverter resistance of 0.09 Ω is placed in parallel with the series field winding.

Solution   Use Eq. (8-43) to find the value of k. Thus

$$150 = \frac{k(250)^2}{\left(0.21 + \dfrac{2\pi}{60}\, 868k\right)^2} = \frac{(250)_k^2}{(0.21 + 90.88k)^2}$$

which leads to

$$k^2 - 0.04583k + 5.34 \times 10^{-6} = 0$$

$$k - 0.04583k \approx 0$$

$$\therefore \quad k = 0.0483$$

Also,

$$\rho = \frac{R_d}{R_s + R_d} = \frac{0.09}{2(0.09)} = 0.5$$

Inserting these values into Eq. (8-44) yields

$$= \frac{1}{\sqrt{0.5}}\, 90.88 + \frac{0.12}{0.04583}\left(\frac{1}{\sqrt{0.5}} - \frac{1}{0.5}\right) + \frac{0.09}{0.04583}\left(\frac{1}{\sqrt{0.5}} - 1\right)$$

$$= 128.52 - 1.534 + 0.8134$$

$$= 127.8 \text{ rad/s}$$

Or

$$n' = \frac{30}{\pi}\,(127.8) = 1220.4 \text{ rpm}$$

Speed Control by External Armature Resistance

A second method of speed adjustment involves the use of an external resistor R_e connected in the armature circuit. The size and cost of this resistor are considerably greater than those of the field rheostat because R_e must be capable of handling the full armature current. Equation (8-26) indicates that the larger R_e is made, the greater will be the speed change. Frequently the external resistor is selected to furnish as much as a 50% drop in speed from the rated value. The chief disadvantage of this method of control is the poor efficiency of operation. For example, a 50% drop in speed is achieved by having approximately half of the terminal voltage V_t appear across R_e. Accordingly, almost 50% of the line input power is dissipated in the form of heat in the resistor R_e. Nonetheless, armature circuit resistance control is often used—especially for series motors.

    Shunt motor.   The circuit arrangement for a shunt motor whose speed can be adjusted by an external armature-circuit resistance R_e as well as by shunt field current control is illustrated in Fig. 8-10. The general expression for the shunt motor speed as a function of torque without the presence of R_e was previously

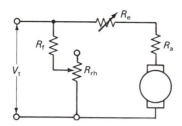

Figure 8-10 Schematic arrangement for speed control of a shunt motor both by pole flux adjustment and armature-voltage control.

found and displayed in Eq. (8-37). It is repeated here with the expressions for the no-load speed (intercept) and the slope of the speed-torque curve included. Thus

$$\omega = \frac{R_f}{K_T K_{sh}}\left(1 + \frac{R_{sh}}{R_f}\right) - R_a \frac{R_f^2}{K_T^2 K_{sh}^2 V_t^2}\left(1 + \frac{R_{sh}}{R_f}\right)^2 T \qquad (8\text{-}45)$$

The inclusion of the external resistance R_e means that effectively the armature circuit resistance is raised from a value of R_a to one of value $(R_a + R_e)$. Hence the expression for the speed-torque characteristic of a shunt motor equipped with an external armature resistance follows from Eq. (8-45) by simply replacing R_a by $(R_a + R_e)$. Accordingly,

$$\omega = \frac{R_f}{K_T K_{sh}}\left(1 + \frac{R_{sh}}{R_f}\right) - (R_a + R_e) \frac{R_f^2}{K_T^2 K_{sh}^2 V_t^2}\left(1 + \frac{R_{sh}}{R_f}\right)^2 T \qquad (8\text{-}46)$$

A study of Eq. (8-46) reveals that R_e does not affect the no-load speed; however, its presence can have a noticeable effect on the *slope* of the drooping straight line represented by Eq. (8-46) especially if R_e is much greater than R_a. This situation is graphically presented in Fig. 8-11. Observe that to deliver power to a constant torque load the insertion of R_e can cause a considerable decrease in speed.

It is helpful to keep in mind for the arrangement depicted in Fig. 8-10 that speeds above the rated or base value can be achieved by pole flux adjustment via R_{sh}, whereas for speeds below this base value armature voltage control through R_e

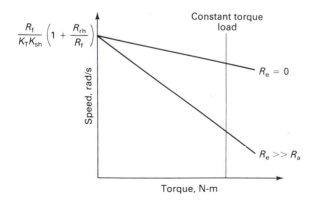

Figure 8-11 Speed-torque characteristic of shunt motor showing the effect of an external armature resistance.

is viable. Overall, by the combination of both schemes, speed control over a range of $4:1$, from the lowest to the highest, is easily managed.

Compound motor. As was the case for speed control by the adjustment of pole flux, speed control of the compound motor by armature-circuit resistance is similar to that which prevails for the shunt motor. Although the presence of the series field winding modifies the results somewhat, the general trends remain the same. This can be best appreciated by examining the general expression for the speed of the compound motor which is

$$\omega = \frac{V_t - I_a(R_a + R_s + R_e)}{K(\Phi_{sh} + \Phi_s)} \tag{8-47}$$

Note that in addition to R_e the numerator also includes the resistance of the series field winding. Moreover, the denominator shows the contribution of the series field mmf on the pole flux. At a given torque an increase in R_e causes the numerator to diminish which in turn drops the speed. Large changes in R_e are followed by correspondingly large changes in speed.

Series motor. Figure 8-12 illustrates the arrangement used to control the speed of a series motor by means of an external resistance. The torque in this case is described by Eq. (8-43) which is repeated here with a modification to account for R_e. Keep in mind that $R = R_a + R_s$. Thus

$$T = \frac{kV_t^2}{(R + R_e + k\omega)^2]} \tag{8-48}$$

where k is defined by Eq. (8-39). Examination of the denominator expression of Eq. (8-48) makes it clear that for a constant torque load, increases in R_e must be accompanied by corresponding decreases in ω in order to keep the expression invariant. Similarly, to preserve constant speed, increases in R_e must produce reduced torque. The net effect of these responses is to cause the speed-torque characteristic with R_e present in the circuit to shift to the left in the manner depicted in Fig. 8-13 relative to its position when no external resistor is used.

The influence of the external resistance in changing the speed *for a constant torque load* can be more dramatically exposed by setting Eq. (8-48) equal to Eq. (8-43). Thus

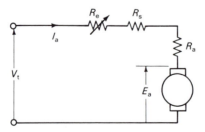

Figure 8-12 Circuit arrangement for series motor speed control by external resistor, R_e.

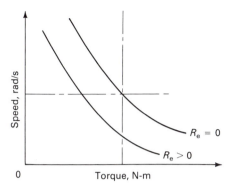

Figure 8-13 The effect of external resistance on the speed-torque characteristic of the series motor.

$$\frac{kV_t^2}{(R + R_e + k\omega')^2} = \frac{kV_t^2}{(R + k\omega)^2} \qquad (8\text{-}49)$$

where ω' denotes the speed when the external resistance is employed. The solution of ω' in the last expression leads directly to

$$\omega' = \omega - \frac{R_e}{k} \qquad (8\text{-}50)$$

Consequently, the insertion of R_e brings about a direct, albeit scaled, decrease in speed when power is delivered to a constant torque load.

Example 8-10

Refer to the motor of Example 8-9 and assume the same initial operating point. Find the external resistance needed to produce the same torque of 150 N-m at a speed of 400 rpm.

Solution The new speed in rad/s is

$$\omega' = \frac{2\pi}{60} n = \frac{2\pi}{60} (600) = 41.89$$

The speed without R_e is

$$\omega = \frac{\pi}{30} 868 = 90.9$$

Then from Eq. (8-50)

$$R_e = k(\omega - \omega') = 0.04583(90.9 - 41.89) = 2.246 \ \Omega.$$

Speed Control by Adjustable Armature Voltage

A third method of speed control involves adjustment of the applied terminal voltage. This scheme is the most desirable from the standpoint of flexibility and high operating efficiency. But it is also the most expensive because it requires its own dc power supply. It means purchasing a motor-generator set or a rectified adjustable voltage source with a capacity at least equal to that of the motor to be controlled. Such expense is not generally justified except in situations where the superior performance achievable with this scheme is indispensable, as is the case in steel mill applications. When the adjustable dc voltage source is obtained from an independent motor-generator set it is referred to as the Ward-Leonard system. Consult Chapter 13 for further details.

Shunt and compound motors.   Figure 8-14 shows the schematic diagram of shunt and compound motors that are operated with adjustable dc voltage sources. The adjustable dc source can be an ac motor driving a dc generator, a dc motor driving a dc generator or a single-phase or a three-phase source that is rectified using thyristors. The latter is the subject matter of Sec. 8-8. A rectified fixed dc source may also be employed for field excitation as indicated.

The manner by which speed control is achieved by adjustable armature voltage is immediately apparent from Eq. (8-3) which at no-load reduces to

$$n = \frac{V_t}{K_E \Phi} \qquad \text{or} \qquad \omega = \frac{V_t}{K_T \Phi} \qquad (8\text{-}51)$$

Clearly, then, for constant field excitation the operating speed is directly proportional to the applied armature voltage. Even under load conditions, which call for accounting for the armature-circuit resistance drop, this proportionality holds within 5% or so. Adjustable armature voltage is invoked to furnish speed control below the base speed (i.e., the speed corresponding to rated armature voltage for full field current), whereas field control is used for speeds above the base value. Adjustable armature voltage achieves the same result for speed control below base speed that is obtained by external armature circuit resistance. However, the former can do it at much higher efficiencies and superior speed regulations albeit at higher initial costs.

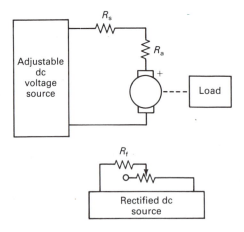

Figure 8-14 Speed control of a dc compound motor by adjustable armature voltage.

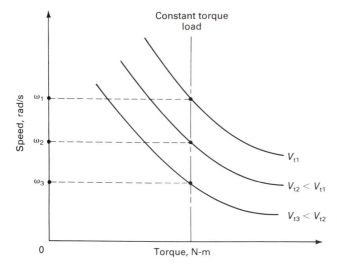

Figure 8-15 The effects of reduced armature voltage on the speed of a series motor at constant load torque.

Series motor.   In the case of the series motor the adjustable dc voltage source is applied to the series connection of the series field and the armature winding. The effect that reductions in the armature voltage applied to the series motor have on speed is easily seen from Eq. (8-18), which defines the approximate speed-torque curve of the motor. It is obvious from this expression that any point on the speed-torque curve can be directly scaled through adjustment of V_t. Thus, for a constant torque load, halving the armature voltage causes a corresponding decrease in speed by one-half. The results are displayed in Fig. 8-15.

Example 8-11

In an industrial application the need arises to provide a constant torque of 105 lb-ft over a speed range of 250 to 1000 rpm. All losses are to be considered negligible for simplicity.

(a) Determine the hp rating and speed rating of a shunt motor to accomplish this task when the control of speed is to be obtained through adjustment of the armature voltage.

(b) Repeat part (a) for the condition where speed control is to be obtained from adjustment of the field current.

(c) Assume that the motors in parts (a) and (b) are designed to operate at 220 V. Find the value of the armature current of both motors when they operate at 1000 rpm delivering the specified torque.

(d) Plot the variation of armature current and armature voltage of both motors over the specified speed range.

(e) Compute the value of the factor $K_T\Phi$ that appears in Eq. (8-23) [or $K_E\Phi$ that appears in Eq. (8-22)] for both motors at a speed of 1000 rpm. Also, plot the variation of this parameter for each motor over the speed range.

(f) Show a plot of the manner in which the delivered hp varies over the specified speed range.

(g) What is the maximum hp that can be delivered by each motor at an operating speed of 250 rpm?

Solution   (a) When armature voltage control is used, the base speed is taken to be the maximum value of the specified speed range. In this case, it is 1000 rpm. Since the motor must deliver a torque of 105 lb-ft = 105(1.356) N-m [see Fig. 8-16(a)], it follows that the required hp rating is determined to be

$$\text{hp} = \frac{(105)(1.356)\left(1000 \times \frac{\pi}{30}\right)}{746} = \frac{(105)(1000)}{5252} = 20$$

Thus the rating for motor A (for armature control) is

A: 20 hp, 1000 rpm

(b) When speed is varied by field current control, the maximum speed of the specified speed range is attained corresponding to minimum field current (i.e., the minimum flux condition). The minimum speed of 250 rpm is then obtained as the field current is advanced to its maximum setting. Because the base speed of a motor is defined as the speed that corresponds to maximum flux at rated armature voltage, it follows that the base speed in this case must be 250 rpm. The required hp rating of this motor can be calculated at the maximum speed value of the speed range. Because the torque is to be constant over the speed range, which includes 1000 rpm, the required hp is the same as for machine A of part (a), namely

$$\text{hp} = \frac{105(1000)}{5252} = 20$$

Hence the complete specification for motor F (for field control) is

F: 20 hp, 250 rpm

A comment is in order at this point concerning motors A and F. First, however, it is useful to cite a general expression for the rating P of electrical machines that puts the emphasis on physical dimensions. This is

$$P = kRD^2Ln \tag{8-52}$$

where D is the rotor diameter, L the effective axial core, n the speed in rpm, and R is an appropriate design parameter. In essence the last expression states that the power rating of a machine is a function of the product of its volume and speed. Accordingly, the greater the speed, the smaller will be the volume. Therefore, motor F in this application will be larger, heavier and so more expensive than motor A. This greater volume of motor F is entirely consistent with the fact that at the lower speeds it must be capable of accommodating values of flux that are much larger than at higher speeds.

(c) Because losses are being neglected, we can set input power equal to output power. Thus

$$I_a(220) = 20(746)$$

or

$$I_a = 67.8 \approx 68 \text{ A}$$

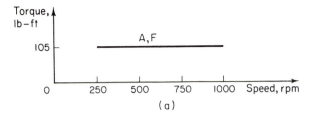

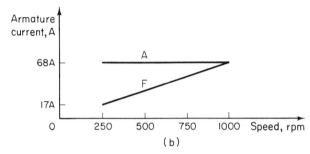

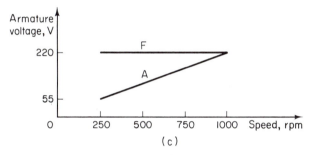

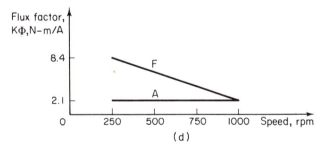

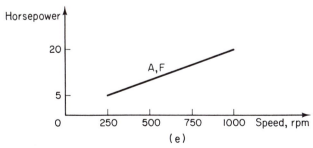

Figure 8-16 Performance features of two motors used to deliver constant torque over a 4 : 1 speed range: (a) constant torque characteristic needed by load; (b) armature current variation at constant torque; (c) armature voltage variation at constant torque; (d) flux factor, KΦ, variation; (e) horsepower variation. A, motor with armature voltage control; F, motor with field current control.

This armature current is the same for both motors at 1000 rpm, the high end of the speed range.

(d) The variation of the armature current for the motor (A) operated with adjustable armature voltage is depicted in Fig. 8-16(b) as curve A. This motor operates with constant field current at all speeds. Hence, its air-gap flux is also constant. The basic torque expression of Eq. (8-23) thus can be written as $T = K_T\Phi I_a = KI_a$. Consequently, a constant torque demands a constant armature current.

The situation is different for the motor (F) that employs field control. In this case a drop in speed is the result of an increase in flux. At a constant torque condition this flux increase means a corresponding decrease in armature current. The variation of this current over the speed range is illustrated as curve F in Fig. 8-16(b).

Appearing in Fig. 8-16(c) are the plots of the armature voltage with speed. Motor F operates at a constant armature voltage. The armature voltage variation of motor A, however, must change with the speed. In fact, the speeds below 1000 rpm are achieved precisely because of the reduction in armature voltage consistent with the equation for the counter emf of the armature which, by the assumption of zero losses, becomes

$$V = K_E\Phi n = K_E' n = E_a$$

Hence, if rated voltage corresponds to operation at 1000 rpm, then operation at 250 rpm is obtained when the armature voltage is reduced to 55 V.

(e) Motor A operates at a constant flux and a constant armature current. The quantity $K_T\Phi$ readily follows from Eq. (8-23) as

$$(K_T\Phi)I_a = T$$

$$K_T\Phi(68) = 105(1.356) = 142.4 \text{ N-m}$$

$$K_T\Phi = 2.1 \text{ N-m/A}$$

Of course, this quantity stays fixed throughout the speed range as indicated by curve A in Fig. 8-16(d).

In the case of motor F part (c) shows that the armature current is also 68 A at 1000 rpm. But because the lower speeds in this motor are the result of increasing flux values, it follows that the quantity $K_T\Phi$ for motor F increases with decreasing speed as illustrated by curve F of Fig. 8-16(d). It is instructive to note that for constant torque the variations of $(K_T\Phi)$ and armature current for motor F are complementary. The increase in $K_T\Phi$ is accompanied by a corresponding decrease in armature current.

In a motor with negligible losses by the law of conservation of energy we can write

$$T\omega_m = E_a I_a$$

$$K_T\Phi I_a \omega_m = K_E\Phi\omega_m I_a$$

or

$$K_T\Phi = K_E\Phi$$

Therefore, the variation of $K_T\Phi$ which bears units of N-m/A is identical to the parameter $K_E\Phi$, which bears units of V/rad/s. As an illustration of the use of this result, we can find the armature voltage of motor A at a speed of 250 rpm (= 26.18 rad/s) by simply multiplying the parameter 2.1 by the speed expressed in radians per second. Thus,

$$2.1(16.18) = 55 \text{ V}$$

(f) The plot for both motors is shown in Fig. 6-23(e). Of course, the hp changes in direct proportion to the speed.

(g) The maximum hp for motor A is 5 because the motor is working as hard electrically and magnetically as it has been designed to work. It is at full capacity electrically since it is drawing rated armature current. It is also at full capacity magnetically since it is operating at its fixed excitation value.

The situation, however, is different with motor F. Although this motor has rated armature voltage applied, it is drawing only 17 A at 250 rpm. Accordingly, this motor could be called on to deliver at 250 rpm an increased torque to the level of 4(105) = 425 lb-ft, corresponding to a fourfold increase in armature current thus raising it to its rated value of 68 A. The corresponding delivered hp is then 20, which is four times larger than that of motor A.

8-4 SPEED REVERSAL

How is the direction of rotation of a dc motor reversed? An examination of the situation that prevails inside the machine reveals that the torque developed on each conductor of the armature winding beneath the field poles can be reversed by reversing either the armature current or the direction of the field flux (by reversal of the direction of flow of the field current). The issue is illustrated in Fig. 8-17. A two-pole motor is shown with a single coil representing the armature winding. The direction of field flux is from left to right. In Fig. 8-17(a) the armature current is assumed to flow out of the plane of the paper on the right side (conductor b) and into the paper on the left side (conductor a). Because a flux reinforcement occurs on the top side of a as well as the bottom side of b, the net torque produces counter clockwise rotation. When the connections to the armature winding are arranged so that the current is reversed as illustrated in Fig. 8-17(b), a flux rein-

Figure 8-17   Speed reversal of a dc motor. (a) Field flux and armature ampere-conductor acting to produce counterclockwise rotations; (b) reversed armature current reverses the developed torque to cause clockwise rotation.

forcement now takes place on the bottom side of a and the top side of b so that the net torque produces a clockwise rotation. If the field direction is reversed in each part of Fig. 8-17 for the specified armature current directions, then the direction of rotation in the configuration of Fig. 8-17(a) becomes clockwise while that of Fig. 8-17(b) becomes counterclockwise.

Is one method of reversal preferred over the other? Let us investigate the advantages and shortcomings of each method. In industrial applications reversal is often called for while the motor is operating under load (e.g., in hoists, punch presses, and so on). Because of the large number of turns that are placed on the field poles of dc motors, the inductance of the field winding is very high as Eq. (1-40) readily indicates. Hence the reversal of the field current through a reversing switch demands a design that will suppress the attendant sparking that would otherwise occur. However, despite such precautions, deterioration of the contact surfaces of the switch will occur over time. It is generally a poor idea to interrupt current in a highly inductive circuit, if it can be avoided. Furthermore, if the motor is equipped with a series field, then reversal of the shunt field must be accompanied by reversal of the series field too otherwise the compounding type will not be the same. Beyond these factors, however, there exists the very serious one of reducing the safety of operation of the motor when a switch is placed in the field circuit. Clearly, a faulty switch here can create an extremely dangerous situation for the motor. This fact alone is sufficient to eliminate speed reversal by field reversal as a viable alternative.

Are there similar disadvantages with armature current reversal? Let us see. When the motor is operating, say, in a clockwise direction under load, it is clearly developing an armature induced voltage, E_a. To reverse the armature current to change the direction of rotation, it is necessary to reverse the applied armature voltage in the manner depicted in Fig. 8-18 through a double-pole double-throw switch. During the instant that the switch is reversed, the armature induced emf still exists. Hence, when the reversed armature voltage appears at the armature terminals, the polarity of V_t is opposed to the original E_a. The effect is to change the speed rather abruptly (called *dynamic breaking*) with little inductance effect. It is only the armature leakage inductance that is active in this state of affairs and it is quite a small quantity compared to the field inductance. Because this feature is such an attractive one and in view of the serious shortcomings associated with field reversal, speed reversal in dc motors in industrial and commercial settings is accomplished exclusively by armature current reversal.

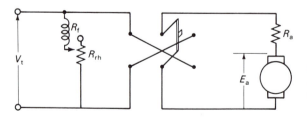

Figure 8-18   Preferred circuit arrangement for reversal of the direction of rotation of a shunt motor via armature current reversal.

8-5 PERMANENT MAGNET (PM) DC MOTORS

For a dc motor to develop torque and thereby attain an operating speed, it is necessary that there exist a flux density distribution along the air gap of the machine. We already know that in conventional dc motors this field distribution is established by the action of a dc current that flows through the field winding which wraps around the salient poles of the stator structure. However, circumstances do arise in the application of dc motors where machines so equipped offer no particular advantage and, in fact, may turn out to be impractical, inconvenient, and/or more expensive than alternative ways of furnishing the machine with flux density. A particularly striking example of this situation is the application of dc motors in countless motor vehicles throughout the world, doing such tasks as driving windshield wipers, fans, windows, air conditioners, radio antennas, and the like. The selection of the dc motor is clearly a natural choice because of the availability of a dc source in the form of a rechargeable battery. The motor size needed to accomplish most of these jobs is very small and in most cases is a small fraction of a horsepower. Accordingly, the use of permanent magnets (PM) in place of field windings offers the advantages of lower cost and simpler construction. It also means higher efficiency because of the absence of the field winding copper losses but in these small sizes this is generally not a significant factor.

Permanent magnet dc motors have also found an application niche in situations requiring integral horsepower sizes exceeding even 100 hp. These motors can be found in systems where motor speed control below base speed only is required and is achieved by armature voltage control through electronic methods like those described in Sec. 8-6. Of course, in applications where control of the motor speed must extend beyond base speed, adjustment of the field flux is necessary and a wound-field machine must be specified because the PM dc motor does not offer this flexibility.

The material used to construct a PM dc motor may be one of three types: Alnico magnets, ceramic (ferrite) magnets, or rare earth magnets. These materials have the distinguishing feature of very high residual flux density and high coercive force. A typical hysteresis loop is shown in Fig. 8-19. Alnico has a residual flux density of the order of 1 T, which is comparable to the flux densities found in wound-field machines, but it falls mostly on the low end of the range. Consequently, for a given hp rating the Alnico-type PM dc motor has a slightly larger armature than its wound-field counterpart. The ceramic magnet has a residual flux density that is usually less than half that of Alnico and it is also the least expensive of the PM materials. As a result ceramic magnets are used for the construction of PM dc motors in the fractional hp range while Alnico magnets are used invariably for motors in the integral hp range.

There are certain disadvantages associated with the use of strong permanent magnets in dc motors especially when the ratings are in the integral hp range. Because the magnetic field is active at all times even when the motor is not being used, the motor enclosure must be more carefully designed to guard against at-

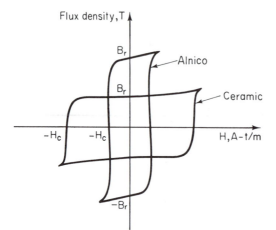

Figure 8-19 General shapes of the hysteresis loops of two permanent magnet materials: (a) Alnico; (b) ceramic (ferrite).

tracting foreign matter that could be harmful to the motor. A more serious drawback concerns demagnetization of the permanent magnets. This can occur especially if the armature current should reach excessive values as might happen under fault conditions such as an armature short-circuit. Equipping the motor with a compensating winding is one effective way to avoid this problem but of course this increases the cost of the motor.

Demagnetization is a problem that is inherently characteristic of permanent magnets. To better understand this statement, let us examine the situation illustrated in Fig. 8-20. Depicted in Fig. 8-20(a) is a permanent magnet attached to a very high permeability magnetic material to form a closed magnetic circuit. If the magnetic material attached to the PM is assumed to offer zero reluctance, the flux

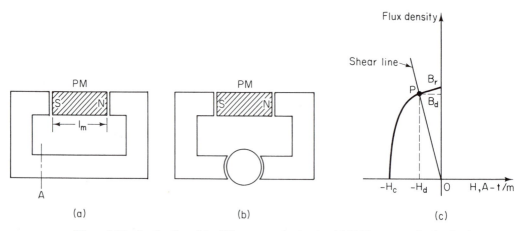

Figure 8-20 Application of the PM to magnetic circuits: (a) PM in a magnetic circuit of zero reluctance; (b) PM in a magnetic circuit with two air gaps; (c) locating the operating flux density value in the magnetic circuit of (b).

density that then penetrates the magnetic circuit is B_r, the residual flux density of the PM. Consider next the situation that is shown in Fig. 8-20(b), which is more in keeping with what is encountered in the actual magnetic circuit of the machine because of the presence of the two air gaps. The insertion of the air gaps into the magnetic circuit of Fig. 8-20(a) necessarily leads to a level of flux density in the magnetic circuit of Fig. 8-20(b) which is less than that found in Fig. 8-20(a). The presence of the air gaps increases the reluctance of the magnetic circuit and the effect may be interpreted in terms of an equivalent demagnetizing field intensity (H_d) associated with the insertion of the air gaps. Expressed mathematically we write

$$-H_d l_m = 2\mathcal{F}_g \qquad (8\text{-}53)$$

where $\mathcal{F}_g$ is the mmf required to force the flux density of the PM to cross one air gap, l_m is the effective length of the PM and H_d is the equivalent demagnetizing field intensity associated with the air gaps. The quantity $H_d l_m$ is that magnetomotive force which would have to be applied on the PM in a direction opposite [hence the negative sign in Eq. (8-53)] to the residual flux to produce the *same* effect as the insertion of the air gaps. The gap mmf is

$$\mathcal{F}_g = \phi_g \mathcal{R}_g \qquad (8\text{-}54)$$

where ϕ_g is now the gap flux associated with a partially demagnetized flux density B_d ($< B_r$) and the cross-sectional area A of the magnetic circuit. Thus

$$\phi_g = B_d A$$

Hence

$$\mathcal{F}_g = \phi_g \mathcal{R}_g = B_d A \left(\frac{g}{\mu_0 A}\right) = \frac{B_d g}{\mu_0}$$

where μ_0 denotes the permeability of air.

Introducing the last expression into Eq. (8-53) and rearranging terms leads to

$$B_d = -\left(\frac{2\mu_0 l_m}{g}\right) H_d \qquad (8\text{-}55)$$

This is clearly the equation of a straight line that passes through the origin and exhibits a negative slope given by the coefficient of H_d. The negative slope plots the line on the demagnetizing side of the B-H curve of the permanent magnet. The intersection of this line with the hysteresis loop of the PM material yields the new operating flux density B_d when the configuration in which the PM magnet is placed changes from that depicted in Fig. 8-20(a) to that of Fig. 8-20(b). Figure 8-20(c) illustrates the result. The straight line of Eq. (8-55) is called the *shear line* of the magnetic circuit in which the PM acts. It is instructive to note that when the gap length is allowed to approach zero, the slope of the shear line becomes infinite. This means that the operating flux density returns to the residual value.

In any practical PM dc motor the operating point is represented by P in Fig.
8-20(c) which means that the actual flux density is less than the residual flux
density that is associated with the *B-H* curve of the PM material. If an Alnico PM
motor is subjected to a severe armature current, a further demagnetization takes
place from the operating point P. Upon subsequent removal of this demagnetizing
force, the effect of a minor hysteresis loop prevents the operating point from
returning to point P. Instead the operating flux density assumes a somewhat
smaller level. It is useful to note, however, that the foregoing situation does not
duplicate itself when the PM material is ceramic. For permanent magnets constructed of ceramic, the operating point P is virtually re-established upon removal
of the demagnetization associated with an excessive armature current.

TABLE 8-1   CHARACTERISTICS AND APPLICATIONS OF DC MOTORS

Type	Starting torque (%)	Max. running torque, momentary (%)
Shunt, constant speed	Medium—usually limited to less than 250 by a starting resistor but may be increased	Usually limited to about 200 by commutation
Shunt, adjustable speed	Same as preceding	Same as preceding
Compound	High—up to 450, depending upon degree of compounding	Higher than shunt—up to 350
Series	Very high—up to 500	Up to 400

Source: By permission from M. Liwschitz-Garik and C. C. Whipple, *Electrical Machinery,* vol. I
(Princeton, N.J.: D. Van Nostrand Co., Inc., 1946).

8-6 OPERATING FEATURES AND APPLICATIONS

The dc motor is often called upon to do the really tough jobs in industry because of
its high degree of flexibility and ease of control. These features cannot easily be
matched by other electromechanical energy-conversion devices. The dc motor
offers a wide range of control of speed and torque as well as excellent acceleration
and deceleration. For example, by the insertion of an appropriate armature circuit resistance, rated torque can be obtained at starting with no more than rated
current flowing. Also, by special design of the shunt-field winding speed adjustments over a range of 4 : 1 above rated speed are readily obtainable. If this is then
combined with armature-voltage control, the range of speed adjustment spreads to

Speed-regulation or characteristic (%)	Speed control (%)	Typical application and general remarks
5–10	Increase up to 200 by field control; de-crease by armature-voltage control	Essentially for constant-speed applications requiring medium starting torque. May be used for adjustable speed not greater than 2 : 1 range. For centrifugal pumps, fans, blowers, conveyors, woodworking machines, machine tools, printing presses
10–15	6 : 1 range by field con-trol, lowered below base speed by arma-ture-voltage control	Same as above, for applications requiring adjustable speed control, either constant torque or constant output
Varying, depending on degree of com-pounding—up to 25–30	Not usually used but may be up to 125 by field control	For drives requiring high starting torque and only fairly constant speed; pulsating loads with flywheel action. For plunger pumps, shears, conveyors, crushers, bending rolls, punch presses, hoists
Widely variable, high at no-load	By series rheostat	For drives requiring very high starting torque and where adjustable, varying speed is satisfactory. This motor is sometimes called the traction motor. Loads must be positively connected, not belted. For hoists, cranes, bridges, car dumpers. To prevent overspeed, lightest load should not be much less than 15 to 20% of full-load torque

8 : 1. In some electronic control devices that are used to provide the dc energy to the field and armature circuits, a speed range of 40 : 1 is possible. The size of the motor being controlled, however, is limited.

Table 8-1 lists some of the salient characteristics and typical applications of the three types of dc motors. It is interesting to note that the maximum torque in the case of the dc motor is limited by commutation and not, as with all other motor types, by heating. Commutation refers to the passage of current from the brushes to the commutator and thence to the armature winding itself. The passage from the brushes to the commutator is an arc discharge. Moreover, as a coil leaves a brush the current is interrupted, which causes sparking. If the armature current is allowed to become excessive, the sparking can become so severe as to cause flashover between brushes. This renders the motor useless.

Another point of interest in the table is the considerably higher starting torque of the compound motor in comparison with the shunt motor. This feature is attributable to the contribution of the series field winding. The same comment is valid as regards the maximum running torque. In each case, of course, the limit for the armature current is the same.

8-7 STARTERS AND CONTROLLERS FOR DC MOTORS

The limitations imposed by commutation as well as voltage-dip restrictions on the source as set forth by the electric utility company make it necessary to use a starter or controller on all dc machines whose ratings exceed 2 hp. A glance at Eq. (8-2) discloses that at starting ($n = 0$) the armature current is limited solely by the armature circuit resistance. Hence, if full terminal voltage is applied, excessive armature currents will flow. This is especially so where large machines are involved because the armature resistance gets smaller as the rating increases. In addition to limiting the armature current, controllers fulfill other useful functions as described in Sec. 4-12.

Depicted in Fig. 8-21 is a simple line starter which is used for small dc motors. The operation is straightforward. Pushing the start button energizes the

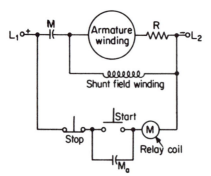

Figure 8-21 Line starter for a low-horsepower dc motor.

main coil which then closes the main contactors M and the interlock M_a. Note that, when the main contactors close, voltage is applied to the armature winding through the starting resistor R and simultaneously to the field winding. This arrangement prevents "shock" starting because some time is needed before the field flux is fully established. Note, too, that this starter has no provisions for removing the starting resistor once the motor has attained its operating speed. In small motors this is of little consequence and it makes for an inexpensive starter. However, the starter resistor in this case does serve another purpose. When the motor is stopped, the field winding is disconnected from the line. The energy stored in the magnetic field then discharges through the starting resistor, preventing possible damage to the field winding.

Three types of magnetic controllers are in use today for controlling the starting current, the starting torque, and the acceleration characteristics of dc motors. One type is the *current-limit* controller, which works on the principle of keeping the current during the starting period between specified minimum and maximum limits. It is not too commonly used because its success depends upon a tricky electrical interlock arrangement which must be kept in excellent operating condition at all times. No further consideration is given to this type. The second type is the *counter-emf* type which is illustrated in Fig. 8-22. Two kinds of relays are used in this controller. One is a light, fast-acting unit called an accelerating relay (AR). The other is the strong, heavy-duty type previously discussed. The accelerating relays appearing across the armature circuit in Fig. 8-22 are voltage-sensitive devices. They are designed to close when the voltage across the coil exceeds a preset value. For the controller under discussion the accelerating relay

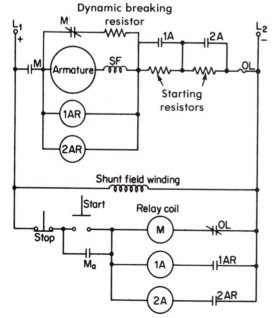

Figure 8-22 Counter-emf magnetic controller for a dc motor.

1AR is usually adjusted to "pick up" at 50% of line voltage and 2AR is adjusted to close at 80% of line voltage.

Pressing the start button energizes coil M, which then closes interlock M_a and the main contactors M. Since the armature is initially stationary, the accelerating relays are de-energized so that both steps of the starting resistor are in the circuit. As the armature gains speed and develops an induced emf exceeding 50% of line voltage, coil 1AR snaps closed, closing contactors 1AR. In turn, coil 1A is energized, closing contactors 1A, which short out the first section of the starting resistor. This then applies increased voltage to the armature, which furnishes further acceleration. When the armature induced emf exceeds 80% of line voltage, accelerating relay 2AR closes. This excites coil 2A, which shorts out the second section of the starting resistor. The motor then accelerates to its full-voltage operating speed.

The controller of Fig. 8-22 is also equipped with a resistor that shunts the armature terminals. It is called a *dynamic breaking resistor* because it serves to bring the motor to a quick halt by dissipating energy obtained from the rotating armature inertia. The diagram of Fig. 8-22 shows that a path is established for the dynamic breaking resistor when the stop button is closed and coil M becomes de-energized. This action closes the normally closed contactor M which is placed in series with the dynamic breaking resistor. Since the motor is still turning at a high speed with stored kinetic energy in its inertia and because the shunt field remains energized, the armature behaves as a generator which in turn delivers current to the dynamic breaking resistor. The subsequent expenditure of energy as heat in the resistor acts quickly to drain the stored kinetic energy of the armature, thus reducing the speed to zero rapidly.

The counter-emf controller has the advantage of providing a contactor closing sequence which adjusts itself automatically to varying load conditions. Furthermore, this is accomplished in a manner that maintains uniform accelerating current and torque peaks. There can be no question about the desirability of such starting performance; however, there is one disadvantage. The contactor closing sequence is based on the assumption that the motor will start on the first step. If it fails to do so, all subsequent operations cannot take place. Furthermore, the starting resistor is in danger of burning up. To avoid such occurrences general-purpose dc motors are most often equipped with *definite time-limit controllers*. Figure 8-23 shows the schematic diagram of such a controller. After a relay coil is energized, the corresponding contactors do not close until the elapse of a preset time delay. The time delay is achieved either by means of magnetic flux decay, a pneumatic device, or a mechanical escapement.

Pressing the start button energizes coil M and closes interlock M_a and the main contactors M, thereby applying voltage to the armature winding through the starting resistor. A definite time after the armature of relay coil M closes, contactors T_M in the control circuit close, regardless of whether the rotor is turning or not. This energizes coil 1A, which closes contactors 1A and thus shorts out the first section of the starting resistor. At a preset time delay after coil 1A is ener-

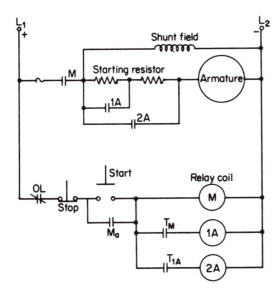

Figure 8-23 Definite time-limit controller for a dc motor.

gized, contactors T_{1A} close. Coil 2A becomes energized, removing the entire starting resistor. The motor then assumes its normal operating speed.

In the definite time-limit controller the preset time intervals between the closing of contactors are adjusted to obtain smooth acceleration and uniform current peaks for average load conditions. If a heavy starting condition occurs and the motor fails to start on the first step, the first accelerating (time-delay) contactors close anyway. This allows an increased starting torque to be developed. Thus the motor is made to "work harder" if it does not start on the first step. Accordingly, whenever a controller must be selected for a general-purpose motor, it is wiser to prescribe the definite time-limit type. The reason is that as a rule for general-purpose applications the starting conditions are not well known.

8-8 SPEED CONTROL BY ELECTRONIC MEANS

The development of reliable thyristors (SCRs) with high current and voltage ratings has made the speed control of dc motors by armature voltage adjustment more viable. Keep in mind that adjustment of the armature voltage offers the advantage of speed control below base speed as well as operation at high efficiency coupled with good speed regulation. Moreover, the control of speed can be achieved whether the motor is operating under load conditions or no-load.

Single-Phase, Half-Wave Thyristor Drive

Figure 8-24 shows the arrangement that is used when a single-phase source is available. It is frequently used for motors which are equipped with permanent magnet fields and rated for 1 hp or less. Because a single thyristor is used in this

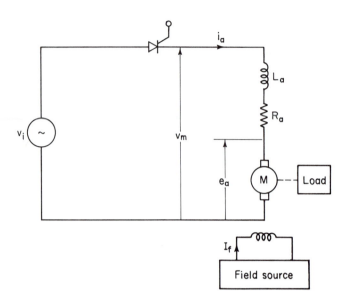

Figure 8-24 Armature voltage control of a dc motor through use of a single-phase, half-wave controlled rectifier. The motor may have either a permanent magnetic field or a separate source of excitation.

circuit, control of the firing point, θ_f, by the gate signal can occur anywhere between 0 and π radians. Illustrated in Fig. 8-25 is a typical situation that can prevail for a firing angle of $\theta_f = 90°$ during steady-state conditions. Time in this diagram is assumed to be measured after steady state is reached. Moreover, the load torque to be supplied by the motor is assumed to be at such a level as to require an average armature current of I_a projected over a full cycle. To produce this average armature current, however, it is necessary for the actual variation of the armature current to have such an instantaneous variation that integration over the conduction period yields an average value of I_a. The conduction period is the time during which the thyristor is conducting and generally this is represented by

$$\theta_c = \theta_e - \theta_f \qquad (8\text{-}56)$$

where θ_f is the angle in the positive excursion of the supply voltage where the gate signal fires the thyristor and θ_e is the extinction angle of the thyristor.

The extinction angle is determined as that point in the cycle where the actual armature current i_a returns to zero, which in turn depends upon the relationship of the instantaneous value of the supply voltage v_i and the variation of the armature counter emf e_a. This e_a must always assume such a position relative to v_i that it produces a time variation in i_a which yields a value of the average armature current that allows a torque to be developed to meet the needs of the load. It is instructive to note here that the instantaneous armature current i_a undergoes an increase in value from θ_f to θ_m, where θ_m corresponds to the time instant where

$v_i = e_a$. In this interval the voltage across the armature winding inductance in Fig. 8-24 is positive. Consequently, an increase in current from zero takes place which can be described by a form of Faraday's law as

$$\Delta i_a = \frac{1}{L_a} \int_{\theta_f}^{\theta_m} (v_i - e_a) \, dt \qquad (8\text{-}57)$$

where L_a denotes the armature winding inductance. Beyond θ_m the quantity $(v_i - e_a)$ becomes negative. By Eq. (8-57) the effect of these negative volt-seconds, appearing across the armature inductance, is to cause the armature current to diminish. In fact, the extinction angle is that point in the cycle where the sum of the negative volt-seconds (i.e., the double-crosshatched area in Fig. 8-25) is equal to the sum of the positive volt-seconds (i.e., the single-cross-hatched area).

When the load calls for an increased developed torque by the motor, the motor responds by reducing its speed (and therefore e_a) by an amount that allows a sufficient increase in positive volt-seconds to yield a greater instantaneous current flow, i_a. Correspondingly, a larger average value of I_a is made to exist over a full cycle thereby meeting the increased torque demand of the load. The variation with time of e_a and the motor speed, ω_m, are represented in Fig. 8-25 by the same curve. It is understood, of course, that the curve is drawn to a different scale. The same applies to the curves marked I_a and T_L. In the interval from θ_f to θ_e observe that the motor is receiving a pulse of energy from the source which manifests itself as a slight increase in speed. The energy stored in the rotating mass, consisting of the rotor of the motor and its attached mechanical load (i.e., the total inertia J), is thereby increased. However, because there is no armature current flow in the interval from θ_e to $(2\pi + \theta_f)$, the motor finds itself coasting along. During this period, energy is delivered to the load by extracting it from the kinetic energy stored in the rotating mass. The result is a dip in speed. When the

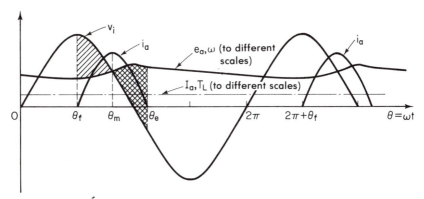

Figure 8-25 Waveshapes of the various electrical quantities and motor speed and load torque associated with the system of Fig. 8-24. The thyristor firing angle is illustrated for $\theta_f = 90°$.

motor is called on to deliver full load torque at low speeds, the speed dip will be greater than when it operates at speeds close to its base value.

Although small dc motors do operate successfully in the circuitry of Fig. 8-24, it is not without notable shortcomings. For example, the need to meet load torque requirements in terms of an armature current averaged over a full cycle of the supply frequency means that large pulses of armature current must flow during the conduction period. This leads to much higher heating losses. Unless special arrangements are made for cooling, it may be necessary to operate the motor at a reduced horsepower rating. Another disadvantage is related to the fact that an average dc current is made to flow through the ac source. If this source should be a single-phase transformer, then it is possible for the transformer to operate in a state of partial saturation. These problems are essentially avoided by operating from a three-phase source.

Three-Phase, Half-Wave Thyristor Motor Drive Circuit

The circuit diagram for this mode of operation is depicted in Fig. 8-26. A three-phase voltage source is assumed to be available with an accessible neutral. It is the line-to-neutral voltages which appear across the series combination of the individual thyristors and the armature winding circuit. The motor field is presumed to be originating from a permanent magnet or from a separate dc source which itself may involve rectification of an ac supply. The application of the three-phase voltage set of Fig. 8-27 in the drive circuit of Fig. 8-26 causes the thyristors S_a, S_b, and S_c to fire in sequence over a potential conduction period of 120°. The actual conduction period depends on when the firing gate signal is

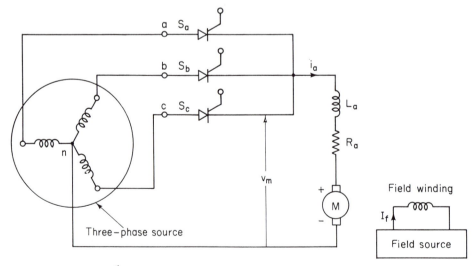

Figure 8-26 Schematic diagram of the three-phase, half-wave controlled rectifier drive circuit for the dc motor.

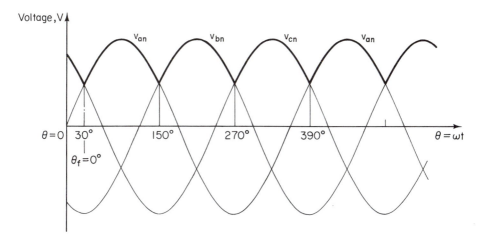

Figure 8-27 Waveforms of the balanced three-phase source. The heavy lines indicate the 120° intervals associated with each phase voltage during which thyristor firing can be initiated.

applied during this 120° interval. A glance at Fig. 8-27 should make it apparent that between $\theta = 30°$ and $\theta = 150°$ a positive potential appears across thyristor S_a. Accordingly, it can act as a switch (hence the notation S for the thyristor) that serves to put a portion or all of v_{an} across the armature terminals of the motor. Simultaneously, during this period, the voltage appearing across the other two thyristors is negative and so S_b and S_c are kept in a cutoff state. Similarly, in the interval from $\theta = 150°$ to $\theta = 270°$, thyristor S_b is prepared for firing because here it now possesses a positive forward voltage while thyristors S_a and S_c have negative voltages on their anode-cathode terminals. In the interval from 270° to 390° thyristor S_c is ready for conduction. This sequenced firing of the thyristors results in a rectification of the three-phase line-to-neutral voltages which has a controllable average value that is determined by the point at which the thyristors are allowed to conduct through action of the gate signals. It is this nonzero average voltage that drives the dc motor.

Let us now obtain an expression that relates this controllable average voltage to the firing angles of the thyristors. Call θ_f the thyristor firing angle and for convenience let $\theta_f = 0°$ correspond to $\theta = \omega t = 30°$ in Fig. 8-27. Because θ_f can be assigned any value from 30° to 150° (or $\pi/6$ to $\pi/6 + \frac{2}{3}\pi$), it follows that the expression for the average voltage over the first conduction period corresponding to firing at angle θ_f is

$$V_m = V_{dc} = \frac{1}{\frac{2}{3}\pi} \int_{\theta_f + \pi/6}^{\theta_f + (5/6)\pi} V_p \sin(\omega t) \, d(\omega t) \qquad (8\text{-}58)$$

where V_p denotes the peak value of the line to neutral voltage and V_m refers to the motor armature voltage. It is important to note in this equation that the upper limit of the integral is written in recognition of the fact that the thyristor stays in the ON state for a period of 120° (i.e., $\frac{2}{3}\pi$) once it is allowed to initiate conduction. Performing the integration and inserting the limits, the desired result is found to be

$$V_{dc} = V_m = 0.827 V_p \cos \theta_f \qquad (8\text{-}59)$$

This equation clearly reveals the controllable aspect of the rectified ac source voltage. A setting of the thyristor gate circuit to yield $\theta_f = 0°$ means that a maximum dc voltage appears across the armature terminals. At the other extreme a setting of $\theta_f = 90°$ reduces this voltage to zero. Accordingly, there is at our disposal a conveniently adjustable armature voltage source to permit speed control below the base speed of the motor.

   An illustration of the speed control capability of the drive circuit of Fig. 8-26 for three values of the thyristor firing angle is shown in Fig. 8-28. Part (a) of this diagram depicts the situation in the drive circuit when the motor is at no load and $\theta_f = 0°$. Thyristor S_a initiates conduction at $\theta = 30°$ and so applies the value of v_{an} to the armature winding of the motor over the period from $\theta = 30°$ to $\theta = 150°$. A similar situation prevails when phases b and c become most positive. The average value of the rectified source is at its maximum. If the armature winding resistance drop is assumed to be zero, then E_a becomes equal to this value and ω_m takes on a corresponding value consistent with the state of its field excitation. Based on the assumption that the maximum rectified voltage is the same as the rated dc motor voltage and the flux field is at its maximum setting, the resulting speed is the base speed. In Fig. 8-28(b) the firing angle is delayed 60°. Consequently, the initial value of v_{an} that appears on the armature circuit is its peak value along with the variation of v_{an} that occurs over the next 120°. In other words, that section of v_{an} is placed across the armature circuit that corresponds to curve def. The average value of this section is $0.413 V_p$ and is consistent with a firing angle of 60°. Of course, this process is repeated for each third of the cycle involving in succession thyristors S_b and S_c. Now the speed is at a level one-half the base speed again based on the assumption of negligible armature winding resistance drop and thyristor drop. Figure 8-28(c) depicts the situation as it exists for $\theta_f = 90°$. Thyristor S_a initiates conduction at $\theta = 120°$ (i.e., $\theta = 30° + \theta_f = 30° + 90°$) and continues in the ON state until $\theta = 240°$. Accordingly, the portion of v_{an} that appears across the motor armature winding is the section represented by $d'e'f'$, which already has an average value of zero. Hence the motor speed becomes zero.

   When the dc motor is operating at a subbase speed under no-load conditions, what is the effect of applying load? The demand for increased torque to be delivered to the attached mechanical load is met by a decrease in speed just as it is in the conventional dc motor. Although the average armature voltage and current are derived from rectification, Eqs. (8-22) to (8-24) continue to be valid provided that the terminal average values are used. As a result in those instances where the armature winding voltage drop is moderate or small, relatively small changes in

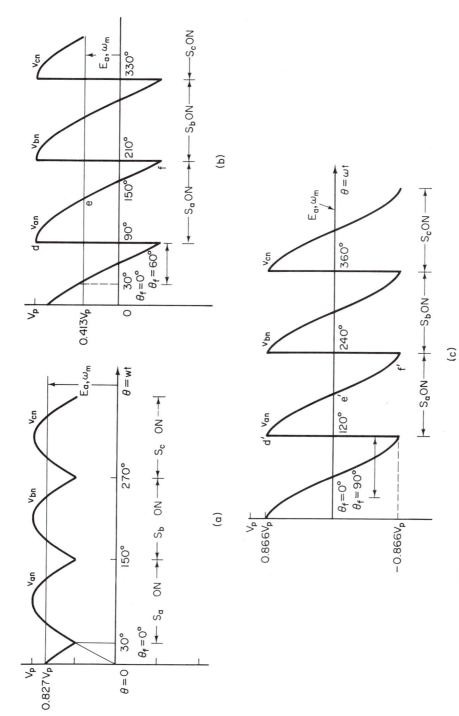

Figure 8-28 Waveshapes of the rectified three-phase source for three values of the thyristor firing angle. Also shown are the average values of the armature induced emf E_a and the motor speed ω_m: (a) full-firing of the thyristors for $\phi_f = 0°$; (b) $\theta_f = 60°$; (c) $\theta_f = 90°$.

the speed accompany the changes in load. This matter is illustrated in the following example.

Example 8-12

A dc motor, which is rated at 550 V and 60 hp, draws 30 A when it delivers rated
torque at speed of 800 rpm. The field is set at a fixed value that gives the motor a
torque and speed parameter of 5.94 N-m/A (or V/rad/s). The motor is supplied from
a three-phase, 60-Hz, 1100 line-to-line source through a half-wave thyristor drive
circuit. The armature circuit resistance is 1 Ω. Assume that the thyristors operate
with negligible voltage drop.

(a) Determine the required firing angle when the motor operates at rated load
conditions.

(b) Compute the change in speed that occurs when the load torque is reduced
to one-half the rated value. What is the percent speed change?

(c) Find the firing angle that is needed when rated torque is to be delivered at a
speed of 500 rpm.

Solution (a) Kirchhoff's voltage law employing average values appearing at the
motor terminals allows us to write

$$V_{dc} = V_m = E_a + I_a R_a$$

$$0.827 V_p \cos \theta_f = (5.94)\omega_m + 30(1)$$

Here

$$\omega_m = 800 \left(\frac{\pi}{30}\right) = 83.78 \text{ rad/s}$$

$$V_P = \sqrt{2} \left(\frac{1100}{\sqrt{3}}\right) = 898.1 \text{ V}$$

Thus

$$742.73 \cos \theta_f = 497.65 + 30 = 527.65$$

$$\theta_f = 44.73°$$

(b) With the load torque reduced to one-half, the required average armature
current is also halved. Since the firing angle is maintained, we have

$$527.65 = 5.94\omega_m + 15$$

or

$$\omega_m = 86.3 \text{ rad/s}$$

Expressed in revolutions per minute, the speed is

$$n_m = \omega_m \left(\frac{30}{\pi}\right) = 86.3 \left(\frac{30}{\pi}\right) = 824.1 \text{ rpm}$$

The change in speed from the rated value expressed in percent is

$$\frac{824.1 - 800}{800} \times 100 = 3.02\%$$

(c) We proceed as in part (a). However, in this part the speed is

$$\omega_m = \frac{5}{8}(83.78) = 52.36 \text{ rad/s}$$

Hence

$$742.73 \cos \theta_f = (5.94)(52.36) + 30 = 341 \text{ V}$$

$$\theta_f = 62.7°$$

Chopper Drives

There is an area of application of dc motors where dc energy is available, often in the form of batteries, and where operation below base speed is extremely important. This is the case for example with forklifts, golf carts, electrically powered trucks, and even rapid-transit cars. Of course, as already pointed out in the preceding section, one workable scheme is to use a series resistor. But we already know that this solution carries with it such serious shortcomings as poor efficiency, poor speed regulation, and high cost. A popular method of providing armature voltage control today in such applications is to use the thyristor in a pulse width modulation mode or a pulse frequency modulation mode. In the former the dc source voltage is switched on and off for a variable part of a cycle in the interest of adjusting the level of average voltage applied to the motor armature winding. In the latter a pulse of voltage of fixed width, which is derived from the dc source, is repeated more or fewer times over a given time period. A typical schematic arrangement of the circuitry is shown in Fig. 8-29. Because the principles are the same, attention is directed here to just one of the schemes, namely, pulse width modulation of the dc source voltage.

The control circuitry that is connected to the thyristor has a twofold pur-

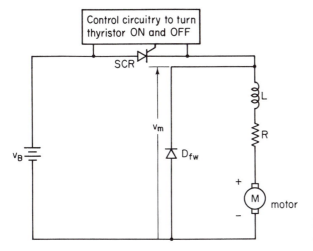

Figure 8-29 Chopper drive circuit for armature control of a dc series motor.

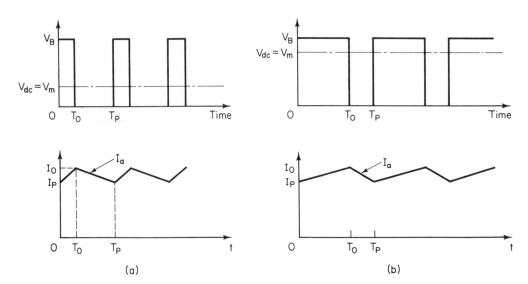

Figure 8-30 Voltage and current waveshapes appearing at the armature terminals of the motor in Fig. 8-25: (a) thyristor fired for low-speed operation; (b) thyristor fired for high-speed operation.

pose. First, it serves to turn the thyristor on over a variable part of a period T_P as illustrated in Figs. 8-30(a) and (b). In the time interval from 0 to T_0 the gate signal puts the thyristor into a conductive state thereby passing the battery voltage to the armature winding. Second, at time T_0, a reverse voltage is instantaneously (measured in the tens of microseconds) applied to the thyristor thereby cutting it off. At the end of the period represented by T_P, the process is repeated. In essence then the control circuitry acts to "chop" the dc supply voltage before it is applied to the motor armature winding. These drive circuits are called chopper drives for obvious reasons. The average value of the dc voltage that is transferred to the motor is quite simply given by

$$V_{\text{dc}} = V_m = \frac{T_0}{T_p} V_B \tag{8-60}$$

where the meaning of the quantities are self-evident from their identities in Fig. 8-30.

The frequency at which the thyristor is pulsed is critical to the design of the control circuitry if large peaks in the armature current are to be avoided. In a properly designed chopper drive circuit, the variation of the armature current looks very much like that illustrated in Fig. 8-30 during steady-state operation. For equilibrium to exist the increase in current during time T_0 is equal to the decrease in current during the time $(T_P - T_0)$. In terms of the quantity of volt-seconds applied to the armature inductance, we can write as the increase in current in period T_0

$$\Delta I_a = \frac{1}{L_a} \int_0^{T_0} [V_B - (V_m - I_a R_a)] \, dt = \frac{[V_B - (V_m - I_a R_a)]T_0}{L_a} \quad (8\text{-}61)$$

since the integrand is essentially constant. The motor voltage V_m is given by Eq. (8-60). Of course, the effect of this increased current is to augment the magnetic energy stored in the armature field. Then, following cutoff of the thyristor at time T_0, the energy is released to the motor through the circuit established by the free-wheeling diode D_{fw}. The rate of decay of the armature current during the time $(T_P - T_0)$ is determined by the time constant of the armature circuit which is often very large compared to the cycle time of the pulses. For this reason the decay is illustrated as a straight line. Because of the important role of the armature inductance in storing sufficient magnetic energy, chopper drives are frequently used with series motors. The series motor has larger armature inductance than the shunt motor, although perfectly acceptable performance can be had when the latter machine is used with an external series inductor.

Example 8-13

Refer to the dc motor described in Example 8-12 and assume that it is to be operated with a chopper drive from a 500-V dc source.

(a) Find the motor speed when the thyristor is made to operate with a ratio of ON time T_0 to cycle time T_P of 0.4. Assume that the motor delivers rated torque.

(b) Determine the required pulse frequency of the thyristors that serves to limit the total change of armature current during the ON period of 5 A. Assume an armature winding inductance of 0.1 H.

(c) Repeat parts (a) and (b) for $T_0/T_P = 0.7$.

Solution   (a) From Eq. (8-60) the average value of the armature terminal voltage is

$$V_m = \frac{T_0}{T_P} V_B = 0.5(550) = 220 \text{ V}$$

The corresponding value of the induced armature voltage then becomes

$$E_a = V_m - I_a R_a = 220 - (30)(1) = 190 \text{ V}$$

Therefore

$$\omega_m = \frac{190}{5.94} = 32 \text{ rad/s} \quad \text{or} \quad 305.4 \text{ rpm}$$

(b) The time T_0 is found directly from Eq. (8-61). Thus,

$$T_0 = \frac{L_a(\Delta I_a)}{V_B - E_a} = \frac{0.1(5)}{550 - 190} = 1.4 \times 10^{-3} \text{ s}$$

Accordingly,

$$T_P = \frac{T_0}{0.4} = \frac{1.4}{0.4} \times 10^{-3} = 3.5 \times 10^{-3} \text{ s}$$

and this leads to the following required pulses per second:

$$f = \frac{1}{T_P} = \frac{10^3}{3.5} = 286 \text{ Hz}$$

(c) Now Eq. (8-60) yields

$$V_m = 0.7(550) = 385 \text{ V}$$

$$E_a = 385 - 30 = 355 \text{ V}$$

$$\omega_m = \frac{355}{5.94} = 64.81 \text{ rad/s or 618.9 rpm}$$

Also, from Eq. (8-61),

$$T_0 = \frac{(0.1)5}{195} = 2.56 \times 10^{-3}$$

$$T_P = \frac{2.56 \times 10^{-3}}{0.7} = 3.66 \times 10^{-3}$$

$$f = \frac{1}{T_P} = 273 \text{ Hz}$$

PROBLEMS

8-1. A 5-hp, 113-V, 1150-rpm, shunt motor has an armature resistance of 0.2 Ω. When delivering rated output the motor draws a line current of 40 A. The field winding has a resistance of 65 Ω. Neglect the demagnetizing effect of the armature mmf.
   (a) Find the motor speed when the motor develops a torque of 15 N-m.
   (b) Determine the speed regulation of this motor. Assume the armature current at no-load is negligible.

8-2. A 15-hp, 230-V, 1750-rpm shunt motor draws 56 A from the line at rated output. The resistance of the armature circuit is 0.16 Ω; the field winding resistance is 153 Ω. The field current is reduced to 80% of its full excitation value by rheostat control.
   (a) If the developed torque is correspondingly reduced in order not to exceed the rated armature current, find the new value of speed. Neglect saturation.
   (b) Determine the new speed when the load demands the same torque.

8-3. The magnetization curve of a 10-hp, 115-V, 1000-rpm shunt motor is shown in Fig. 7-17. Each vertical diversion represents 10 V and each horizontal unit represents 0.4 A. The armature circuit has a resistance of 0.07 Ω and the field winding resistance is 48 Ω.
   At a given load condition the motor armature current is 60 A. The demagnetizing effect of the associated armature winding mmf is found to be 0.4 A expressed in equivalent field amperes. Find the motor speed.

8-4. A 115-V dc machine is equipped with both a shunt and series field winding the resistances of which are respectively 60 Ω and 0.15 Ω. The armature winding resistance is 0.2 Ω. When operating as a shunt motor, it draws a line current of 40 A and develops a torque of 31 N-m to a shaft load. When operating as a compound motor supplying the same torque, the series field contribution to the pole flux is known to be 25%.

(a) Find the speed as a compound motor.

(b) Find the speed as a shunt motor.

8-5. A 40-hp, 550-V, 500-rpm dc machine is equipped with a shunt and series field winding the resistances of which are 225 Ω and 0.25 Ω, respectively. The resistance of the armature circuit is 0.36 Ω. When the motor is operated as a series motor with an armature current of 59 A, the motor speed is found to be 1120 rpm.

The machine is next operated as a cumulatively compounded motor drawing the same armature current. At this load condition the shunt field contributes 1.5 times as much pole flux as does the series field.

(a) Find the speed of the compound motor.

(b) Compute the torque developed by the compound motor.

(c) What is the developed horsepower?

8-6. A 200-hp, 550-V, 450-rpm dc series motor draws a line current of 295 A at rated output. The resistances of the armature and series field windings are 0.07 Ω and 0.04 Ω, respectively. The torque is reduced to one-fourth the rated value. Find the new operating speed.

8-7. Refer to the motor of Prob. 8-6.

(a) Evaluate the parameters that appear in Eq. (8-17) and show that the equation that relates the speed ω to torque T for this motor is

$$\omega = \frac{2843.6}{\sqrt{T}} - 2.95$$

(b) Next, evaluate K in Eq. (8-20) and compare the result with the result found in part (a).

(c) Use the result of part (a) to find the speed corresponding to a torque of 3166 N-m.

(d) Repeat part (c) using the result of part (b).

(e) Compute the percent error caused by neglecting the winding resistances for the series motor.

8-8. A 200-hp, 550-V, 450-rpm dc series motor draws a line current of 295 A at rated output. The resistance of the series field winding is 0.11 Ω. Determine both the speed and torque when the load changes so that it draws a current of 180 A.

8-9. A 10-hp, 230-V shunt motor has an armature circuit resistance of 0.5 Ω and a field resistance of 115 Ω. At no-load and rated voltage the speed is 1200 rpm and the armature current is 2 A. If load is applied, the speed drops to 1100 rpm. Determine:

(a) The armature current and the line current.

(b) The developed torque.

(c) The horsepower output assuming the rotational losses are 500 W.

Neglect armature reaction.

8-10. A 230-V, 50-hp, dc shunt motor delivers power to a load drawing an armature current of 200 A and running at a speed of 1100 rpm. The magnetization curve is given by Fig. 7-17, where each vertical unit represents 20 V and each horizontal unit represents 2 A. Neglect armature reaction. Also $R_a = 0.02$ Ω.

(a) Find the value of the armature induced emf at this load condition.

(b) Compute the motor field current.

(c) Compute the value of the load torque. The rotational losses are 600 W.

(d) At what efficiency is the motor operating?

(e) At what percentage of rated power is it operating?

8-11. Refer to Prob. 8-10, and assume the load is reduced so that an armature current of 75 A flows.

(a) Find the new value of speed.

(b) What is the new horsepower being delivered to the load?

8-12. The magnetization curve of Fig. 7-17 applies to a 230-V, 15-hp motor, where each ordinate unit represents 20 V and each abscissa unit represents 0.2 A. At no-load, the field circuit draws a current of 1.9 A from the line, while the armature current is negligibly small. The speed at no-load is 1000 rpm. A mechanical load is then applied to the motor shaft. The measured armature current is found to be 40 A, causing an armature circuit resistance drop of 16 V. Moreover, the demagnetizing effect of the armature mmf is found to be 0.3 equivalent field ampere.

(a) Compute the load speed in rpm, neglecting the demagnetizing effect of the armature reaction.

(b) Find the speed for the specified load when the demagnetizing effect is accounted for. (*Hint:* The procedure is similar to that outlined for generator behavior, except that now the voltage equation for the motor is used in place of that for the generator. Also, keep in mind that the magnetization curve is valid only for a single speed, so that adjustments are needed.)

(c) Compute the percentage of change in speed from no-load to full-load for parts (a) and (b); compare and explain the difference.

8-13. A 230-V, 15-hp, dc shunt motor has the magnetization curve shown in Fig. 7-17, where each ordinate unit represents 20 V and each abscissa unit represents 0.2 A. This curve is valid for operation at 1000 rpm.

(a) Rated voltage is applied to this motor, and at no-load the speed is found to be 900 rpm. Negligible armature current flows at this condition. Find the total field circuit resistance for this point of operation.

(b) Load is applied to this motor, corresponding to an armature current of 40 A. If the armature circuit resistance drop is 16 V and the demagnetizing effect of the armature mmf is 0.3 equivalent field ampere, determine the new speed in rpm.

8-14. A 115-V, 15-hp, dc cumulative-compound motor has the magnetization curve shown in Fig. 7-17, where each ordinate unit represents 10 V and each abscissa unit represents 1 A. The motor speed at no-load and rated voltage applied is 1000 rpm.

(a) Assuming negligible armature current at no-load, find the field circuit resistance for this operating point.

(b) The motor delivers power consistent with an armature current of 100 A, which is also made to flow through the series field winding. The magnetization effect of the series field is represented by 3 equivalent field amperes. The demagnetizing effect of the armature winding is denoted by 0.7 equivalent field ampere. The armature circuit resistance drop (including the series field) is 6 V. Find the speed under load.

8-15. A 20-hp, 230-V, 1150-rpm, four-pole, dc shunt motor has a total of 620 conductors arranged in two parallel paths and yielding an armature circuit resistance of 0.2 Ω. When it delivers rated power at rated speed, the motor draws a line current of 74.8 A and a field current of 3 A. Compute:

(a) The flux per pole.
(b) The developed torque.
(c) The rotational losses.
(d) The total losses expressed as a percentage of the rated power.
Assume negligible armature reaction.

8-16. A 230-V, 10-hp, dc series motor draws a line current of 36 A when delivering rated power at its rated speed of 1200 rpm. The armature circuit resistance is 0.2 Ω, and the series field winding resistance is 0.1 Ω. The magnetization curve may be considered linear. Effects of armature reaction may be neglected.
   (a) Find the speed of this motor when it draws a line current of 20 A.
   (b) What is the developed torque at the new conditions?
   (c) How does this torque compare with the original value? Why?

8-17. A 250-V, 50-hp, 1000-rpm, dc shunt motor drives a load that requires a constant torque regardless of the speed of operation. The armature circuit resistance is 0.04 Ω. When this motor delivers rated power, the armature current is 160 A.
   (a) If the flux is reduced to 70% of its original value, find the new value of armature current.
   (b) What is the new speed?

8-18. When a 250-V, 50-hp, 1000-rpm, dc shunt motor is used to supply rated output power to a constant-torque load, it draws an armature current of 160 A. The armature circuit has a resistance of 0.04 Ω, and the rotational losses are equal to 2 kW. An external resistance of 0.5 Ω is inserted in series with the armature winding. For this condition compute:
   (a) The speed.
   (b) The developed power.
   (c) The efficiency assuming the field loss is 1.6 kW.
Armature reaction is negligible.

8-19. A dc shunt motor has the magnetization curve shown in Fig. 7-17, where one vertical unit represents 20 V and one horizontal unit represents 2 field amperes. The armature circuit resistance is 0.05 Ω. At a specified load and a speed of 1000 rpm, the field current is found to be 12 A when a terminal voltage of 245 V is applied to the motor. The rotational losses are 2.5 kW, and armature reaction is negligible.
   (a) Find the armature current.
   (b) Compute the developed torque.
   (c) What is the efficiency?

8-20. Answer whether the following statements are true or false. When the statement is false, provide the correct answer.
   (a) The emf induced in a single coil of the armature of a dc machine is necessarily unidirectional.
   (b) Putting a suitable limit on the amount of heat dissipated is the single most important design factor for all electric machines.
   (c) A dc motor responds to the need for increased power at its output shaft by changing its ampere-conductor distribution relative to its field distribution.

8-21. A 230-V, dc shunt-wound motor has an armature circuit resistance of 0.5 Ω. At full-load the armature winding draws 40 A and the speed is measured at 1100 rpm corresponding to a field circuit resistance of 115 Ω.
   (a) Find the developed motor torque in newton-meters.

(b) The field circuit resistance is increased to 144 Ω. Find the new operating speed assuming the developed torque remains constant in demand to load requirements. Neglect saturation.

(c) Calculate the efficiency in part (b). Assume that rotational losses amount to 600 W.

8-22. A 230-V, dc shunt motor has an armature circuit resistance of 0.15 Ω and a field winding resistance of 68 Ω. The pole flux at full field current is 0.1 Wb. The rated torque is 100 N-m. This motor undergoes a change in speed of 6% from no-load to rated load.

(a) Find the value of the torque constant K_T.

(b) Determine the no-load speed.

(c) What is the speed at rated load?

8-23. A 230-V, dc shunt motor has an armature winding resistance of 0.2 Ω and a field winding resistance of 125 Ω. At no-load it operates at a speed of 1500 rpm.

Find the speed when it delivers a full-load rated torque of 75 N-m.

8-24. Refer to the motor of Prob. 8-23. While operating at no-load, a field rheostat resistance of 75 Ω is inserted in the field winding circuit.

(a) Find the new no-load speed.

(b) What is the value of torque that the motor can develop without exceeding the rated armature current?

(c) Determine the speed at the load condition of part (b).

(d) Determine the percent change encountered in part (c) relative to the no-load value.

(e) Find the speed regulation when the motor operates with the specified rheostat resistance and delivers rated torque.

8-25. A 230-V, dc shunt motor has a field winding resistance of 115 Ω, an armature resistance of 0.18 Ω, and a no-load speed of 1250 rpm. Find the value of field rheostat resistance that permits the motor to deliver a torque of 40 N-m at a speed of 2000 rpm.

8-26. A 460-V, 50-hp series motor delivers rated torque at a speed of 600 rpm. The armature circuit resistance is 0.06 Ω and the series field winding resistance is 0.04 Ω. Determine the value of diverter resistance that is needed to permit the motor to develop rated torque at 900 rpm.

8-27. Refer to the motor of Prob. 8-23 and answer the following questions.

(a) Determine the value of external armature resistance that is needed to allow the motor to develop rated torque at 750 rpm. The field rheostat resistance is set at zero.

(b) At this load condition how much larger is the loss in the external armature resistance than in the armature winding resistance?

(c) If the rated armature winding copper loss is 5% of the rated output and is also equal to the fixed losses, find the efficiency of the motor in part (a).

8-28. A 230-V, 75-hp, 450-rpm shunt motor has an armature resistance of 0.043 Ω and a field winding resistance of 42 Ω. When operating at no-load with full excitation, the speed is measured to be 1512 rpm.

(a) Determine the motor speed when it delivers rated torque and an external resistance of 0.2 Ω is placed in the armature circuit.

(b) Find the speed regulation.

(c) Repeat parts (a) and (b) for an external resistance equal to 1 Ω.

8-29. A 550-V, 100-hp, 400-rpm shunt motor has an armature resistance of 0.19 Ω and a field winding resistance of 250 Ω. The no-load speed at full excitation is 420 rpm.

(a) An external armature resistance of 1.6 Ω is inserted into the armature circuit. Determine the motor speed at rated load torque.

(b) What is the speed regulation for the condition of part (a)?

8-30. A 150-hp, 550-V, 690-rpm series motor has a series field and armature winding resistances of 0.065 Ω and 0.1 Ω, respectively. While the motor is delivering rated output to a constant torque load, an external armature resistance is inserted to reduce the speed to 450 rpm. Find the ohmic value of the external resistance.

8-31. Refer to the motor of Prob. 8-30. An external armature resistance of 1.2 Ω is used when the motor delivers one-half of its rated torque to a coupled load. Find the motor speed.

8-32. A 15-hp, 230-V, 1750-rpm shunt motor draws 56 A from the line at rated output. The resistance of the armature circuit is 0.16 Ω and the field winding resistance is 153 Ω. It is desired that 80% of the rated torque be delivered at a speed of 1000 rpm. Determine the value of the external armature resistance that permits the motor to operate at this condition.

8-33. A 230-V, dc shunt motor is used as an adjustable speed drive over a range of zero to 2400 rpm. Speeds from zero to 1600 rpm are obtained by adjusting the armature terminal voltage from zero to 230 V, with the field current kept fixed at full field value. Speeds from 1600 rpm to 2400 rpm are obtained by increasing the field current, with the armature terminal voltage maintained at 230 V. Ignore machine losses and armature reaction effects.

(a) The torque required by the load remains constant over the entire speed range. Show the general form of the curve for armature current versus speed, over the full speed range.

(b) Instead of keeping the load torque constant, suppose that the armature current is not to exceed a specified value. Show the general form of the curve for allowable load torque versus speed, over the entire speed range.

8-34. A customer is interested in buying a dc shunt motor to supply a load requiring a constant torque of 525 lb-ft continuously, over a speed range of 500 to 2000 rpm.

(a) List three ways that this job can be done.

(b) For each method of part (a), specify the base speed and horsepower rating of the motor.

(c) List the outstanding advantage associated with each method of control.

8-35. Two adjustable-speed, dc shunt motors have maximum speeds of 2000 rpm and minimum speeds of 500 rpm. Motor A drives a load requiring constant horsepower over the speed range. Motor B drives one requiring constant torque. Neglect all losses and armature reaction.

(a) If speed adjustment is obtained by field control and if the horsepower outputs are equal at 500 rpm and the armature currents are each 100 A, compute the armature currents at 2000 rpm.

(b) Specify the nameplate horsepower and speed ratings of the individual motors to be purchased to do these jobs. Assume that at 500 rpm the torque for both loads is the same, and equal to 1050 lb-ft.

8-36. Figure P8-36 shows a dc series motor, with a rheostat in parallel with the field winding. The resistance of this rheostat is equal to two-thirds of the field winding resistance. The motor is operating in the steady state, with the rheostat switch open. The terminal voltage is 250 V, the armature current 40 A, the speed 700 rpm. When the load torque is increased by 50% and the rheostat switch is closed, find the new steady-state values of speed and armature current. Given values:

Armature resistance 0.2 Ω, series field resistance 0.15 Ω.

Neglect saturation, armature reaction, and rotational power losses.

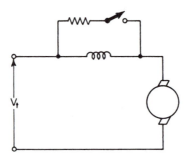

Figure P8-36

8-37. A dc series motor runs at 300 rpm, and draws a current of 75 A from the 500-V line. The total resistance of the armature and field circuits is 0.4 Ω. Find the new steady-state values of armature current and speed if the line voltage is increased from 500 to 600 V. Assume that the developed torque is to remain constant.

8-38. Read the following sentence carefully, and correct any wrong statement that you may find in it:

> A dc shunt motor operates under load in the steady state. At the time $t = 0$, the field rheostat resistance is reduced; this increases the flux and thereby the torque. So the motor accelerates and reaches a new steady-state speed which is higher than it was before.

8-39. A dc shunt motor draws 10.0 A from its 220-V supply line. The resistance of its armature circuit (including brushes, brush contacts, and interpoles) is $R_a = 1$ Ω. The field winding resistance is $R_f = 300$ Ω. (There is no field rheostat.) The rotational losses (hysteresis, eddy currents, friction, and windage) are 10% of the output power. Find the efficiency, as accurately as the given information and the use of a calculator permit.

8-40. A dc shunt motor is rated at 3 hp, 115 V, 1000 rpm. At rated operating conditions, the efficiency is 82.5%. The armature resistance (including brush contact) is 0.44 Ω, the field circuit resistance is 145 Ω.

  (a) Calculate the value of the induced voltage E_a at rated operating conditions.
  (b) If this motor is tested at *no-load,* with the supply voltage adjusted to the value calculated under (a) and the speed kept at 1000 rpm (so that the *rotational losses* remain the same as at full load), how much is the armature current under this condition? (*Note:* The $I_a r_a$ drop at no-load can be neglected.)

8-41. Two identical dc machines are tested in "opposition" as shown in Fig. P8-41 (armatures in parallel, shafts coupled, losses supplied from an electrical source). The letters *G* and *M* stand for *generator* and *motor*.
  (a) Which machine has the higher armature copper losses?
  (b) Which machine has the higher field copper losses?
  (c) Which machine has the higher core losses?
  To each answer, add a brief explanation.

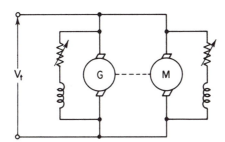

Figure P8-41

8-42. The magnetization curve of Fig. 7-17 shows the no-load characteristic for a dc shunt generator at 900 rpm, where each abscissa unit represents 0.5 A and each ordinate unit represents 20 V. The armature circuit resistance is 0.2 Ω, the field circuit resistance (without field rheostat) 30 Ω. The effect of armature reaction is equivalent to four demagnetizing ampere-turns per pole for every ampere of armature current. The field winding has 1250 turns per pole.
  (a) Find the resistance of the field rheostat (in series with the field winding) for the machine operating as a generator driven at 900 rpm and carrying an armature current of 125 A at a terminal voltage of 220 V.
  (b) The field rheostat remains as for part (a). The load is now disconnected. The speed remains unchanged. What is the new value of the terminal voltage?
  (c) Find the field rheostat resistance for a no-load voltage of 420 V at a speed of 1800 rpm.

8-43. Depicted in Fig. P8-43 is the reversing controller circuitry for a dc shunt motor. Identify the unmarked armature contacts and explain the reversing operation. Explain, too, why the auxiliary interlock contacts F_a and R_a are placed in series with the Rev and Fwd switches.

8-44. A current-limit type controller for a dc shunt motor is shown in Fig. P8-44. The accelerating relays (AR) respond to armature current. These relays have a lightweight armature so that they easily pick up and open their contacts whenever the armature current exceeds the rated value. By means of a step-by-step procedure, describe the operation of this controller. Be careful to identify the proper closing and opening sequence whenever two or more contactors appear in series.

8-45. In the three-phase half-wave controlled rectifier circuit of Fig. 8-26, sketch the waveform of the source voltage that appears across the armature circuit when the firing angle of the SCRs is set at 45°. What is the average value of this voltage?

8-46. Repeat Prob. 8-45 for the case where the firing angle of the SCRs is set at 75°.

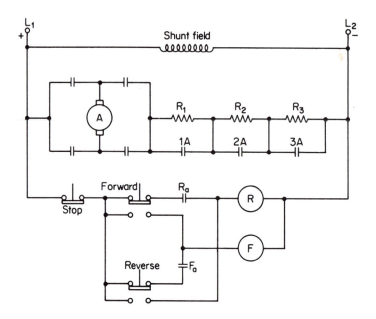

Figure P8-43

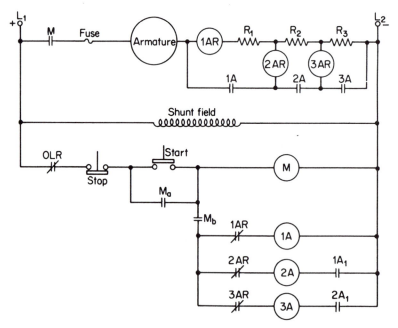

Figure P8-44

8-47. A 25-hp, 250-V, dc motor is driven from a three-phase, 550-V source through a half-wave thyristor drive circuit as displayed in Fig. 8-26. The armature circuit resistance is 0.1 Ω and the developed torque parameter of this motor is known from design data to be 6.25 N-m/A. The thyristor voltage drops and losses are negligible.
   (a) When the firing angle of the SCRs is set at 30° and the motor is operating at no-load, find the speed in rpm.
   (b) Rated load torque of 575 N-m is applied to the motor shaft. Determine the new operating speed in rpm.
   (c) Calculate the speed regulation of this electronically controlled dc motor.

8-48. It is desirable that the motor of Prob. 8-47 be made to deliver its rated torque of 575 N-m at a speed of 250 rpm.
   (a) Find the required firing angle of the SCRs.
   (b) Compute the speed regulation.

8-49. The motor of Prob. 8-47 is driven by the chopper circuitry of Fig. 8-29 from a 250-V dc source. The thyristor control circuitry is arranged to provide a ratio of ON time to cycle time of 0.5.
   (a) Find the operating speed when rated torque is delivered by the motor.
   (b) Determine the no-load speed and the speed regulation for the specified switching ratio.
   (c) Repeat parts (a) and (b) for $T_0/T_p = 1.0$.

8-50. The motor of Prob. 8-47 is driven by a chopper drive of a 250-V source where the ratio of ON-time to cycle time is 0.5. Determine the required pulse frequency of the thyristors in order that the change in armature current during the ON period be limited to 5 A when operating at rated load. The armature inductance is 0.05 H.

8-51. Repeat Prob. 8-50 for the case where the ratio of ON time to cycle time is 0.8.

$8\text{-}8\ a)$ $\dfrac{T'}{T_R} = \left(\dfrac{I_a'}{I_a}\right)^2 \Rightarrow T' = \left(\dfrac{180}{295}\right)^2 3166 = 1178\ N\text{-}m$

$h_p = \dfrac{(T(RPM)) \times \frac{\pi}{30}}{746} = 200 \Rightarrow T_R = 3166\ N\text{-}M$

$b)\ E_a' = 550 - 180(.11) = 530.2$
$E_R = 550 - 295(.11) = 517.55$

$\dfrac{E_a'}{E_R} = \dfrac{K_T K_f I_a' \omega'}{K_T K_f I_a \omega} = \dfrac{530}{517}$ $\omega = 450 \times \dfrac{17}{30}$

$\Rightarrow \omega' = \dfrac{530}{517}(47.124)\dfrac{295}{180} = 79.173$

$N = \dfrac{120 f}{2} = (120)\dfrac{\frac{79.173}{2\pi}}{2} = 756\ rpm$

$8\text{-}12)\ a)$ from curve
$13.05 \times 20 = 261$ $V_E = \dfrac{230}{20} = 11.5 = E_a$ at no load
$11.5 \times 20 = 230$

$N = 1000(13.05/11.5) = 1135\ rpm$
$N_2 = N_1(B_0 P_0/A_0 P_0) = 1135(10.7/13.05) = 930\ rpm$
$b)\ N_2' = N_1(B_0 P_0/F_d) = 1135(10.7/12.55) = 968\ rpm$
$c)\ \Delta\% = 100(1000 - 930)/930 = 7.53\%$ $\phi_2 > \phi_2'$ $\downarrow E_{a_3} = E_{a_2}'$
$\quad '' = 100(1000 - 968)/968 = 3.31\%$ $\therefore N_2' \rightarrow (\phi_2/\phi_2')N_2 > N_2$

9

Single-Phase Induction Motors

By far the vast majority of single-phase induction motors are built in the fractional-horsepower range. Single-phase motors are found in countless applications doing all sorts of jobs in homes, shops, offices, and on the farm. An inventory of the appliances in the average home in which single-phase motors are used would probably number beyond a dozen. An indication of the volume of such motors can be had from the fact that the sum total of all fractional-horsepower motors in use today far exceeds the total of integral-horsepower motors of all types.

9-1 HOW THE ROTATING FIELD IS OBTAINED

In its pure and simple form the single-phase motor usually consists of a distributed stator winding (not unlike one phase of a three-phase motor) and a squirrel-cage rotor. The ac supply voltage is applied to the stator winding, which in turn creates a field distribution. Since there is a single coil carrying an alternating current, a little thought reveals that the air-gap flux is characterized by being fixed in space and alternating in magnitude. If hysteresis is neglected, the flux is a maximum when the current is instantaneously a maximum and it is zero when the current is zero. Such an arrangement gives the single-phase motor no starting torque. To understand this in terms of the concepts discussed in Sec. 3-1 refer to Fig. 9-1. As

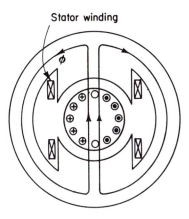

Stator winding

Figure 9-1 Simplified diagram of the single-phase motor.

a matter of convenience the distributed single-phase winding is represented by a coil wrapped around protruding pole pieces; some motors do in fact use this configuration. Assume that instantaneously the flux-density wave is increasing in the upward direction as shown. Then by transformer action a voltage is induced in the rotor having that distribution which enables the corresponding rotor mmf to oppose the changing flux. To accomplish this, current flows out of the right-side conductors and into the left-side conductors as illustrated in Fig. 9-1. Note that this resulting ampere-conductor distribution corresponds to a space phase angle of $\psi = 90°$. By Eq. (3-27) the net torque is therefore zero. Of course what this means is that beneath each pole piece there are as many conductors producing clockwise torque as there are producing counterclockwise torque. This condition, however, prevails only at standstill. If by some means the rotor is started in either direction, it will develop a nonzero net torque in that direction and thereby cause the motor to achieve normal speed. The problem therefore is to modify the configuration of Fig. 9-1 in such a way that it imparts to the rotor a nonzero starting torque.

The answer to this problem lies in so modifying the motor that it closely approaches the conditions prevailing in the *two-phase* induction motor. In accordance with the general results developed in Sec. 4-1 we know that to obtain a revolving field of constant amplitude and constant linear velocity in a two-phase induction motor two conditions must be satisfied. One, there must exist two coils (or windings) whose axes are space-displaced by 90 electrical degrees. Two, the currents flowing through these coils must be time-displaced by 90 electrical degrees and they must have such magnitudes that the mmf's are equal. If the currents are less than 90° apart in time but greater than 0°, a rotating field will still be developed but the locus of the resultant flux vector will be an ellipse rather than a circle. Hence in such a case the linear velocity of the field varies from one point in time to another. Also, if the currents are 90° apart but the mmf's of the two coils are unequal, an elliptical locus for the rotating field again results. Finally, if the currents are neither 90° time-displaced nor of a magnitude to furnish equal mmf's, a rotating magnetic field will continue to be developed but now the locus will be more elliptical than in the previous cases. However, the important aspect of all this is that a revolving field can be so obtained even if its amplitude is not constant during its time history, and satisfactory performance can be achieved with such a revolving field. Of course such performance items as power factor and efficiency will be poorer than for the ideal case, but this is not too serious because the motors are of relatively small power.

Appearing in Fig. 9-2 is the schematic diagram which shows the modifications needed to give the single-phase motor a starting torque. A second winding

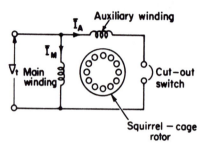

Figure 9-2 Schematic diagram of the resistance split-phase motor.

called the *auxiliary* winding is placed in the stator with its axis in quadrature with that of the main winding. Usually the main winding is made to occupy two-thirds of the stator slots and the auxiliary winding is placed in the remaining one-third. In this way the space-displacement condition is met exactly. The time displacement of the currents through the two windings is obtained at least partially by designing the auxiliary winding for high resistance and low leakage reactance. This is in contrast to the main winding, which has low resistance and higher leakage reactance. Figure 9-3 depicts the time displacement existing between the

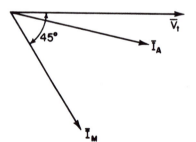

Figure 9-3 Phasor diagram showing the time-phase displacement between the auxiliary and main winding currents at standstill.

auxiliary winding current $\bar{I}_A$ and the main winding current $\bar{I}_M$ at standstill. Frequently in motors of this design the $\bar{I}_A$ and $\bar{I}_M$ phasors are displaced by about 45° in time. Thus with the arrangement of Fig. 9-2 a revolving field results and so the motor achieves normal speed. Because of the high-resistance character of the auxiliary winding, this motor is called the *resistance-start split-phase* induction motor. Also, the auxiliary winding used in these motors has a short time power rating and therefore must be removed from the line once the operating speed is reached. To do this a cut-out switch is placed in the auxiliary winding circuit which, by centrifugal action, removes the auxiliary winding from the line when the motor speed exceeds 75% of synchronous speed.

9-2 EQUIVALENT CIRCUIT BY THE DOUBLE-REVOLVING FIELD THEORY

In this analysis the focus is on the operation of the single-phase induction motor with only the main winding energized by a single-phase source. When this voltage is applied to the coil at standstill, there is no starting torque because the pulsating field distribution produced by this distributed winding merely produces transformer action as already described. It is useful at this point to introduce an alternative explanation of this condition by replacing the alternating flux field by oppositely revolving fields having half the amplitude of the alternating field and each traveling at the same frequency at which the pulsating field alternates. We resort to mathematics to illustrate this equivalence in precise terms.

As described in Sec. 4-1 the equation for an alternating magnetic field whose axis is fixed in space is given by

$$b(\alpha) = B_m \sin \omega t \cos \alpha \tag{9-1}$$

where B_m denotes the amplitude of the sinusoidally distributed air-gap flux density produced by a properly distributed stator winding carrying an alternating current of frequency ω, and α denotes a space-displacement angle measured from the axis of the stator (or field) winding. If the trigonometric identity

$$\sin x \cos y = \tfrac{1}{2} \sin (x - y) + \tfrac{1}{2} \sin (x + y) \tag{9-2}$$

is introduced into Eq. (9-1), the expression for the magnetic field becomes alternatively

$$b(\alpha) = \frac{B_m}{2} \sin (\omega t - \alpha) + \frac{B_m}{2} \sin (\omega t + \alpha) \tag{9-3}$$

The first term on the right side of Eq. (9-3) is the equation of a revolving field that moves in the positive α direction (see Sec. 4-1) and has an amplitude equal to one-half that of the original alternating field. The second term is the equation of a revolving field whose amplitude is also one-half that of the alternating field and whose direction of rotation is in the negative α direction. The field that moves in the positive α direction is sometimes referred to as the *forwardly rotating field* or the positive sequence. Similarly, the backwardly rotating field corresponds to the negative-sequence field. By definition the forward direction is that direction in which the single-phase motor is initially started.

The absence of a starting torque in a single-phase motor that is not equipped with an auxiliary starting device may now be understood in terms of effects produced by the forward and backward fields. The amplitudes of the two revolving fields in Eq. (9-3) are the same and thereby yield no net revolving field to produce a starting torque. The two revolving fields in the motor produce effects that are equal and opposite.

The equivalent circuit of the single-phase induction motor at standstill is easy to identify because it is behaving simply as a transformer with the secondary short-circuited. Accordingly, drawing from our experience with transformers (see Fig. 2-12), we can draw the equivalent circuit as shown in Fig. 9-4. Appearing in Fig. 9-4(a) is the conventional form of the exact equivalent circuit of a transformer. The exact version is cited here because the presence of the air gap in the motor calls for appreciable magnetizing current and consequently its effect is no longer small and cannot be ignored. Also, note that the core loss resistor, which is found in parallel with the magnetizing reactance, is here omitted for simplicity. It can always be added later, if it is needed. Moreover, r_2' denotes the rotor winding resistance per phase referred to the stator and x_2' is the rotor winding leakage reactance per phase, stator referred. Figure 9-4(b) depicts the equivalent circuit of Fig. 9-4(a) modified to bring into evidence the interpretation based on the double-revolving field representation of the alternating field produced by the energized main winding, which is designated by subscript m. In recognition of the fact that the forward and backward revolving fields have amplitudes that are one-half the value of the alternating field, it is necessary to apply a factor of 0.5 to the parameters appearing between the terminals where the emf induced by these fields appears. This is necessary to preserve the equivalence of $\bar{E}_f + \bar{E}_b$ in Fig. 9-4(b) to $\bar{E}_m$ in Fig. 9-4(a). Of course, $\bar{E}_f$ denotes the emf produced by the forward field in the rotor winding referred to the stator, and, similarly, $\bar{E}_b$ is the emf generated by the oppositely rotating field also referred to the stator winding.

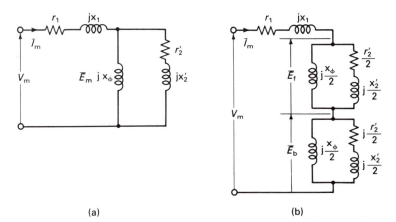

(a)   (b)

Figure 9-4 Equivalent circuit of the single-phase induction motor at standstill with only the main winding energized. (a) Conventional configuration; (b) modified configuration to stress the double-field equivalence of the alternating field caused by the main winding.

When the single-phase motor is given a starting torque and allowed to reach a speed n in the forward direction, the rotor is operating at a slip s which is defined in the usual way as

$$s = \frac{n_s - n}{n_s} = 1 - \frac{n}{n_s} \tag{9-4}$$

In the part of the equivalent circuit associated with the forwardly revolving field the effect of s is to put into evidence the resistance $r_2'/2s$ and the leakage reactance $x_2'/2$. The rationale that yields this result is the same as that for the balanced three-phase induction motor since the situation between the rotor and the forward field is the same. However, the situation is different between the forwardly turning rotor and the backwardly revolving field. Now the slip of the rotor relative to this field is

$$s_b = \frac{n - (-n)}{n_s} = 1 + \frac{n}{n_s} \tag{9-5}$$

which also reveals that this slip is greater than unity. By employing Eq. (9-4) in Eq. (9-5) it is possible to express the backward slip s_b in terms of the forward slip s. Thus

$$s_b = 1 + 1 - s = 2 - s \tag{9-6}$$

Accordingly, the parameters associated with the section of the equivalent circuit where the effects of the backward field are indicated must be expressed in terms of resistance $(r_2')/2(2 - s)$ to represent the power transferred across the air gap along with a leakage reactance of $x_2'/2$. Thus, the equivalent circuit of the single-phase motor operating at a slip s assumes the final form shown in Fig. 9-5. The current I_{2f}' is the forward rotor current per phase referred to the stator and I_{2b}' is the backward rotor current per phase referred to the stator.

In normal operation of the single-phase motor the slip can be expected to lie in the range $0 < s \le 0.15$. This means that the impedance of the parallel combina-

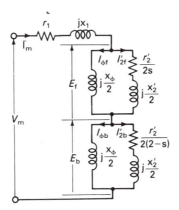

Figure 9-5 Equivalent circuit of a single-phase induction motor operating at a slip s base on the double-revolving field theory.

tion associated with the forward field, namely

$$\frac{\bar{Z}_f}{2} = \frac{1}{2} \frac{jx_\phi \left(\dfrac{r_2'}{s} + jx_2'\right)}{\dfrac{r_2'}{s} + j(x_\phi + x_2')} \tag{9-7}$$

will be much larger than the impedance associated with the backward field

$$\frac{\bar{Z}_b}{2} = \frac{1}{2} \frac{jx_\phi \left(\dfrac{r_2'}{2 - s} + jx_2'\right)}{\dfrac{r_2'}{2 - s} + j(x_\phi + x_2')} \tag{9-8}$$

Consequently, since each of these impedances carries the same current, the magnitude of $\bar{E}_f \gg \bar{E}_b$. In turn the magnitude of the forward field, ϕ_f, which generates $\bar{E}_f$ is much greater than the magnitude of the backward field, ϕ_b, which is responsible for $\bar{E}_b$. Figure 9-6 shows a simplified representation of Fig. 9-5 expressed in terms of the forward and backward impedance symbols introduced in Eqs. (9-7) and (9-8).

Figure 9-7 depicts the forward and backward fields moving in opposite directions but frozen in time at a specific instant. This representation shows how ϕ_f and ϕ_b cooperate to produce a resultant pole flux ϕ_r. It is especially instructive to note that operation of the single-phase motor at $s \neq 1$ does in fact result in a pole flux that rotates with time. The magnitude of this resultant is not constant. For example, as time elapses in Fig. 9-7, an instant is reached when ϕ_f and ϕ_b will lie along the horizontal opposing one another to yield a resultant that is the difference, $(\phi_f - \phi_b)$. Ninety degrees later in time the two fluxes will again be collinear but now the resultant is their sum. Accordingly, the resultant flux is a revolving one with an elliptical locus.

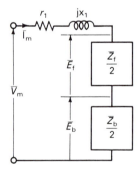

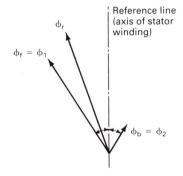

Figure 9-6 Simplified representation of Figure 9-5 through use of Eqs. (9-7) and (9-8).

Figure 9-7 Illustrating how a resultant revolving field is obtained when the motor operates at a slip $\neq 1$.

It is useful to note too that at standstill when $s = 1$ and $\phi_f = \phi_b$, the resultant flux ϕ_r always lies along the vertical in Fig. 9-7 and so yields an alternating field rather than a revolving one.

9-3 EQUIVALENT CIRCUIT VIA SYMMETRICAL COMPONENTS

The single-phase motor can be analyzed as a special case of an unbalanced two-phase motor where in the extreme case the second phase is entirely missing. In turn, unbalanced two-phase motors can be treated in terms of *balanced* two-phase systems through the use of symmetrical components. The underlying theory of symmetrical components for this purpose is fully exposed in the next chapter and the interested reader is encouraged to read Secs. 10-1 to 10-3 prior to continuing with this section if the intention is not to proceed further without first seeing the derivation of the results cited here.

Chapter 10 demonstrates that any two-phase unbalanced set of voltages, $\bar{V}_a$ and $\bar{V}_b$, can be replaced by two sets of balanced two-phase voltages of opposite sequence. Here subscript a refers to phase a and subscript b to phase b. The first set is called the *positive sequence* set and the reversed set is identified as the *negative sequence*. (Respectively, these are analogous to the forward and backward fields described in Sec. 9-2.) The balanced positive sequence set is represented by $\bar{V}_1$ and $-j\bar{V}_1$ where subscript 1 is used to denote the positive sequence and the $-j$ attached to the second component of the balanced set stresses the 90° lag in time between the two balanced voltages. Similarly, the second balanced set is represented by $\bar{V}_2$ and $j\bar{V}_2$ where subscript 2 denotes the negative sequence which, of course, is assured by attaching j to the second component. By this scheme the unbalanced voltages of phases a and b can now be expressed in terms of the two balanced sets as

$$\bar{V}_a = \bar{V}_1 + \bar{V}_2 \tag{9-9}$$

$$\bar{V}_b = -j\bar{V}_1 + j\bar{V}_2 \tag{9-10}$$

The theory of symmetrical components also provides the instructions on how to calculate the positive and negative sequence components from the original unbalanced set. The results are

$$\bar{V}_1 = \tfrac{1}{2}(\bar{V}_a + j\bar{V}_b) \tag{9-11}$$

$$\bar{V}_2 = \tfrac{1}{2}(\bar{V}_a - j\bar{V}_b) \tag{9-12}$$

The results contained in Eqs. (9-9) through (9-12) lead to a very useful and interesting physical interpretation of the analysis of the unbalanced two-phase motor. Each phase voltage consists of a component of a balanced two-phase set, one positive and the other negative. The positive-sequence components in each unbalanced phase voltage, $\bar{V}_1$ and $-j\bar{V}_1$, can be considered to cooperate in creating a revolving magnetic field that interacts with the rotor winding to produce a positive-sequence torque T_1. At the same time the negative-sequence components in each phase, $\bar{V}_2$ and $-j\bar{V}_2$, can be considered to combine in creating a revolving magnetic field of reversed direction that interacts with the rotor winding to produce a negative-sequence torque T_2. Of course the rotor responds to the resultant torque, which is the difference between T_1 and T_2. A schematic representation of this situation is depicted in Fig. 9-8. Note that for convenience the rotor is treated as consisting of two rotors that are mechanically coupled. The positive-sequence voltage set ($\bar{V}_1$ and $-j\bar{V}_1$) produces T_1, and the negative-sequence set produces T_2. All the known techniques of balanced operation can now be used to determine T_1 and also T_2, and the results can then be superposed to obtain the resultant. Unbalanced operation is thereby treated entirely in terms of the techniques of balanced operation. Accordingly, no new theory is needed to treat the motor analysis beyond this point.

The equivalent circuit of the unbalanced two-phase motor actually consists of two parts: one representing the motor response to the positive-sequence voltage set and the other representing the response to the negative-sequence set. Because each voltage sequence is a balanced set, the equivalent circuit expressed on a per-phase basis is similar to that used for the balanced three-phase motor [see Fig. 4-7(f)], which is repeated in Fig. 9-9(a). The applied stator voltage is the positive-sequence voltage $\bar{V}_1$. The quantity r_1 is the stator winding resistance. The quantity x_ϕ is the total magnetizing reactance associated with the magnetic circuit of the motor. The stator referred rotor leakage reactance per phase is denoted as x_2', and r_2' is the rotor resistance per phase referred to the stator. Use of r_2'/s implies that the rotor travels in the direction of the positive-sequence rotating field.

The equivalent circuit of the unbalanced two-phase motor in response to the balanced negative-sequence set is shown in Fig. 9-9(b). However, since the rotor travels opposite in direction to the negative sequence field, the backward slip is given by Eq. (9-6) and so the quantity $r_2'/2 - s$ appears in Fig. 9-9(b). By superposition the same negative-sequence voltage set is assumed applied to the same stator windings and producing its own reactions in the same rotor winding. This is the meaning that is intended in the schematic diagram of Fig. 9-8. There is really only one pair of stator coils and one rotor, but the effect is the same as if there were two identical pairs of stator coils and two identical rotors mechanically coupled.

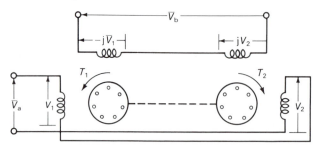

Figure 9-8 Representation of unbalanced two-phase motor by a positive-sequence motor coupled to an identical negative-sequence motor.

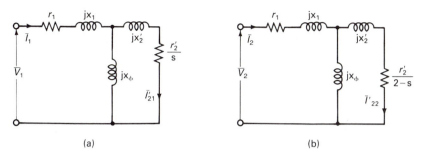

Figure 9-9 Equivalent circuit of unbalanced two-phase motor. (a) Circuit for computing motor performance of the positive-sequence voltage set; (b) circuit for the negative sequence.

In accordance with the theory of symmetrical components the phase currents of the unbalanced two-phase motor can now be found by using the equivalent circuits of Fig. 9-9 to find the positive-sequence current $\bar{I}_1$ and the negative-sequence current $\bar{I}_2$. Then

$$\bar{I}_a = \bar{I}_1 + \bar{I}_2 \tag{9-13}$$

$$\bar{I}_b = -j\bar{I}_1 + j\bar{I}_2 \tag{9-14}$$

It is at this juncture that we can now extend the analysis of the unbalanced two-phase motor to the single-phase induction motor. The latter can be treated as that special case of the two-phase motor where only *one* phase is used. In other words, we have a situation where the current $\bar{I}_b$ of Eq. (9-14) is necessarily zero because of the absence of phase *b*. By Eq. (9-14) this forces the positive- and negative-sequence currents in the single-phase motor to be identical. It then follows from Eq. (9-13) that

$$\bar{I}_1 = \bar{I}_2 = \frac{\bar{I}_a}{2} \tag{9-15}$$

Figure 9-10(a) depicts the equivalent circuit of the single-phase motor in terms of the positive- and negative-sequence voltage sets. But, because the positive- and negative-sequence currents are the same, *a series connection of these individual equivalent circuits is now permissible.* The result is shown in Fig. 9-10(b) and it is notable that now both the current and voltage equations for the stator winding can be represented by a single circuit employing appropriate positive- and negative-sequence components. The impedance looking in at the terminals at which $\bar{V}_1$ appears is the positive sequence impedance $\bar{Z}_1$, which specifically is given by

$$\bar{Z}_1 = r_1 + jx_1 + \bar{Z}_{p1} \tag{9-16}$$

where
$$\bar{Z}_{p1} = \frac{jx_\phi \left(\dfrac{r_2'}{s} + jx_2' \right)}{\dfrac{r_2'}{s} + j(x_\phi + x_2')} \tag{9-16a}$$

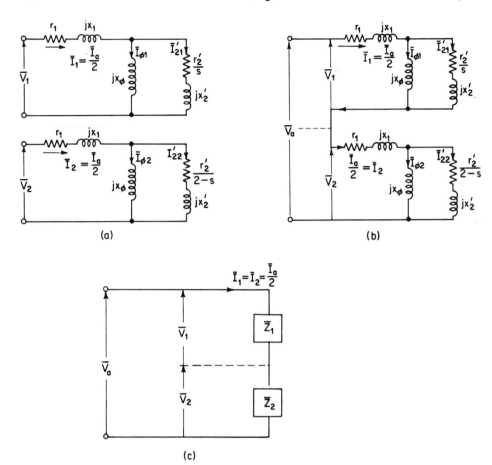

Figure 9-10 Equivalent circuit of the single-phase induction motor, based on symmetrical components: (a) the equivalent circuits associated with the positive- and negative-sequence fields; (b) series connection representation that is permitted because the positive- and negative-sequence currents are always the same; (c) simplified representation in terms of the positive-sequence input impedance $\bar{Z}_1$ and the negative-sequence input impedance $\bar{Z}_2$. Core loss is neglected throughout.

Similarly, the negative-sequence impedance is described as

$$\bar{Z}_2 = r_1 + jx_1 + \bar{Z}_{p2} \qquad (9\text{-}17)$$

where

$$\bar{Z}_{p2} = \frac{jx_\phi \left(\dfrac{r_2'}{2 - s} + jx_2' \right)}{\dfrac{r_2'}{2 - s} + j(x_\phi + x_2')} \qquad (9\text{-}17a)$$

Examination of Fig. 9-10(b) discloses that the values of $\bar{V}_1$ and $\bar{V}_2$ are influenced by the slip at which the motor is operating. Varying slip values yield varying values for $\bar{Z}_1$ and $\bar{Z}_2$. Because $\bar{Z}_1$ and $\bar{Z}_2$ can be series connected (see Fig. 9-10(c)) for the single-phase motor, it follows that the positive- and negative-sequence voltages are directly related to $\bar{Z}_1$ and $\bar{Z}_2$ through voltage division. Accordingly, we can write

$$\bar{V}_1 = \frac{\bar{Z}_1}{\bar{Z}_1 + \bar{Z}_2}\, \bar{V}_a \qquad (9\text{-}18)$$

$$\bar{V}_2 = \frac{\bar{Z}_2}{\bar{Z}_1 + \bar{Z}_2}\, \bar{V}_a \qquad (9\text{-}19)$$

It should be noted from these expressions that the values of the positive- and negative-sequence components change with each value of slip.

Several simplifications of the equivalent circuit of Fig. 9-10(b) can be introduced. The first involves combining the stator impedance drop into a single total value which is permissible for the single-phase induction motor because the positive- and negative-sequence currents are the same. Thus,

total impedance drop $= \bar{I}_1(r_1 + jx_1) + \bar{I}_2(r_1 + jx_1)$

$$= \frac{\bar{I}_a}{2}\,(r_1 + jx_1) + \frac{\bar{I}_a}{2}\,(r_1 + jx_1) = \bar{I}_a(r_1 + jx_1) \qquad (9\text{-}20)$$

The second simplification comes about upon noticing that the voltage drop across $\bar{Z}_{p1}$ [the parallel combination of jx_ϕ and $(r_2'/s + jx_2')$] and $\bar{Z}_{p2}$ can be written as

$$\frac{\bar{I}_a}{2}\,\bar{Z}_{p1} = \bar{I}_a\,\frac{\bar{Z}_{p1}}{2} \qquad (9\text{-}21)$$

and

$$\frac{\bar{I}_a}{2}\,\bar{Z}_{p2} = \bar{I}_a\,\frac{\bar{Z}_{p2}}{2} \qquad (9\text{-}22)$$

Whenever the stator impedance drop is expressed by the result of Eq. (9-20), the preferred way of expressing the voltage drops appearing across the parallel elements of the positive- and negative-sequence circuits is by the right side of the last two equations. With the formulations represented by Eqs. (9-20), (9-21), and (9-22), the equivalent circuit of Fig. 9-10(b) can be replaced by the simpler version depicted in Fig. 9-11. Note that it is now the total stator current $\bar{I}_a$ that flows in the series arrangement of the circuit elements, $r_1 + jx_1$, $\bar{Z}_{p1}/2$, and $\bar{Z}_{p2}/2$ in response to the driving force represented by $\bar{V}_a$. The manipulation involved in Eqs. (9-21) and (9-22) calls for the use of one-half the values of the actual rotor winding resistance and leakage reactance per phase as well as one-half the actual value of the magnetizing reactance.

Once the single-phase induction motor is started and achieves normal speed, the slip frequently lies in the range $0 < s \le 0.15$. Moreover, for conventional single-phase motors $x_\phi > |r_2' + jx_2'|$. Accordingly, very little error is made if the

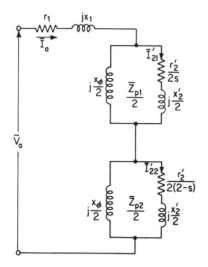

Figure 9-11 Simplified approximate equivalent circuit of the single-phase induction motor based on symmetrical components.

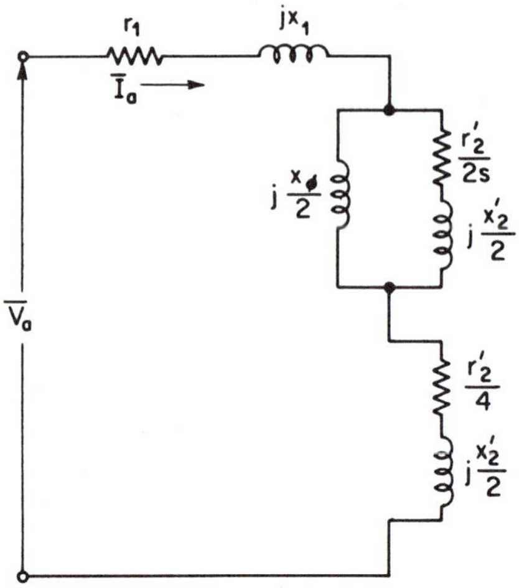

Figure 9-12 A modified simplified equivalent circuit applies for $0 \leq s \leq 0.15$.

magnetizing reactance is entirely eliminated from the impedance $\bar{Z}_{p2}/2$. Also, $r_2'/[2(2 - s)] \approx r_2'/4$ and, since this resistor carries the same current as r_1, the former may be combined with the latter. Similarly, $jx_2'/2$ may be combined with jx_1. Imposing these assumptions on the circuit of Fig. 9-11 leads to a still simpler form as depicted in Fig. 9-12.

This form of the configuration of Fig. 9-12 is like that of the per-phase equivalent circuit of a balanced polyphase induction motor. A considerable and useful simplification has clearly been made possible in a situation that at first appeared quite complex.

A comparison of the configurations appearing in Figs. 9-5 and 9-11 reveals these circuits to be identical except for the notation. The comparison makes it obvious that phase a is the same as the main winding m; and subscript 1 for positive sequence plays the role of f for forward field and subscript 2 for the negative sequence corresponds to b for backward field. Finally, $\bar{E}_f$ and $\bar{V}_1$ are corresponding quantities as well as $\bar{E}_b$ and $\bar{V}_2$.

9-4 TORQUE-SPEED CHARACTERISTIC

We have seen that the curve relating torque as a function of speed is a very important characteristic of electromechanical energy-conversion devices operating as motors. The single-phase motor is no exception. It is to be expected, however, that the resultant torque-speed curve that applies over the entire operating range is not as straight-forward to obtain as it was for the three-phase induction motor. The reason is that the positive- and negative-sequence voltage components change in value with slip [refer to Eqs. (9-18) and (9-19)]. Accordingly, when the resultant torque-speed curve is found, care must be taken to compute T_1 and T_2 at a given slip by first calculating those values of $\bar{V}_1$ and $\bar{V}_2$ that are valid for the slip concerned.

To emphasize this point further, refer to the plot appearing in Fig. 9-13(a). It shows the plot of the positive-sequence torque-speed curve T_1 corresponding to $\bar{V}_1 = \bar{V}_a/2$, and the negative-sequence torque-speed curve T_2 corresponding to $\bar{V}_2 = \bar{V}_a/2$. When this plot is made to apply to the single-phase induction motor, the only point on the resultant curve shown that has any validity is the one for $s = 1$. It indicates that the resultant torque is zero—a familiar result by now. When the motor is assumed to be operating at a normal slip, say s_n, it is incorrect to identify the corresponding torque as T'_n in Fig. 9-13(a), because $\bar{V}_1$ and therefore T_1 are much larger than the values prevailing at standstill. Rather, the correct value of the resultant torque at slip s_n is determined by first finding the T_1 and T_2 quantities using the appropriate values of $\bar{V}_1$ and $\bar{V}_2$ and then subtracting to obtain

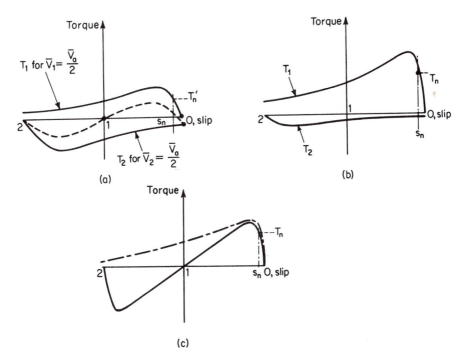

Figure 9-13 Torque-speed curves of the single-phase induction motor: (a) valid representation of resultant torque at $s = 1$ only; (b) valid representation of resultant torque at $s = s_n$ only; (c) resultant curve that is valid over the entire speed range (solid line only).

T_n as illustrated in Fig. 9-13(b). The true resultant torque-speed curve of the single-phase induction motor assumes a form like that depicted in Fig. 9-13(c).

The broken-line curve of Fig. 9-13(c) denotes the torque-speed curve of the two-phase motor operated under balanced conditions. A comparison with single-phase operation reveals two interesting facts: (1) operation as a single-phase motor at small slips is quite acceptable in spite of increased rotor losses; (2) the single-phase motor will run in the direction in which it is started.

Effect of Rotor Resistance on the Torque-Speed Curve

In the case of the polyphase induction motor it was shown that neither the shape nor the maximum value of the torque-speed curve was influenced by the value of rotor resistance. Refer to Eq. (4-33). However, this is not so for the single-phase induction motor and the reason is obvious from a glance at the equivalent circuit of Fig. 9-12. Recall that the slip at which the maximum torque is developed corresponds to the conditions for which maximum power is delivered to the resistor $r'_2/2s$. The slip at which this occurs is found by setting $r'_2/2s$ equal to the magnitude of the impedance looking back into the equivalent circuit from the terminals of this resistance. Clearly, the ensuing value for s_m will be found to depend upon r'_2. Accordingly, the resultant torque too is so influenced by r'_2, and the manner in which this occurs is depicted in Fig. 9-14. Note that it not only causes a reduction in the peak developed torque but it brings about a flattening of the curve as well. This behavior of the torque-speed curve restricts the use of external rotor resistance for controlling the speed of the single-phase induction motor.

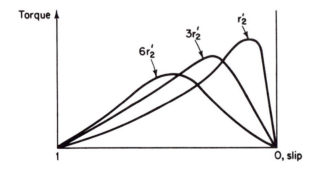

Figure 9-14 The effect of rotor resistance on the torque-speed curve of a single-phase induction motor.

9-5 PERFORMANCE ANALYSIS

The calculation of the performance of the single-phase induction motor can be readily determined through the use of the equivalent circuit and the appropriate power-flow diagram. The approximate equivalent circuit of Fig. 9-12 may be used whenever it is desired to compute the behavior of the motor in the slip range of $0 < s \leq 0.15$. Otherwise use is made of the circuit depicted in Fig. 9-11.

The flow of power from the line to the shaft in the single-phase motor is a bit more complicated than it is in the balanced polyphase motor because of the need to work with the positive- and negative-sequence fields. The complete power-flow diagram appears in Fig. 9-15. The electrical input power is applied to the motor at the stator winding. After removing a small part of this power to supply the stator winding copper loss and the core loss produced by the resultant field, the remainder is the power transferred across the air gap. However, note that the total gap power is now expressed in two parts, one associated with the positive-sequence field, P_{g1}, and the other associated with the negative-sequence field, P_{g2}. Because the rotor travels in the same direction as the positive-sequence field, the negative-sequence gap power is necessarily consumed entirely as rotor copper loss. It is worthwhile to note too that this gap power P_{g2} only partially supplies the total copper loss associated with the rotor resistance of the negative-sequence impedance. From the equivalent circuit of Fig. 9-11 it is seen that the negative-sequence gap power is the power associated with $r_2'/2(2 - s)$ and is expressed by

$$P_{g2} = (I_{22}')^2 \frac{r_2'}{2(2 - s)} \tag{9-23}$$

Upon rearrangement of this equation, it follows that the expression for the total copper loss in the rotor resistance of the negative sequence becomes

$$(I_{22}')^2 \frac{r_2'}{2} = (2 - s)P_{g2} = P_{g2} + (1 - s)P_{g2} \tag{9-24}$$

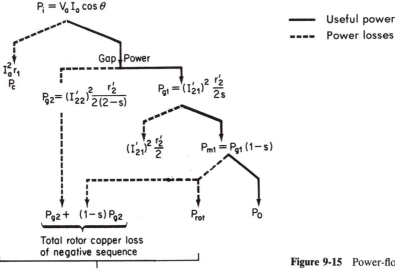

Figure 9-15 Power-flow diagram of the single-phase induction motor.

The right side of Eq. (9-24) is written in two parts to emphasize that a portion of this copper loss, P_{g2}, is supplied directly from the line through the medium of the negative-sequence field, while the remaining portion, $(1 - s)P_{g2}$, is supplied mechanically by the rotor in response to the positive-sequence field. At normal operating slips the positive-sequence gap power is many times greater than that of the negative-sequence. Upon subtracting the copper loss in the rotor winding associated with the positive-sequence [i.e., $(I'_{21})^2(r'_2/2)$ from P_{g1}], there results the useful mechanical developed power P_{m1}. Specifically,

$$P_{m1} = P_{g1}(1 - s) \tag{9-25}$$

and this quantity represents the force that drives the load. Further examination of the power-flow diagram indicates, however, that before useful power can be delivered to the load the mechanical power developed by the positive-sequence field must first supply the rotational losses as well as the power needed to drive the rotor against the opposing action of the negative-sequence field. This latter quantity is already denoted in Eq. (9-24) as $(1 - s)P_{g2}$.

Example 9-1

A $\frac{1}{4}$-hp, 110-V, 60-Hz, four-pole single-phase induction motor has a rotational loss of 15 W at normal speeds. The equivalent circuit parameters are known to be as follows:

$$r_1 = 1.3 \ \Omega, \qquad r'_2 = 3.2 \ \Omega$$
$$x_1 = 2.5 \ \Omega, \qquad x'_2 = 2.2 \ \Omega$$
$$x_\phi = 48 \ \Omega$$

Determine the performance (i.e., line current, line power factor, horsepower output, and efficiency) of this motor when it operates at a slip of 4%.

Solution   The approximate equivalent circuit of Fig. 9-12 is used. It is repeated in Fig. 9-16 with all parameter values shown. The impedance of the parallel circuit is given by

$$\frac{\bar{Z}_{p1}}{2} = \frac{j\frac{x_\phi}{2}\left(\frac{r'_2}{2s} + j\frac{x'_2}{2}\right)}{\frac{r'_2}{2s} + j\left(\frac{x_\phi}{2} + \frac{x'_2}{2}\right)} = \frac{j24(40 + j1.1)}{40 + j25.1} = 10.8 + j17.2 = 20.4\underline{/57.9°} \tag{9-26}$$

The total input impedance is therefore

$$\bar{Z}_T = \left(r_1 + \frac{r'_2}{4}\right) + j\left(x_1 + \frac{x'_2}{2}\right) + \frac{\bar{Z}_{p1}}{2} = 2.1 + j3.6 + 10.8 + j17.2$$

$$= 12.9 + j20.8 = 24.4\underline{/58.4°} \tag{9-27}$$

Accordingly, the line current is

$$\bar{I}_a = \frac{\bar{V}_a}{\bar{Z}_T} = \frac{110}{24.4\underline{/58.4°}} = 4.51\underline{/-58.4°} \ \text{A} \tag{9-28}$$

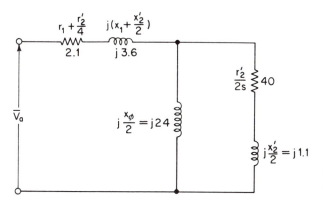

Figure 9-16 Equivalent circuit for Example 9-1.

and the line power factor is

$$\text{power factor} = \cos 58.4° = 0.525 \text{ lagging} \tag{9-29}$$

Such a low value for the line power factor is typical of single-phase motor operation. The input power is found to be

$$P_i = V_a I_a \cos \theta = 110(4.51)0.525 = 261 \text{ W} \tag{9-30}$$

The positive-sequence gap power is found from

$$P_{g1} = (I_a)^2 \frac{1}{2} R_{p1} = (4.51)^2 10.8 = 220 \text{ W} \tag{9-31}$$

Also, the negative-sequence gap power is determined to be

$$P_{g2} = (I'_{22}) \frac{r'_2}{4} \approx I_a^2 \frac{r'_2}{4} = (4.51)^2 0.8 = 16.3 \text{ W} \tag{9-32}$$

From the power-flow diagram the expression for the output power is

$$P_0 = P_{m1} - P_{\text{rot}} - (1 - s)P_{g2}$$
$$= P_{g1} - sP_{g1} - P_{\text{rot}} - (1 - s)P_{g2} \tag{9-33}$$

Inserting values

$$P_0 = 220 - (0.04)220 - 15 - (0.96)16.3$$
$$= 220 - 8.8 - 15 - 15.6$$
$$= 182.4 \text{ W} \tag{9-34}$$

The horsepower delivered to the load is therefore

$$\text{horsepower} = \frac{P_0}{746} = \frac{182.4}{746} = 0.245 \tag{9-35}$$

This result indicates that the motor is delivering almost rated output power. The efficiency is found to be

$$\eta = \frac{P_0}{P_i} = \frac{182.4}{261} = 0.70, \text{ or } 70\% \tag{9-36}$$

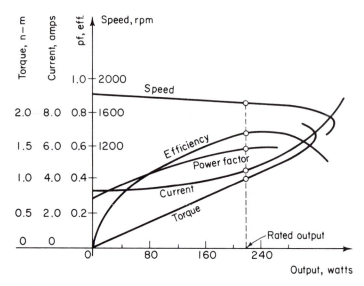

Figure 9-17 Typical performance characteristics for a single-phase induction motor, rated at $\frac{1}{4}$ hp, 60 Hz, four poles.

By repeating the calculations described in the foregoing example for various values of slip, the performance characteristics of the motor can be determined over the full operating range. Curves for a typical $\frac{1}{4}$-hp, 1725-rpm motor are depicted in Fig. 9-17.

9-6 APPROXIMATE DETERMINATION OF THE EQUIVALENT CIRCUIT PARAMETERS

The data obtained in a blocked-rotor test and a no-load test can be used to furnish information about the parameters appearing in the equivalent circuit. The blocked-rotor test data are used to reveal information about the leakage reactances and the stator-referred rotor winding resistance. The determination of these parameters is approximate because the associated procedure involves simplifying assumptions. One assumption is to consider that for $s = 1$ the impedance of $x_\phi/2$ in the equivalent circuit of Fig. 9-11 is so large compared with $(r_2'/2) + j(x_2'/2)$ that for all practical purposes it may be eliminated entirely from the equivalent circuit. Accordingly, the equivalent circuit of Fig. 9-11 reduces to the form depicted in Fig. 9-18.

If we let the measured values of voltage, current, and power obtained in the rotor-blocked test be denoted respectively by V_{sc}, I_{sc}, and P_{sc}, it then follows that the equivalent impedance referred to the stator coil is

$$Z_e = \frac{V_{sc}}{I_{sc}} \qquad (9\text{-}37)$$

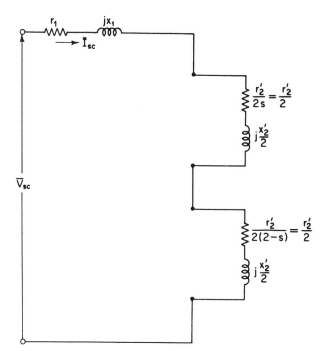

Figure 9-18 Equivalent circuit of the single-phase induction motor as it applies at standstill.

Moreover, the equivalent resistance component of Eq. (9-37) is readily determined from the wattmeter and ammeter readings. Thus

$$R_e \equiv r_1 + r_2' = \frac{P_{sc}}{I_{sc}^2} \qquad (9\text{-}38)$$

To determine r_2' from the result of Eq. (9-38), it is necessary to introduce a second important assumption, i.e.,

$$r_1 = r_2' \qquad (9\text{-}39)\dagger$$

This assumption is valid for all well-designed machines, i.e., those machines for which a minimum amount of iron and copper is used for a specified output rating. Unless otherwise stated, the assumption stated in Eq. (9-39) is always made in analyzing the single-phase motor. Therefore, the rotor winding resistance term appearing in the equivalent circuit can be identified as

$$\frac{r_2'}{2} = \frac{R_e}{4} \qquad (9\text{-}40)$$

where R_e is determined by Eq. (9-38).

---

†An alternative procedure is to find r_1 in the manner described for the three-phase induction motor. See Sec. 4-7.

The value of the equivalent leakage reactance is obtained from

$$X_e = \sqrt{Z_e^2 - R_e^2} \tag{9-41}$$

where Z_e and R_e are found from Eqs. (9-37) and (9-38). The stator-referred rotor winding leakage reactance is then found by assuming that

$$x_1 = x_2' \tag{9-42}$$

and this leads directly to

$$\frac{x_2'}{2} = \frac{X_e}{4} \tag{9-43}$$

The sole parameter of the equivalent circuit that remains to be determined is the magnetizing reactance x_ϕ. Its value can be established from the no-load test data and knowledge of x_1, x_2', and r_2' as derived from the rotor-blocked test. The procedure is illustrated by referring to the equivalent circuit of Fig. 9-19 which is derived from Fig. 9-11 upon imposing simplifying assumptions. Recall that at no-load the slip is very close to zero; it then follows that the quantity $r_2'/2s$ represents a very high resistance and so may be omitted without much error. Therefore, at no-load $\bar{Z}_{p1}/2$ is assumed to be equal simply to $jx_\phi/2$. Similarly, the impedance $\bar{Z}_{p2}/2$ associated with the negative-sequence field can be simplified to the low impedance represented by $(r_2'/4) + j(x_2'/2)$—which is many times smaller than $jx_\phi/2$. Certainly this assumption is even more valid than the corresponding assumption used in the rotor-blocked test.

Determination of x_ϕ essentially requires finding the voltage across a and b in Fig. 9-19. By letting the instrument readings for voltage, current, and power in the no-load test be denoted respectively by V_n, I_n, and P_n, the power factor angle of the current can be found from

$$\cos \theta_n = \frac{P_n}{V_n I_n} \tag{9-44}$$

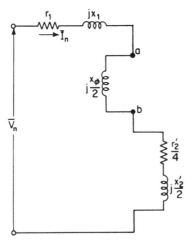

Figure 9-19 Equivalent circuit of the single-phase induction motor at no-load ($s \approx 0$).

Then,

$$\bar{V}_{ab} = \bar{V}_n - I_n \underline{/-\theta_n} \left[\left(r_1 + \frac{r_2'}{4} \right) + j \left(x_1 + \frac{x_2'}{2} \right) \right] \tag{9-45}$$

Since all the quantities on the right side of this equation are known, the magnetizing reactance is readily computed as

$$\frac{x_\phi}{2} = \frac{V_{ab}}{I_n} \tag{9-46}$$

The foregoing procedure can be somewhat simplified by using an approximate expression for the voltage across a and b that uses magnitudes instead of phasors. Thus

$$V_{ab} \approx V_n - I_n \left(x_1 + \frac{x_2'}{2} \right) \tag{9-47}$$

The validity of this assumption is based on the fact that the single-phase induction motor operates at very poor power factor at no-load. A glance at Fig. 9-20 indicates that the three quantities involved in Eq. (9-47) are practically collinear. When Eq. (9-47) is used to determine V_{ab}, there is no need to use Eq. (9-44); x_ϕ is found directly through the application of the last two equations only.

Example 9-2

Laboratory tests on a single-phase induction motor yield the following data:

Blocked-rotor test:	$V_{sc} = 110$ V,	$I_{sc} = 14.8$ A,	$P_{sc} = 1130$ W
No-load test:	$V_n = 110$ V,	$I_n = 2.8$ A,	$P_n = 60$ W
Rotational losses:	17 W		

Determine the five parameters of the equivalent circuit and the value of the core loss in watts.

Solution   From Eq. (9-38) the equivalent resistance is

$$R_e = r_1 + r_2' = \frac{P_{sc}}{I_{sc}^2} = \frac{1130}{14.8^2} = 5.15 \ \Omega$$

Hence,

$$r_1 = r_2' = 2.65 \ \Omega$$

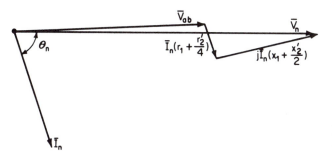

Figure 9-20   Phasor diagram for the circuit of Fig. 9-19.

and

$$\frac{r_2'}{2} = 1.33 \ \Omega$$

Also,

$$Z_e = \frac{V_{sc}}{I_{sc}} = \frac{110}{14.8} = 7.44 \ \Omega$$

$$X_e = \sqrt{7.44^2 - 5.15^2} = 5.36 \ \Omega$$

Therefore,

$$x_1 = x_2' = \frac{5.36}{2} = 2.68 \ \Omega$$

and

$$\frac{x_2'}{2} = 1.34 \ \Omega$$

Moreover, by Eq. (9-47) the voltage across the magnetizing reactance at no-load is

$$V_{ab} = V_n - I_n \left(x_1 + \frac{x_2'}{2} \right) = 110 - 2.8(2.68 + 1.34) = 110 - 11.2 = 98.8 \ V$$

Hence,

$$\frac{x_\phi}{2} = \frac{V_{ab}}{I_n} = \frac{98.8}{2.8} = 35.3 \ \Omega$$

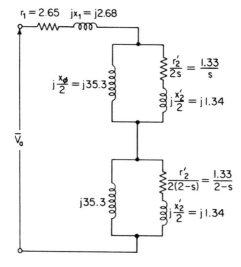

Figure 9-21 Equivalent circuit showing parameters computed in Example 9-2.

The core loss is found from

$$P_c = P_n - I_n^2 \left(r_1 + \frac{r_2'}{4} \right) - P_{\text{rot}}$$

$$= 60 - 2.8^2(3.17) - 17 = 18.2 \text{ W} \qquad (9\text{-}48)$$

The equivalent circuit showing the computed parameters appears in Fig. 9-21.

9-7 SINGLE-PHASE MOTOR TYPES

Many different types of single-phase motors have been developed primarily for two reasons. One, the torque requirements of the appliances and applications in which they are used vary widely. Two, it is desirable to use the lowest-priced motor that will drive a given load satisfactorily. For example, a high-torque version of the split-phase induction motor discussed in Sec. 9-1 is designed almost exclusively for washing-machine applications. This motor is available in a single speed rating and in just two horsepower ratings. This, coupled with the large volume of sales, enables it to be the least expensive motor in its category. See line two of Table 9-1 for more details.

Capacitor-Start Induction-Run Motor

The chief difference between the various types of single-phase motors lies in the method used to start them. In the case of this motor a capacitor is placed in the auxiliary winding circuit so selected that it brings about a 90-degree time displacement between $\bar{I}_A$ and $\bar{I}_M$. See Fig. 9-22. The result is a much larger starting torque than is achievable with resistance split-phase starting, because the starting torque is essentially proportional to the magnitudes of $\bar{I}_M$ and $\bar{I}_A$ as well as the sine of the phase angle between these quantities i.e., $T_S \alpha \bar{I}_M \bar{I}_A \sin < \frac{\bar{I}_M}{\bar{I}_A}$.

A typical torque-speed curve for this motor is depicted in Fig. 9-22(c). The cut-out switch serves to remove the auxiliary winding and the starting capacitor from the line once the motor has reached operating speed. Both items have intermittent ratings and could easily be damaged if allowed to remain energized. The starting capacitor is of the electrolytic type and generally has a value in the range of 100 μF for motor ratings of $\frac{1}{8}$ to $\frac{3}{4}$ hp. The capacitor can be easily damaged by a faulty cut-out switch; so proper maintenance of this switch is important.

Permanent-Split Capacitor Motor

At normal speeds the capacitor-start induction-run motor behaves as a conventional single-phase motor which means it has a resultant revolving field that has an elliptical locus. The result is a motor that operates less smoothly and in a noisier

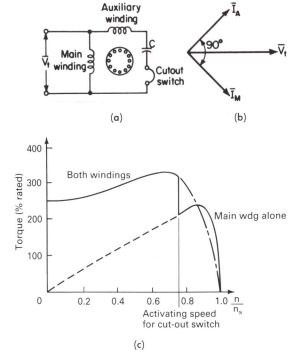

(a)

(b)

(c)

Figure 9-22 Capacitor-start induction-run motor: (a) schematic diagram; (b) phasor diagram at starting; (c) typical torque-speed characteristic of capacitor-start induction-run motor.

fashion than one where the resultant field has a constant amplitude with time. This latter condition can be closely duplicated by designing the single-phase motor with an auxiliary winding that is identical to the main winding and then placing in series with the auxiliary winding a full-duty, oil-filled capacitor whose value is selected to provide the constant amplitude rotating field in the vicinity of rated output. The schematic arrangement appears in Fig. 9-23 and for obvious reasons this motor is called the permanent-split capacitor motor. Basically, this motor is designed to operate as a balanced two-phase motor near rated load.

A typical torque-speed curve is illustrated in Fig. 9-24. There is no need for a cut-out switch because of the continuous-duty rating of the auxiliary winding. Part of the price paid for the improved running performance is the much lower torque at starting and breakdown. The starting torques are usually 60% to 75% of rated value. This is attributable to the fact that the capacitor value that gives balanced performance at normal slips is far different from the value needed for large starting torques. This is entirely understandable in view of the large difference between the input impedance at standstill compared to running conditions.

Besides the quiet, smooth running performance and the increased operating reliability (owing to the absence of a cut-out switch), the permanent-split capacitor motor also makes *reversible* operation a routine matter since basically it is a two-phase motor. Reversible operation of the capacitor-start induction-run motor is trickier to achieve. (Why?)

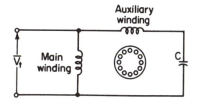

Figure 9-23   Permanent-split capacitor motor.

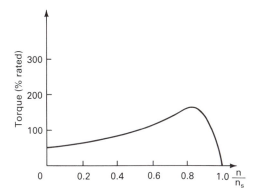

Figure 9-24   Typical torque-speed characteristic of permanent-split capacitor motor. No cut-out switch.

Capacitor-Start Capacitor-Run Motor

The high starting torque of the capacitor-start motor can be combined with the good running performance of the permanent-slit motor by employing two capacitors in the manner shown in Fig. 9-25. Of course, the auxiliary winding must carry a continuous rating as in the case of the permanent-split motor and C_1 must also have a continuous-duty rating. Capacitor C_2, which is used only during starting, can be a less expensive, intermittently rated element. A typical torque-speed curve for this motor is illustrated in Fig. 9-26.

Shaded-Pole Motor

The shaded-pole motor is used extensively in applications that require $\frac{1}{20}$ hp or less. The construction is very rugged as a study of Fig. 9-27 reveals. It consists essentially of copper wound on iron, a squirrel-cage rotor, and not much else. It contains no cut-out switch, which can be a potential source of trouble.

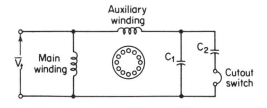

Figure 9-25   Capacitor-start capacitor-run motor.

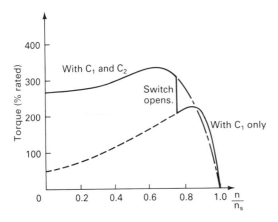

Figure 9-26 Typical torque-speed characteristic of capacitor-start capacitor-run motor. See Fig. 9-25 for identity of C_1 and C_2.

Figure 9-27 Construction features of the shaded-pole motor.

The manner of obtaining a moving flux field in this motor is different than it is in those considered previously. Here use is made of a copper ring that is embedded in each salient-pole piece as depicted in Fig. 9-27. The action of this ring is to oppose flux changes; and, in the process, over the course of the cyclic variation of the coil's exciting current, it causes the flux to move in a sweeping fashion across the face of the poles from the unshaded to the shaded sections. Reference to Fig. 9-28 illustrates the procedure.

In the discussion that follows hysteresis is neglected; hence flux and current are in time phase. Appearing in Fig. 9-28(a) is the situation that prevails at the N-pole flux at time ωt_1 in the cyclic variation of the magnetizing coil current i_ϕ [see Fig. 9-28(d)]. At this time instant flux is increasing, but the reaction in the copper ring is to produce a current which acts, by Lenz's law, to prevent the increase in

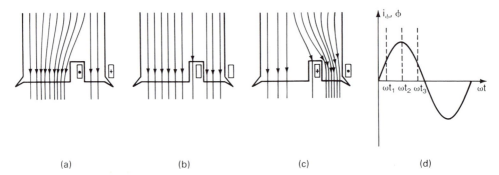

(a)   (b)   (c)   (d)

Figure 9-28 The action of the copper ring in producing a moving flux across the pole face. (a) Field distribution at time ωt_1; (b) time ωt_2; (c) time ωt_3; (d) coil current.

flux. This calls for a current direction in the copper ring that is into the plane of the paper on the right side and out on the left side. The result is to crowd the flux to the unshaded portion. Later, at ωt_2, the induced emf in the copper ring is zero because $d\phi/dt$ is zero. Hence there occurs a uniform distribution of flux across the pole face as illustrated in Fig. 9-28(b). Next, at ωt_3, the pole flux is decreasing in response to a decreasing magnetizing current. Since the reaction in the ring is always to prevent change, which at this instant means to prevent a decrease in flux, the induced ring current now takes the direction shown in Fig. 9-28(c). The result is a crowding of flux in the shaded portion of the pole. The net effect, therefore, is to produce a sweeping flux motion in the air gap beneath the pole face which in turn induces currents in the squirrel-cage rotor that react with the pole flux to produce torque.

A typical torque-speed characteristic is shown in Fig. 9-29. Note the modest value of starting torque of these motors. The breakdown torque is also of moderate value.

Universal Motor

The notion of a universal motor is tied to the capability of the motor to operate with dc or ac voltages applied, i.e., to operate over the range of frequencies from zero to 60 Hz. By way of introducing this topic we ask the question: Can a *shunt motor,* designed for dc operation, be made to deliver power with a 60-Hz voltage connected to its input line terminals? Recourse to the basic torque equation reminds us that we need to identify a flux field, an ampere-conductor distribution and a proper field orientation of the two that is maintained over time. The ampere-conductor distribution is ensured by the flow of armature current through the armature winding. There will be a pole flux but its magnitude will be very small for two reasons. One, since the field winding is equipped with many turns and the inductance is proportional to the number squared [see Eq. (1-40)], the inductive reactance of the field winding is extremely high thus producing very little field current. Two, the air gap of dc machines, unlike ac machines, is quite large. Hence the net effect is to produce a very small pole flux. Furthermore, although the brushes are correctly placed along the interpolar axis, still during every cyclic

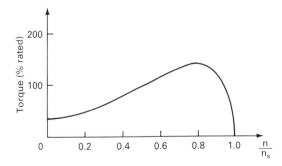

Figure 9-29 Typical torque-speed characteristic of a shaded-pole motor. Note low starting and breakdown torques.

variation of the field and armature currents, the flux field and the armature ampere-conductor distribution will not produce a unidirectional torque on each conductor. This is because the armature current is not in time phase with the field current; and the reason is that the inductance of the armature circuit is not as large as that of the field circuit. Therefore, despite the fact that the ac operated shunt motor produces some starting and running torque, its performance would be far from acceptable.

What happens when the single-phase ac voltage is applied to a dc series motor? The situation now is very much improved. First, because the series field and armature winding carry the same current in the same circuit, the field orientation for producing torque is optimum. Second, the pole flux is now greater because the series field inductance is low owing to the use of a small number of series turns compared to the number of field turns in a shunt motor. The series field mmf is chiefly dependent on the flow of a large armature current. Accordingly, acceptable performance can be expected for this mode of operation. Core losses, however, can be exorbitant if the stator iron is unlaminated. Also, commutation is likely to be poor because of the induced currents that occur through transformer action when a coil undergoes commutation.

The term *universal motor* is applied to a dc series motor the design of which has been modified to mitigate the shortcomings associated with ac operation. These modifications include the following:

1. A completely laminated stator structure including the poles.
2. High resistance connections from the coils to the commutator segments to reduce the transformer induced voltages.
3. A smaller number of series field turns to reduce the reactance drop in the armature circuit.
4. An increased number of armature conductors to compensate for the reduced pole flux associated with the smaller number of series field turns.

It should come as no surprise that the universal motor is an excellent dc series motor.

The torque developed by the universal motor is proportional to the square of the armature current just as it is for the conventional dc series motor. But, because the armature current in the universal motor is sinusoidally varying, the instantaneous torque as a function of time varies in accordance with double the line frequency as illustrated in Fig. 9-30. Although the average torque is responsible for the useful work at the motor shaft, this double frequency variation of the instantaneous torque makes the universal motor more prone to noise and vibration.

Typical torque-speed curves for the universal motor and the conventional dc series motor are depicted in Fig. 9-31. Observe that at any given torque the speed of the universal motor is less than that of the dc series motor. The reason lies in

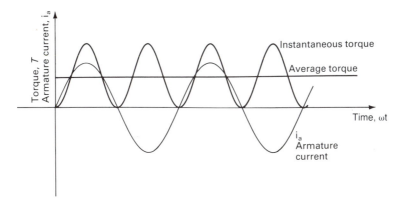

Figure 9-30   Instantaneous torque of the universal motor.

the reactance drops that are associated with the series field and armature windings.

It is useful to note that the universal motor develops more torque for a given size than any other single-phase motor because the torque is proportional to the armature current squared. Hence, it finds considerable popularity in those applications where space is at a premium and high torque is needed. The whole area of home appliances is typical of such applications. Furthermore, speed control of such motors is easily achieved either through the use of a series resistor (for the smaller ratings) or through the use of an adjustable rectified source employing silicon controlled rectifiers.

Hysteresis Motors

This topic is included here primarily because the hysteresis motor is most often used on single-phase power supplies in competition with the other motors described in this section. However, it is sometimes designed for three-phase operation and it always runs at synchronous speed in steady state.

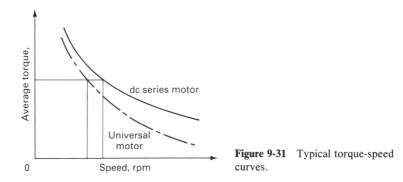

Figure 9-31   Typical torque-speed curves.

The single-phase hysteresis motor depends for its success on a rotor that has no teeth or windings but is constructed of high-retentivity steel in a smooth, cylindrical form. The high-retentivity feature of the rotor material essentially furnishes the motor with the equivalent of permanent magnet poles that permits operation at synchronous speed. The rotating field of the stator is usually obtained by employing two well-distributed windings with a fixed capacitor permanently connected to one of the windings to provide the time displacement needed to generate a rotating field of constant amplitude. Care is taken in the stator design to ensure a nearly sinusoidal field distribution in order to minimize losses.

Starting torque.   When power is applied to the stator windings at standstill, the resulting rotating stator mmf generates hysteresis loops in the rotor steel with a cyclic variation of line frequency. The magnetized rotor exhibits induced magnetic poles that lag behind the magnetizing stator mmf by the hysteretic angle γ.† Materials of high retentivity have large values of γ. Accordingly, at standstill, these induced poles follow the rotating stator mmf but do so with a lag of γ electrical degrees. The torque produced in this fashion by hysteresis can be described by

$$T_h = K_h \phi_S \phi_R \sin \gamma$$

where K_h is an appropriate design parameter, ϕ_S denotes the flux produced by the stator mmf, ϕ_R is the induced flux per pole in the rotor caused by the hysteresis effect and γ is the space displacement between these two fields expressed in electrical degrees. Observe that T_h is a fixed quantity for a given machine.

Accelerating torque.   If the starting torque exceeds the load torque, the excess torque accelerates the load toward synchronous speed. During this time the frequency of generation of the hysteresis loop in the rotor steel is diminishing but neither ϕ_R nor γ are affected because these quantities depend on the retentivity property of the hysteresis loop of the rotor material. Therefore, T_h remains constant from zero to synchronous speed. This is indicated in Fig. 9-32 by the horizontal line. The resultant torque-speed curve over this range is actually sloping negatively because of the contribution to the net rotor developed torque that is associated with eddy-currents that are produced in the rotor steel during the cyclic variation of the rotor flux. Of course, once synchronous speed is attained, the cyclic variation of flux in the rotor ceases and so too does the eddy-current torque contribution. However, the hysteresis torque remains because despite the absence of cyclic variation of flux in the rotor, the high-retentivity feature of the

†See p 55.

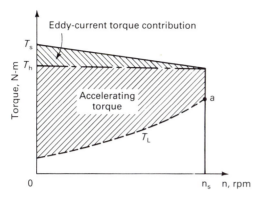

Figure 9-32 Torque-speed characteristic of the hysteresis motor.

rotor steel provides the rotor with what can now be considered to be permanent magnets. Moreover, if at this synchronous speed the hysteretic torque exceeds that required by the load, the rotor will readjust the physical lag between the axis of the stator poles and the axis of the rotor poles to ensure equilibrium. This operating point is indicated by a in Fig. 9-32.

Two outstanding features of the hysteresis motor are quiet operation and good accelerating torque. Also, the capability of this motor to operate at synchronous speed is advantageous especially where the electric power system furnishes power at virtually constant frequency. In such circumstances the hysteresis motor is ideal for driving clocks and other speed sensitive devices such as record, tape, and compact disc players.

9-8 OPERATING FEATURES AND TYPICAL APPLICATIONS

The principal performance features as well as typical applications of single-phase ac motors appear in Table 9-1. Note that for applications involving less than $\frac{1}{20}$ hp the shaded-pole motor is invariably used. On the other hand, for applications involving $\frac{1}{20}$ hp to $\frac{3}{4}$ hp the choice of the motor depends upon such factors as the starting and breakdown torque and even quietness of operation. Where a minimum of noise is desirable and low starting torque is adequate, such as for driving fans and blowers, then the motor to choose is the permanent-split type. If quietness is to be combined with high starting torque, as might be the case where a compressor drive is needed, then the suitable choice would be the capacitor-start capacitor-run motor. Of course, in circumstances where the compressor is located in a noisy environment, clearly the choice should be the capacitor-start induction-run motor because it is less expensive.

TABLE 9-1   CHARACTERISTICS AND APPLICATIONS OF SINGLE-PHASE A-C MOTORS

Type designation	Starting torque (% of normal)	Approx. comparative price (%)	Breakdown torque (% of normal)	Starting current at 115 V
General-purpose split-phase motor	90–200 Medium	85	185–250 Medium	23 1/4 hp
High-torque split-phase motor	200–275 High	65	Up to 350	32 High 1/4 hp
Permanent-split capacitor motor	60–75 Low	155	Up to 225	Medium
Permanent-split capacitor motor	Up to 200 Normal	155	260	
Capacitor-start general-purpose motor	Up to 435 Very high	100	Up to 400	
Capacitor-start capacitor-run motor	380 High	190	Up to 260	
Shaded-pole motor	50	—	150	

Source: By permission from M. Liwschitz-Garik and C. C. Whipple, *Electric Machinery,* vol. II (Princeton, N.J.: D. Van Nostrand Co., Inc., 1946).

PROBLEMS

9-1. Two field coils are space displaced by 45 electrical degrees. Moreover, sinusoidal currents that are time displaced by 45 electrical degrees flow through these windings. Will a revolving flux result? Explain.

9-2. Two coils are placed in a low reluctance magnetic circuit with their axes in quadrature—the first along the vertical and the second along the horizontal. Sinusoidal currents that are time displaced by 90 electrical degrees and of the same frequency ω are made to flow through the coils. Moreover, the mmf of the first coil is twice that of the second.

(a) Assuming that at time ωt_0 the mmf of the first coil is at its positive peak value, determine the relative magnitude and direction of the resultant mmf.

(b) Repeat part (a) for a time ωt_1 that is 45° later than ωt_0.

TABLE 9-1   (Continued)

Power Factor (%)	Efficiency (%)	Horsepower range	Application and general remarks
56–65	62–67	1/20 to 3/4	Fans, blowers, office appliances, food-preparation machines. Low- or medium-starting torque, low-inertia loads. Continuous-operation loads. May be reversed.
50–62	46–61	1/6 to 1/3	Washing machines, sump pumps, home work-shops, oil burners. Medium- to high-starting torque loads. May be reversed.
80–95	55–65	1/20 to 3/4	Direct-connected fans, blowers, centrifugal pumps. Low-starting torque loads. Not for belt drives. May be reversed.
80–95	55–65	1/6 to 3/4	Belt-driven or direct-drive fans, blowers, centrifugal pumps, oil burners. Moderate-starting torque loads. May be reversed.
80–95	55–65	1/8 to 3/4	Dual voltage. Compressors, stokers, conveyors, pumps. Belt-driven loads with high static friction. May be reversed.
80–95	55–65	1/8 to 3/4	Compressors, stokers, conveyors, pumps. High-torque loads. High power factor. Speed may be regulated.
30–40	30–40	1/300 to 1/20	Fans, toys, hair dryers, unit heaters. Desk fans. Low-starting torque loads.

    (c) Repeat part (a) for a time ωt_2 that is 90° later than ωt_0.

    (d) What conclusion can you draw from parts (a), (b), and (c)?

9-3. The two coils of Prob. 9-2 are now energized with sinusoidal currents that are 45 electrical degrees out of phase but yield equal mmfs.

    (a) Assuming that at time ωt_0 the mmf of the first coil is at its positive peak value, find the relative magnitude and direction of the resultant mmf.

    (b) Repeat part (a) for a time ωt_1 that is 45° later than ωt_0.

    (c) Repeat part (a) for a time ωt_2 that is 90° later than ωt_0.

    (d) What conclusion can you draw from parts (a), (b), and (c)?

9-4. In Fig. P9-4, I, II, and III denote the three coils of a balanced three-phase winding which are spaced displaced by 120° from one another. If a single-phase voltage is applied to the three-phase, squirrel-cage motor in the manner shown, will the motor operate? Justify your answer.

9-5. The stator windings of a two-phase, two-pole induction motor are in space quadrature. When the effective values of the phase currents are $\bar{I}_a = 10\underline{/0°}$ and $\bar{I}_b = 10\underline{/90°}$ A, the resulting stator mmf wave rotates clockwise, and its peak value is 2000 A-t per

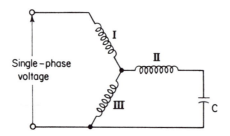

Figure P9-4

pole. If the phase currents are changed to $\bar{I}_a = 5\underline{/0°}$ and $\bar{I}_b = 10\underline{/53.1°}$, find the peak values of the clockwise and counterclockwise rotating mmf waves.

9-6. The phase currents in the stator windings of a two-phase induction motor are $\bar{I}_a = 8\underline{/0°}$ and $\bar{I}_b = 8\underline{/-90°}$ A. Together they produce an mmf wave rotating counterclockwise and having a peak value of 1000 A-t per pole. These currents are then changed to the following: $\bar{I}_a = 10\underline{/0°}$ and $\bar{I}_b = 6\underline{/+36.9°}$. Find the peak values of the resulting clockwise and counterclockwise rotating mmf waves.

9-7. At standstill, the currents in the main and auxiliary windings of a single-phase induction motor are given by $\bar{I}_M = 10\underline{/0°}$ and $\bar{I}_A = 5\underline{/60°}$. There are five turns per pole on the auxiliary winding for every four turns on the main winding. Moreover, these windings are in space quadrature. What should be the magnitude and phase of the current in the auxiliary winding to produce an mmf wave of constant amplitude and velocity?

9-8. The armature mmf of a machine is described by the equation

$$\mathcal{F}_a = K_1 \cos \omega t \cos \alpha + K_2 \cos \omega t \sin \alpha$$

Is this a single-phase or a polyphase machine? Explain your answer.

9-9. An unbalanced two-phase voltage system is described by

$$\bar{V}_a = 120\underline{/0°}$$
$$\bar{V}_b = 80\underline{/-53°}$$

  (a) Find the positive-sequence voltage set of this system.
  (b) What is the negative-sequence set?
  (c) Draw the complete phasor diagram of this two-phase system, showing each positive- and negative-sequence component.

9-10. Repeat Prob. 9-9 for the case where $\bar{V}_a = 120\underline{/0°}$ and $\bar{V}_b = 80\underline{/-90°}$.

9-11. Find the unbalanced two-phase system that is represented by a positive-sequence component of $95.2\underline{/14.6°}$ and a negative-sequence component of $31\underline{/40.6°}$ expressed in volts.

9-12. The results of a load test on a $\frac{1}{4}$-hp, 115-V, 60-Hz, four-pole, single-phase induction motor are as follows:

applied voltage	= 115 V
main winding current	= 3.7 A
power input	= 270 W
stator core loss	= 10 W
slip	= 90 rpm

The stator winding resistance is 2 Ω and the stator referred rotor resistance is 1.6 Ω. The starting winding was disconnected during this test. Calculate the power output during the test. Assume that the rotational losses are equal to the stator core losses.

9-13. A 110-V, 60-Hz, single-phase induction motor has the following parameters:

$$r_1 = 1.86 \ \Omega, \qquad r_2' = 1.78 \ \Omega$$
$$x_1 = 2.56 \ \Omega, \qquad x_2' = 1.28 \ \Omega$$
$$x_\phi = 25 \ \Omega$$

At a slip of 5% find:
(a) Input line current and power factor.
(b) Horsepower output.
(c) The efficiency.

9-14. A 115-V, $\frac{1}{4}$-hp, 60-hz, single-phase induction motor has the following design parameters:

$$r_1 = 1.48 \ \Omega, \qquad r_2' = 1.68 \ \Omega$$
$$x_1 = 3.22 \ \Omega, \qquad x_2' = 1.6 \ \Omega$$
$$x_\phi = 25 \ \Omega$$

Also, stator core loss is negligible.
(a) Determine the operating performance of this motor when it runs at a slip of 10%.
(b) Would you permit continuous operation of this motor under the conditions of part (a)?

9-15. The rotor of the machine described in the preceding problem is blocked and rated voltage is applied. Compute the readings of a line wattmeter and ammeter. Make approximations where proper.

9-16. A $\frac{1}{2}$-hp, 115-V, four-pole, single-phase induction motor presents the following test data:

> At no-load: 115 V, 4.81 A, 105 W
>
> With rotor blocked: 115 V, 27.2 A, 1308 W

The stator winding resistance is 0.83 Ω. The stator core losses are equal to the rotational losses. Determine the parameters of the equivalent circuit.

9-17. The following no-load and blocked rotor test data are available for a single-phase induction motor:

> Blocked rotor: $I_{sc} = 6.6$ A, $V_{sc} = 53$ V, $P_{sc} = 210$ W
>
> No-load: $I_n = 3.85$ A, $V_n = 208$ V, $P_n = 160$ W

The stator resistance is 1.48 Ω and the rotational losses are 60 W.
(a) Determine the parameters of the equivalent circuit.
(b) What is the value of the core loss?
(c) Compute the value of resistance that can be used to represent this core loss in the equivalent circuit. Show its location.

9-18. A $\frac{1}{4}$-hp, 110-V, 60-Hz, four-pole, single-phase induction motor has the following parameters:

$$r_1 = 1.5 \ \Omega, \qquad r_2' = 1.4 \ \Omega$$
$$x_1 = 2.4 \ \Omega, \qquad x_2' = 1.2 \ \Omega$$
$$x_\phi = 40 \ \Omega$$

The stator core loss is 20 W.

(a) Calculate the developed torque at a slip of 5%.

(b) If the rotational losses are equal to 20 W, what is the shaft torque?

9-19. A $\frac{1}{2}$-hp, 220-V, 60-Hz, four-pole, single-phase induction motor has the following equivalent circuit parameters:

$$r_1 = 2.8 \; \Omega, \qquad r_2' = 2.4 \; \Omega$$

$$x_1 = 5.5 \; \Omega, \qquad x_2' = 2.8 \; \Omega$$

$$x_\phi = 90 \; \Omega$$

The core loss is 30 W and the rotational losses are 46 W. Determine for $s = 0.05$:

(a) Line current and line power factor.

(b) Horsepower output.

(c) Efficiency.

9-20. The following data apply to a $\frac{1}{4}$-hp, 110-V, 60-Hz, four-pole, single-phase induction motor.

No-load test: $V_n = 110 \; V$, $I_n = 2.73 \; A$, $P_n = 56 \; W$

Blocked rotor test: $V_{sc} = 110 \; V$, $I_{sc} = 16.6 \; A$, $P_{sc} = 1260 \; W$

The resistance of the main winding is 1.5 Ω. Assume the rotational losses are equal to the stator core loss. On a carefully drawn equivalent circuit, show the value of each parameter for a slip of 5%.

9-21. A two-phase, 60-Hz, squirrel-cage induction motor has a blocked resistance per phase of 6 Ω and a reactance of 8 Ω. When connected as a normal two-phase motor, its starting torque is 200%. It is desired to use this motor on a single-phase circuit.

(a) A resistance of 10 Ω is inserted in series with one phase and the machine is operated as a split-phase motor. Determine the developed torque at standstill. Consider that the torque is proportional to the product of the phase currents as well as the sine of the phase angle between them.

(b) A capacitor is to be used in place of the resistor. Calculate the microfarad rating of this capacitor so that the phase angle between the currents is 90°.

(c) Determine the starting torque for the condition of part (b).

9-22. Discuss in some detail the capability offered by the following motors in permitting reversible operation. Be sure to cite the ease or difficulty that is likely to be encountered in each case as well as any precautions to be observed. In those instances where reversible operation is deemed practical show a complete schematic diagram of the motor including the circuit that allows the reversible mode.

(a) Permanent-split capacitor motor.

(b) Capacitor-start capacitor-run motor.

(c) General-purpose split-phase motor.

(d) Universal motor.

(e) Hysteresis motor.

9-23. The following questions relate to the single-phase shaded-pole induction motor:

(a) Can the direction of rotation be reversed by reversing the line leads? Explain.

(b) Devise a scheme that permits reversible operation of the single-phase shaded-pole motor.

10

Unbalanced Two-Phase Motors: Servomotors

The two-phase servomotor has gained widespread use as an output actuator in feedback control systems. A high rotor resistance is one of its distinguishing design characteristics. This feature ensures a negative slope for the torque-speed characteristic over its entire operating range and thereby furnishes the motor with positive damping for good stability. The ac servomotor often operates with unbalanced two-phase voltages. A fixed reference voltage is applied to its main winding and a varying voltage is applied to its control winding. The control winding is frequently fed from the output terminals of a servoamplifier with a signal that is usually displaced 90° from that on the main winding.

The analytical treatment of the two-phase servomotor supplied with unbalanced voltages can be conveniently accomplished by the use of symmetrical components. This technique is described in the first section of this chapter for the more general case of an unbalanced four-phase system and is subsequently applied to the two-phase case. The symmetrical components approach allows the unbalanced voltages to be replaced by two or more sets of balanced voltages. In this way the unbalanced machine is treated in terms of theoretical considerations that have already been fully exposed. (The computation of the performance of polyphase motors is described in Chapter 4.)

10-1 SYMMETRICAL COMPONENTS

The method of replacing an unbalanced four-phase voltage system with balanced voltage sets is described first chiefly because a more complete formulation of the problem can be made. If the unbalanced two-phase voltage system were treated directly, there would be no need to deal with zero-sequence voltages or balanced single-phase voltage sets so that some generality would thereby be lost. In Fig. 10-1(a) is shown an unbalanced four-phase voltage set. In a balanced system the phase voltages would be equal in amplitude and exactly 90° displaced from one another. Neither condition is satisfied by the set depicted in Fig. 10-1(a). Figure 10-1(b) shows that the phasor sum is not zero but adds, rather, to a resultant quantity which is identified as $4\bar{V}_0$. The quantity $\bar{V}_0$ is called the *zero-sequence voltage* and the factor 4 states that this unbalance can be assumed to have been brought about by the existence of an equal unbalance in each phase voltage. It is called zero-sequence because the $\bar{V}_0$ in any one phase voltage is *in-phase* with the $\bar{V}_0$ quantity of the other phase voltages.

For any given unbalanced system the zero-sequence component can be readily found from

$$\bar{V}_0 = \tfrac{1}{4}(\bar{V}_a + \bar{V}_b + \bar{V}_c + \bar{V}_d) \qquad (10\text{-}1)$$

The removal of $\bar{V}_0$ from each of the phase voltages results in a four-phase voltage system that adds up to zero. Thus

$$(\bar{V}_a - \bar{V}_0) + (\bar{V}_b - \bar{V}_0) + (\bar{V}_c - \bar{V}_0) + (\bar{V}_d - \bar{V}_0) = 0 \qquad (10\text{-}2)$$

For convenience let

$$\bar{V}_A \equiv \bar{V}_a - \bar{V}_0$$
$$\bar{V}_B \equiv \bar{V}_b - \bar{V}_0$$
$$\bar{V}_C \equiv \bar{V}_c - \bar{V}_0 \qquad (10\text{-}3)$$
$$\bar{V}_D \equiv \bar{V}_d - \bar{V}_0$$

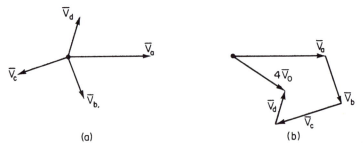

(a)   (b)

Figure 10-1   Unbalanced four-phase voltage system: (a) star notation; (b) identifying the zero-sequence component $\bar{V}_0$.

Equation (10-2) then becomes

$$\bar{V}_A + \bar{V}_B + \bar{V}_C + \bar{V}_D = 0 \qquad (10\text{-}4)$$

The foregoing is graphically represented in Fig. 10-2. It is important to keep in mind that, although the phasor sum of $\bar{V}_A$, $\bar{V}_B$, $\bar{V}_C$, and $\bar{V}_D$ is zero, these quantities still represent an unbalanced four-phase system in the sense that the voltages are unequal in magnitude and have unequal phase-displacement angles.

Attention is next directed to the conditions that must be imposed in order generally to represent the unbalanced voltage system of Eq. (10-4). Simplicity is served if we can succeed in representing this unbalanced system by groups of balanced voltages. Specifically, because we are dealing with a four-phase system, it is desirable to make this representation in terms of *balanced four-phase systems* to the extent possible. In this connection then a good starting point is to assume that each of the phase voltages of the system of Eq. (10-4) contains an appropriate component of a *balanced positive sequence set* as illustrated in Fig. 10-3. The quantity $\bar{V}_1$ is assumed to be a component of $\bar{V}_A$. The subscript 1 is used to denote the *positive sequence*. Note that in a positive sequence set of four-phase balanced voltages, the phase voltages are equal in magnitude and 90 degrees displaced from one another and the rotation is counterclockwise (i.e., positive). The quantity $a\bar{V}_1$ is the corresponding component of $\bar{V}_B$ where the symbol a denotes the rotational operator defined by

$$a = 1\underline{/-90^\circ} \qquad (10\text{-}5)$$

Similarly, $a^2\bar{V}_1$ is the positive sequence component of $\bar{V}_C$ and $a^3\bar{V}_1$ is the positive sequence component of $\bar{V}_D$. Observe that $a^2 = -1$ and $a^3 = 1\underline{/90^\circ}$.

What can we now add to the positive-sequence four-phase voltage set of Fig. 10-3 to ensure at least some unbalance in the resulting phase voltages? The inclusion of a second set of balanced positive-sequence voltages is fruitless, as a glance

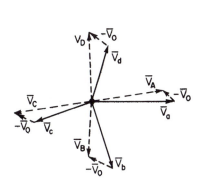

Figure 10-2 Removing the zero-sequence voltage $\bar{V}_0$ from each phase voltage.

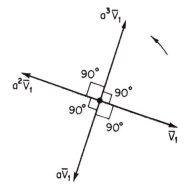

Figure 10-3 Balanced positive-sequence four-phase voltage set.

at Fig. 10-4 reveals. The result of adding two positive-sequence voltage sets still yields a balanced four-phase system. It therefore seems worthwhile at this point to investigate the effect of adding instead a balanced *negative-sequence* four-phase voltage set to the system of Fig. 10-3. Now the result is that illustrated in Fig. 10-5. Note that the selection of the opposite sequence has the effect of unbalancing two pairs of voltages while still preserving a balanced system, i.e., yielding a zero resultant. The quantity $\bar{V}_2$ is the negative-sequence component of $\bar{V}_A$. The subscript 2 is used to denote the negative sequence. Note too that for the negative-sequence set it is $a^3\bar{V}_2$ that is a component of $\bar{V}_B$, and $a\bar{V}_2$ is a component of $\bar{V}_D$. This marks a further step in achieving our goal of representing an unbalanced voltage system in terms of suitable balanced sets.

It is now desirable to add a third set of voltages to each phase of the system depicted in Fig. 10-5 but in such a way as to cause each of the phase voltages (and not just pairs) to be different in phase as well as in magnitude. In this connection let us explore the effect of adding a balanced single-phase set, keeping in mind that the resultant of the four components is to yield zero in order to preserve balance. Accordingly, if we assume that a third component $\bar{V}_3$ is added to $\bar{V}_1$ and $\bar{V}_2$ to yield $\bar{V}_A$, then an equal and opposite quantity, $a^2\bar{V}_3$, is assumed added to $a\bar{V}_1$ and $a^3\bar{V}_2$ to yield $\bar{V}_B$. Similarly, the third component of $\bar{V}_C$ is taken to be $\bar{V}_3$, while the third component of $\bar{V}_D$ is taken to be $a^2\bar{V}_3$. The results of these additions are depicted in Fig. 10-6. Note that this single-phase reversed-phase set differs from the zero-sequence set in that the former set taken together adds to zero. An inspection of Fig. 10-6 discloses that, with the use of the indicated balanced voltage sets, it becomes possible to represent any four-phase system of unbalanced voltages. The reason is that with such a composition there exists sufficient flexibility to make each phase voltage different in angle and magnitude from the others.

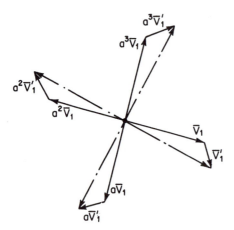

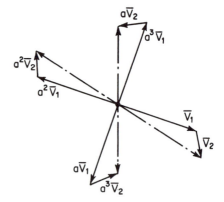

Figure 10-4 The addition of two positive-sequence voltage sets yields a balanced resultant set.

Figure 10-5 Addition of positive- and negative-sequence balanced four-phase voltage sets.

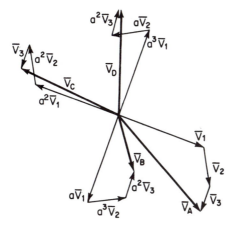

Figure 10-6 Illustrating how an unbalanced set of four-phase voltages can be represented generally by balanced sets.

However, our interest really focuses on the reverse situation, i.e., finding the balanced voltage components in each phase when an unbalanced system is specified. The foregoing treatment was described as presented in order to emphasize better how each component affects the total voltage.

The unbalanced phase voltages of the four-phase system can now be expressed mathematically as follows:

$$\bar{V}_A = \bar{V}_1 + \bar{V}_2 + \bar{V}_3 \tag{10-6}$$

$$\bar{V}_B = a\bar{V}_1 + a^3\bar{V}_2 + a^2\bar{V}_3 \tag{10-7}$$

$$\bar{V}_C = a^2\bar{V}_1 + a^2\bar{V}_2 + \bar{V}_3 \tag{10-8}$$

$$\bar{V}_D = a^3\bar{V}_1 + a\bar{V}_2 + a^2\bar{V}_3 \tag{10-9}$$

The first term on the right side of each equation represents a component of the balanced positive-sequence set, the second term represents a component of the balanced negative-sequence set, and the third term represents a component of the balanced single-phase set. Since each component set is a balanced set, together they yield a resultant that is zero. This is not surprising in view of the fact that the unbalanced component was previously removed. [Refer to Eq. (10-3).]

If Eq. (10-3) is inserted into Eqs. (10-6) to (10-9), expressions for the original unbalanced voltages in terms of the appropriate components are obtained. Thus

$$\bar{V}_a = \bar{V}_1 + \bar{V}_2 + \bar{V}_3 + \bar{V}_0 \tag{10-10}$$

$$\bar{V}_b = a\bar{V}_1 + a^3\bar{V}_2 + a^2\bar{V}_3 + \bar{V}_0 \tag{10-11}$$

$$\bar{V}_c = a^2\bar{V}_1 + a^2\bar{V}_2 + \bar{V}_3 + \bar{V}_0 \tag{10-12}$$

$$\bar{V}_d = a^3\bar{V}_1 + a\bar{V}_2 + a^2\bar{V}_3 + \bar{V}_0 \tag{10-13}$$

The manner of finding the zero-sequence component $\bar{V}_0$ in terms of the given unbalanced four-phase system is indicated by Eq. (10-1). An appropriate matter of concern at this point is consideration of how the remaining three balanced components can be determined in terms of the given unbalanced system. Note that, if the positive-sequence component $\bar{V}_1$ of phase a is determined, then the corresponding components in each of the remaining phases are also known because they differ by a rotation factor only. A similar situation obtains for the negative-sequence and the balanced single-phase components.

To find the positive-sequence component, Eqs. (10-10) through (10-13) must be manipulated in such a way as to isolate $\bar{V}_1$. A study of these equations reveals that this can be accomplished by multiplying Eq. (10-11) by a^3, Eq. (10-12) by a^2, and Eq. (10-13) by a and then summing each of these equations with Eq. (10-10). Thus

$$\bar{V}_a = \bar{V}_1 + \bar{V}_2 + \bar{V}_3 + \bar{V}_0 \tag{10-14}$$

$$a^3\bar{V}_b = \bar{V}_1 + a^2\bar{V}_2 + a\bar{V}_3 + a^3\bar{V}_0 \tag{10-15}$$

$$a^2\bar{V}_c = \bar{V}_1 + \bar{V}_2 + a^2\bar{V}_3 + a^2\bar{V}_0 \tag{10-16}$$

$$a\bar{V}_d = \bar{V}_1 + a^2\bar{V}_2 + a^3\bar{V}_3 + a\bar{V}_0 \tag{10-17}$$

$$\overline{\bar{V}_a + a^3\bar{V}_b + a^2\bar{V}_c + a\bar{V}_d = 4\bar{V}_1 + 0 + 0 + 0} \tag{10-18}$$

Keep in mind that the quantities $(1 + a + a^2 + a^3)$ and $(1 + a^2)$ are identically equal to zero. From the last equation it follows that the value of the positive-sequence component of voltage is

$$\boxed{\bar{V}_1 = \tfrac{1}{4}(\bar{V}_a + j\bar{V}_b - \bar{V}_c - j\bar{V}_d)} \tag{10-19}$$

since $a^3 = +j$, $a^2 = -1$, and $a = -j$.

Applying a similar procedure to the negative-sequence and the balanced single-phase set leads to the following results:

$$\boxed{\bar{V}_2 = \tfrac{1}{4}(\bar{V}_a - j\bar{V}_b - \bar{V}_c + j\bar{V}_d)} \tag{10-20}$$

and

$$\boxed{\bar{V}_3 = \tfrac{1}{4}(\bar{V}_a - \bar{V}_b + \bar{V}_c - \bar{V}_d)} \tag{10-21}$$

Therefore by means of Eqs. (10-1), (10-19), (10-20), and (10-21) the values of balanced voltage sets can be found to replace an unbalanced four-phase system of voltages.

Example 10-1

The phase voltages of an unbalanced four-phase system are given by

$$\bar{V}_a = 120\underline{/0°} = 120 + j0 \tag{10-22}$$

$$\bar{V}_b = 75\underline{/-60^\circ} = 37.5 - j65 \tag{10-23}$$

$$\bar{V}_c = 100\underline{/-150^\circ} = -86.6 - j50 \tag{10-24}$$

$$\bar{V}_d = 50\underline{/80^\circ} = 8.7 + j49 \tag{10-25}$$

(a) Find the zero-sequence, positive-sequence, negative-sequence voltages and the balanced single-phase set.

(b) Draw the phasor diagram showing the manner by which the individual phase components add to yield the given unbalanced voltages.

Solution   (a) By Eq. (10-1) the zero-sequence component for each phase is found to be

$$\bar{V}_0 = \tfrac{1}{4}(\bar{V}_a + \bar{V}_b + \bar{V}_c + \bar{V}_d)$$

$$= \tfrac{1}{4}(79.6 - j66) = 19.9 - j16.5 = 25.8\underline{/-39.8^\circ} \tag{10-26}$$

The positive-sequence component as obtained from Eq. (10-19) is

$$\bar{V}_1 = \tfrac{1}{4}(\bar{V}_a + j\bar{V}_b - \bar{V}_c - j\bar{V}_d)$$

$$= \tfrac{1}{4}(120 + j0 + j37.5 + 65 + 86.6 + j50 - j8.7 + 49)$$

$$= 80.2 + j19.7 = 82.6\underline{/13.8^\circ} \tag{10-27}$$

The negative-sequence component is determined from Eq. (10-20) to be

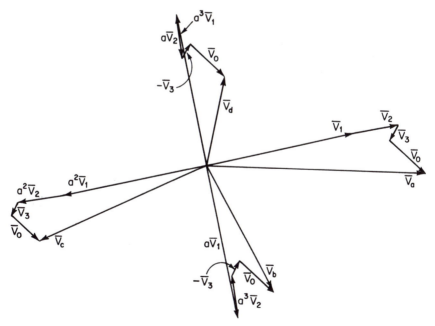

Figure 10-7   Unbalanced four-phase voltage system of Example 10-1, and the balanced voltage components.

$$\bar{V}_2 = \tfrac{1}{4}(\bar{V}_a - j\bar{V}_b - \bar{V}_c + j\bar{V}_d)$$

$$= \tfrac{1}{4}(120 + j0 + j37.5 - 65 + 86.6 + j50 + j8.7 - 49)$$

$$= 23.2 + j5.3 = 23.7\underline{/12.9^\circ} \tag{10-28}$$

Finally, the $\bar{V}_3$ component is computed to be

$$\bar{V}_3 = \tfrac{1}{4}(\bar{V}_a - \bar{V}_b + \bar{V}_c - \bar{V}_d)$$

$$= \tfrac{1}{4}(120 + j0 - 37.5 + j65 - 86.6 - j50 - 8.7 - j49)$$

$$= -3.2 - j8.5 = 8.9\underline{/-107.8^\circ} \tag{10-29}$$

(b) The required phasor diagram appears in Fig. 10-7.

10-2 SYMMETRICAL COMPONENTS APPLIED TO UNBALANCED TWO-PHASE MOTORS

The unbalanced two-phase voltage system can be treated as a special case of the unbalanced symmetrical four-phase system. The unbalanced symmetrical four-phase system is one in which the voltages $\bar{V}_c$ and $\bar{V}_d$ are respectively equal to $-\bar{V}_a$ and $-\bar{V}_b$. Thus in the phasor diagram $\bar{V}_c$ and $\bar{V}_d$ are mirror images of $\bar{V}_a$ and $\bar{V}_b$. Appearing in Fig. 10-8(a) is the schematic representation of an unbalanced two-phase system. Keep in mind that for a balanced two-phase system $|\bar{V}_a| = |\bar{V}_b|$; also these two quantities are exactly out of phase by 90 degrees. If either condition is not satisfied, the system is unbalanced. Depicted in Fig. 10-8(b) is the schematic diagram of the corresponding symmetrical four-phase system. The use of any two successive phases yields a two-phase system.

What are the balanced components that can now be used to represent an unbalanced two-phase system of voltages? Do we need as many components as

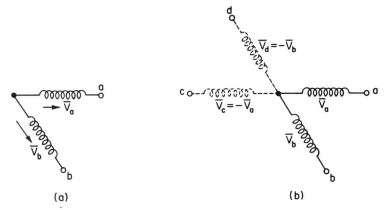

(a)   (b)

Figure 10-8 Two-phase system as a special case of the four-phase system: (a) unbalanced two-phase system; (b) unbalanced symmetrical four-phase system.

were required to represent the unbalanced four-phase system? The answers to these questions follow immediately by treating the two-phase system in terms of the symmetrical four-phase system and then applying Eqs. (10-1) and (10-19) to (10-21). Introducing $\bar{V}_c = -\bar{V}_a$ and $\bar{V}_d = -\bar{V}_b$ into Eq. (10-1) leads to the conclusion that the zero-sequence voltage is zero. This is entirely consistent with the fact that, as a symmetrical system, the resultant of the four-phase voltages must be identically equal to zero. Also, the insertion of these same quantities into Eq. (10-19) discloses that the value of the positive-sequence component of phase a is given by

$$\bar{V}_1 = \tfrac{1}{4}(\bar{V}_a + j\bar{V}_b + \bar{V}_a + j\bar{V}_b)$$

or

$$\boxed{\bar{V}_1 = \tfrac{1}{2}(\bar{V}_a + j\bar{V}_b)} \tag{10-30}$$

Similarly, by Eq. (10-20) the negative-sequence component is found to be

$$\bar{V}_2 = \tfrac{1}{4}(\bar{V}_a - j\bar{V}_b + \bar{V}_a - j\bar{V}_b)$$

or

$$\boxed{\bar{V}_2 = \tfrac{1}{2}(\bar{V}_a - j\bar{V}_b)} \tag{10-31}$$

The value of $\bar{V}_3$ is determined from Eq. (10-21). Thus

$$\bar{V}_3 = \tfrac{1}{4}(\bar{V}_a - \bar{V}_b - \bar{V}_a - \bar{V}_b) = 0 \tag{10-32}$$

On the basis of the foregoing analysis we reach the conclusion that *any unbalanced two-phase system can be completely represented by a two-phase balanced positive-sequence component and a two-phase balanced negative-sequence component.* The unbalanced phase voltages can then be expressed simply as

$$\boxed{\begin{aligned} \bar{V}_a &= \bar{V}_1 + \bar{V}_2 \\ \bar{V}_b &= a\bar{V}_1 + a^3\bar{V}_2 = -j\bar{V}_1 + j\bar{V}_2 \end{aligned}}$$

$$(10\text{-}33)$$
$$(10\text{-}34)$$

where $\bar{V}_1$ and $\bar{V}_2$ are given by Eqs. (10-30) and (10-31).

Example 10-2

An unbalanced two-phase voltage system is described by

$$\bar{V}_a = 120\,\underline{/0^\circ} = 120 + j0 \tag{10-35}$$

$$\bar{V}_b = 75\,\underline{/-60^\circ} = 37.5 - j65 \tag{10-36}$$

Find the positive- and negative-sequence components that can be used to represent the unbalanced voltage system.

Solution   The positive-sequence component of phase a is found from Eq. (10-30) to be

$$\bar{V}_1 = \tfrac{1}{2}(\bar{V}_a + j\bar{V}_b) = \tfrac{1}{2}(120 + j0 + j37.5 + j65)$$

$$= 9.25 + j18.8 = 94.4\underline{/11.5°} \qquad (10\text{-}37)$$

Figure 10-9(a) depicts a simple graphical procedure for finding $\bar{V}_1$. It first requires that $\bar{V}_b$ be rotated by 90° in the positive direction to yield $j\bar{V}_b$, then finding the phasor sum of $\bar{V}_a$ and $j\bar{V}_b$, and finally taking one-half of this resultant. The desired quantity is indicated by OA.

The analytical determination of the negative-sequence component follows from Eq. (10-31). Thus

$$\bar{V}_2 = \tfrac{1}{2}(\bar{V}_a - j\bar{V}_b) = \tfrac{1}{2}(120 - j0 - j37.5 - 65)$$

$$= 27.5 - j18.8 = 33.3\underline{/-34.4°} \qquad (10\text{-}38)$$

The graphical determination of $\bar{V}_1$ is illustrated in Fig. 10-9(b). It first requires that $\bar{V}_b$ be rotated in the negative direction by $-90°$ to yield $-j\bar{V}_b$, then adding $\bar{V}_a$ and $-j\bar{V}_b$, and finally taking one-half of this sum. The desired quantity is identified as OB and is equivalent to the result expressed in Eq. (10-38).

The complete expressions for the phase voltages in terms of the positive- and negative-sequence components are then

$$\bar{V}_a = \bar{V}_1 + \bar{V}_2 = 94.4\underline{/11.5°} + 33.3\underline{/-34.4°} \qquad (10\text{-}39)$$

$$\bar{V}_b = a\bar{V}_1 + a^3\bar{V}_2 = -j\bar{V}_1 + j\bar{V}_2 = 94.4\underline{/-78.5°} + 33.3\underline{/55.6°} \qquad (10\text{-}40)$$

The graphical representation of the last two equations appears in Fig. 10-9(c).

The results embodied in Eqs. (10-33) and (10-34) lead to a very useful and interesting physical interpretation of the analysis of the unbalanced two-phase motor. Each phase voltage consists of a component of a balanced two-phase set, one positive and the other negative. The positive-sequence components in each phase, $\bar{V}_1$ and $a\bar{V}_1$, can be considered to cooperate in creating a revolving magnetic field that interacts with the rotor winding to produce a positive-sequence

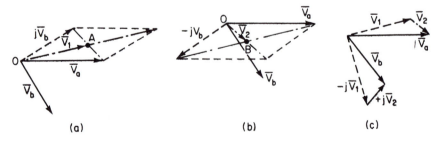

      (a)   (b)   (c)

Figure 10-9   Phasor diagrams for Example 10-2: (a) graphical determination of the positive-sequence component; (b) graphical determination of the negative-sequence component; (c) unbalanced voltages with balanced components.

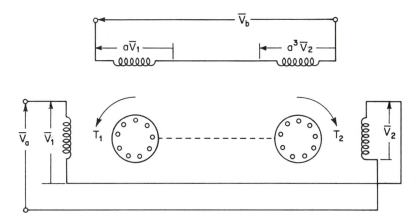

Figure 10-10   Representation of the unbalanced two-phase servomotor in terms of positive- and negative-sequence voltages.

torque T_1. At the same time the negative-sequence components in each phase, $\bar{V}_2$ and $a^3\bar{V}_2$, can be considered to combine in creating a revolving magnetic field of reversed direction that interacts with the rotor winding to produce a negative-sequence torque T_2. Of course the rotor responds to the resultant torque, which is the difference between T_1 and T_2. A schematic representation of this situation is depicted in Fig. 10-10. Note that for convenience the rotor is treated as consisting of two rotors that are mechanically coupled. The positive-sequence voltage set—$\bar{V}_1$ and $a\bar{V}_1$—produces T_1, and the negative-sequence set produces T_2. All the known techniques of balanced operation can now be used to determine T_1 and also T_2, and the results can then be superposed to obtain the resultant. Unbalanced operation is thereby treated entirely in terms of the techniques of balanced operation. Accordingly, no new theory is needed to handle the two-phase motor beyond this point.

10-3 EQUIVALENT CIRCUIT AND PERFORMANCE BASED ON SYMMETRICAL COMPONENTS

The equivalent circuit of the unbalanced two-phase motor actually consists of two parts, one representing the motor response to the positive-sequence voltage set and the other representing the response to the negative-sequence set. Because each voltage sequence is a balanced set, the equivalent circuit configuration expressed on a per-phase basis is similar to that shown in Fig. 4-7(f). The circuit is repeated in Fig. 10-11(a). The applied stator voltage for this case is the positive-sequence voltage $\bar{V}_1$. If the main winding (phase a) of the motor is assumed identical with the control winding (phase b), then the equivalent circuit for each winding is identical. In this chapter we restrict attention to motors that have identical stator phase windings. However, in those instances where this condition

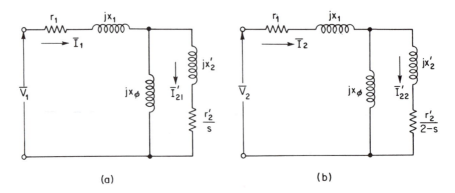

Figure 10-11 Two-part equivalent circuit of the two-phase servomotor with unbalanced applied voltages: (a) equivalent circuit per phase of the balanced positive-sequence set; (b) the per-phase equivalent circuit of the negative-sequence balanced set.

is not met, the use of a suitable transformer can be made to bring about the balanced condition.

The parameters appearing in the equivalent circuit of Fig. 10-11(a) have the same meanings as described in Chapter 4. The quantity r_1 denotes the stator winding resistance per phase. For the motor with balanced windings this represents the winding resistance of either phase. The total magnetizing reactance per phase is denoted by x_ϕ, and the total leakage reactances per phase of the stator and rotor windings are respectively denoted by x_1 and x_2'. The prime notation indicates that the last quantity is referred to the stator winding. The total rotor resistance per phase is identified as r_2'. Use of the quantity r_2'/s in the equivalent circuit implies that the rotor travels in the direction of the positive-sequence rotating field. Finally, note that a core loss resistor is not included. Hence the assumption throughout is that core loss is negligible.

The equivalent circuit of the motor in response to the balanced negative-sequence set is depicted in Fig. 10-11(b). Each circuit parameter has the same meaning as it has for the equivalent circuit of the positive-sequence set. By superposition the negative-sequence voltage set is assumed to be applied to the same stator windings and producing its own reactions in the same rotor winding. This is the meaning that is intended in the schematic diagram of Fig. 10-10. There is really only one pair of stator coils and one rotor, but the effect is the same as if there were two identical pairs of stator coils and two identical rotors mechanically coupled. A comparison of Fig. 10-11(b) with 10-11(a) shows differences occurring in two places. One, the stator applied voltage now has the value $\bar{V}_2$ instead of $\bar{V}_1$. Two, the slip of the actual rotor rotation with respect to the revolving field of the negative-sequence voltage set is $2 - s$ instead of s. Keep in mind that even with unbalanced voltages the rotor can revolve at only *one* speed. The slip of the rotor with respect to the positive-sequence field is then clearly

$$s = \frac{n_s - n}{n_s} \tag{10-41}$$

Because the negative-sequence voltage set establishes a field that revolves in the opposite direction to that of the positive sequence, it follows that the slip of the rotor with respect to the negative-sequence field, s_2 is given by

$$s_2 = \frac{n_s - (-n)}{n_s} = 1 + \frac{n}{n_s} \tag{10-42}$$

In the interest of dealing solely in terms of s—the slip with respect to the positive-sequence field—use is made of Eq. (10-41) in Eq. (10-42) to yield

$$\boxed{s_2 = 1 + \frac{n}{n_s} = 2 - s} \tag{10-43}$$

Therefore in all future work that involves the use of positive- and negative-sequence fields, the rotor slip with respect to the negative-sequence field will always be denoted by $2 - s$.

The computation of performance can be readily determined through the use of our knowledge of the operation of a polyphase induction motor in response to balanced applied voltages. Let us find the expression for the resultant torque and the individual phase currents when the motor operates at a slip s with unbalanced stator voltages. First we find the behavior of the motor in response to the positive-sequence voltages. This is easily accomplished by means of the equivalent circuit of Fig. 10-11(a). In a typical problem $\bar{V}_1$, s, and the circuit parameters are specified. As the initial step the input impedance occurring with the positive-sequence voltage is found. Calling this impedance Z_1 we have

$$\bar{Z}_1 = \bar{z}_1 + \bar{Z}_{p1} = r_1 + jx_1 + \frac{jx_\phi[(r_2'/s) + jx_2']}{(r_2'/s) + j(x_\phi + x_2')} \tag{10-44}$$

where Z_{p1} obviously denotes the impedance of the parallel arrangement of the circuit elements. The second subscript, 1, identifies the positive sequence. With the positive-sequence input impedance known, the positive-sequence component of the input current in phase a is given by

$$\bar{I}_1 = \frac{\bar{V}_1}{\bar{Z}_1} \tag{10-45}$$

The rotor current produced by the action of the positive-sequence rotating field follows from

$$\bar{I}_{21}' = \frac{\bar{I}_1 \bar{Z}_{p1}}{(r_2'/s) + jx_2'} \tag{10-46}$$

where the second subscript denotes the positive sequence. A little thought discloses that an alternative expression for $\bar{I}_{21}'$ can be had by applying the current divider rule for the parallel circuit. However, Eq. (10-46) is usually preferred because the quantities on the right side are frequently available in a convenient form.

The gap power associated with the positive-sequence rotating field, P_{g1}, is then found from

$$P_{g1} = q(I'_{21})^2 \frac{r'_2}{s} = 2(I'_{21})^2 \frac{r'_2}{s} \tag{10-47}$$

The corresponding developed torque is then

$$T_1 = \frac{P_{g1}}{\omega} \qquad \text{N-m} \tag{10-48}$$

where ω is the synchronous angular frequency expressed in radians per second.

The performance of the motor in response to the negative-sequence set is found in a similar fashion. Thus the negative-sequence input impedance can be expressed by

$$\bar{Z}_2 = \bar{z}_1 + \bar{Z}_{p2} = r_1 + jx_1 + \frac{jx_\phi[(r'_2/2 - s) + jx'_2]}{(r'_2/2 - s) + j(x_\phi + x'_2)} \tag{10-49}$$

where $\bar{Z}_{p2}$ is the impedance of the parallel arrangement appearing in the circuit of Fig. 10-11(b). The negative-sequence component of the current in stator phase a becomes

$$\bar{I}_2 = \frac{\bar{V}_2}{\bar{Z}_2} \tag{10-50}$$

The associated rotor current takes on the value

$$\bar{I}'_{22} = \frac{\bar{I}_2 \bar{Z}_{p2}}{(r'_2/2 - s) + jx'_2} \tag{10-51}$$

so that the expression for the power transferred across the air gap by the negative-sequence flux field becomes

$$P_{g2} = q(I'_{22})^2 \frac{r'_2}{2 - s} = 2(I'_{22})^2 \frac{r'_2}{2 - s} \tag{10-52}$$

and

$$T_2 = \frac{P_{g2}}{\omega} \tag{10-53}$$

The resultant torque produced by the simultaneous effects of the positive- and negative-sequence voltages expressed in newton-meters is

$$\boxed{T = T_1 - T_2 = \frac{1}{\omega}(P_{g1} - P_{g2})} \tag{10-54}$$

For servomotors operating as unbalanced two-phase motors used in instrument servomechanisms the value of torque in newton-meters is a very small quantity. It is customary in such instances to refer rather to the torque in terms of its proportional quantity, power. Hence it is not uncommon for control engineers to refer to a servomotor torque as "so many units of synchronous watts." We too adopt this practice.

The resultant line current in phase a is

$$\boxed{\bar{I}_a = \bar{I}_1 + \bar{I}_2} \tag{10-55}$$

and that in phase b is similarly obtained as

$$\boxed{\bar{I}_b = -j\bar{I}_1 + j\bar{I}_2} \tag{10-56}$$

Information about the power factor of each stator phase current is implied in the polar expression of the phase currents. In normal applications of the servomotor, efficiency is not very important and so this matter receives no further attention here.

Example 10-3

A 5-W, 60-Hz, 120-V, two-pole, two-phase servomotor has the following parameters:

$$r_1 = 285\ \Omega, \qquad r_2' = 850\ \Omega$$

$$x_1 = 60\ \Omega, \qquad x_2' = 60\ \Omega$$

$$x_\phi = 995\ \Omega$$

When it is operating at a slip of 0.6, determine (a) the resultant torque in synchronous watts and (b) the stator phase currents. Assume the servomotor operates with the following unbalanced two-phase voltages: $\bar{V}_a = 120\ \underline{/0°}$ and $\bar{V}_b = 75\ \underline{/-60°}$.

Solution (a) The positive- and negative-sequence voltages for the specified unbalanced voltages are computed in Example 10-2 to be as follows:

$$\bar{V}_1 = 94.4\ \underline{/11.5°} \quad \text{and} \quad \bar{V}_2 = 33.3\ \underline{/-34.4°}$$

We first compute the motor response to the positive-sequence set. Thus

$$\bar{Z}_{p1} = \frac{j995[(850/0.6) + j60]}{(850/0.6) + j1055} \approx \frac{j955(1415)}{1760\ \underline{/36.8°}} = 800\ \underline{/53.2°} \tag{10-57}$$

and

$$\bar{Z}_1 = \bar{z}_1 + \bar{Z}_{p1} = 285 + j60 + 481 + j644 = 1038\ \underline{/42.6°} \tag{10-58}$$

Hence

$$\bar{I}_1 = \frac{\bar{V}_1}{\bar{Z}_1} = \frac{94.4\ \underline{/11.5°}}{1038\ \underline{/42.6°}} = 0.091\ \underline{/-31.1°} \tag{10-59}$$

and

$$\bar{I}'_{21} = \frac{\bar{I}_1 \bar{Z}_{p1}}{1415 + j60} = \frac{0.091 \underline{/-31.1°} \, (800 \underline{/53.2°})}{1415} = 0.0514 \underline{/22.1°} \qquad (10\text{-}60)$$

The positive-sequence gap power then becomes

$$P_{g1} = 2(0.0514)^2 1415 = 2(3.7) = 7.4 \text{ W} \qquad (10\text{-}61)$$

The response to the negative-sequence voltage set is computed to be as follows:

$$\bar{Z}_{p2} = \frac{j995[(850/1.4) + j60]}{(850/1.4) + j(995 + 60)} \approx \frac{j995(606)}{1218 \underline{/60.1°}} = 496 \underline{/29.9°} \qquad (10\text{-}62)$$

$$\bar{Z}_2 = \bar{z}_1 + \bar{Z}_{p2} = 285 + j60 + 431 + j248 = 779 \underline{/23.3°} \qquad (10\text{-}63)$$

$$\bar{I}_2 = \frac{\bar{V}_2}{\bar{Z}_2} = \frac{3.3 \underline{/-34.4°}}{779 \underline{/23.3°}} = 0.042 \underline{/-57.7°} \qquad (10\text{-}64)$$

$$\bar{I}'_{22} = \frac{\bar{I}_2 \bar{Z}_2}{606 + j60} = \frac{0.042 \underline{/-57.7°} \, (496 \underline{/29.9°})}{606} = 0.034 \underline{/-33.5°} \qquad (10\text{-}65)$$

$$P_{g2} = 2(0.0344)^2 \frac{r'_2}{2 - s} = 2(0.0344)^2 (606) = 2(0.716) = 1.43 \text{ W} \qquad (10\text{-}66)$$

Hence the resultant gap power is

$$P_g = P_{g1} - P_{g2} = 7.4 - 1.43 = 5.97 \text{ synchronous watts} \qquad (10\text{-}67)$$

(b) By Eq. (10-55) the expression for the current in stator phase a is

$$\bar{I}_a = \bar{I}_1 + \bar{I}_2 = 0.09 \underline{/-31.1°} + 0.042 \underline{/-57.7°}$$

$$= 0.0778 - j0.047 + 0.0224 - j0.0355 \qquad (10\text{-}68)$$

$$= 0.1002 - j0.0825 = 0.13 \underline{/-39.5°}$$

Similarly by Eq. (10-56) the phase current in the control winding b is

$$\bar{I}_b = -j\bar{I}_1 + j\bar{I}_2 = -j0.0778 - 0.047 + j0.0224 + 0.0355$$

$$= -0.0115 - j0.0554 = 0.0566 \underline{/-101.7°} \qquad (10\text{-}69)$$

Attention is next turned to the derivation of an oftentimes useful equation for computing the resulting torque. The insertion of Eqs. (10-45) and (10-46) into Eq. (10-47) leads to the following expression for the gap power of the positive-sequence field

$$P_{g1} = 2 \left(\frac{V_1}{Z_1}\right)^2 \left[\frac{Z_{p1}}{(r'_2/s) + jx_2}\right]^2 \left(\frac{r'_2}{s}\right) \qquad (10\text{-}70)$$

Let us now determine the expression for the gap power when the motor is operated with balanced two-phase voltages, i.e., $V_a = V_b$ and each displaced from the other by 90 degrees. In this instance the phase current is given by

$$\bar{I}_a = \frac{\bar{V}_a}{\bar{Z}_1} \tag{10-71}$$

where $\bar{Z}_1$ is described by Eq. (10-44). Moreover the rotor current expression is

$$\bar{I}_2' = \frac{\bar{I}_a \bar{Z}_{p1}}{(r_2'/s) + jx_2'} \tag{10-72}$$

where $\bar{Z}_{p1}$ is specified as before. The expression for the total gap power for balanced operation, P_g, becomes therefore

$$P_g = 2(I_2')^2 \frac{r_2'}{s} = 2\left(\frac{V_a}{Z_1}\right)^2 \left(\frac{Z_{p1}}{(r_2'/s) + jx_2'}\right)^2 \left(\frac{r_2'}{s}\right) \tag{10-73}$$

By formulating the ratio of Eq. (10-70) to Eq. (10-73) there results

$$P_{g1} = \left(\frac{V_1}{V_a}\right)^2 P_g \tag{10-74}$$

By proceeding in a similar fashion it can be shown that

$$P_{g2} = \left(\frac{V_2}{V_a}\right)^2 P_g' \tag{10-75}$$

where P_g' denotes the *gap power computed for balanced two-phase operation at a slip of* $2 - s$. The resultant gap power is accordingly expressed as

$$P_{gr} = P_{g1} - P_{g2} = \left(\frac{V_1}{V_a}\right)^2 P_g - \left(\frac{V_2}{V_a}\right)^2 P_g' \tag{10-76}$$

Correspondingly, the resultant torque can then be written as

$$T = \frac{1}{\omega}\left[\left(\frac{V_1}{V_a}\right)^2 P_g - \left(\frac{V_2}{V_a}\right)^2 P_g'\right] \tag{10-77}$$

or

$$\boxed{T = \left(\frac{V_1}{V_a}\right)^2 T_{B1} - \left(\frac{V_2}{V_a}\right)^2 T_{B2}} \tag{10-78}$$

where $T_{B1} = P_g/\omega$ and denotes the developed torque computed at slip s for balanced two-phase operation at rated voltage, and $T_{B2} = P_g'/\omega$ and denotes the developed torque computed at slip $2 - s$ for balanced two-phase operation at rated voltage.

The usefulness of Eq. (10-78) lies in the fact that the resultant torque at any slip and for any condition of unbalanced two-phase voltages may be computed in terms of the torque for full-voltage balanced operation. This information is usually supplied by the motor manufacturer. Hence the torque-speed characteristic for various values of the control winding voltage is readily determined from Eq. (10-78) without resorting to the equivalent-circuit computational procedure. This matter is further elaborated in the next section.

Example 10-4

Compute the resultant torque of the servomotor of Example 10-3 by employing Eq. (10-78).

Solution   Under conditions of balanced two-phase operation the phase current becomes

$$\bar{I}_a = \frac{\bar{V}_a}{\bar{Z}_1} = \frac{120}{1038} \underline{/-42.6°} = 0.116 \underline{/-42.6°} \qquad (10\text{-}79)$$

The corresponding rotor current is

$$\bar{I}_2' = \frac{\bar{I}_a \bar{Z}_{p1}}{(r_2'/s) + jx_2'} \approx \frac{0.116 \underline{/-42.6°} \; (800 \underline{/53.2°})}{1415} = 0.0655 \underline{/10.6°}$$

$$\therefore \quad P_g = 2(I_2')^2 \frac{r_2'}{s} = 2(0.0655)^2(1415) = 12.1 \text{ W}$$

Therefore the gap power of the positive-sequence set becomes by Eq. (10-61)

$$P_{g1} = \left(\frac{V_1}{V_a}\right)^2 P_g = \left(\frac{94.4}{120}\right)^2 (12.1) = 7.45 \qquad (10\text{-}80)$$

A comparison with Eq. (10-61) indicates favorable agreement.

To find P_g', the motor is assumed to be operating at a slip of $2 - s = 1.4$ with balanced two-phase voltages applied. Then

$$\bar{I}_a = \frac{\bar{V}_a}{\bar{Z}_2} = \frac{120}{779 \underline{/23.3°}} = 0.152 \underline{/-23.3°} \qquad (10\text{-}81)$$

and

$$\bar{I}_2' = \frac{\bar{I}_a \bar{Z}_{p2}}{(r_2'/2 - s) + jx_2'} \approx \frac{0.152 \underline{/-23.3°} \; (496 \underline{/29.9°})}{606} = 0.1245 \underline{/6.6°}$$

$$P_g' = 2(0.1245)^2 \frac{r_2'}{2 - s} = 2(0.0155)606 = 18.8 \text{ W} \qquad (10\text{-}82)$$

Hence by Eq. (10-75) the gap power of the negative-sequence field is

$$P_{g2} = \left(\frac{V_2}{V_a}\right)^2 P_g' = \left(\frac{33.3}{120}\right)^2 (18.8) = 1.45 \qquad (10\text{-}83)$$

which compares favorably with Eq. (10-66).

The resultant torque is therefore

$$T = \frac{1}{\omega} (7.45 - 1.45) = \frac{1}{375} (6) = 0.0159 \text{ N-m} \qquad (10\text{-}84)$$

10-4 SERVOMOTOR TORQUE-SPEED CURVES

The torque-speed curve of a conventional two-phase induction motor supplied by a balanced two-phase voltage has a shape similar to that of the three-phase induction motor. This characteristic is depicted as curve a in Fig. 10-12. The use of such a motor in control systems is intolerable because of the positive slope that prevails over most of the operating speed range. The positive slope represents negative damping in the control system, which in turn can lead to a condition of instability. Therefore for control systems applications the motor must be modified in a way that ensures positive damping over the full speed range. A convenient way to achieve this result is to design the rotor with very high rotor resistance. The torque-speed characteristic then assumes the shape shown by curve b of Fig. 10-12. It is sketched over a range of slip from zero to 2 because of the pertinence of that portion of the curve lying between $s = 1$ and $s = 2$ when negative-sequence voltages exist in unbalanced operation. Whenever the two-phase induction motor includes the high rotor resistance design feature, it is referred to as a two-phase *servomotor*.

In many applications of the servomotor in feedback control systems, phase a is energized with fixed rated voltage (often called the *reference voltage*), while phase b is energized with a varying control voltage that is usually obtained from a preceding amplifier stage. Moreover, the arrangement in these configurations is such that the control voltage is frequently adjusted to be exactly 90° out of phase with the voltage applied to phase a. In these instances unbalanced operation is really an unbalance in voltage magnitude only. If we proceed under the assumption that the reference voltage $\bar{V}_a$ and the control voltage $\bar{V}_b$ are always 90° apart in phase and if we define the quantity ρ such that

$$\rho \equiv \frac{V_b}{V_a} \tag{10-85}$$

then the phasor expression for the control voltage becomes

$$\bar{V}_b = -j\rho\bar{V}_a \tag{10-86}$$

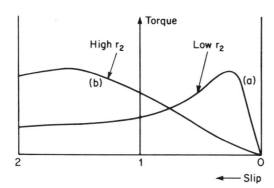

Figure 10-12 Torque-speed curves: (a) for the balanced two-phase induction motor with low rotor resistance; (b) for the same motor with high rotor resistance.

The $-j$ factor accounts for the 90° lag between the two stator phase voltages, and ρ expresses the magnitude of V_b as a per-unit ratio of the reference voltage. Upon inserting Eq. (10-86) into Eq. (10-30), the expression for the positive-sequence voltage becomes

$$\bar{V}_1 = \tfrac{1}{2}(\bar{V}_a + j\bar{V}_b) = \frac{\bar{V}_a}{2}(1 + \rho) \qquad (10\text{-}87)$$

Similarly

$$\bar{V}_2 = \tfrac{1}{2}(\bar{V}_a - j\bar{V}_b) = \frac{\bar{V}_a}{2}(1 - \rho) \qquad (10\text{-}88)$$

Appearing in Fig. 10-13 is a set of actual torque-speed curves for a two-phase servomotor corresponding to various values of the control voltage expressed as a fraction (ρ) of the reference voltage. The curve marked 115 V (or $\rho = 1$) represents the characteristic for balanced operation. However, note that, as the control voltage is reduced, the associated torque-speed characteristic becomes severely restricted. If information about the torque-speed characteristic of the servomotor is supplied by the manufacturer for balanced rated operation over the full range of slip from zero to 2, then the torque-speed characteristic for any value of ρ can be easily derived by means of Eqs. (10-87), (10-88), and (10-78). The procedure is illustrated in Fig. 10-14.

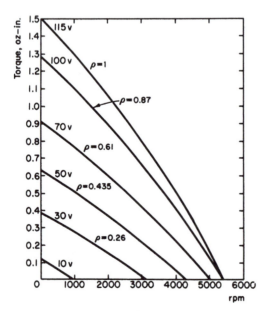

Figure 10-13 Torque-speed curves of the two-phase servomotor for various values of ρ.

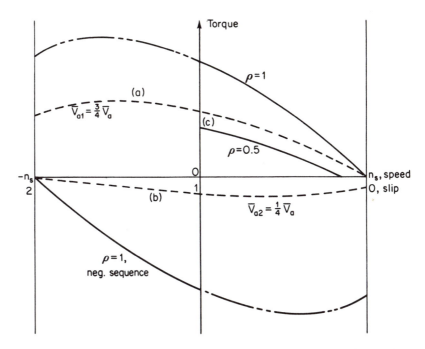

Figure 10-14   Graphical determination of the servomotor torque-speed curves for $\rho = 0.5$.

Consider that the servomotor operates with the control voltage fixed at one-half the value of the reference voltage, i.e., $\rho = \frac{1}{2}$. Then by Eq. (10-87) the positive-sequence voltage has a magnitude of

$$\bar{V}_1 = \frac{\bar{V}_a}{2}(1 + \rho) = \frac{\bar{V}_a}{2}(1 + \tfrac{1}{2}) = \tfrac{3}{4}\bar{V}_a \tag{10-89}$$

and the negative-sequence voltage is found by Eq. (10-88) to be

$$\bar{V}_2 = \frac{\bar{V}_a}{2}(1 - \rho) = \frac{\bar{V}_a}{2}(1 - \tfrac{1}{2}) = \tfrac{1}{4}\bar{V}_a \tag{10-90}$$

Inserting these results into Eq. (10-78) leads to the following expression for the resultant torque:

$$T_r = \left(\frac{V_1}{V_a}\right)^2 T_{B1} - \left(\frac{V_2}{V_a}\right)^2 T_{B2} = \tfrac{9}{16}T_{B1} - \tfrac{1}{16}T_{B2} \tag{10-91}$$

The quantity $\tfrac{9}{16}T_{B1}$ is readily obtained over the entire speed range by taking nine-sixteenths of the ordinate value of the characteristic for full-voltage balanced operation at all values of slip. This result is identified as curve a in Fig. 10-14. In like manner the quantity $\tfrac{1}{16}T_{B2}$ is easily determined over the entire speed range by taking one-sixteenth of the value of the characteristic supplied by the manufacturer for full-voltage balanced operation. This is represented by curve b in Fig. 10-14 and is drawn below the abscissa axis because the negative-sequence torque is opposed to the positive-sequence torque. The resultant torque-speed characteristic over the useful speed range is found by adding curves a and b, which leads to curve c. So long as the servomotor is operated with a voltage $\bar{V}_b = -j0.5\,\bar{V}_a$, curve c describes its operating behavior. By repeating the foregoing procedure the resultant torque-speed characteristic for other values of ρ can be determined. It is interesting to note that these characteristics are obtainable without the necessity of making repeated calculations in conjunction with the equivalent circuit as outlined in Example 10-3. However, the alternative procedure described here does require having the torque-speed characteristic for balanced operation at full voltage over the slip range from 0 to 2.

Another useful characteristic of the servomotor is a description of the developed torque as a function of the control voltage with speed as a parameter. These curves, too, can be derived with the use of Eq. (10-78) and the availability of the

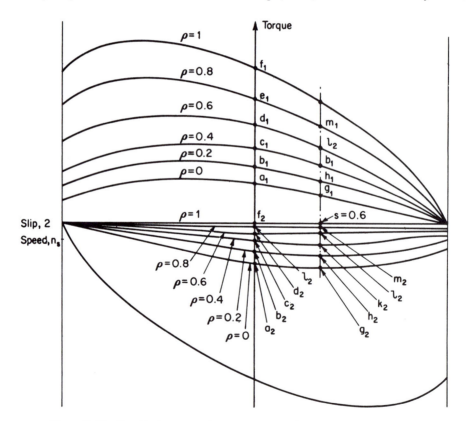

Figure 10-15 Graphical construction for deriving the curves of Fig. 10-16. The abscissa axis is slip or speed.

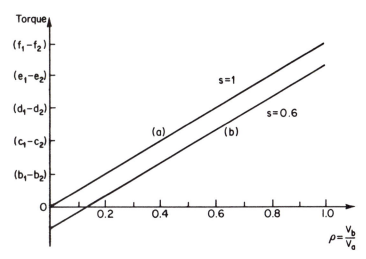

Figure 10-16   Variation of resultant torque versus control voltage with slip as a parameter.

torque-speed characteristic for balanced operation at rated voltage. Appearing in Fig. 10-15 are the torque-speed characteristics corresponding to various values of ρ for both the positive- and negative-sequence voltages. The curves for $\rho = 0$ are identical but merely reversed in position consistent with their association with the positive- and negative-sequence voltage sets. As ρ is increased toward unity the positive-sequence contribution becomes increasingly larger than it is for $\rho = 0$, while the corresponding negative-sequence contribution becomes increasingly smaller than it is for $\rho = 0$. To obtain a plot of resultant torque as a function of control voltage expressed as a ratio of the rated reference voltage at zero speed, it is merely necessary to subtract the negative-sequence contribution from the positive-sequence torque. Thus at zero speed (or $s = 1$) the resultant torque is $a_1 - a_2 = 0$ for $\rho = 0$. For $\rho = 0.2$ the resultant torque is $b_1 - b_2$, and this quantity is greater than zero. The net torque for $\rho = 0.4$ is $c_1 - c_2$, and it should be apparent that this quantity is greater than $b_1 - b_2$. The complete plot is depicted in Fig. 10-16 as curve (*a*). Repeating the procedure at a second value of slip leads to curve (*b*). Note the region of negative torque.

PROBLEMS

10-1. An unbalanced four-phase system is described by the following voltage set:

$$\bar{V}_a = 110\underline{/15°}$$

$$\bar{V}_b = 110\underline{/-50°}$$

$$\bar{V}_c = 110\underline{/-135°}$$

$$\bar{V}_d = 110\underline{/90°}$$

(a) What is the value of the zero-sequence voltage?
(b) Find the phasor expression of the positive-sequence voltage set, and draw the corresponding phasor diagram.
(c) Repeat part (b) for the negative-sequence set.
(d) Repeat part (b) for the balanced single-phase set.
(e) Draw the complete phasor diagram, showing how the individual components combine to yield the given unbalanced voltage set.

10-2. A four-phase voltage system is described by the following phasors:

$$\bar{V}_a = 110\,\underline{/15°}$$

$$\bar{V}_b = 60\,\underline{/-50°}$$

$$\bar{V}_c = 140\,\underline{/-135°}$$

$$\bar{V}_d = 40\,\underline{/90°}$$

(a) Determine that set of balanced components that can replace the unbalanced system.
(b) Draw the complete phasor diagram of this four-phase system, showing all components.

10-3. Show the derivation that leads to Eq. (10-20).

10-4. Show the derivation that leads to Eq. (10-21).

10-5. The positive-sequence component of phase a of a four-phase voltage system is known to be equal to $60\,\underline{/-30°}$ V
(a) What is the phasor expression of the positive-sequence component of phase c?
(b) What is the positive-sequence component of phase b?
(c) Draw the complete phasor diagram for the positive-sequence set.

10-6. Under what conditions is it possible for an unbalanced four-phase voltage system to yield a resultant voltage that is zero?

10-7. One phase of a four-phase voltage system has the following components:

$$\bar{V}_1 = 60\,\underline{/-60°}$$

$$\bar{V}_2 = 30\,\underline{/-90°}$$

$$\bar{V}_3 = 20\,\underline{/-110°}$$

$$\bar{V}_0 = 0$$

(a) Find the phasor expression of phase b voltage.
(b) Is this a balanced four-phase system? Explain.

10-8. A four-phase system has the positive-, negative-, and zero-sequence voltages specified in Prob. 10-7. Draw to scale the complete phasor diagrams of the resulting four-phase system. A graphical solution is permissible.

10-9. A four-phase voltage system is described as follows:

$$\bar{V}_a = 100\,\underline{/0°}$$

$$\bar{V}_b = 50\,\underline{/-90°}$$

$$\bar{V}_c = 100\,\underline{/-180°}$$

$$\bar{V}_d = 50\,\underline{/90°}$$

(a) Is there a need for a balanced single-phase set when replacing this voltage system with balanced sets? Explain.

(b) Find the positive-sequence voltage set.

(c) Find the negative-sequence voltage set.

10-10. The positive-sequence component of phase a of a four-phase system is $75\underline{/-30°}$, and the negative-sequence component is $25\underline{/-30°}$. The zero-sequence and the balanced single-phase sets are both zero. Determine the expressions for the phase voltages of the original four-phase system.

10-11. A four-phase voltage system is given by:

$$\bar{V}_a = 100\underline{/0°}$$
$$\bar{V}_b = 60\underline{/-90°}$$
$$\bar{V}_c = 60\underline{/-180°}$$
$$\bar{V}_d = 60\underline{/90°}$$

(a) Determine the phasor expressions of that four-phase system which has a zero resultant.

(b) Find the positive-sequence voltage set for the given system.

(c) Is the positive-sequence set the same for the system of part (a)? Explain.

(d) Find the negative-sequence voltage set.

(e) Compute the balanced single-phase set.

(f) Draw the phasor diagram for phase b, clearly identifying each component.

10-12. Compute the zero-, positive-, and negative-sequence sets as well as the balanced single-phase set that can be used to replace the following four-phase system:

$$\bar{V}_a = 100\underline{/0°}$$
$$\bar{V}_b = 60\underline{/-65°}$$
$$\bar{V}_c = 100\underline{/-180°}$$
$$\bar{V}_d = 60\underline{/115°}$$

What generalization can you draw concerning the component sets that make up the original voltage system?

10-13. Find the unbalanced two-phase system that is represented by a positive-sequence component of $95.2\underline{/14.6°}$ and a negative-sequence component of $31\underline{/40.6°}$.

10-14. An unbalanced two-phase voltage system is described by:

$$\bar{V}_a = 120\underline{/0°}$$
$$\bar{V}_b = 80\underline{/-53°}$$

(a) Find the positive-sequence voltage set of this system.

(b) What is the negative-sequence set?

(c) Draw the complete phasor diagram of this two-phase system, showing each positive- and negative-sequence component.

10-15. Repeat Prob. 10-14 for the case where $\bar{V}_a = 120\underline{/0°}$ and $\bar{V}_b = 80\underline{/-90°}$.

10-16. A two-phase, four-pole servomotor is energized by an auxiliary winding voltage that is 90° out of phase with the main winding voltage. The main winding voltage rating at 60 Hz is 75 V. The motor parameters are as follows:

$$r_1 = 50 \ \Omega, \qquad r_2' = 100 \ \Omega$$

$$x_1 = 120 \ \Omega, \qquad x_\phi = 100 \ \Omega$$

The rotor leakage reactance is negligible. Compute the developed torque at a slip of 0.4 when the auxiliary voltage is 37.5 V.

10-17. Appropriate no-load tests on a two-phase servomotor rated at 115 V, 60 Hz, two poles, reveal the following parameters:

$$r_1 = 302 \ \Omega, \qquad r_2' = 1380 \ \Omega$$
$$x_\phi = 695 \ \Omega$$
$$x_1 = 385 \ \Omega, \qquad x_2' = 385 \ \Omega$$

This servomotor is operated with $\bar{V}_a = 115 \underline{/0°}$ and $\bar{V}_b = 80 \underline{/-90°}$, and a rotor slip of 0.25.

  (a) Draw the appropriate equivalent circuit for the positive- and negative-sequence voltage sets. Show the proper applied voltage in each case.
  (b) Compute the developed torque, expressed in synchronous watts and newton-meters.
  (c) What is the rms value of the current that flows through the control winding (phase b)?

10-18. Refer to the servomotor of Prob. 10-17 and assume that it continues to operate at a slip of 25%.
  (a) Compute the torque developed by this motor when it operates as a balanced two-phase motor, i.e., $\bar{V}_a = 115 \underline{/0°}$ and $\bar{V}_b = 115 \underline{/-90°}$.
  (b) Compute the developed torque when the motor operates as a balanced two-phase motor at a slip of 1.75 per unit.
  (c) By employing the results of parts (a) and (b), compute the torque developed by this motor when $\bar{V}_a = 115 \underline{/0°}$ and $\bar{V}_b = 80 \underline{/-90°}$.

10-19. Refer to the servomotor of Prob. 10-17. When this motor operates at a slip of 0.5, the positive-sequence current is found to be $80.5 \underline{/-51.1°}$ ma and the negative-sequence current is $38.5 \underline{/-98.4°}$ ma.
  (a) Determine the voltages applied to the main and control windings.
  (b) Find the control winding current.

10-20. A two-phase servomotor has $115 \underline{/0°}$ V applied to the main winding. At a particular point of operation corresponding to a control winding voltage of $75 \underline{/-80°}$, the positive- and negative-sequence impedances are, respectively, $220 \underline{/67°}$ and $175 \underline{/57°}$.
  (a) Determine the positive- and negative-sequence components of the main and control winding currents.
  (b) Find the total values of the main and control winding currents.

10-21. The servomotor of Prob. 10-17 is operated with the main and control winding voltages always in quadrature.
  (a) Compute the stall torque when the control winding voltage equals the main winding voltage, i.e., $\rho = 1$.
  (b) Determine the stall torque for $\rho = 0.5$.

(c) Assuming the no-load speed is 3400 rpm, calculate the ratio of stall torque to speed, in radians per second, for $\rho = 1$.

(d) Of what usefulness is the quantity computed in part (c)?

10-22. A two-phase, two-pole, ac servomotor equipped with a drag-cup rotor has the following parameters at 60 Hz:

$$r_1 = 360 \ \Omega, \qquad r_2' = 260 \ \Omega$$

$$x_1 = 50 \ \Omega, \qquad x_2' = 50 \ \Omega$$

$$x_\phi = 890 \ \Omega$$

The main winding is identical to the control winding, and both are rated at 115 V. The source of the control winding voltage is arranged to provide a quadrature voltage at all times. Moreover, the internal impedance of this source is negligibly small.

(a) Determine the developed torque in synchronous watts for $\rho = 0.8$ and $s = 0.3$.

(b) How much power is supplied to the main winding in part (a)?

(c) What is the power in the control winding in part (a)?

10-23. (a) For the servomotor of Prob. 10-22 compute the value of the developed torque corresponding to operation with $\rho = 0.1$ and $s = 0.05$.

(b) Repeat this calculation for the servomotor of Prob. 7-17. Compare and comment.

10-24. Refer to the servomotor of Prob. 10-22.

(a) Determine the developed torque in synchronous watts for $\rho = 1$ and $s = 0.3$.

(b) Repeat part (a) for $\rho = 1$ and $s = 1.7$.

(c) Find the developed torque for $\rho = 0.7$ and a rotor slip of 30%.

10-25. For the operating condition specified in Prob. 10-17, compute the ratio of the positive-sequence flux wave to the negative-sequence flux wave. Neglect saturation.

10-26. For the operating condition specified in Prob. 10-22(a) determine the ratio of the positive-sequence flux wave to the negative-sequence flux wave. What significance can you draw from this result?

11

Stepper Motors

The explosive growth of the computer industry in recent years has meant an enormous growth too for stepper motors because these motors provide the driving force in many computer peripheral devices. They can be found, for example, driving the paper feed mechanism in line printers and printing terminals. Stepper motors are also used exclusively in floppy disk drives where they provide precise positioning of the magnetic head on the disks. The *x*- and *y*-coordinate pens in expensive plotters are driven by stepper motors. Here they offer good dynamic performance while eliminating the need for a heavy maintenance schedule associated with alternative drive schemes that employ gears, slide wires, and brushes. The stepper motor is especially suited for such applications because essentially it is a device which serves to convert input information in *digital form* to an output that is mechanical. It thereby provides a natural interface with the digital computer.

The stepper motor, however, can be found performing countless tasks outside the computer industry as well. In many commercial, military, and medical applications, the stepper motors perform such functions as mixing, cutting, stirring, metering, blending, and purging. They provide many supporting roles in the manufacture of packaged foodstuffs, commercial end products, and even the production of science fiction movies. The advantages offered by the stepper motor in these applications can easily be exploited with the use of two pieces of auxiliary equipment: the microprocessor and controlled switches in the form of transistors.

Although the stepper motor is driven by a properly sequenced set of digital signals, the motor itself exhibits the characteristics of a synchronous motor. In one of its most popular forms, it acts as a doubly excited machine, which in turn means it produces a steady-state torque at one speed.

11-1 CONSTRUCTION FEATURES

The construction of the stepper motor is quite simple. It consists of a slotted stator equipped with two or more individual coils and a rotor structure that carries no winding. The classification of the stepper motor is determined by how the rotor is designed. If it is provided with a permanent magnet attached to its shaft, it is called a permanent magnet (PM) stepper motor. If the permanent magnet is not included, it is simply classified as a reluctance-type stepper motor. Attention here is directed first to the PM stepper motor. Of course, the presence of the permanent magnet furnishes the motor with the equivalent of a constant dc excitation. Thus, when one or more of the stator coils are energized, the machine behaves as a synchronous motor.

The basic construction details of the PM stepper motor are illustrated in Fig. 11-1(a) for a four-pole stator and a five-pole rotor structure. Observe that the action of the permanent magnet on the particular orientation of the rotor structure is to magnetize each of the poles (or slots) at one end of the rotor with an N-pole polarity and each of the poles at the other end with an S-pole polarity. Moreover, the N-pole set of rotor poles is arranged to be displaced from the S-pole set of rotor poles by one-half of a pole-pitch and is readily evident by a comparison of Figs. 11-1(b) and (c). The symmetry that exists between the stator poles and each set of rotor poles makes it apparent that each rotor pole set behaves in an identical fashion. For example, if coil A-A' is assumed to be energized to yield an N-pole polarity at A and an S-pole polarity at A', then the relationship between N of the stator and S1, S2, and S5 of the rotor in Fig. 11-1(b) corresponds exactly to that of S of the stator and N1, N2, and N5 of the rotor in Fig. 11-1(c). A similar statement can be made for the remaining sets of poles.

11-2 METHOD OF OPERATION

The manner in which the PM stepper motor can be used to perform precise positioning is explained by examining the sequence of diagrams that appear in Fig. 11-2. The starting point is represented in Fig. 11-2(a) with coil A-A' energized to carry positive dc current so that the upper stator pole is given an N-pole polarity and the corresponding lower stator pole is made to behave with an S-pole polarity. Coil B-B' is assumed to be de-energized. The rotor maintains this orientation

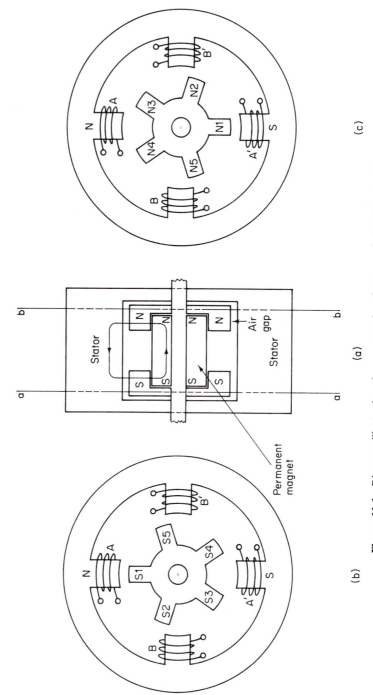

Figure 11-1 Diagram illustrating the construction features of the **PM** stepper motor: (a) axial view showing the permanent magnet attached to the rotor shaft; (b) cross-sectional view at *a-a* showing rotor poles of S-pole polarity; (c) cross-sectional view at *b-b* showing poles of N-pole polarity.

as long as the stator coil currents remain unchanged. At this position, the net torque developed by the stepper motor is zero. This conclusion is readily apparent in this case from symmetry considerations. For example, at the S-pole end of the rotor it is seen that S1 is in alignment with the N-pole of the stator. The torque produced by the action of the field distributions of the stator pole and the rotor pole S1 is dependent on the sine of the angular displacement of these two flux fields. With exact alignment the angle is clearly zero and so too is the torque between these two fields. Proceeding on the basis of superposition, we note that a torque of attraction exists between the stator N-pole field and the field associated with S5, which acts to cause counterclockwise (ccw) rotation. The magnitude of this torque is proportional to the sin 72°, which is the rotor pole-pitch angle. However, the action between the N-pole stator field and S2 is also one of attraction but it is exactly equal and opposite to that produced between the stator N-pole and S5. A similar result is obtained upon examining the situation that exists under the S-pole stator field. It is useful to note here that a difference in detail does exist in that the torques involved are of a repulsive nature rather than attractive as occurs beneath the N-pole stator field for this orientation. Because the repulsive force has its maximum effect when a rotor flux field is in exact alignment with a stator flux field, the cosine of the displacement angle between the axes of two flux fields of opposite angle must be used. Accordingly, there is a torque existing between the stator S-pole and the rotor S3 pole which is proportional to the strength of these fields as well as the cos 36°, which acts to produce a clockwise rotation. But the action between S4 and the stator S-pole is exactly equal and opposite, thus yielding a resultant of zero torque produced by conditions beneath the stator S-pole. The net torque is zero and so the rotor remains in a position of equilibrium.

Let us now examine the effect of alternately energizing the two quadrature stator coils in sequence through positive and negative values of dc excitation. The first step in the sequence is represented in Fig. 11-2(b) by placing coil A-A' in a de-energized state and coil B-B' in a fully energized state. The rotor is shown in the final steady-state position that results as long as coil B-B' is energized with positive current (i.e., so that the left pole becomes an N-pole and the right pole assumes an S-pole polarity). The net rotor displacement for this switch in coil excitation in this case is 18°. Once the rotor assumes the position depicted in Fig. 11-2(b), the orientation of the rotor poles relative to the newly located stator poles is identical to that appearing in Fig. 11-2(a) and so the rotor is again in equilibrium. It is instructive, however, to examine the torque conditions that develop when coil A-A' is de-energized and coil B-B' is energized while the rotor is still in the position corresponding to Fig. 11-2(a). We proceed on the assumption that the currents in these coils change instantaneously, i.e., the time constants of the windings are very small compared to the mechanical time constants involved. This is not an unrealistic assumption. Because the magnitude of the flux field produced by a given coil excitation in the fixed geometry of a given machine is determinable and since the permanent magnet flux field of each rotor pole is also

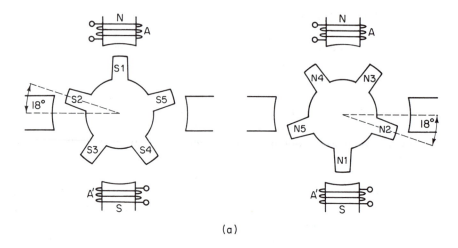

(a)

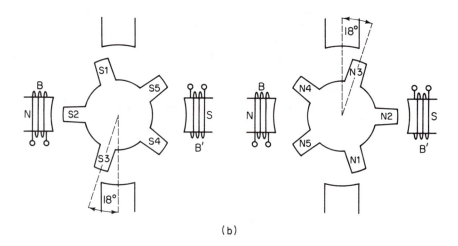

(b)

Figure 11-2 Illustrating the indexing of the rotor in a ccw direction for a full cycle of coil *A-A'*/coil *B-B'* excitation in sequence through positive and negative values. Observe that the orientation between rotor and stator in (e) exactly corresponds to the starting point in (a) except for rotation. (a) Coil *A-A'* energized for an N-S orientation; coil *B-B'* deenergized. Comment: starting point. (b) Coil *A-A'* deenergized; coil *B-B'* energized for an N-S orientation. Result: Rotor advances ccw 18° from its position in (a). (c) Coil *A-A'* energized with reversed current for an S-N orientation; coil *B-B'* de-energized. Result: Rotor advances ccw an additional 18° from its position in (b). (d) Coil *A-A'* de-energized; coil *B-B'* energized with reverse current for an S-N orientation. Result: Rotor advances ccw an additional 18° from its position in (c). (e) Coil *A-A'* energized with positive current for an N-S orientation; coil *B-B'* de-energized. Result: Rotor advances an additional 18° for a total of 72° or one rotor pole pitch corresponding to one cycle of ± excitation of stator coils.

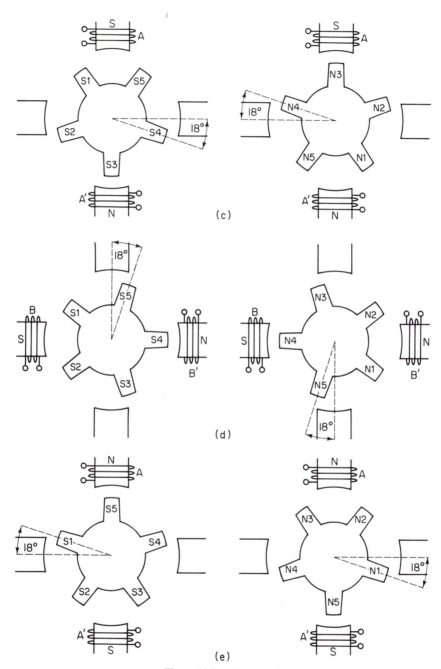

(c)

(d)

(e)

Figure 11-2 (*Continued*)

calculable, we can represent the maximum torque produced by these fields as T_m. When the rotor orientation is such that the torque developed is other than maximum, the appropriate sine or cosine function is introduced depending upon whether the torque is one of attraction or one of repulsion. Thus with coil B-B' energized and the rotor still in the position depicted in Fig. 11-2(a), the expression for the torque employing superposition is

$$T = T_m \left[\sin \left\langle \begin{matrix} S2 \\ N \end{matrix} \right. + \sin \left\langle \begin{matrix} S1 \\ N \end{matrix} \right. - \sin \left\langle \begin{matrix} S3 \\ N \end{matrix} \right. + \cos \left\langle \begin{matrix} S5 \\ S \end{matrix} \right. - \cos \left\langle \begin{matrix} S4 \\ S \end{matrix} \right. \right] \quad (11\text{-}1)$$

where the plus sign is used if the torque acts to produce ccw movement and the negative sign is used if the movement would be cw. Inserting the proper angles corresponding to the stator pole orientation of Fig. 11-2(b) and the rotor position of Fig. 11-2(a), Eq. (11-1) becomes

$$T = T_m(\sin 18° + \sin 90° - \sin 54° + \cos 18° - \cos 54°)$$

$$= T_m(0.309 + 1.0 - 809 + 0.951 - 0.588)$$

$$= 0.863 T_m \quad (11\text{-}2)$$

Because this is a positive torque, the rotor in Fig. 11-2(a) moves to the new equilibrium position shown in Fig. 11-2(b).

Consider next that coil B-B' is switched off and coil A-A' is again energized but this time with current flowing in the opposite direction. The upper stator pole assumes an S-pole polarity and the lower stator pole takes on an N-pole polarity. Accordingly, the new stator flux configuration [in Fig. 11-2(c)] exerts a torque on the rotor orientation of Fig. 11-2(b) which is again described by Eq. (11-2). The ensuing torque displaces the rotor by an additional 18° counterclockwise in order to reach the new equilibrium position shown in Fig. 11-2(c). In the subsequent step in the sequence, the current in coil A-A' is switched off while the current in coil B-B' is switched on but in the reversed direction. The stator poles then assume the polarity depicted in Fig. 11-2(d), which in turn produces a torque on the rotor to yield a further 18° displacement in the ccw direction. In other words, the rotor moves to the new equilibrium position that finds S4 in alignment with the N-pole associated with coil B-B'.

It is instructive to observe here that once a sequence is initiated and maintained through the proper positive and negative alternate excitation of the stator coils, the action of the torque produced by the subsequent coil excitation is to displace the rotor in the established direction. This is done in a manner to ensure that the rotor pole having the least angular alignment with a stator pole of opposite polarity is brought into exact alignment with that pole. Accordingly, when coil A-A' is next energized, the result is to move S5 into alignment with the upper stator N-pole. The new equilibrium position is shown in Fig. 11-2(e). Observe that this last orientation is identical to the first one except that the entire rotor structure has been indexed through a total displacement equal to one rotor pole pitch, or 72° in this case. Note too that for a *two*-coil (or two-phase) stator with *two* states (positive and negative), it requires *four* switching operations to complete a switching cycle. Thus, this motor is made to *step* four times during each cycle of excitation which in turn produces a net displacement of one rotor pole pitch. Or, stated differently, each switch of the stator excitation produces a rotor displacement of one-fourth a rotor pole pitch. The *stepping* action exhibited by this motor in response to alternate sequenced excitation of stator coils is the reason it is called a *stepper motor*.

The direction of rotation of the stepper motor can be conveniently reversed by merely reversing the switching sequence. Thus, if in Fig. 11-2(b) the excitation of coil *B-B'* were reversed, then the action of the developed torque would be to displace S5 in Fig. 11-2(a) into a position of exact alignment with the N-pole flux which would now be associated with coil *B'*. The result is a clockwise displacement of 18° in contrast to the counterclockwise displacement that occurs when coil *B* is made to produce an N-pole polarity.

11-3 DRIVE AMPLIFIERS AND TRANSLATOR LOGIC

The foregoing description of the method of operation of the stepper motor makes it plain that the stepping action of the motor is dependent on a specific switching sequence that serves to energize and de-energize the stator coils. A practical scheme that achieves this objective is the one of a drive amplifier consisting of transistors that are switched on and off sequentially in accordance with the signals that originate from an appropriate translator logic control circuit. A block diagram of the arrangement is shown in Fig. 11-3. The drive amplifier provides the dc excitation to coils *A-A'* and *B-B'* through switched circuitry that is controlled by the translator logic device. Appearing in Fig. 11-4 is a detailed description of the composition of one type of drive amplifier, namely, the *bipolar* type. A total of four switching transistors is needed for each phase coil whenever a single power

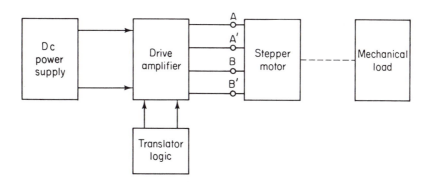

Figure 11-3 Block diagram of the stepper motor and its switched supply.

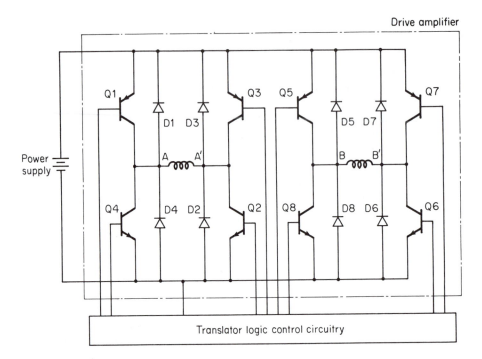

Figure 11-4 Diagram illustrating the details of a bipolar drive amplifier.

supply is used and current reversal in the phase coils is desired. The term bipolar is used for such a circuit because the circuitry permits either end of the phase coil to be connected to the positive side of the power supply. There is a drive circuit available that employs two power supplies in which case it becomes necessary to use only two transistors per phase coil. In addition to the eight transistors, the drive amplifier of Fig. 11-4 also includes eight diodes. Of course, the diodes are required to provide paths for the inductive coil currents to flow whenever the transistors are switched off by logic control circuitry. For example, if transistors Q1 and Q2 are ON during the first quarter of a switching cycle thus causing current to flow in the phase coil from A to A', the transition into the next quarter of the cycle calls for Q1 and Q2 to be switched OFF. Because of the magnetic storage associated with the current in coil A-A', the current cannot be reduced to zero instantaneously. But, as it attempts to do so, a large emf is induced in the coil with terminal A' positive and A negative. This induced emf puts a positive voltage across diodes D3 and D4 thus furnishing a path for the switched-off current in coil A-A'. The current decays rapidly because of the small time constant of the discharging circuit.

How is the logic schedule of the translator control circuit determined? This is best explained by developing the logic required to perform the operations of the stepper motor described in connection with Fig. 11-2. In this case, a complete switching cycle required that the stator coils be energized in accordance with the schedule in the table below for a counterclockwise rotation.

Coil A-A'	Coil B-B'
ON+	OFF
OFF	ON+
ON−	OFF
OFF	ON−

Here ON+ is distinguished from ON− by a current reversal in the coil. The logic schedule that achieves this switching sequence for the drive amplifier of Fig. 11-4 expressed in binary language where 1 denotes ON and 0 denotes OFF is as follows:

Coil A-A'				Coil B-B'			
Q1	Q2	Q3	Q4	Q5	Q6	Q7	Q8
1	1	0	0	0	0	0	0
0	0	0	0	1	1	0	0
0	0	1	1	0	0	0	0
0	0	0	0	0	0	1	1

Each repetition of this switching cycle provides an advancement of the stepper motor equal to one rotor pole pitch. Hence in this case five cycles of this switching sequence are needed to complete one full revolution of the rotor.

A very popular circuit for the drive amplifier of stepper motors is the one illustrated in Fig. 11-5. This circuit clearly has the advantage of fewer electronic components and simpler bias circuitry for the switching transistors. Each phase coil of the stepper motor that is used with these drives is equipped with a center tap and the switching logic for the associated transistors arranged so that current flows only in one section at a time. Reversal of the magnetic pole polarity is achieved by directing the current either to one section or to the other. The disadvantage of such a scheme is that the copper is not made to work continuously in producing useful output. This arrangement is called the bifilar unipolar drive. It is referred to as unipolar because the ends of both coil sections stay connected to the negative side of the power supply during operation unlike the bipolar case where the connection alternates between the high and low side of the source.

The logic schedule to be furnished by the translator control circuitry for each full switching cycle of the operation of a stepper motor such as the one illustrated in Fig. 11-2 is the following:

Coil A		Coil B	
Q1	Q2	Q3	Q4
1	0	0	0
0	0	1	0
0	1	0	0
0	0	0	1

It is assumed here that coil A is equipped with a center tap by using the point of connection of coil A and coil A' in Fig. 11-1. A similar arrangement is assumed to be used for coil B.

   The diodes D1 through D4 appearing in the circuitry of the drive amplifier shown in Fig. 11-5 are for the purpose of maintaining a current path following the cutoff of a particular transistor switch. For example, when current is flowing through the left-half section of coil A in Fig. 11-5 and Q1 is subsequently cut off, a circuit is established by transformer action in the second section of coil A which allows the current in the first section to drop to zero as the current in the second section instantaneously rises to the level of the current in the first section. The latter current then flows through a path that involves the power supply and diode D2. Diode D2 is made conductive since a positive voltage appears across it as a

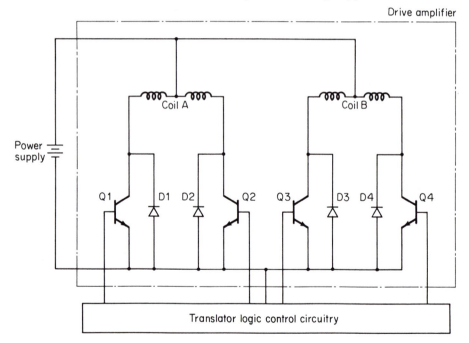

Figure 11-5 Drive amplifier of the bifilar, unipolar type. Each phase coil of the stepper motor comes equipped with a center tap.

result of the transformer voltage induced in the right-half section of coil *A*. The transfer of current from the left section to the right section of coil *A* is allowed to occur instantaneously because there is *no change* in magnetic energy in this process. The mmf of both coils acts on the same magnetic circuit.

11-4 HALF-STEPPING AND THE REQUIRED SWITCHING SEQUENCE

The description of the operation of the stepper motor so far has been limited to situations where either one stator coil or the other was allowed to be energized at any one time. What would be the consequence if a switching schedule is devised which permits both stator coils to be excited simultaneously at certain periods in a cycle? To examine this matter, let us return to the configuration depicted in Fig. 11-2(a). Recall that a full 18° ccw displacement resulted from switching coil *A-A'* off and coil *B-B'* on. However, suppose that coil *B-B'* is switched on while maintaining the excitation on coil *A-A'*. In other words, coil *A-A'* is not de-energized when excitation is applied to coil *B-B'*. Intuition, based on the symmetry preva-

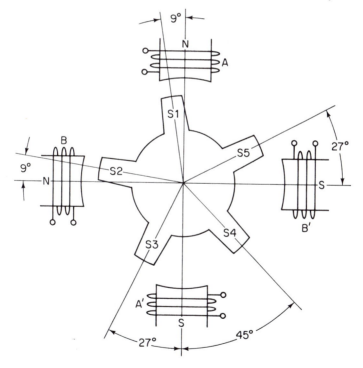

Figure 11-6 Half-step equilibrium position that occurs when the rotor is initially at the position illustrated in Fig. 11-2(a) and both coils *A-A'* and *B-B'* are energized to produce the field orientation shown here. The result is a 9° displacement of the rotor as indicated here.

lent in this situation, leads us to suspect that the motor takes a *half-step,* i.e., it advances in the ccw direction by one-eighth of a rotor pole-pitch. An examination of the resultant torque on the rotor structure at this displacement angle reveals the motor to be in equilibrium. This is readily verified by writing the expression for the net torque on the rotor corresponding to this condition. The situation is illustrated in Fig. 11-6 for a half-step angle of 9°. Note that all four stator poles are energized with the polarities indicated. Applying superposition, the expression for the net torque can be written as

$$T_R = T_m \left[-\sin \overset{N_A}{\underset{S1}{<}} - \sin \overset{N_A}{\underset{S2}{<}} + \sin \overset{N_A}{\underset{S5}{<}} - \cos \overset{S_{A'}}{\underset{S3}{<}} + \cos \overset{S_{A'}}{\underset{S4}{<}} \right.$$

$$\left. + \sin \overset{N_B}{\underset{S2}{<}} + \sin \overset{N_B}{\underset{S1}{<}} - \sin \overset{N_B}{\underset{S3}{<}} + \cos \overset{S_{B'}}{\underset{S5}{<}} - \cos \overset{S_{B'}}{\underset{S4}{<}} \right]$$

where the first line represents the torque produced by the stator poles created by coil *A-A'* interacting with the S-pole section of the rotor and the terms on the second line represent the torque contributions associated with the stator poles of coil *B-B'* and the S-pole section of rotor. Upon inserting the angular displacements existing between the axes of the indicated poles, we get

$$T_R = T_m(-\sin 9° - \sin 81° + \sin 63° - \cos 27° + \cos 45°$$

$$+ \sin 9° + \sin 81° - \sin 63° + \cos 27° - \cos 45°)$$

$$= 0$$

Hence, this half-step displacement of the rotor represents an equilibrium position.

If coil *A-A'* is now de-energized while the excitation of coil *B-B'* is maintained, the motor steps an additional 9°, thus yielding the originally expected 18° that results when one coil is switched off and the other is switched on. An extension of this analysis through a complete switching cycle executed by half-step increments leads to the following switching sequence for counterclockwise rotation:

Coil *A-A'*	Coil *B-B'*	
ON+	OFF	
ON+	ON+	
OFF	ON+	
ON−	ON+	One complete
ON−	OFF	switching cycle
ON−	ON−	
OFF	ON−	
ON+	ON−	
ON+	OFF	
.	.	
.	.	
.	.	

A full switching cycle now includes a total of eight switching states, which yield eight half-steps, to give a total displacement of one rotor pole pitch. The switching schedule required to half-step the motor of Fig. 11-1 arranged to operate in the bifilar mode and driven by the unipolar drive amplifier of Fig. 11-5 is detailed in the following table:

Coil *A*		Coil *B*	
Q1	Q2	Q3	Q4
1	0	0	0
1	0	1	0
0	0	1	0
0	1	1	0
0	1	0	0
0	1	0	1
0	0	0	1
1	0	0	1

The advantage of half-stepping of course is that it improves the precision of the stepper motor in applications where more precise positioning is demanded.

11-5 RELUCTANCE-TYPE STEPPER MOTOR

The reluctance-type stepper motor is not equipped with a permanent magnet on the rotor. It develops an indexing torque in response to the sequenced dc excitation of stator coils by virtue of the large difference in magnetic reluctance that exists between a direct-axis path and a quadrature-axis path. The resulting stationary flux fields produced by the dc current in selected stator coils always exert a force which acts to minimize the reluctance of the flux path in accordance with the characterization of Eq. (1-65). A typical configuration of the reluctance-type stepper motor is illustrated in Fig. 11-7. The stator in this case is designed for eight poles and is equipped with four phase coils each of two sections. The rotor is slotted to produce the effect of six poles corresponding to which is a pole pitch of 60°. The rotor is at a position of equilibrium which occurs when coil *A-A'* (i.e., phase A) is energized and the remaining phases are left unexcited. Observe that the constant flux field produced by phase A encounters a path of minimum reluctance by passing through rotor poles 1 and 4 which are aligned with the stator pole structures associated with phase A.

For the combination of stator and rotor poles that is designed into this motor, it is instructive to note that the axes of both rotor poles 2 and 5 are 15° displaced from the axes of the two stator poles associated with phase B. Hence, if phase A is de-energized and phase B is energized to produce an N-pole polarity at coil *B* and an S-pole polarity at coil *B'*, a reluctance torque is developed that acts

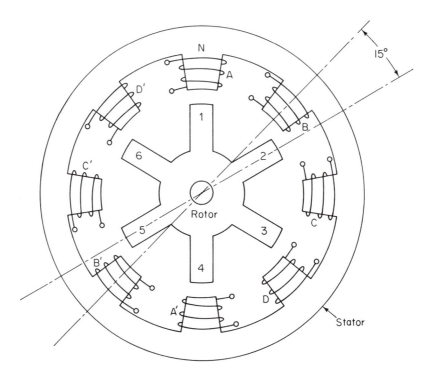

Figure 11-7 Construction details of a reluctance-type stepper motor equipped with four stator phases (i.e., eight poles) and six rotor poles.

to align the axes of rotor poles 2 and 5 with the axes of the stator poles at *B-B'*. The rotor advance is 15° ccw which in this machine too is one-quarter of a rotor pole pitch. Successive excitations of phases C, D, and A yield a total advance in the ccw direction equal to a rotor pole pitch. A sixfold repetition of this switching sequence for the excitation of the stator phases results in one revolution of the motor shaft.

The equilibrium position of the rotor in Fig. 11-7 also prepares the motor to move in a clockwise direction. However, it now requires that coil *D'* be energized for an N-pole polarity and coil *D* for an S-pole polarity following the removal of excitation on phase A. A reluctance torque now develops that pulls rotor teeth 6 and 3 into alignment with the poles produced by coil *D'-D*. Clearly this results in a clockwise rotation.

The design of the drive amplifiers and the translator logic for the reluctance-type stepper motor differ in some details from those reported for the PM stepper motor. This is necessary in order to account for differences in the design aspects of the two types of motors. The principles, however, remain the same and are easily transferable from one motor type to the other.

11-6 RATINGS AND OTHER CHARACTERISTICS

On the basis of the material that is related in the preceding pages concerning the stepper motor, it should come as no surprise that these motors are used in circumstances where the need for large horsepower is not a factor. A motor's output power is dependent upon the extent to which the copper is worked electrically and the iron is worked magnetically. The stepper motor ranks low on both counts. To achieve the characteristics which make the stepper motor unique, it is necessary that during each complete switching cycle each of the phase coils spends time in a de-energized state and that some parts of the magnetic flux paths be essentially free of field flux. Consequently, stepper motors are found to range in output power from about 1 W to a maximum size of 3 hp. In terms of physical dimensions, the largest machines are built with diameters up to seven inches while the smallest ones are constructed in a pancake form with diameters as small as 1 in.

Many step sizes are also available with stepper motors. They can be purchased with step sizes as small as 0.72° or as large as 90°. The most common step sizes are 1.8°, 7.5°, and 15°. To procure a step size of 1.8° in a PM type, it is necessary to design the stator for 40 poles and the rotor for 50 poles (or slots).

The shape of the torque-speed curve of the stepper motor takes on the general form depicted in Fig. 11-8. The speeds indicated in this particular motor are achieved as a result of programming the switching sequence for appropriate repetition. For example, to realize one revolution per second, it is necessary for the translators' logic to switch a 1.8° step motor at the rate of 200 steps per second. Any attempt to operate this motor at a rate of five revolutions per second fails because the torque drops to zero. This decrease in torque is attributable to the effects of the speed voltages induced in the phase coils at the higher speeds as well as to the inductance of these coils. Compensation circuits are available

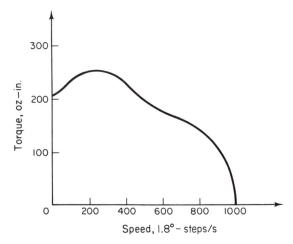

Figure 11-8 Torque-speed curve of a PM stepper motor operated at full step.

which can double the useful speed range but this is at the cost of increased circuit complexity for the drive amplifiers.

PROBLEMS

11-1. Consider the situation corresponding to the rotor position depicted in Fig. 11-2(b). Assume that coil A-A' is energized so that the magnetic material on which coil A' is wrapped is of an N-pole polarity and that of coil A is of an S-pole polarity.
   (a) Write the expression for the torque produced on the rotor by the S-pole section of the rotor. Assume sinusoidal flux distributions throughout.
   (b) How many degrees of rotation occur before a position of equilibrium is reached?

11-2. Repeat Prob. 11-1 for the conditions that exist at the N-pole section of the rotor.

11-3. Let T_m denote the maximum torque produced by the cooperation of a single PM rotor pole with the stator pole in the configuration of Fig. 11-2. Calculate the net torque produced on the rotor shaft for the condition represented by Fig. 11-2(c).

11-4. The schematic diagram of a four-phase, two-pole PM stepper motor is shown in Fig. P11-4. The phase coils are excited in sequence by means of a translator logic circuit.
   (a) Write the logic schedule for full-stepping of this motor.
   (b) What is the displacement angle of the full step?
   (c) Write the logic schedule for operating this motor in the half-step mode.
   (d) Make a sketch of the stator poles and the relative rotor position when coils B and C are simultaneously energized.
   (e) What is the total rotor displacement in part (d) measured from the position depicted in Fig. P11-4?

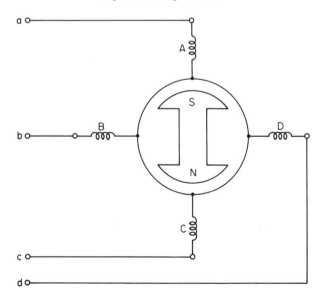

Figure P11-4

11-5. The excitation signals used to drive the coils in the motor of Fig. P11-4 originate from digital signals that drive the supply amplifier. A continuous sequencing of these signals will cause the stepper motor to run at a constant speed (thus behaving as a synchronous motor). Determine the frequency rate of these digital signals if the speed of rotation is to be 360 rpm. Assume a half-stepping pattern.

11-6. Determine the frequency rate at which the digital signals that drive the amplifier of the motor of Fig. 11-2 must be set in the half-step mode if the motor is to run at a speed of 360 rpm.

11-7. The drive amplifier of Fig. 11-4 is used to drive the stepper motor of Fig. 11-2 in the full-step mode. Write the logic schedule for switching the transistors in a manner such that it yields a *clockwise* direction of rotation.

11-8. A PM stepper motor is designed to provide a half-step size of 0.9°.
   (a) Determine the rotor pole pitch in degrees.
   (b) Find the number of stator poles.

11-9. Draw the diagram that illustrates the orientation of the stator poles with the N-pole section of the rotor for the half-step case illustrated in Fig. 11-6.

12

Synchros

Synchros are ac electromagnetic devices that find wide applications in servomechanisms.† However, they can also be used as indicating instruments for registering mechanical displacements remotely. In this chapter attention is first directed to a description of the construction features of the synchros as well as an explanation of their theory of operation. This is then followed by a discussion of the types of errors that are found in such devices which include static errors, dynamic errors, and residual voltages.

12-1 CONSTRUCTION FEATURES

There are four basic types of synchros: the control transmitter (denoted by cx), the control transformer (cT), the control differential (cD), and the control receiver (cR). The essential construction characteristics of the control transmitter are represented in Fig. 12-1(a). It consists of a stator and a rotor. The inner surface of the stator is slotted to accommodate a balanced three-phase winding, which is usually of the concentric-coil type. The rotor is of the dumbbell construction with

†A servomechanism is an error-actuated feedback control system in which the controlled variable is mechanical.

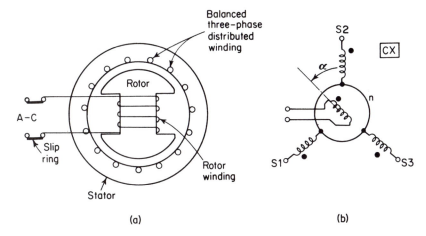

Figure 12-1   Synchro control transmitter: (a) construction features; (b) schematic diagram.

many turns of wire wrapped around the stem. An ac voltage is applied to the rotor winding through a pair of slip rings, which causes an excitation current to flow and produce an alternating flux field with respect to the rotor structure. By definition the electrical zero position of the rotor corresponds to maximum coupling with the turns of phase winding 2 of the stator. Moreover, since there are two rotor positions for maximum coupling, a unique electrical zero is identified by matching dots as shown in Fig. 12-1(b). Figure 12-1(b) depicts the rotor axis displaced from the electrical zero by the angle α. Note the similarity in construction to the three-phase synchronous generator; the synchro can be aptly described as a miniature generator. It is important, however, to understand the distinction in the way the two machines are used. The synchronous generator is excited with a dc voltage and driven at constant speed, and produces a three-phase voltage at the stator terminals. In contrast, the synchro is excited with an ac voltage and is often merely displaced by finite amounts from the electrical zero, and thus induces *single-phase* voltages through transformer action in the stator winding. Sometimes, in control system applications, the synchros are driven at low tracking speeds.

The construction details of the synchro control transformer differ somewhat from those of the control transmitter. One important difference is that the air gap is uniform owing to the rotor's cylindrical or umbrella-like construction. Refer to Fig. 12-2(a) and (b). This feature is included to keep the magnetizing current drawn by the control transformer to a minimum. Another difference is the way in which the electrical zero is defined. For the control transformer it is defined as that position of the rotor which makes the coupling with winding 2 of the stator zero. This particular configuration is depicted in Fig. 12-2(c). The stator winding is also a balanced three-phase winding; however, it differs from the transmitter in

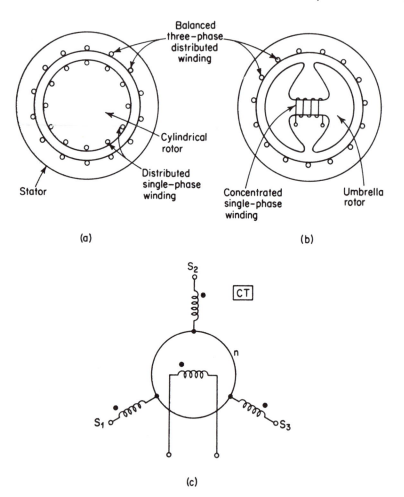

Figure 12-2 Control transformer: (a) construction features of cylindrical rotor type; (b) construction features of umbrella rotor type; (c) schematic diagram with rotor winding at electrical zero.

that the transformer has a higher impedance per phase. This latter feature permits several control transformers to be fed from a single transmitter.

The differential synchro has a balanced three-phase distributed winding in both the stator and the rotor. Moreover, the rotor is cylindrically shaped as shown in Fig. 12-3(a). Although three-phase windings are involved, it is important to keep in mind that these units deal solely with single-phase voltages.

The basic structure of the control receiver is the same as that of the control transmitter. It has a balanced three-phase stator winding and a salient-pole rotor. The one detail in which the receiver differs from the transmitter is the inclusion of a mechanical viscous damper on its shaft. In normal use both the rotor and

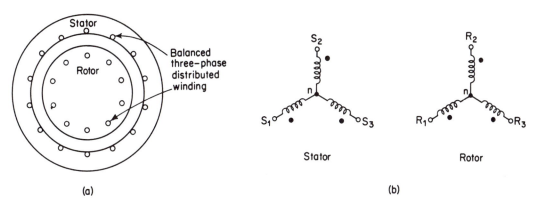

Figure 12-3 Differential synchro: (a) construction features; (b) schematic representation.

stator windings are excited with single-phase currents. Accordingly, the field distribution of the rotor interacts with the ampere-conductor distribution of the stator to produce torque and hence rotation. The purpose of the damper is to permit the receiver rotor to respond to the changing stator ampere-conductor distribution without causing the rotor to overshoot its mark. If the overshoot is large, average torque may be produced such that the receiver runs as a single-phase motor.

12-2 VOLTAGE RELATIONS

A description of how the stator phase voltages vary with rotor displacement is useful in understanding the various applications of synchros. In this connection refer to the schematic diagram of the control transmitter shown in Fig. 12-4. Assume that the rotor winding is excited with a single-phase ac voltage and that the rotor is displaced from its electrical zero position by the angle α and held fast. By transformer action single-phase voltages having the same frequency as the rotor applied voltage will be induced in each stator phase. The value of the induced voltage depends upon the coupling of each phase with the rotor winding. To describe these relationships mathematically let

$$a \equiv \frac{\text{effective stator turns}}{\text{effective rotor turns}} \tag{12-1}$$

$$\bar{E}_r = \text{rms value of the induced emf in the rotor winding} \tag{12-2}$$

and

$$\bar{E} = a\bar{E}_r \tag{12-3}$$

The last expression denotes the maximum value of rms voltage that can be induced in any balanced stator phase.

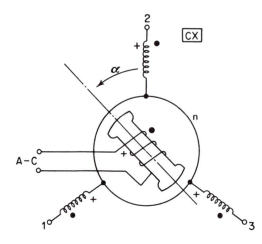

Figure 12-4 Circuit for deriving the stator phase and line voltage variations as a function of rotor displacement α.

The rms value of the emf induced in winding 2 of the stator when the rotor is at the angle α is described by

$$\bar{E}_{2n} = \bar{E} \cos \alpha \qquad (12\text{-}4)$$

Note that, when α is zero, the rotor is completely linked with winding 2 and so yields the maximum rms voltage $\bar{E}$. This situation is entirely consistent with the definition of the electrical zero for the control transmitter. The expression for the instantaneous value of this voltage can be written as

$$e_{2n}(t) = \sqrt{2}\, E \sin \omega t \cos \alpha \qquad (12\text{-}5)$$

where ω is the frequency of the applied rotor voltage. A study of the last expression should make it apparent that the sin ωt term accounts for the transformer action that takes place in the synchro and that cos α describes the coupling between the rotor winding and winding 2 of the stator. The use of a properly distributed and pitched stator winding that yields a quasisinusoidally distributed field is the reason for the cosine function in Eq. (12-4). If the winding were distributed differently, a different functional relationship would be needed.† However, throughout this chapter, we assume the use of sinusoidally distributed windings only.

The value of the rms voltage induced in winding 1 is given by

$$\bar{E}_{1n} = \bar{E} \cos (\alpha - 120°) \qquad (12\text{-}6)$$

The argument is modified by $-120°$ to account for the fact that the axis of winding 1 is located 120° ahead of the axis of winding 2. The corresponding expression for the instantaneous voltage then becomes

$$e_{1n}(t) = \sqrt{2}\, E \sin \omega t \cos (\alpha - 120°) \qquad (12\text{-}7)$$

†Some synchros are designed to provide a linear output with rotor displacement.

Note that the sin ωt factor remains the same as in Eq. (12-5), which emphasizes that this induced emf is of the same frequency and of the same phase as that occurring in winding 2. Only the amplitudes differ because of the difference in coupling.

Finally the emf induced in winding 3 can be expressed as

$$\bar{E}_{3n} = \bar{E} \cos (\alpha + 120°) \tag{12-8}$$

The presence of 120° in the argument accounts for the fact that winding 3 is located behind the axis of winding 2 by this amount. The expression for the instantaneous voltage in winding 3 is accordingly

$$e_{3n}(t) = \sqrt{2}\, E \sin \omega t \cos (\alpha + 120°) \tag{12-9}$$

Equations (12-4) to (12-9) describe the stator *phase* voltages. Information about the associated line voltages is readily obtained from the phase voltages. Thus the induced voltage appearing between terminals 1 and 2 in Fig. 12-4 can be written as

$$\bar{E}_{12} = \bar{E}_{1n} + \bar{E}_{n2} = \bar{E}_{1n} - \bar{E}_{2n} \tag{12-10}$$

Inserting Eqs. (12-4) and (12-6) into Eq. (12-10) yields

$$\bar{E}_{12} = \bar{E} \cos \alpha \cos 120° + \bar{E} \sin \alpha \sin 120° - \bar{E} \cos \alpha$$

$$= \left(-\frac{3}{2} \cos \alpha + \frac{\sqrt{3}}{2} \sin \alpha \right) \bar{E} \tag{12-11}$$

Or combining the sine and cosine functions into an equivalent cosine function leads to

$$\bar{E}_{12} = \sqrt{3}\, \bar{E} \cos (\alpha - 150°) \tag{12-12}$$

The expression for the line voltage between terminals 2 and 3 can be similarly determined. Thus,

$$\bar{E}_{23} = \bar{E}_{2n} + \bar{E}_{n3} = \bar{E}_{2n} - \bar{E}_{3n} \tag{12-13}$$

Substituting Eqs. (12-4) and (12-8) leads to

$$\bar{E}_{23} = \left(\frac{3}{2} \cos \alpha + \frac{\sqrt{3}}{2} \sin \alpha \right) \bar{E} \tag{12-14}$$

or

$$\bar{E}_{23} = \sqrt{3}\, \bar{E} \cos (\alpha - 30°) \tag{12-15}$$

The line voltage between terminals 1 and 3 is

$$\bar{E}_{31} = \bar{E}_{3n} + \bar{E}_{n1} = \bar{E}_{3n} - \bar{E}_{1n} \tag{12-16}$$

Upon inserting Eqs. (12-6) and (12-8) there results

$$\bar{E}_{31} = \bar{E} \cos (\alpha + 120°) - \bar{E} \cos (\alpha - 120°)$$
$$= - \sqrt{3}\, \bar{E} \sin \alpha \qquad\qquad (12\text{-}17)$$

or

$$\bar{E}_{31} = \sqrt{3}\, \bar{E} \cos (\alpha + 90°) \qquad\qquad (12\text{-}18)$$

Equation (12-17) is put in the cosine form to facilitate comparison with Eqs. (12-12) and (12-15) to show that these line voltages are such that the arithmetic sum of two of them must be equal to the third.

As a result of the variations in the line voltages that can be effected by displacement of the rotor, it is possible to feed these voltages to the stator of a control transformer, a receiver, or even a differential synchro and thereby produce useful effects. Such applications are discussed in the next section.

12-3 APPLICATIONS

The use of synchros in servomechanisms represents one of the biggest areas of applications. Shown in Fig. 12-5 is the arrangement most often found. This configuration constitutes that part of the servomechanisms called the *error detector*. Its chief function is to convert a command in the form of a mechanical displacement of the cx rotor to an electrical signal appearing as a voltage at the rotor winding terminals of the control transformer. It offers the distinct advantage of furnishing information about shaft displacement by means of wires, which allows the control transformer to be located remotely from the control transmitter.

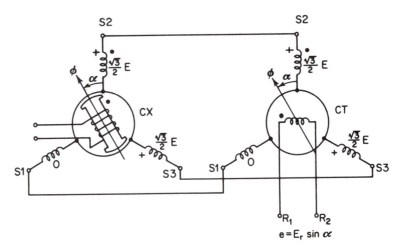

Figure 12-5 Synchro arrangement for converting a CX displacement to a corresponding voltage signal at the rotor winding of the CT.

What is the mechanism by which this is brought about? Keep in mind that the rotor of the control transmitter is excited with an ac voltage so that an alternating field is created whose axis coincides with the CX rotor axis. The time-varying nature of the air-gap flux and its linkage with the stator coils induce in them corresponding voltages. Thus, if α equals 30° measured in the counterclockwise direction, it follows that the voltages induced in the stator phases are as specified in Fig. 12-5. When the CX stator terminals are connected to the CT stator terminals in the fashion shown in Fig. 12-5, the transmitter is made to supply magnetizing currents in the CT stator windings, which in turn create an alternating flux field in its own air gap. The values of the CT stator phase currents must be such that the resulting air-gap flux induces voltages that are equal and opposite to those prevailing in the stator of the transmitter. This denotes the equilibrium condition satisfying Kirchhoff's voltage law in the circuit between the stator windings of the two synchros. Accordingly, the direction of the resultant flux produced by the CX stator phase currents is forced to take a position of exact correspondence to that of the rotor axis of the control transmitter.

If the CT rotor is assumed to be held fast at its electrical zero position as depicted in Fig. 12-5, then the rms voltage induced in the rotor can be expressed as

$$e = E_r \sin \alpha \qquad (12\text{-}19)$$

where E_r denotes the maximum rms voltage induced by the CT air-gap flux when the coupling with the rotor winding is greatest and α is the displacement angle of the rotor of the control transmitter. The sinusoidal function appears because the distributed stator winding produces a sinusoidally distributed flux density along the air gap. The polarity markings are used to indicate those sides of the stator coils that have the same instantaneous polarity. For equilibrium to be established in the stator circuits it is therefore necessary that both phase and magnitude conditions be satisfied.

It is important to note that the value of the CT rotor voltage is in essence an indication of a lack of correspondence between the rotor positions of the transmitter and the transformer from their electrical zeros. Thus, if the CT rotor were manually turned counterclockwise in Fig. 12-5 by an amount $\alpha = 30°$, then clearly the rotor emf would become zero. In this manner exact correspondence is provided with the transmitter measured from the respective electrical zero positions of each synchro.

A general expression for the instantaneous value of the voltage (e) induced in the CT rotor when the displacement of the CX is α_X and that of the rotor of the CT is α_T is

$$e = E_r \sin \omega t \sin (\alpha_X - \alpha_T) \qquad (12\text{-}19\text{a})$$

Figure 12-6 is the schematic diagram showing how the synchro pair of Fig. 12-5 is used in a servomechanism. If it is desirable here for the load to be moved a specified amount in response to a command applied in the form of a displacement

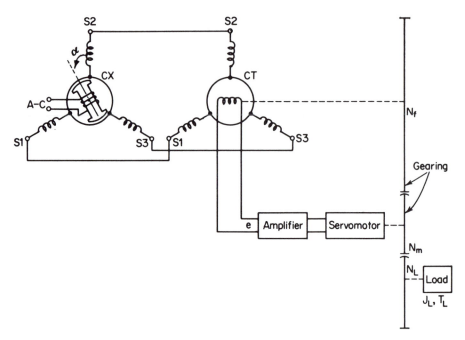

Figure 12-6 Servomechanism illustrating the use of synchros as the error detector. Dashed lines denote mechanical connections.

of the cx rotor, an error voltage e should initially exist because of the inertia on the output shaft associated with the servomotor and the load. As time elapses, the servomotor will begin to respond to this error voltage and thus displace the load. As the output displacement increases, the mechanical connection between the servomotor shaft and that of the CT rotor will cause the CT rotor to advance in the direction to reduce the error voltage e. The servomotor will cease to turn when a position of exact correspondence occurs—at which time e is zero. It is worthwhile to note that with such a control system a large load (e.g., a 100-ft radar dish antenna) can be controlled with very little effort applied at the rotor of the cx synchro.

A second application of synchros is for *torque transmission* over an appreciable distance without the use of a rigid mechanical connection. The connection diagram is illustrated in Fig. 12-7. The arrangement requires the use of a transmitter and a receiver unit. The manner in which the system works is quite straightforward. When the rotor of the transmitter is displaced, say, by $\alpha = 30°$, the stator behaves like the secondary of a transformer and delivers current to the three-phase stator winding of the receiver; this produces an ampere-conductor distribution whose axis is fixed by α. By exciting the rotor of the receiver from the same ac source as the rotor of the transmitter, a flux field is created which interacts with the stator ampere-conductor distribution to produce an electromagnetic torque. The action of this torque brings the rotor axis of the receiver, which

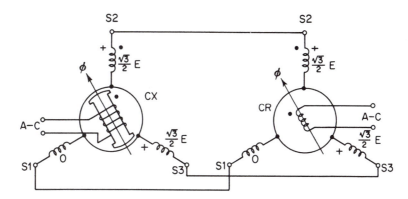

Figure 12-7 Use of synchros for remote position indication.

is free to turn, to a position of correspondence with the rotor of the transmitter. Hence the CR rotor revolves by α degrees in the counterclockwise direction. Note that at this position the induced voltages in the stator windings of the receiver ideally bear the same phase and magnitude as those prevailing in the stator coils of the transmitter.

The use of a differential synchro to add or subtract an angle in the torque transmission system is illustrated in Fig. 12-8. In part (a) of this figure the voltage distribution in the stator phase windings of each synchro is shown corresponding to a CX rotor displacement of $\alpha_X = 30°$ in the positive (or counterclockwise) direction. The differential synchro is assumed to be kept at the electrical zero position in this instance and so does not alter the transmission characteristic from that represented by Fig. 12-7. The rotor of the receiver, of course, responds by turning counterclockwise by an amount equal to an α_X of 30°. It is helpful to note that a counterclockwise displacement of the CX rotor yields a counterclockwise displacement of the CR rotor.

Depicted in Fig. 12-8(b) is the effect of introducing a displacement of α_D in the differential synchro superimposed upon the situation represented in Fig. 12-8(a). For the convenience of illustration α_D is taken equal to −60°. The minus sign means that the rotor of the differential synchro is turned clockwise. Since the axis of the flux field in the differential synchro is determined by α_X, the displacement of the CD rotor by −60° means that the induced emf in rotor winding 2 must be zero, while in the remaining windings the rms value of the induced emf is ($\sqrt{3}/2)E$ having the polarities indicated. Immediately after the displacement of the CD rotor, a voltage unbalance exists between the rotor of the differential and the stator of the receiver. This unbalance causes currents to flow so that the resulting ampere-conductor distribution in the CR stator produces an electromagnetic torque that brings the CR rotor to the position that ensures voltage balance. An examination of Fig. 12-8(b) discloses that the resulting equilibrium position of the CR rotor is a position that is 90° displaced counterclockwise from the electrical zero. Note that a *clockwise* displacement of the CD rotor yields a *counterclock-*

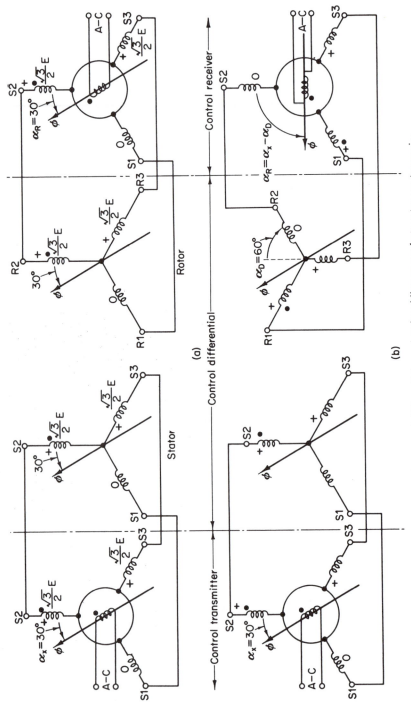

Figure 12-8 Use of a control differential synchro in adding an angle to a torque transmission system.

wise displacement of the CR rotor. A mathematical description of the transmitter-differential-receiver configuration of Fig. 12-8 is given by

$$\alpha_R = a_X - \alpha_D \qquad (12\text{-}20)$$

where α_R is the net displacement of the CR synchro corresponding to an α_X displacement of the transmitter and an α_D displacement of the differential synchros. Positive values of the angles expressed in degrees are used for counterclockwise displacements and negative values for clockwise displacements.

12-4 THE ADVANTAGE OF SINUSOIDALLY DISTRIBUTED WINDINGS

Rather than consider separately the various schemes that are used for synchro windings, attention is directed in this section to show that no harmonic induced emf's can occur in a sinusoidally distributed winding even though such harmonics do exist in the field distribution created by the field winding. Synchro residual voltages originating from this source can accordingly be kept to a minimum by employing winding arrangements that approximate the sinusoidal distribution. The concentric type of winding is such a scheme and so is frequently employed in synchro design.

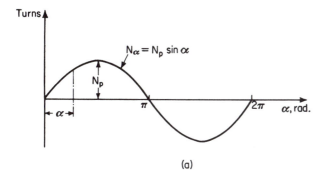

(a)

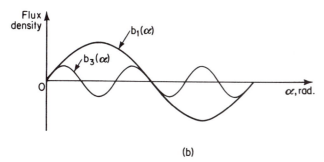

(b)

Figure 12-9   (a) Sinusoidally distributed armature winding; (b) air-gap field distribution, showing the fundamental and the third harmonic components.

Depicted in Fig. 12-9 is the sinusoidally distributed armature winding† of a synchro. Let

$$N = \text{turns/pole} \tag{12-21}$$

$$N_\alpha = N_p \sin \alpha = \text{sinusoidal distribution of turns} \tag{12-22}$$

These quantities are related by

$$N = \int_0^\pi N_\alpha \, d\alpha \tag{12-23}$$

Performing the indicated integration then yields

$$N = N_p[-\cos \alpha]_0^\pi = 2N_p \tag{12-24}$$

Hence the distribution of turns may be expressed in turns per pole as

$$N_\alpha = \frac{N}{2} \sin \alpha \tag{12-25}$$

In general the field distribution produced by the CT synchro field winding contains a large fundamental component plus odd harmonics. No even harmonics exist because the field patterns beneath the north and south poles of the synchros are mirror images of one another. The expression for the nth harmonic flux density is given by

$$b_n(\alpha) = B_n \sin (n\alpha - \gamma) \tag{12-26}$$

where B_n is the amplitude of the nth harmonic field component. In the case of the transmitter synchro, B_n is a sinusoidal function of the line frequency ω applied to the rotor winding. Also, α denotes a space angle measured in the manner illustrated in Fig. 12-9. Moreover, the quantity γ is proportional to the CX rotor speed ω_r. A comparison of Eq. (12-26) with Eq. (4-8) discloses that both are expressions for traveling waves.

Because the induced emf in the synchro armature winding is determined by the time rate of change of the flux linkage, it is worthwhile to find the expression for the flux linkage for both the fundamental field component and the harmonic field components. But first the equation for the flux is needed. For the nth harmonic this is

$$\phi_n = \int_\alpha^{\alpha + \pi} b_n(\alpha) lr \, d\alpha \tag{12-27}$$

where r denotes the rotor radius and l is the axial length. For simplicity a unit axial length is considered. Equation (12-27) may then be written as

$$\phi_n = \int_\alpha^{\alpha + \pi} B_n \sin (n\alpha - \gamma) r \, d\alpha = \frac{2B_n r}{n} \cos (n\alpha - \gamma) \tag{12-28}$$

---

†Armature winding is the stator for the CX synchro and the rotor for the CT synchro.

The associated expression for the flux linkage then becomes

$$\lambda_n = \int_0^\pi N_\alpha \, d\alpha \, \phi_n(\alpha) = \frac{NB_n r}{n} \int_0^\pi \sin \alpha \cos(n\alpha - \gamma) \, d\alpha \qquad (12\text{-}29)$$

Inserting the trigonometric identity

$$\sin \alpha \cos(n\alpha - \gamma) = \tfrac{1}{2} \sin[(n+1)\alpha - \gamma] - \tfrac{1}{2} \sin[(n-1)\alpha - \gamma] \qquad (12\text{-}30)$$

leads to

$$\lambda_n = \frac{NB_n r}{2n} \int_0^\pi \{\sin[(n+1)\alpha - \gamma] - \sin[(n-1)\alpha - \gamma]\} \, d\alpha \qquad (12\text{-}31)$$

The flux linkage of the fundamental field component is found upon substituting $n = 1$ into Eq. (12-31) and then performing the indicated integration. This leads to

$$\lambda_1 = \frac{B_1 N \pi r}{2} \sin \gamma \qquad (12\text{-}32)$$

Keep in mind that γ varies with the angular position of the CX rotor and that superimposed upon this is a sinusoidal line-frequency variation of B_1, the amplitude of the fundamental field component. It is also worthwhile to note that the fundamental flux linkage by Eq. (12-32) is not identically equal to zero.

A more useful expression for the flux linkage of the nth harmonic is obtained by integrating Eq. (12-31) and then inserting the specified limits. Thus

$$\lambda_n = \frac{NB_n r}{2n} \left(\left\{ \frac{-1}{n+1} \cos[(n+1)\alpha - \gamma] \right\}_0^\pi + \left\{ \frac{1}{n-1} \cos[(n-1)\alpha - \gamma] \right\}_0^\pi \right)$$

$$(12\text{-}33)$$

or

$$\lambda_n = \frac{NB_n r}{2n} \left[\frac{-1}{n+1} (\cos \gamma - \cos \gamma) + \frac{1}{n-1} (\cos \gamma - \cos \gamma) \right] \equiv 0 \qquad (12\text{-}34)$$

This last result is a significant one. It states that, for a sinusoidally distributed armature winding, the induced emf's of all the field harmonics are zero despite their presence in the field distribution.†

12-5 STATOR MMF OF THE CONTROL TRANSFORMER IN THE CX-CT MODE

In Sec. 12-3 it is pointed out that the magnetizing currents that flow in the transformer of the CX-CT configuration are such that the axis of the resultant air-gap flux follows the position of the CX rotor axis. This conclusion was reached on the

---

†Another way of describing this result is to state that all the pitch and distribution factors of the harmonics are zero.

basis of satisfying Kirchhoff's voltage law in the circuit involving the three stator phases of each synchro. Now attention is directed to a verification of this result in appropriate mathematical formulations. For the sake of generality the analysis is carried out for the nth harmonic. This procedure offers the advantage of making the general result available in the next section to determine the static errors caused by the CT space harmonics.

To focus attention solely on the stator mmf harmonics, several simplifying assumptions are made concerning the control transformer. These are: no saturation of the iron, uniform air gap between rotor and stator, and balanced phase windings. In addition, the control transmitter is assumed to be perfect, which means that it is equipped with a sinusoidally distributed winding as well as balanced stator windings and a uniform air gap. In Fig. 12-10 is a typical stator mmf wave for the actual winding distribution employed in synchros. Specifically only the mmf curve of winding 2 is illustrated here. A Fourier-series analysis applied to this stepwise function makes it clear that mmf space harmonics do exist and must be of odd order because of half-wave symmetry. Also depicted in Fig. 12-10 is the location of the rotor axis of the transmitter measured from a zero reference line. Note that displacements measured in the CT air gap are denoted by Γ, while the CX displacements are represented by α.†

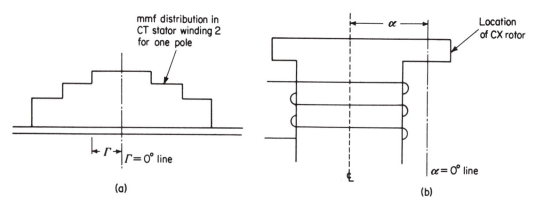

Figure 12-10 (a) Magnetomotive force distribution in phase 2 of the control transformer, showing one pole; (b) transmitter rotor position that produces the magnetizing current leading to the mmf distribution of (a).

The expression for the mmf per pole for winding 2 in the stator of the control transformer for the nth harmonic is

$$f_{n2} = F_n \sin \omega t \cos n\Gamma \qquad (12\text{-}35)$$

where F_n is the amplitude of the mmf harmonic contained within the stepped wave shape. The $\sin \omega t$ factor represents the alternating nature of the mmf wave; the

†This α is not the same as that used in Sec. 12-4.

angle Γ permits finding the value of the mmf at any fixed point in space for a specified time instant. The harmonic amplitude F_n can be expressed in terms of the winding constants and the effective value of the stator phase current. In balanced polyphase machines used in conjunction with three-phase voltages this effective current is a fixed quantity. However, in the case of the control transformer this current is a variable quantity dependent upon the value of α. If I denotes the maximum rms current in winding 2 corresponding to the maximum applied voltage and if N_e denotes an equivalent number of effective turns per pole, then Eq. (12-35) can be more specifically written as

$$f_{n2} = (N_e I)\cos \alpha \sin \omega t \cos n\Gamma \tag{12-36}$$

Keep in mind that Γ and α are independent quantities. For given values of α and ωt we may choose different values of Γ.

By introducing the fact that windings 1 and 3 are located, respectively, ahead and behind winding 2 by 120°, the expressions for the mmf of windings 1 and 3 are correspondingly

$$f_{n1} = (N_e I) \sin \omega t \cos (\alpha + 120°) \cos n(\Gamma + 120°) \tag{12-37}$$

and

$$f_{n3} = (N_e I) \sin \omega t \cos (\alpha - 120°) \cos n(\Gamma - 120°) \tag{12-38}$$

To find the resultant mmf in the stator of the control transformer, it is merely necessary to add the last three equations. Thus

$$f_{nr} = (N_e I) \sin \omega t \left[\cos \alpha \cos n\Gamma + \cos (\alpha - 120°) \cos (n\Gamma - 120°n)\right.$$
$$\left. + \cos (\alpha + 120°) \cos (n\Gamma + 120°n)\right] \tag{12-39}$$

Inserting the trigonometric identity

$$\cos x \cos y = \tfrac{1}{2} \cos (x - y) + \tfrac{1}{2} \cos (x + y) \tag{12-40}$$

into Eq. (12-39) yields

$$f_{nr} = (N_e I) \sin \omega t \{ \tfrac{1}{2} \cos (\alpha - n\Gamma) + \tfrac{1}{2} \cos (\alpha + n\Gamma)$$
$$+ \tfrac{1}{2} \cos [(\alpha - n\Gamma) - (n - 1)120°] + \tfrac{1}{2} \cos [(\alpha + n\Gamma) + (n + 1)120°]$$
$$+ \tfrac{1}{2} \cos [(\alpha - n\Gamma) + (n + 1)120°] + \tfrac{1}{2} \cos [(\alpha + n\Gamma) - (n + 1)120°]\} \tag{12-41}$$

After some manipulation the last equation reduces to

$$f_{nr} = \frac{N_e I}{2} \sin \omega t [A \cos (\alpha - n\Gamma) + B \cos (\alpha + n\Gamma)] \tag{12-42}$$

where

$$A = 1 + 2 \cos (n - 1)120° \tag{12-43}$$

$$B = 1 + 2 \cos (n + 1)120° \tag{12-44}$$

Equation (12-42) can be further simplified by determining the values of A and B for various values of n. In this connection recall that n is restricted to odd values. Accordingly,

for $n = 1$, $A = 3$

                                                            $B = 0$

        $n = 3$, $A = 0$

                                                            $B = 0$

        $n = 5$, $A = 0$

                                                            $B = 3$

or more succinctly

$$A = 3 \qquad \text{for } n = 3k + 1 \qquad\qquad (12\text{-}45)$$

where $k = 0, 2, 4, 6, \ldots ,$
and

$$B = 3 \qquad \text{for } n = 3k - 1 \qquad\qquad (12\text{-}46)$$

where $k = 2, 4, 6, \ldots .$

Therefore, the expression for the resultant mmf may be written as

$$f_{nr} = \tfrac{3}{2}(N_e I) \sin \omega t \cos (\alpha - n\Gamma) \qquad \text{for } n = 3k + 1 \qquad (12\text{-}47a)$$
$$= \tfrac{3}{2}(N_e I) \sin \omega t \cos (\alpha + n\Gamma) \qquad \text{for } n = 3k - 1 \qquad (12\text{-}47b)$$

where the k's are defined as in Eqs. (12-45) and (12-46). Examination of Eq. (12-47) discloses that a single expression can be used provided that k is allowed to take on negative as well as positive values. Accordingly, we may write

$$\boxed{f_{nr} = \tfrac{3}{2}(N_e I) \sin \omega t \cos (\alpha - n\Gamma)} \qquad\qquad (12\text{-}48a)$$

where

$$n = 3k + 1 \qquad\qquad (12\text{-}48b)$$

and

$$k = 0, \pm 2, \pm 4, \pm 6, \pm \cdots \qquad\qquad (12\text{-}48c)$$

When a negative number is used for k, it yields a negative value for n. This sign must be carried along with the magnitude of n. When it is, Eq. (12-47b) immediately results.

The use of the negative sign offers still another advantage. If for the moment the cx rotor is assumed to be driven at a constant velocity ω_r, it follows that $\alpha = \omega_r t$. Insertion of this result into Eq. (12-48a) yields the equation of a traveling wave having a fixed amplitude of $\frac{3}{2}(N_eI)$. The use of $k = -2$ in Eq. (12-48a) leads to

$$f_{5r} = \tfrac{3}{2}(N_eI) \sin \omega t \cos (\omega_r t + 5\Gamma) \tag{12-49}$$

Examination of the last expression discloses that it is the equation of a wave traveling in the *negative* Γ direction. Therefore the minus sign that results from Eq. (12-48b) corresponding to $k = -2$ can be used to denote those harmonic components of the mmf wave that travel in the negative Γ direction.

How can Eq. (12-48) be used to show that the fundamental component of the resultant field in the air gap of the control transformer follows exactly the cx rotor displacements? First let us write the expression for this fundamental term. Thus,

$$f_{1r} = \tfrac{3}{2}(N_eI) \sin \omega t \cos (\alpha - \Gamma) \tag{12-50}$$

Now assume that an observer is placed at $\Gamma = 0$ degrees in the air gap of the control transformer and stays fixed there. Then with the cx rotor at its zero position (i.e., $\alpha = 0°$), Eq. (12-50) states that the observer at $\Gamma = 0$ degrees sees the amplitude of the field mmf. Of course, this amplitude will be at its maximum value when the ct stator phase current is a maximum. Next assume that the cx rotor is displaced by $\alpha = 30°$. In accordance with Eq. (12-50) the observer at $\Gamma = 0$ degrees now sees a value of the mmf which is given by

$$f_{r1} = \tfrac{3}{2}(N_eI) \sin \omega t \cos \alpha = \tfrac{3}{2}(N_eI) \sin \omega t \cos 30° \tag{12-51}$$

A comparison of Eq. (12-51) with Eq. (12-50) reveals a reduction by the factor cos α. Hence one can conclude that the axis of the mmf in the air gap of the control transformer (i.e., the location of its amplitude) has revolved in the positive direction by the amount $\alpha = 30°$. *Therefore the stator field of the control transformer follows exactly the position of the transmitter rotor axis.*

Perhaps a more convincing demonstration of this conclusion can be had by placing the observer at a point $\Gamma = 30°$. Then when α is 0°, the largest mmf seen by the observer is $\sqrt{3}/2$ times the peak value. However, when α is increased to 30°, Eq. (12-48) indicates that the observer now sees the peak mmf. Hence the axis of the air-gap flux has moved in the positive† Γ direction by the amount α.

12-6 DISPLACEMENT ERRORS CAUSED BY CT SPACE HARMONICS

The presence of harmonics in the stator mmf of the ct synchro is responsible for the existence of a displacement error in the cx-ct configuration. The variation of this error can be shown to be a sixth-harmonic function of the cx rotor displacement angle. In this section consideration is first given to a physical demonstration

---

†If the wave would have moved in the negative Γ direction, the observer would see only one-half the peak mmf.

of why the presence of harmonics leads to such a displacement error. A mathematical description then follows.

The physical explanation requires the use of a suitable zero reference from which the cx and ct rotor displacements are measured. For convenience this reference is taken as the previously defined electrical zero positions. At this point it is assumed that the rotors of the cx and ct synchros are in exact correspondence. This situation can be readily illustrated through a graphical interpretation of Eq. (12-48). In the interests of simplifying the representation, only the fundamental component and the fifth and seventh harmonics are assumed present in the ct stator mmf wave. At the electrical zero position, α is zero. Similarly, to place the ct rotor at electrical zero it is assumed that Γ is also zero. Equation (12-48) then yields the following results:

$n = 1,$

$$f_{1r} = \frac{3}{2}(N_e I) \sin \omega t \cos (\alpha - \Gamma) = \frac{3}{2}(N_e I) \sin \omega t \cos 0°$$

$$= \frac{3}{2}(N_e I) \sin \omega t \equiv F_1 \tag{12-52}$$

$n = -5,$

$$f_{5r} = \frac{3}{2}\left(\frac{N_e I}{5}\right) \sin \omega t \cos (0° + 5 \times 0°)$$

$$= \frac{3}{2}\left(\frac{N_e I}{5}\right) \sin \omega t \equiv F_5 \tag{12-53}$$

$n = +7,$

$$f_{7r} = \frac{3}{2}\left(\frac{N_e I}{7}\right) \sin \omega t \cos (0° - 7 \times 0°)$$

$$= \frac{3}{2}\left(\frac{N_e I}{7}\right) \sin \omega t \equiv F_7 \tag{12-54}$$

where F_1, F_5, and F_7 are defined as indicated. A glance at these expressions shows that all three components are in *space* phase as well as in *time* phase. The space-phase condition is graphically illustrated in Fig. 12-11(a). Note that the line of action of the resultant mmf F_R coincides with that of the fundamental F_1 because the harmonic terms are in phase at this point too. The coupling with the ct rotor winding is therefore zero so that no emf occurs. The control transformer is accordingly at the null† position.

Next consider the case where the cx rotor is displaced by 20° in the positive counterclockwise direction. Then imagine that the rotor of the ct synchro is

---

†The null position is the position of minimum voltage in the ct rotor winding, which in this case is zero.

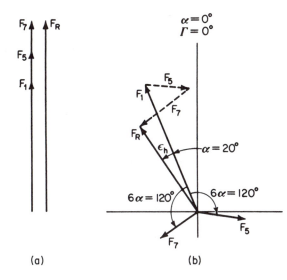

Figure 12-11 Graphical representation of the displacement error caused in a CX-CT synchro configuration by the CT mmf harmonics: (a) condition at electrical zero; (b) location of the resultant mmf when $\alpha = 20°$ and the CT rotor is displaced by $\Gamma = \alpha = 20°$.

displaced by a like amount in the same direction, i.e., $\Gamma = 20°$. If the resultant field mmf continues to be in space phase with the fundamental component, then a null voltage occurs at $\Gamma = 20°$ and so no displacement error exists. On the other hand, if the resultant mmf, which produces the net field flux, does not lie along the $20°$ line, then some coupling with the CT rotor winding exists and so causes a small voltage to appear between its terminals. To reduce this voltage to zero it will then be necessary to displace the CT rotor by an additional amount, ε_h. The CX-CT synchro configuration is then said to have a displacement error of ε_h at $\alpha = 20°$. By resorting once again to Eq. (12-48) we can ascertain which condition prevails. Accordingly, we get

$n = 1$,

$$f_{1r} = \frac{3}{2}(N_e I) \sin \omega t \cos(\alpha - \Gamma) = F_1 \cos(0° - 0°)$$

$$= F_1 \cos 0° \tag{12-55}$$

$n = 5$,

$$f_{5r} = \frac{3}{2}\left(\frac{N_e I}{5}\right) \sin \omega t \cos(\alpha + 5\Gamma) = F_5 \cos[20° + 5(20°)]$$

$$= F_5 \cos 120° \tag{12-56}$$

$n = 7$,

$$f_{7r} = \frac{3}{2}\left(\frac{N_e I}{7}\right) \sin \omega t \cos(\alpha - 7\Gamma) = F_7 \cos[20° - 7(20°)]$$

$$= F_7 \cos(-120°) \tag{12-57}$$

A graphical representation of these results is shown in Fig. 12-11(b). The fifth and seventh harmonics are now no longer in phase with the fundamental component as can be seen by the difference in the argument of the cosine terms in the foregoing equations. Keep in mind that the angle Γ in these equations denotes the amount by which the rotor of the CT synchro is displaced. To determine whether or not a null results, Γ is set equal to α. Hence for the fundamental component the argument of the cosine term becomes identically zero. The argument of the cosine term can thus be considered to be an angle that is measured relative to the position of the CX rotor axis. The zero-degree value for the fundamental means that there is exact correspondence between the axis of the fundamental CT field and the CX rotor axis. In the case of the fifth harmonic the argument is 120°, which means that the axis of the fifth-harmonic mmf lies 120° displaced from the location of the fundamental component. In the diagram, F_5 is depicted *behind* F_1 by 120° because the fifth harmonic is known to rotate in a direction opposite to that of the fundamental component. However, note that F_5 is displaced 100° from the electrical zero reference line, which is consistent with the value $n\Gamma = -5(20°) = -100°$. For the seventh harmonic, which travels in the same direction† as the fundamental, F_7 is shown ahead of F_1 by 120°. Note again that this is consistent with the fact that F_7 is displaced by $n(20°) = 7(20°) = 140°$ in the positive direction from the zero reference line. It is also worthwhile to note that the positions of F_5 and F_1 are *sixth*-harmonic functions of the CX rotor displacement angle α.

In general the sum of the fundamental mmf component and the harmonics yields a resultant mmf that is displaced from the fundamental by the angle ε_h. Accordingly, if the CX-CT arrangement is used as an error detector in a servomechanism and a command is applied at the transmitter in the form of $\alpha = 20°$, the servomechanism achieves a position of equilibrium (i.e., a null position) when the CT synchro is displaced through the mechanical feedback connection by an amount equal to $\alpha + \varepsilon_h$. Therefore, the output follows the command but with an error of ε_h.

Mathematical Analysis

An equation for the displacement error ε_h can be found by first determining the expression for the null voltage. This can be conveniently done by finding the voltage induced in the transformer rotor winding by each harmonic component of the revolving mmf. Keeping in mind that at the electrical zero position the axis of the CT rotor winding is displaced by 90 electrical degrees from the assumed zero position of the stator mmf axis (i.e., the $\Gamma = 0$ degree point), the voltage induced in the rotor winding by the fundamental field component is

$$E_1 = K_1 \cos \left[\alpha - \left(\Gamma - \frac{\pi}{2} \right) \right] \qquad (12\text{-}58)$$

---

†The minus sign in the argument of the cosine term of Eq. (12-48) represents positive rotation.

where

$$K_1 = 4.44 f\, N_r K_{wr} \Phi_1 \qquad (12\text{-}59)$$

Note that K_1 actually denotes the rms voltage induced in the rotor winding by the fundamental field component corresponding to maximum coupling. Accordingly, Eq. (12-58) describes the rms value of the voltage induced in the rotor winding for a given CX rotor displacement α and a CT rotor displacement Γ. The cosine term of this equation is simply a factor that indicates the degree of coupling that exists between the fundamental field flux Φ_1 produced by the fundamental field mmf and the CT rotor winding. Equation (12-58) may be rewritten as

$$E_1 = -K_1 \sin(\alpha - \Gamma) \qquad (12\text{-}60)$$

This result can now be used to find the voltage induced in the rotor winding when the CT rotor is placed at the null position. It is important to keep in mind that in general the coupling of the fundamental flux component with the rotor winding will not be zero at the null point. This is graphically illustrated in Fig. 12-11, where the rotor winding is at null when its axis is in quadrature with the resultant mmf. However, note that there is some small amount of coupling of the winding with the fundamental field component even though the coupling with the resultant field is zero. Since the null position is found by displacing the CT rotor by $\Gamma = \alpha + \varepsilon_h$ rather than by $\Gamma = \alpha$, it follows that the rms value of the voltage induced by the fundamental field at null is

$$E_{1\text{ null}} = -K_1 \sin(\alpha - \Gamma) = -K_1 \sin(\alpha - \alpha - \varepsilon_h) = K_1 \sin \varepsilon_h \qquad (12\text{-}61)$$

Because such displacement errors are often very small, the last expression may be written as

$$E_{1\text{ null}} = K_1 \varepsilon_h \qquad (12\text{-}62)$$

A similar procedure can be followed to find the emf induced in the CT rotor by the fifth harmonic of the field mmf. Thus the general expression for the rms value of this voltage is

$$E_5 = K_5 \cos\left[\alpha + 5\left(\Gamma - \frac{\pi}{2}\right)\right] \qquad (12\text{-}63)$$

where

$$K_5 = 4.44 f\, N_r K_{wr5} \Phi_5 \qquad (12\text{-}64)$$

Here K_{wr5} is the rotor winding factor for the fifth harmonic and Φ_5 denotes the fifth-harmonic flux per pole. In Eq. (12-63) the $\pi/2$ quantity is treated in the same way as Γ because it is a space angle associated with the CT rotor. By means of an appropriate trigonometric identity Eq. (12-63) can be rewritten as

$$E_5 = K_5 \sin(\alpha + 5\Gamma) \qquad (12\text{-}65)$$

When the CT rotor is assumed to be placed at its null position, Γ takes on the value

$\alpha + \varepsilon_h$, which upon insertion into Eq. (12-65) yields

$$E_{5\,\text{null}} = K_5 \sin (\alpha + 5\Gamma) = K_5 \sin (\alpha + 5\alpha + 5\varepsilon_h) = K_5 \sin (6\alpha + 5\varepsilon_h) \qquad (12\text{-}66)$$

An important observation to make at this point is that the voltage induced by the fifth-harmonic field mmf is of the *same* frequency as that of the fundamental component. Moreover, it is in *time phase* with the fundamental. Therefore the fundamental and the harmonic voltage contributions can be added directly to obtain the resultant voltage.

The expression for the rms value of the induced voltage for the seventh harmonic is found to be

$$E_7 = K_7 \sin (\alpha - 7\Gamma) \qquad (12\text{-}67)$$

where K_7 is defined in a manner similar to that for K_5. The corresponding equation for the null voltage associated with this harmonic is then

$$E_{7\,\text{null}} = -K_7 \sin (6\alpha + 7\varepsilon_h) \qquad (12\text{-}68)$$

It is interesting to note that the argument in Eqs. (12-66) and (12-68) involves a sixth-harmonic function of α. It is reasonable to expect therefore that extension of the foregoing procedure to the eleventh and thirteenth mmf harmonics leads to expressions that involve twelfth-harmonic functions of α. Specifically it can be shown that

$$E_{11\,\text{null}} = K_{11} \sin (12\alpha + 11\varepsilon_h) \qquad (12\text{-}69)$$

$$E_{13\,\text{null}} = K_{13} \sin (12\alpha + 13\varepsilon_h) \qquad (12\text{-}70)$$

Of course, as higher order harmonics are considered, their importance diminishes because the magnitudes of the K coefficients vary inversely with the order of the harmonic. Hence the harmonics that bear the greatest influence in causing displacement errors are the fifth and seventh.

The complete expression for the null voltage that occurs when the CT rotor is placed at the position $\Gamma = \alpha + \varepsilon_h$ may be written as

$$E_{\text{null}} = E_{1\,\text{null}} + E_{5\,\text{null}} + E_{7\,\text{null}} + \cdots \qquad (12\text{-}71)$$

An algebraic sum is permitted because these voltages are of the same frequency and in time phase. Inserting Eqs. (12-62), (12-66), (12-68), (12-69), and (12-70) yields

$$E_{\text{null}} = \varepsilon_h K_1 + K_5 \sin (6\alpha + 5\varepsilon_h) - K_7 \sin (6\alpha + 7\varepsilon_h)$$
$$+ K_{11} \sin (12\alpha + 11\varepsilon_h) - K_{13} \sin (12\alpha + 13\varepsilon_h) + \cdots \qquad (12\text{-}72)$$

This expression can be simplified by noting that the quantity $n\varepsilon_h$ appearing in the arguments of the sine terms can be neglected with little or no error. Accordingly, we have

$$E_{\text{null}} = \varepsilon_h K_1 + (K_5 - K_7) \sin 6\alpha + (K_{11} - K_{13}) \sin 12\alpha + \cdots \qquad (12\text{-}73)$$

However, with the CT rotor placed at the null position the resultant emf induced is zero. Hence

$$E_{null} = 0 = \varepsilon_h K_1 + (K_5 - K_7) \sin 6\alpha + (K_{11} - K_{13}) \sin 12\alpha + \cdots \qquad (12\text{-}74)$$

from which it follows that the expression for the displacement error is

$$\varepsilon_h = \frac{K_5 - K_7}{K_1} \sin 6\alpha + \frac{K_{11} - K_{13}}{K_1} \sin 12\alpha + \cdots \qquad (12\text{-}75)$$

An examination of Eq. (12-75) discloses that the magnitude of the displacement error caused by the space mmf harmonics in a CX-CT synchro configuration can be controlled by properly manipulating the K coefficients. Thus, if K_5 and K_7 are made equal, the sixth harmonic displacement error is entirely eliminated. This result can be achieved in some synchros through the judicious selection of the winding factors for these harmonics. When cancellations cannot be accomplished, the winding design is chosen so that the winding factors for the harmonics are kept as small as possible.

12-7 STATIC ERRORS AND RESIDUAL VOLTAGES

The accuracy of a servomechanism can be no better than the synchro pair that is used as the error detector. Hence the static displacement errors must be kept as small as manufacturing tolerances will allow, consistent of course with economical production. A typical static error curve for a CX-CT synchro error detector is depicted in Fig. 12-12. This curve is obtained by displacing the rotor of the transmitter by an amount α and then displacing the CT rotor to that position which

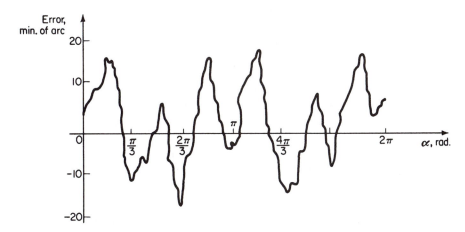

Figure 12-12 Typical displacement error curve for a CX-CT synchro pair. Note the predominance of a six-cycle and a two-cycle error variation.

yields a minimum voltage across its rotor winding terminals. The difference between the CT rotor displacement Γ and the CX rotor displacement α constitutes the static displacement error. In well-designed synchros this error rarely exceeds $\frac{1}{2}$ degree. A study of Fig. 12-12 reveals the predominance of a six-cycle error. As already described in the preceding section, this is due to the effect of the mmf harmonics in the stator of the control transformer. Note that elimination of this source of error goes a long way towards increasing the accuracy of the CX-CT synchro pair.

Another dominant component that can readily be distinguished in the static error curve is a double frequency error of the CX rotor displacement α. The source of this second harmonic term lies in the slight unbalance that necessarily occurs in the stator phase windings of the synchros. It can be shown that, if the resistive and reactive components of the phase impedances are not identical, then a displacement error exists which varies as a second harmonic function of α. The nonidentical nature of the windings is very often attributable to manufacturing difficulties. These often involve one or more of the following factors: slight deviations in the turns used for each phase, varying tensions on the wire spools during the winding process, slight deviations in the cross-sectional area of the wires obtained from different spools or different manufacturers, slight differences in the leakage reactance of each phase winding, and finally the effect of tooth iron saturation. The relationship between the static error and the unbalanced impedances discloses that unbalances in the reactive parts of the impedances exert a greater influence on the static error than unbalance in the resistive parts.

Another source of a double frequency error in the error curve is the absence of a perfectly uniform air gap. This can happen for several reasons: ellipsing of the rotor, ellipsing of the stator, an eccentrically placed rotor, or a skewed rotor axis. Usually production methods are such that there is no difficulty in getting a perfectly round rotor. However, the same is not true for the stator structure; there is a greater tendency for the stator to come off the production line slightly ellipsed. Actually the ellipsing need not be very large to cause a significant static error. For example, a difference between the major and minor axes of the stator of only 0.0001 in. can cause a static error spread of as much as 3 to 4 minutes of arc.

Besides static displacement errors, the CX-CT synchro pair has the additional shortcoming of residual voltages at null. Whenever residual voltages do in fact exist, they cannot contain a component of fundamental frequency that is in phase with the applied line voltage. After all, the null position is achieved whenever the in-phase fundamental frequency component is zero. Hence, any voltage that exists at null must be due to other sources. Figure 12-13 depicts a typical curve showing how the residual voltage varies with α (as the transformer is placed at the null position for each value of α).

Analysis shows that the residual voltage contains two components: time harmonics of the fundamental frequency and quadrature voltages. The time harmonics in turn arise from two sources, i.e., the saturation of the iron used in the

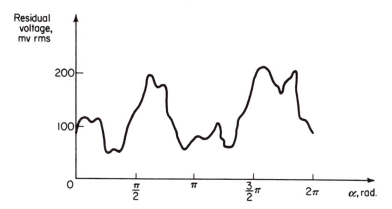

Figure 12-13 Typical variation of a residual voltage content in a CX-CT configuration at null.

magnetic circuit of the synchros as well as the nonuniform field distribution in the air gap of the control transmitter. The quadrature voltage is defined as the minimum voltage of fundamental frequency existing at null. Its presence is caused either by unbalanced stator phase windings or a nonuniform air gap or by a combination of both. In each instance, however, it can be shown that the quadrature voltage is a double-frequency function of α. A glance at Fig. 12-13 shows the prevalence of the second harmonic term. Moreover, analysis also shows that resistance unbalance in the synchro phase windings exerts a greater influence than reactance unbalance does. It is for this reason that switches in the stator line connections of the CX-CT synchro pair should be avoided if at all possible.

Good synchro design focuses on keeping the residual voltages as small as is economically possible especially when these units are used as the error detectors in servo systems. Unusually large residual voltages can easily cause saturation of servoamplifiers which provide the control voltages that drive the output devices. When saturation sets in, the system gain is sharply reduced, which adversely affects the total system behavior. Moreover, although the residual voltages are incapable of producing any useful output in such systems, they can nevertheless cause currents to flow in the output devices that create unnecessary heat losses.

PROBLEMS

12-1. Explain why an umbrella-like rotor construction keeps to a minimum the magnetizing current drawn by the control transformer.

12-2. Prepare a rough sketch to show how the magnetizing current of a control transformer equipped with a salient-pole rotor varies with rotor displacement.

12-3. When the rotor of the synchro depicted in Fig. P12-3 is at the electrical zero position, the rms value of the phase voltage induced in winding 2 is measured to be 52 V. Determine the voltage induced in each winding from neutral-to-line and from line-to-line when the rotor is at a position 60° from electrical zero in a counterclockwise direction. Clearly indicate the polarity markings for each voltage.

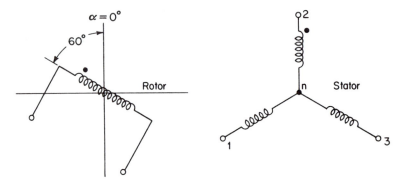

Figure P12-3

12-4. Repeat Prob. 12-3 for the instance where the rotor occupies the following positions:
   (a) 180° counterclockwise.
   (b) 90° clockwise.

12-5. A control transmitter, a control differential, and a control receiver are connected in the manner shown in Fig. 12-8(a). The control transmitter is displaced 60° counterclockwise and fastened. The control receiver is displaced 30° clockwise and fastened. In which direction and by what amount does the differential synchro move? In your solution, redraw the diagram for the new position of the synchro, and clearly indicate the rms value of induced emf in each synchro winding. Assume that the maximum rms voltage from line-to-neutral in each unit is E.

12-6. The output voltage of a control transmitter is fed to the stator of a differential synchro whose rotor output is then applied to the stator of a control transformer. Assume that, initially, each synchro is at its proper electrical zero position. If the rotor of the transmitter is displaced 90° counterclockwise and the rotor of the differential synchro is turned 60° counterclockwise, determine the value of the induced voltage at the rotor terminals of the control transformer, expressed in terms of its maximum voltage E.

12-7. The output voltage of a control transmitter is fed to the stator of a differential synchro whose output is then applied to the stator of a control transformer.
   (a) Draw the schematic diagram of the system with each unit at the electrical zero positions.
   (b) If the rotor of the transmitter is displaced 30° counterclockwise and that of the differential synchro is turned 60° clockwise, find the value of the induced voltage at the rotor terminals of the CT in terms of its maximum voltage E.

12-8. A synchro transmitter has its stator connected to the stator of a differential synchro whose rotor, in turn, is connected to the stator of a synchro control transformer. The rotor of the transmitter is properly energized and all three synchros are adjusted to their zero positions. The transmitter rotor is then turned 15° clockwise, and the differential rotor is turned 75° counterclockwise.

(a) Find the angle and direction of the flux axis in each synchro.

(b) Determine the voltage across the rotor terminals of the control transformer in terms of the maximum possible value of this voltage E.

12-9. Repeat Prob. 12-8 for the case where the rotor of the CX is turned 45° clockwise and that of the CD is displaced 90° counterclockwise.

12-10. With the aid of Eq. (12-48a) demonstrate that the resultant mmf associated with the fifth harmonic travels in the negative Γ direction. Assume that an observer is placed at $\Gamma = 30°$.

12-11. Determine the maximum value of the sixth harmonic error in a CX-CT synchro transmission unit where the effect of the fifth space harmonic is 20% and that of the seventh harmonic is 14% of the fundamental. Assume that all other errors are negligible.

12-12. The mmf distribution in the air gap of the control transformer in a CX-CT synchro transmission configuration is known to have a fundamental component as well as fifth and seventh harmonics. The fifth harmonic flux per pole is one-fifth that of the fundamental, and the seventh harmonic flux per pole is one-seventh that of the fundamental. Moreover, the rotor winding factor is one-fourth and one-fifth for the fifth and seventh harmonics, respectively. Assume for convenience that the displacement error due to the space harmonics is zero at the electrical zero position.

(a) For a CX rotor displacement of 30°, compute the value of displacement error that occurs when the CT is turned to the null position.

(b) Repeat part (a) for a CX displacement of 240°.

(c) Prepare a rough diagram of the conditions prevailing in part (b).

12-13. An ideal control transmitter is used with a control transformer in a typical CX-CT voltage indicating system. The air-gap flux of the CT contains a fifth and seventh harmonic component in addition to the fundamental. The magnitude of the flux per pole of the harmonics is inversely proportional to the order of the harmonic. A rotor winding is used for the CT unit, which causes a 5% reduction in effective induced voltage. This winding also yields identical winding factors for the harmonics. Compute the harmonic winding factor that corresponds to a sixth harmonic displacement error which never exceeds 1°.

12-14. The predominant space harmonics in the control transformer of a CX-CT arrangement are known to be the 5th, 7th, 11th, and 13th. The winding factors for the fundamental and the 11th and 13th harmonics are each 0.96, while for the 5th and 7th they are 0.25 and 0.2, respectively. Assume that the harmonic displacement error is zero at the electrical zero of the CT.

(a) Compute the value of the harmonic displacement error for the following values of the CX rotor displacement: 5°, 10°, 15°, 20°, 25°, and 30°.

(b) Sketch the general nature of the curve that passes through these points.

12-15. Refer to the CX-CT synchros of Prob. 12-11, and assume that the harmonic displacement function passes through zero at the electrical zero position of the CT unit. Also, assume that the maximum rms voltage induced by the fundamental flux is E_1 V.

  (a) What is the value of the voltage induced in the rotor of the CT by the fundamental flux, when the CT is at the null position corresponding to a CX rotor displacement of 10°?
  (b) What is the value of the fifth harmonic induced voltage in part (a)?
  (c) Is the quantity computed in part (b) of the same frequency as the fundamental? Explain.

12-16. Repeat Prob. 12-15 for the conditions specified in Prob. 12-12.

13

Dynamics of Electric Machines

In this chapter we study the dynamic behavior of electromechanical devices. In all the preceding chapters the analysis was concerned with steady-state performance only. However, nowadays, a knowledge of the behavior of electromechanical devices in the transient state is at least equally important because this equipment is often found as part of a total system. For example, generators and motors are frequently found in the same interconnected electric power grid system. As loads change and shift, it is the dynamic characteristic of the individual machines taken together that will determine whether the system can react in a stable manner to such disturbances. Similarly, in many feedback control systems such as the servomechanism, it is not uncommon to find dc motors, two-phase servomotors, pilot generators, amplidynes, and other electromechanical devices as important components of such systems. The overall behavior and stability of the entire control system is often greatly influenced by the dynamic characteristics of these devices. An understanding of the steady-state characteristics alone is not sufficient to provide a useful study of total performance in these situations. After all, if the system of interest fails to achieve steady-state operation because of poor dynamics, an understanding of the steady-state performance is really only academic.

The *transfer function* and the state variable method are the vehicles by which the dynamic characteristics of electromechanical devices are described in this treatment. However, the development proceeds on the assumption that the reader already has a background in Laplace transforms.

13-1 THE TRANSFER FUNCTION

As the name implies, the *transfer function* is a mathematical formulation that relates the output variable of a device to the input variable. For linear devices the transfer function is dependent solely upon the parameters of the device plus any operations of time, such as differentiation and integration, which it may possess. It is independent of the input quantity—both its magnitude and variation.

The general procedure for deriving the transfer function is always the same irrespective of the type component or engineering device under study. It involves the following three steps: (1) determine the governing equation for the device expressed in terms of the output and input variables; (2) Laplace transform the governing equation, assuming all initial conditions are zero; and (3) rearrange the equation to formulate the ratio of the output to input variable.

An example is useful at this point to illustrate the procedure and to show how the dynamic behavior is thereby described. Refer to Fig. 13-1. Assume that

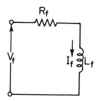

Figure 13-1 Field winding circuit of an electromechanical energy converter. Example of a first-order system.

the input quantity is the voltage applied to the field winding of a machine, V_f, and that the desired output quantity is the field winding current, I_f. The equation that relates I_f to V_f in terms of the appropriate parameters of the circuit is by Kirchhoff's voltage law

$$V_f = I_f R_f + L_f \frac{dI_f}{dt} \qquad (13\text{-}1)$$

Equation (13-1) is the governing equation and specifically is a first-order differential equation because it involves one energy-storing element, the inductance L_f. Laplace transforming Eq. (13-1) for zero initial field winding current leads to the following transformed differential equation:

$$V_f(s) = I_f(s)R_f + L_f s I_f(s) \qquad (13\text{-}2)$$

where $V_f(s)$ denotes the Laplace transform of V_f, $I_f(s)$ denotes the Laplace transform of I_f, and the Laplace operator s denotes the first derivative of time. Upon formulating the ratio of output to input variable in the last equation, the desired transfer function results. Thus

$$\boxed{\frac{I_f(s)}{V_f(s)} = \frac{1}{R_f + sL_f} = \frac{1}{R_f}\frac{1}{1 + s\tau_f}} \tag{13-3}$$

where

$$\tau_f = \frac{L_f}{R_f} \tag{13-4}$$

and is the time constant of the field winding.

Inspection of the right side of Eq. (13-3) discloses that the transfer function is solely dependent on the circuit parameters R_f and L_f and the derivative operator s. Moreover, the dynamic behavior of this circuit is completely determined by the form of the *denominator of the transfer function*. This is because the denominator of the transfer function is the *characteristic equation* of the governing differential equation when set equal to zero. Reference to Eq. (13-1) makes this readily apparent. To obtain the characteristic equation by the classical method of solving differential equations, it is necessary first to set the forcing function V_f equal to zero and then to put the remaining terms in operator form. Thus

$$0 = R_f I_f + L_f s I_f \tag{13-5}$$

or

$$R_f + sL_f = 0 \tag{13-6}$$

A comparison of Eq. (13-6) with the denominator of Eq. (13-3) reveals the two expressions to be identical.

The complete time description of the output variable can always be found from the transfer function. This includes steady-state as well as transient behavior. In the case of the *R-L* circuit, e.g., the expression for the transformed solution of the field current can be written from Eq. (13-3) as

$$I_f(s) = \frac{1}{R_f + sL_f} V_f(s) \tag{13-7}$$

If V_f is now identified for convenience as a step voltage of magnitude V_f applied to the field winding, the corresponding Laplace transform is

$$V_f(s) = \frac{V_f}{s} \tag{13-8}$$

Inserting this expression into Eq. (13-7) yields

$$I_f(s) = \frac{V_f}{L_f s(s + R_f/L_f)}\frac{1}{} \tag{13-9}$$

A partial fraction expansion allows Eq. (13-9) to be written as

$$I_f(s) = \frac{V_f}{L_f s(s + R_f/L_f)} = \frac{K_0}{s} + \frac{K_1}{s + R_f/L_f} \tag{13-10}$$

Evaluation of K_0 and K_1 then leads to

$$I_f(s) = \frac{V_f}{R_f}\left(\frac{1}{s} - \frac{1}{s + R_f/L_f}\right) \tag{13-11}$$

By employing a table of Laplace-transform pairs the corresponding time solution is found to be

$$i_f(t) = \frac{V_f}{R_f}(1 - \varepsilon^{-t/\tau_f}) \tag{13-12}$$

where $\tau_f = L_f/R_f$.

A comparison of Eq. (13-12) with Eq. (13-11) shows that the transient term $-\varepsilon^{-t/\tau_f}$ is directly related to that term of the transformed version of the solution that results from the transfer function [the second term of Eq. (13-11)].

As a matter of convenience the transfer function is frequently represented graphically by the block diagram notation depicted in Fig. 13-2(a). Sometimes the *signal flow diagram* shown in Fig. 13-2(b) is used. The output quantity of the block is always the Laplace transformed version of the output variable and the input quantity is the Laplace transformed version of the input variable.

Systems often are represented by a block diagram showing the interconnected transfer functions of the individual components of which they are comprised. Large complex systems contain many direct transfer functions of the type depicted in Fig. 13-2, but they may also contain feedback transfer functions. Whenever a feedback transfer function appears around a direct transfer function,† the total transfer function of the combination is readily found by applying the feedback formula which is now derived. Refer to the block diagram appearing in Fig. 13-3. The quantity $G(s)$ denotes the direct transmission function between the reference input signal $R(s)$ and the output signal $C(s)$. The quantity $H(s)$ represents the feedback function. Note that the output quantity $C(s)$ is operated on by this function and then *subtracted* from the reference input $R(s)$ to generate the

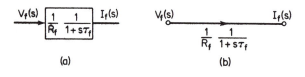

Figure 13-2 Graphical representation of the transfer function of the circuit of Fig. 13-1: (a) block diagram form; (b) signal flow diagram form.

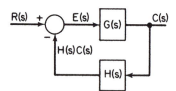

Figure 13-3 Block diagram for a feedback system.

†A direct transfer function of a system component is one that appears in the direct path linking the input and output variables.

actual input signal $E(s)$ to the direct transfer function $G(s)$. The quantity $E(s)$ is often called the *actuating signal* of the feedback system. The reference input signal and the feedback signal must be of opposite signs. This is a desirable condition for suitable operation of the feedback system and is commonly termed *negative* feedback. Whenever a feedback path such as $H(s)$ is wrapped around a direct path such as $G(s)$, the significant transfer function is the one that relates the output $C(s)$ to the input $R(s)$. The ratio of these two quantities can be readily expressed in terms of $H(s)$ and $G(s)$. A glance at Fig. 13-3 makes it apparent that the output quantity can be expressed as

$$E(s)G(s) = C(s) \tag{13-13}$$

But

$$E(s) = R(s) - H(s)C(s) \tag{13-14}$$

so that upon its insertion into Eq. (13-13) there results

$$[R(s) - H(s)C(s)]G(s) = C(s) \tag{13-15}$$

This last equation now involves just the input and output variables. Accordingly, after terms are collected and the ratio of output to input is formulated, the desired *closed-loop transfer function $T(s)$* results. Thus

$$\boxed{T(s) = \frac{G(s)}{1 + H(s)G(s)}} \tag{13-16}$$

The plus sign appears in the denominator whenever negative feedback is used. The quantity $H(s)G(s)$ is called the *loop transfer function* since it is obtained by taking the product of the transfer functions encountered in traversing a closed loop.

In dealing with systems that involve several energy-storing elements identified in the form of Eq. (13-3), it is sometimes useful to represent the known transfer function in terms of an equivalent transfer function having a direct transfer function that is pure integration (as denoted by the Laplace operator $1/s$). The advantage of such a block diagram representation is that the respective output quantities of all the pure integrators in the direct transmission path are then in fact the *state variables*† of the total system in terms of which the complete system

---

†When dealing with systems containing energy-storing elements, it is convenient to use as the dependent variables those variables by which the system's energy state can be described at any specified time. For example, in an R-L circuit the particular state of the circuit can always be conveniently identified in terms of the current irrespective of the past history of the circuit. This is because a knowledge of the current immediately yields the energy state by the relationship $\frac{1}{2}Li^2$. A similar situation prevails for the capacitor. Here, however, it is convenient to choose the instantaneous value of the voltage across the capacitor v_c as the state variable. Once v_c is known, the state of the network containing C is known through the relationship $\frac{1}{2}Cv_c^2$. The state function depends solely upon the final state of the element.

The principal advantage of working with state variables is that it makes it possible to represent

behavior can be described. In the simple case of the first-order system that leads to Eq. (13-3), it is possible through appropriate manipulation to rewrite the direct transfer function in the form of Eq. (13-16). Keeping in mind that it is desirable to identify $G(s)$ as $1/s$ in the equivalent formulation of Eq. (13-16), it is helpful to divide numerator and denominator of Eq. (13-3) by $s\tau$. This leads to

$$\frac{1}{R_f} \frac{1}{1 + s\tau_f} = \frac{1}{R_f} \frac{1/s\tau_f}{(1/s\tau_f) + 1} = \frac{1}{R_f\tau_f} \left[\frac{1/s}{1 + (1/s)(1/\tau_f)} \right] \quad (13\text{-}17)$$

A block diagram representation of the bracketed term of Eq. (13-17) appears in Fig. 13-4(b). Note that it is identical to the representation depicted in Fig. 13-4(a), but the former does identify a pure integration block in the direct transmission path.

The transformed equation associated with the block diagram of Fig. 13-4(b) by inspection can be written as

$$V_f(s) \frac{1}{R_f\tau_f} - \frac{1}{\tau_f} I_f(s) = sI_f(s) \quad (13\text{-}18)$$

The corresponding time equation is obtained by replacing s with d/dt and by dropping the "of s" notation.† Accordingly, we get

$$\frac{dI_f}{dt} = \frac{V_f}{R_f\tau_f} - \frac{1}{\tau_f} I_f \quad (13\text{-}19)$$

Equation (13-19) is called the *state equation* for this system since it involves a relationship between the state variable I_f and its first derivative. A little reflection of course discloses that this equation is identical with the original governing differential equation [see Eq. (13-1)].

13-2 DYNAMIC BEHAVIOR OF THE DC MOTOR WITH CONNECTED MECHANICAL LOAD

As already pointed out in the preceding section, the dynamic behavior of electromechanical devices is determined by first finding the appropriate transfer function

---

the state of a system containing n independent energy-storing elements in terms of n *state equations*, which are always first-order differential equations. Thus, if a system contains three independent energy-storing elements, the total system behavior can be described by three state equations. In each instance the state equation is a first-order differential equation involving the state variables and possibly an external forcing function. In contrast, if the total system behavior were to be represented by a single governing equation, it would be a third-order differential equation. An outstanding advantage of the state approach to system analysis is that it lends itself naturally to easy simulation on the analog and digital computers.

†Although it is customary to use lower-case letters for time functions and capital letters for the Laplace-transformed functions, that practice is not followed in this chapter because of the desire to preserve the identity of the variables with the notation used in preceding chapters. As a substitute this distinction is made by the absence or presence of the "of s" notation.

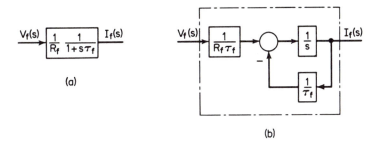

(a)

(b)

Figure 13-4 Alternative block diagram representations of transfer functions: (a) direct representation; (b) equivalent representation, bringing into evidence pure integration in the direct transmission path.

and then applying the techniques of the Laplace transform theory. In the process suitable block diagram representations are developed and the proper state equations identified in order to enhance a computer study of the machine dynamics for those readers who are so inclined. A study of the dynamics of the dc motor is made for two cases—one where the armature winding leakage inductance[†] is negligibly small and the other where it is large enough to be considered.

Figure 13-5 gives the schematic diagram of the dc motor with its connected load. The combined inertia of the load and the rotor of the motor is denoted by J.[‡] The equivalent viscous friction of the motor and the load is denoted by F. The opposing load torque is called T_L. The quantity L_a is the armature leakage inductance which is initially assumed equal to zero; R_a is the armature winding resistance; and ω denotes the speed of the rotor expressed in radians per second.

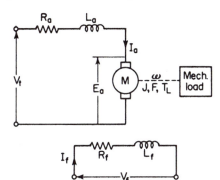

Figure 13-5 Schematic diagram of the dc motor with mechanically connected load. Field current is maintained constant.

Assume we wish to find the manner in which the motor speed responds to changes in the applied armature winding voltage for constant field current. In

[†] This refers to the part of the flux that can be considered produced by the armature winding mmf and not linked with the field winding.

[‡] A convenient set of units is assumed throughout this discussion for J, F, and K. For example, J would be expressed in lb-ft-sec², F in lb-ft-sec, and K in lb-ft/rad.

short, we want to find the transfer function that relates ω to V_t. To do this, it is necessary to establish a governing equation that relates ω to V_t. At the motor shaft the electromagnetic torque of the motor must be equal to the sum of the opposing torques. Thus

$$T = J\frac{d\omega}{dt} + F\omega + T_L \tag{13-20}$$

or, expressed more conveniently,

$$T - T_L = J\frac{d\omega}{dt} + F\omega \tag{13-21}$$

It is helpful to note at this point that Eq. (13-21) is a first-order differential equation; the sole energy-storing element is denoted by the total inertia J. Furthermore, by Eq. (8-23) the electromagnetic torque is expressed as

$$T = K_T\Phi I_a \tag{13-22}$$

where Φ denotes the air-gap flux. Since flux is held fixed in the system of Fig. 13-5, Eq. (13-22) can be more succinctly written as

$$T = K_t I_a \tag{13-23}$$

where K_t is a torque constant expressed in units of torque per ampere. Also, the armature current is related to the terminal voltage by

$$I_a R_a = V_t - E_a \tag{13-24}$$

which follows from Eq. (8-24) for R_s equal to zero. Moreover, the armature induced emf is directly proportional to the speed as shown by Eq. (8-22). Thus

$$E_a = K_E\Phi n = K_\omega\omega \tag{13-25}$$

where K_ω is a speed constant expressed in units of volts per radian per second. Inserting Eq. (13-25) into Eq. (13-24) yields

$$I_a R_a = V_t - K_\omega\omega \tag{13-26}$$

It is interesting to note that the right-hand side of the last equation involves both the input variable V_t and the output variable ω of the transfer function being sought. Accordingly, Eq. (13-26) lends itself to the partial block diagram representation of Fig. 13-6, expressed as a negative feedback arrangement (see solid portion of the diagram). Note further that, if the quantity $I_a R_a$ is multiplied by $1/R_a$, the armature current I_a results. Subsequent multiplication by K_t then yields the developed electromagnetic torque. The last two statements are represented in Fig. 13-6 by the dashed-line blocks.

In order to complete the block diagram, it is necessary to relate torque to speed. Since this relationship is already available in Eq. (13-21), the desired result is found by Laplace transformation. Thus

$$T(s) - T_L(s) = Js\omega(s) + F\omega(s) \tag{13-27}$$

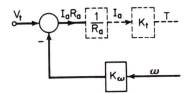

Figure 13-6 The block diagram representation of Eq. (13-26) is drawn with solid lines.

or

$$\frac{\omega(s)}{T(s) - T_L(s)} = \frac{1}{sJ + F} = \frac{1}{F} \frac{1}{1 + s\tau_m} \tag{13-28}$$

where

$$\tau_m = \text{mechanical time constant} = \frac{J}{F} \tag{13-29}$$

Upon adding the block diagram representation of Eq. (13-28) to the partial block diagram of Fig. 13-6, the complete block diagram shown in Fig. 13-7 is obtained. Finally, by applying the feedback formula of Eq. (13-6) to the configuration of Fig. 13-7, the transfer function relating speed to terminal voltage results. Accordingly,

$$\frac{\omega(s)}{V_t(s)} = \frac{(K_t/R_aF)[1/(1 + s\tau_m)]}{1 + (K_tK_\omega/R_aF)[1/(1 + s\tau_m)]} \tag{13-30}$$

Simplifying and collecting terms yields

$$\boxed{\frac{\omega(s)}{V_t(s)} = \frac{K_t}{R_aF + K_tK_\omega} \frac{1}{1 + s\tau'_m}} \tag{13-31}$$

where

$$\tau'_m = \frac{JR_a}{R_aF + K_tK_\omega} \tag{13-32}$$

It is possible of course to derive the desired transfer function without first developing the block diagram. A single equation relating the input and output

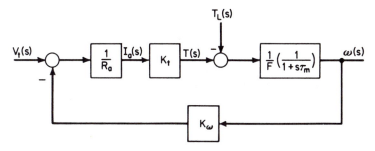

Figure 13-7 Complete block diagram of the system of Fig. 13-5.

variables can be obtained by substituting the expression for I_a from Eq. (13-26) into Eq. (13-23) and then inserting the latter expression into Eq. (13-21). This leads to

$$\frac{K_t}{R_a} V_t = J \frac{d\omega}{dt} + F\omega + \frac{K_t K_\omega}{R_a} \omega + T_L \tag{13-33}$$

If we now let T_L be zero for convenience and then Laplace transform Eq. (13-33), we get

$$\frac{K_t}{R_a} V_t(s) = Js\omega(s) + F\omega(s) + \frac{K_t K_\omega}{R_a} \omega(s) \tag{13-34}$$

Hence

$$\frac{\omega(s)}{V_t(s)} = \frac{K_t/R_a}{sJ + F + (K_t K_\omega/R_a)} = \frac{K_t}{FR_a + K_t K_\omega} \frac{1}{1 + s\tau'_m} \tag{13-35}$$

A comparison of Eq. (13-35) with Eq. (13-31) shows the two to be identical.

What is the dynamic response of the motor speed to a step change in the applied armature voltage V_t for zero load torque? Assuming the step change in applied voltage is V_t, it follows that $V_t(s) = V_t/s$. From the transfer function of Eq. (13-31), the Laplace transformed solution for the speed becomes

$$\omega(s) = \frac{K_t V_t}{R_a F + K_t K_\omega} \frac{1}{s} \frac{1}{1 + s\tau'_m} = \frac{K_t V_t}{JR_a} \frac{1}{s} \frac{1}{s + 1/\tau'_m} \tag{13-36}$$

or

$$\omega(s) = \frac{K_t V_t}{JR_a} \frac{1}{s} \frac{1}{s + 1/\tau'_m} = \frac{K_0}{s} + \frac{K_1}{s + 1/\tau'_m} \tag{13-37}$$

Evaluation of the coefficients of the partial fraction expansion leads to

$$\omega(s) = \frac{K_t V_t}{FR_a + K_t K_\omega} \left(\frac{1}{s} - \frac{1}{s + 1/\tau'_m} \right) \tag{13-38}$$

The corresponding time solution is found to be

$$\omega(t) = \frac{K_t V_t}{FR_a + K_t K_\omega} (1 - \varepsilon^{-t/\tau'_m}) \tag{13-39}$$

Examination of the last expression reveals that the speed rises as depicted in Fig. 13-8 in accordance with a time constant τ'_m to a steady-state speed given by the coefficient on the right side of Eq. (13-39).

How does one obtain the dynamic response of the motor speed to a step change in load torque for a constant applied terminal voltage? The answer is readily obtained once the transfer function between speed and load torque is known. A glance at Fig. 13-7 provides this information. Note that between $T_L(s)$ and $\omega(s)$ there is a direct transmission function of $(1/F)[1/(1 + s\tau_m)]$. The corre-

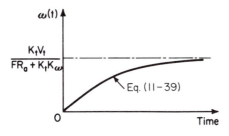

Figure 13-8 Response of motor speed to a step change in V_t for the system of Fig. 13-5.

sponding feedback function in this instance is merely $K_t K_\omega / R_a$. Hence the use of Eq. (13-16) yields as the appropriate transfer function

$$\frac{\omega(s)}{-T_L(s)} = \frac{(1/F)[1/(1 + s\tau_m)]}{1 + (K_t K_\omega / R_a F)[1/(1 + s\tau_m)]} = \frac{R_a}{R_a F + K_t K_\omega}\left(\frac{1}{1 + s\tau_m'}\right) \qquad (13\text{-}40)$$

where τ_m' is defined by Eq. (13-32). Then by following a procedure similar to that employed for a change in V_t the complete expression for $\omega(t)$ is obtained. The details are left as an exercise for the reader.

The block diagram of Fig. 13-7 can be redrawn by employing the manipulation illustrated in Eq. (13-17) to bring into evidence pure integration in the direct transmission path of the system block diagram. This leads to the configuration shown in Fig. 13-9. Since the output of a pure integrator is a state variable, it follows that the motor speed ω is the state variable. The corresponding state equation is found by writing the equation that applies at the summing junction of the integrator. Thus for Fig. 13-9 we get (neglecting T_L for convenience)

$$\frac{T(s)}{F\tau_m} - \frac{1}{\tau_m}\omega(s) = s\omega(s) \qquad (13\text{-}41)$$

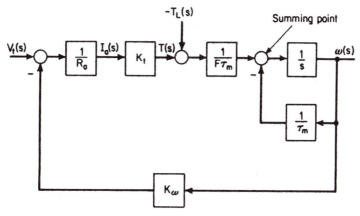

Figure 13-9 Modification of the block diagram of Fig. 13-7, to bring into evidence pure integration in the direct transmission path.

The associated time-domain equation is

$$\frac{d\omega(t)}{dt} = \frac{T}{F\tau_m} - \frac{1}{\tau_m}\,\omega(t) \tag{13-42}$$

Inserting the appropriate expression for T leads to

$$\frac{d\omega(t)}{dt} = \frac{V_t - K_\omega\omega(t)}{R_a}\,K_t\,\frac{1}{F\tau_m} - \frac{1}{\tau_m}\,\omega(t)$$

$$= \frac{K_t}{R_aF\tau_m}\,V_t - \left(\frac{1}{\tau_m} + \frac{K_\omega K_t}{R_aF\tau_m}\right)\omega(t) \tag{13-43}$$

The last expression is the state equation. Note that it involves the state variable $\omega(t)$ and its first derivative. Of course the solution of this equation has already been shown to be Eq. (13-39).

Motor Dynamics with Armature Leakage Inductance Not Negligible

Since the general solution procedure has already been just outlined, attention here is focused only on those parts of the analysis where the presence of L_a bears an influence. Actually this occurs in just one place, Kirchhoff's voltage law for the armature circuit. Thus, instead of the difference between the terminal voltage and the armature induced emf being equal to the I_aR_a drop alone, it is now equal to the armature impedance drop. Expressed mathematically,

$$I_aR_a + L_a\frac{dI_a}{dt} = V_t - E_a \tag{13-44}$$

Laplace transformed, this becomes

$$I_a(s)R_a(1 + s\tau_a) = V_t(s) - K_\omega\omega(s) \tag{13-45}$$

where

$$\tau_a \equiv \frac{L_a}{R_a} \tag{13-46}$$

or

$$\frac{I_a(s)}{V_t(s) - K_\omega\omega(s)} = \frac{1}{R_a(1 + s\tau_a)} \tag{13-47}$$

Equation (13-47) discloses that, to obtain the armature current from the difference between the terminal voltage and the counter armature emf, this difference must be operated upon, not by $1/R_a$, but rather by $1/[R_a(1 + s\tau_a)]$. Hence the block diagram of Fig. 13-7 becomes that illustrated in Fig. 13-10 when $L_a \neq 0$.

Application of the feedback formula of Eq. (13-16) to Fig. 13-10 discloses that the motor speed is now related to a change in terminal voltage by

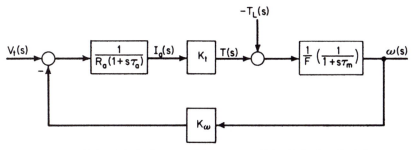

Figure 13-10   Block diagram of the system of Fig. 13-5 when $L_a \neq 0$.

$$\frac{\omega(s)}{V_t(s)} = \frac{\dfrac{K_t}{R_aF(1 + s\tau_a)(1 + s\tau_m)}}{1 + \dfrac{K_tK_\omega}{R_aF(1 + s\tau_a)(1 + s\tau_m)}} \tag{13-48}$$

which simplifies to

$$\frac{\omega(s)}{V_t(s)} = \frac{K_t}{\tau_a\tau_mR_aF} \frac{1}{s^2 + \dfrac{\tau_a + \tau_m}{\tau_a\tau_m}s + \dfrac{1}{\tau_a\tau_m} + \dfrac{K_tK_\omega}{\tau_a\tau_mR_aF}} \tag{13-49}$$

A glance at Eq. (13-49) indicates that now the characteristic equation is second order. This is not surprising in view of the presence of an additional energy-storing element in the form of the armature leakage inductance. The nature of the dynamic response to a step change in applied voltage depends upon the character of the roots of the denominator quadratic in Eq. (13-49). If they are complex conjugate roots with negative real parts, the response is a damped oscillation. If the roots are both negative real, the solution consists of a fixed term plus two exponentially decaying terms.

Appearing in Fig. 13-11 is the block diagram of Fig. 13-10 redrawn to bring into evidence the two pure integration terms associated with the second-order system of Fig. 13-5 when $L_a \neq 0$. The two state variables are the output quantities of the two integrators, armature current I_a and motor speed ω. The two state equations, which are linear first-order differential equations, are obtained by writing the expression for the input signal to each integrator, multiplying by $1/s$, and then equating to the specific state variable. Thus at the first integrator in Fig. 13-11 the transformed version of the state equation is

$$sI_a(s) = \frac{1}{R_a\tau_a}[V_t(s) - K_\omega\omega(s)] - \frac{1}{\tau_a}I_a(s) \tag{13-50}$$

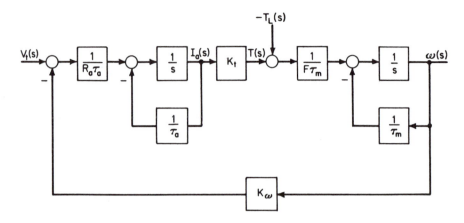

Figure 13-11 Alternative form of the block diagram of Fig. 13-10, to bring into evidence pure integrators in the direct transmission path. The output quantities of the integrators are the state variables of the system.

At the second integrator the result is

$$s\omega(s) = \frac{1}{F\tau_m}[K_t I_a(s) - T_L(s)] - \frac{1}{\tau_m}\omega(s) \qquad (13\text{-}51)$$

The corresponding time-domain form of the state equations is clearly

$$\frac{dI_a}{dt} = -\frac{1}{\tau_a}I_a - \frac{K_\omega}{R_a L_a}\omega + \frac{1}{R_a \tau_a}V_t \qquad (13\text{-}52)$$

$$\frac{d\omega}{dt} = \frac{K_t}{F\tau_m}I_a - \frac{1}{\tau_m}\omega - \frac{1}{F\tau_m}T_L \qquad (13\text{-}53)$$

In both equations, I_a and ω are functions of time. In matrix notation the time-domain formulation of the state equations may be written as follows:

$$\begin{bmatrix} \dfrac{dI_a}{dt} \\[2ex] \dfrac{d\omega}{dt} \end{bmatrix} = \begin{bmatrix} -\dfrac{1}{\tau_a} & -\dfrac{K_\omega}{R_a L_a} \\[2ex] \dfrac{K_t}{F\tau_m} & -\dfrac{1}{\tau_m} \end{bmatrix} \begin{bmatrix} I_a \\[2ex] \omega \end{bmatrix} + \begin{bmatrix} \dfrac{1}{R_a \tau_a} & 0 \\[2ex] 0 & -\dfrac{1}{F\tau_m} \end{bmatrix} \begin{bmatrix} V_t \\[2ex] T_L \end{bmatrix} \qquad (13\text{-}54)$$

Equation (13-54) is called the *normal* form of the state equations describing the complete behavior of the system depicted in Fig. 13-5.

13-3 DC GENERATOR DYNAMICS

How does the armature induced emf of a dc generator respond to a change in field current? How does the armature winding current of the dc generator respond to changes in field current? These are questions that are treated in this section. As

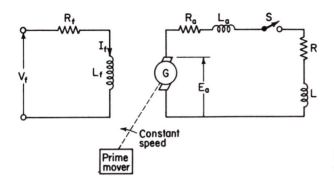

Figure 13-12 Direct-current generator with connected load.

with motor dynamics, the answers lie in establishing the appropriate transfer function relating the variables of interest. Two cases of generator dynamic behavior are treated: one where no external load is applied to the generator (i.e., switch S in Fig. 13-12 is open), and the other where external load is applied (S closed).

No External Generator Load

Consider the generator of Fig. 13-12 driven at constant speed by the prime mover. Let it be desirable to find the transfer function relating the armature induced voltage to the field winding voltage. The differential equation for the field winding circuit is given by

$$V_f = I_f R_f + L_f \frac{dI_f}{dt} \tag{13-55}$$

In general I_f is a function of time whenever V_f or R_f is changing. The Laplace transformation of Eq. (13-55) for zero initial conditions yields

$$V_f(s) = I_f(s)R_f + L_f s I_f(s) \tag{13-56}$$

where $I_f(s)$ denotes the Laplace transform of the time function I_f and $V_f(s)$ is the Laplace transform of V_f. Upon formulating the ratio of $I_f(s)$ to $V_f(s)$, the transfer function between these quantities results. Thus

$$\frac{I_f(s)}{V_f(s)} = \frac{1}{R_f(1 + s\tau_f)} \tag{13-57}$$

where

$$\tau_f \equiv \frac{L_f}{R_f} = \text{field winding time constant} \tag{13-58}$$

The relationship between the field current and the armature induced emf is described by Eq. (7-10), which is repeated here for convenience. Thus

$$E_a = K_E \Phi n = K_f I_f \tag{13-59}$$

The right side of Eq. (13-59) is a valid equivalence provided that no saturation occurs. For consistency the units of K_f are volts per field ampere. Expressed in Laplace notation, Eq. (13-59) becomes merely

$$E_a(s) = K_f I_f(s) \tag{13-60}$$

so that the transfer "function" relating field current to the corresponding induced emf is just a constant. Accordingly,

$$\frac{E_a(s)}{I_f(s)} = K_f \tag{13-61}$$

The complete transfer function is then obtained by combining Eqs. (13-61) and (13-57). Thus

$$\boxed{\frac{E_a(s)}{V_f(s)} = \left[\frac{I_f(s)}{V_f(s)}\right]\left[\frac{E_a(s)}{I_f(s)}\right] = \frac{K_f}{R_f(1 + s\tau_f)}} \tag{13-62}$$

The block diagram representation of the last equation appears in Fig. 13-13.

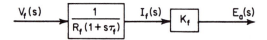

Figure 13-13   Block diagram of the system of Fig. 13-12, with load switch open.

    Depicted in Fig. 13-14 is the form of the block diagram that brings into evidence the pure integrator in the direct transmission path. The output of this integrator is the state variable, namely, field current. The corresponding state equation in transformed notation is found by taking the input to this integrator and setting it equal to $sI_f(s)$. Thus

$$sI_f(s) = \frac{1}{R_f\tau_f} V_f(s) - \frac{1}{\tau_f} I_f(s) \tag{13-63}$$

The corresponding time-domain form is then clearly

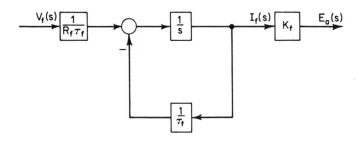

Figure 13-14   Block diagram of Fig. 13-13 modified to show the pure integrator. I_f is the state variable.

$$\frac{dI_f}{dt} = \frac{1}{R_f \tau_f} V_f - \frac{1}{\tau_f} I_f \tag{13-64}$$

Because state equations are only associated with energy-storing elements, E_a does not appear in Eq. (13-64). However, the transition from the state variable I_f to the desired output variable E_a is readily achieved through multiplication by a constant, K_f.

Generator Dynamics Including the Effect of the Load

This situation is represented in Fig. 13-12 by closing switch S. The generator load is assumed to contain inductance as well as resistance. Also, R_a and L_a are respectively the armature winding resistance and leakage inductance.

The transfer function that relates $E_a(s)$ to $V_f(s)$ is identical with Eq. (13-62). Always keep in mind that transfer functions are dependent upon the system parameters—in this instance K_f, R_f, and L_f—and these have not been altered. If it is now desirable to describe how the armature current is influenced by changes in field voltage, then an additional expression is needed that relates E_a to I_a. Clearly this is readily available through the application of Kirchhoff's voltage law to the armature circuit. In differential equation form we have

$$E_a = I_a R_a + L_a \frac{dI_a}{dt} + I_a R + L \frac{dI_a}{dt} \tag{13-65}$$

where R and L are the resistance and inductance of the load respectively. Laplace transforming and collecting terms yield

$$E_a(s) = I_a(s)(R_a + R)(1 + s\tau_A) \tag{13-66}$$

where

$$\tau_A = \text{armature circuit time constant} = \frac{L_a + L}{R_a + R} \tag{13-67}$$

The transfer function then becomes

$$\frac{I_a(s)}{E_a(s)} = \frac{1}{(R_a + R)(1 + s\tau_A)} \tag{13-68}$$

It is worthwhile to note here that the existence of s in the last equation is occasioned by the presence of a second energy-storing element in the system of Fig. 13-12 with the load connected.

The total transfer function relating the armature current to the field voltage is determined by formulating the product of Eqs. (13-62) and (13-68). Accordingly,

$$\boxed{\frac{I_a(s)}{V_f(s)} = \frac{K_f}{R_f(R_a + R)} \frac{1}{1 + s\tau_f} \frac{1}{1 + s\tau_a}} \tag{13-69}$$

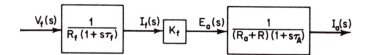

Figure 13-15 Block diagram of the system of Fig. 13-12 when the load switch is closed.

Again note that the denominator of the complete transfer function is now second order in s, which is consistent with the fact that there are now two independent energy-storing elements in the system. The block diagram of the system is depicted in Fig. 13-15.

The modified form of the block diagram of Fig. 13-15 is illustrated in Fig. 13-16. Two pure integrators appear because the system is second order. Hence there are two state variables, which a glance at Fig. 13-16 discloses to be the field current and the armature current. The two state equations are found by employing the procedure previously described. In the s notation these are

$$sI_f(s) = -\frac{1}{\tau_f} I_f(s) + \frac{1}{R_f \tau_f} V_f(s) \tag{13-70}$$

$$sI_a(s) = -\frac{1}{\tau_A} I_a(s) + \frac{K_f}{(R_a + R)\tau_A} I_f(s) \tag{13-71}$$

The corresponding time-domain versions are

$$\frac{dI_f}{dt} = -\frac{1}{\tau_f} I_f + \frac{1}{R_f \tau_f} V_f \tag{13-72}$$

$$\frac{dI_a}{dt} = \frac{K_f}{(R_a + R)\tau_A} I_f - \frac{1}{\tau_A} I_a \tag{13-73}$$

In matrix notation these state equations are written as

$$\begin{bmatrix} \dfrac{dI_f}{dt} \\[2mm] \dfrac{dI_a}{dt} \end{bmatrix} = \begin{bmatrix} -\dfrac{1}{\tau_f} & 0 \\[2mm] \dfrac{K_f}{(R_a + R)\tau_A} & -\dfrac{1}{\tau_A} \end{bmatrix} \begin{bmatrix} I_f \\[2mm] I_a \end{bmatrix} + \begin{bmatrix} \dfrac{1}{R_f \tau_f} \\[2mm] 0 \end{bmatrix} V_f \tag{13-74}$$

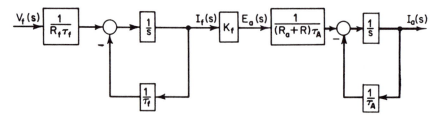

Figure 13-16 Modified version of the block diagram of Fig. 13-15 to bring into evidence the state variables, I_f and I_a, as the outputs of pure integrators.

Example 13-1

The dc generator depicted in Fig. 13-12 has a field winding resistance of 40 Ω and a field inductance of 8 H. The generated emf per field ampere is 100, and the magnetization curve is linear. Moreover, the armature winding resistance and leakage inductance are respectively 0.1 Ω and 0.2 H. The load resistance is 5 Ω and the associated load inductance is 2.35 H.

Determine the time solution for the armature current when a field voltage of 102 V is applied to the field winding. Assume that the prime mover is running at rated speed and that the load switch is closed.

Solution The transfer function that relates the armature current to the field voltage is given by Eq. (13-69). Thus

$$\frac{I_a(s)}{V_f(s)} = \frac{K_f}{R_f(R_a + R)} \frac{1}{1 + s\tau_f} \frac{1}{1 + s\tau_A}$$

$$= \frac{100}{40(0.1 + 5)} \frac{1}{1 + (s/5)} \frac{1}{1 + (s/2)} \tag{13-75}$$

where

$$\tau_f = \frac{L_f}{R_f} = \frac{8}{40} = \frac{1}{5} \text{ s} \tag{13-76}$$

and

$$\tau_A = \frac{L_a + L}{R_a + R} = \frac{0.2 + 2.35}{0.1 + 5} = \frac{1}{2} \text{ s} \tag{13-77}$$

The Laplace transform of the field voltage for a step change of 102 V is described by

$$V_f(s) = \frac{V_f}{s} = \frac{102}{s} \tag{13-78}$$

Hence the transformed form of the solution for the armature current can be written as

$$I_a(s) = \frac{102}{s} \left(\frac{1000}{204} \frac{1}{s + 5} \frac{1}{s + 2} \right) = \frac{500}{s(s + 5)(s + 2)} \tag{13-79}$$

Evaluation of the last equation is facilitated by a partial fraction expansion. Accordingly,

$$I_a(s) = \frac{500}{s(s + 5)(s + 2)} = \frac{K_0}{s} + \frac{K_1}{s + 5} + \frac{K_2}{s + 2} \tag{13-80}$$

These coefficients are found to be

$$K_0 = \left[\frac{500}{(s + 5)(s + 2)} \right]_{s=0} = 50 \tag{13-81}$$

$$K_1 = \left[\frac{500}{s(s + 2)} \right]_{s=-5} = +33.3 \tag{13-82}$$

$$K_2 = \left[\frac{500}{s(s + 5)} \right]_{s=-2} = -83.3 \tag{13-83}$$

Insertion into Eq. (13-80) allows the transformed solution to be written as

$$I_a(s) = \frac{50}{s} + \frac{33.3}{s + 5} - \frac{83.3}{s + 2} \tag{13-84}$$

The corresponding time solution as obtained from a suitable table of Laplace-transform pairs becomes

$$I_a(t) = 50 + 33.3\varepsilon^{-5t} - 83.3\varepsilon^{-2t} \tag{13-85}$$

which is the desired solution. This last expression completely describes the process of armature current buildup when a field voltage of 102 V is suddenly applied to the field circuit. Note that the steady-state (or final-state) value of the armature current is 50A—a result that can be verified through use of the procedure described in Chapter 7 for finding steady-state performance.

13-4 DYNAMICS OF A GENERATOR-MOTOR SYSTEM

A method of speed adjustment that is employed in industry whenever control over a wide speed range is required is shown in Fig. 13-17. Speed adjustment above base speed† is obtained by field current (I_{fm}) variation with rated voltage applied to the armature terminals. Below base speed the speed is adjusted by changing the armature voltage below its rated value. A convenient way to control the armature voltage is to control the field current of a supplying generator as depicted in Fig. 13-17. Large amounts of power at the load can be controlled by easy adjustment of the generator field current. This method of speed control is referred to as the *Ward-Leonard system.*

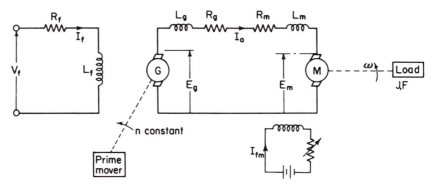

Figure 13-17 System diagram for studying dynamic behavior of the Ward-Leonard generator and motor system of speed control.

A little thought should make it apparent that adjustments of the field current in the system of Fig. 13-17 do not translate immediately into appropriate speed

---

†Base speed corresponds to operation with rated armature voltage and rated field current at rated motor output.

changes. The change in speed must occur *smoothly* from one commanded level to another. The change cannot occur abruptly because of the presence of energy-storing elements. Examination of this system reveals the presence of three energy-storing elements, each of which influences system dynamics. These are generator field inductance, combined generator and motor armature winding leakage inductances, and the inertia of the load and the motor.

A description of the system's dynamic behavior lies in identifying the transfer function that properly relates the motor speed ω to the generator field voltage V_f. As a first step in deriving this transfer function, we write the differential equation for the generator field winding circuit. Thus

$$V_f = I_f R_f + L_f \frac{dI_f}{dt} \tag{13-86}$$

Laplace transforming yields

$$V_f(s) = I_f(s) R_f (1 + s\tau_f) \tag{13-87}$$

where $\tau_f = L_f/R_f$. Formulating the ratio of field current to field voltage leads to

$$\frac{I_f(s)}{V_f(s)} = \frac{1}{R_f(1 + s\tau_f)} \tag{13-88}$$

The generated armature voltage is related to the field current by

$$\frac{E_g(s)}{I_f(s)} = K_f \qquad \text{V/A} \tag{13-89}$$

In turn the generated emf is related to armature current by Kirchhoff's voltage law. Specifically,

$$E_g = I_a R_g + L_g \frac{dI_a}{dt} + I_a R_m + L_m \frac{dI_a}{dt} + E_m \tag{13-90}$$

where R_g and R_m are the armature winding resistances of the generator and motor, L_g and L_m denote the leakage inductances of the armature windings, and E_m denotes the counter emf of the motor. More explicitly E_m can be written as

$$E_m = K_\omega \omega \tag{13-91}$$

where K_ω is expressed in volts per radian per second and the motor speed ω is expressed in radians per second. Inserting the last expression into Eq. (13-90) and Laplace transforming gives

$$I_a(s) = \frac{E_g(s) - K_\omega \omega(s)}{R(1 + s\tau_A)} \tag{13-92}$$

where

$$R = R_g + R_m \tag{13-93}$$

$$\tau_A = \frac{L_g + L_m}{R_g + R_m} \tag{13-94}$$

It is helpful at this point to rewrite Eq. (13-92) as

$$\frac{I_a(s)}{E_g(s) - K_\omega \omega(s)} = \frac{1}{R(1 + s\tau_A)} \qquad (13\text{-}95)$$

This is in the form of a transfer function that lends itself to block diagram representation. In particular, note that the input to the transfer function block consists then of the difference between the generator source voltage $E_g(s)$ and a quantity that is a function of the output variable $K_\omega \omega(s)$. Hence the latter quantity can be considered as coming from a feedback path originating from the output.

Continuing the development in a logical stepwise fashion, we next note that the armature current combines with the motor field flux to produce an electromagnetic torque that can be expressed as

$$T = K_t I_a \qquad (13\text{-}96)$$

where K_t bears units of torque per armature ampere. The transfer function notation is simply

$$\frac{T(s)}{I_a(s)} = K_t \qquad (13\text{-}96a)$$

Finally, to relate the developed torque to the motor speed, we note that the former must be equal to the sum of all opposing torques. If we now let J denote the total inertia of the load and the motor and F denote the total viscous friction, we can write

$$T = J\frac{d\omega}{dt} + F\omega \qquad (13\text{-}97)$$

The load torque is omitted in this analysis because it is assumed to be constant. Our interest here, remember, is to find the manner in which the speed changes in response to changes in the generator field current (or field voltage). Upon Laplace transforming Eq. (13-97) and formulating the ratio of speed to torque, we have

$$\frac{\omega(s)}{T(s)} = \frac{1}{F(1 + s\tau_m)} \qquad (13\text{-}98)$$

where

$$\tau_m = \text{mechanical time constant} = \frac{J}{F} \qquad (13\text{-}99)$$

A successive block diagram representation of Eqs. (13-88), (13-89), (13-95), (13-96), and (13-98) taken in order leads to the complete system block diagram depicted in Fig. 13-18(a). By applying the feedback formula of Eq. (13-16) to the feedback loop, the complete expression of speed to generated voltage is obtained.

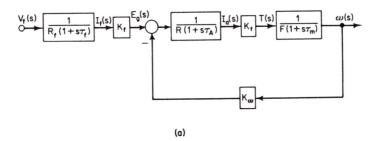

(a)

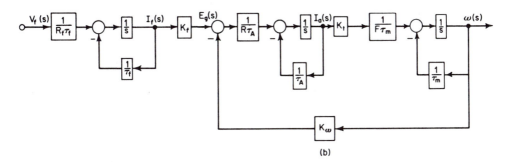

(b)

Figure 13-18 (a) Block diagram of the generator motor system of Fig. 13-17. (b) Alternate form of the block diagram of Fig. 13-17, to show explicitly the pure integrators of the system. There are three; hence the number of state variables is also three.

Thus

$$\frac{\omega(s)}{E_g(s)} = \frac{\dfrac{K_t}{RF(1 + s\tau_A)(1 + s\tau_m)}}{1 + \dfrac{K_t K_\omega}{RF(1 + s\tau_A)(1 + s\tau_m)}} = \frac{K_t}{RF(1 + s\tau_A)(1 + s\tau_m) + K_\omega K_t}$$

$$= \frac{K_t}{RF + K_\omega K_t} \left[\frac{1}{s^2 \dfrac{\tau_A \tau_m RF}{RF + K_\omega K_t} + s \dfrac{(\tau_A + \tau_m)RF}{RF + K_\omega K_t} + 1} \right] \qquad (13\text{-}100)$$

This result can also be found analytically without the use of the block diagram by inserting Eq. (13-92) into Eq. (13-96) and then substituting for $T(s)$ in Eq. (13-98). Accordingly,

$$\frac{\omega(s)}{\dfrac{E_g(s) - K_\omega \omega(s)}{R(1 + s\tau_A)} K_t} = \frac{1}{F(1 + s\tau_m)} \qquad (13\text{-}101)$$

or

$$RF(1 + s\tau_A)(1 + s\tau_m)\omega(s) = K_t E_g(s) - K_\omega K_t \omega(s) \qquad (13\text{-}102)$$

from which

$$\frac{\omega(s)}{E_g(s)} = \frac{K_t}{RF(1 + s\tau_A)(1 + s\tau_m) + K_\omega K_t} \qquad (13\text{-}103)$$

Comparison with the intermediate step of Eq. (13-100) discloses the two expressions to be identical.

The complete transfer function is determined of course by multiplying Eq. (13-100) by Eqs. (13-88) and (13-89). Hence

$$\boxed{\frac{\omega(s)}{V_f(s)} = \frac{K_f K_t}{R_f(RF + K_\omega K_t)} \frac{1}{1 + s\tau_f} \left[\frac{1}{s^2 \dfrac{\tau_A \tau_m RF}{RF + K_\omega K_t} + s \dfrac{(\tau_A + \tau_m)RF}{RF + K_\omega K_t} + 1} \right]}$$

$$(13\text{-}104)$$

The characteristic equation of the governing differential equation of this system is obtained by setting the denominator of Eq. (13-104) equal to zero. This yields a third-order equation which indicates the presence of three time lags in the system, a result which is entirely consistent with the presence of three energy-storing elements.

Appearing in Fig. 13-18(b) is the modified version of the block diagram of Fig. 13-18(a) in order to show explicitly the pure integrators, the outputs of which identify the state variables. Accordingly, the state variables of this system are field current, armature current, and motor speed. The state equations are found as described in the preceding sections. This leads to the following transformed forms:

$$sI_f(s) = -\frac{1}{\tau_f} I_f(s) + \frac{1}{R_f \tau_f} V_f(s) \qquad (13\text{-}105)$$

$$sI_a(s) = \frac{K_f}{R\tau_A} I_f(s) - \frac{1}{\tau_A} I_a(s) - \frac{K_\omega}{R\tau_A} \omega(s) \qquad (13\text{-}106)$$

$$s\omega(s) = \frac{K_t}{F\tau_m} I_a(s) - \frac{1}{\tau_m} \omega(s) \qquad (13\text{-}107)$$

Expressed in the time domain the equations are

$$\frac{dI_f}{dt} = -\frac{1}{\tau_f} I_f + \frac{1}{R_f \tau_f} V_f \qquad (13\text{-}108)$$

$$\frac{dI_a}{dt} = \frac{K_f}{R\tau_A} I_f - \frac{1}{\tau_A} I_a - \frac{K_\omega}{R\tau_A} \omega \tag{13-109}$$

$$\frac{d\omega}{dt} = \frac{K_t}{F\tau_m} I_a - \frac{1}{\tau_m} \omega \tag{13-110}$$

Finally, the matrix formulation becomes

$$\begin{bmatrix} \dfrac{dI_f}{dt} \\[2ex] \dfrac{dI_a}{dt} \\[2ex] \dfrac{d\omega}{dt} \end{bmatrix} = \begin{bmatrix} -\dfrac{1}{\tau_f} & 0 & 0 \\[2ex] \dfrac{K_f}{R\tau_A} & -\dfrac{1}{\tau_A} & -\dfrac{K_\omega}{R\tau_A} \\[2ex] 0 & \dfrac{K_t}{F\tau_m} & -\dfrac{1}{\tau_m} \end{bmatrix} \begin{bmatrix} I_f \\[2ex] I_a \\[2ex] \omega \end{bmatrix} + \begin{bmatrix} \dfrac{1}{R_f\tau_f} \\[2ex] 0 \\[2ex] 0 \end{bmatrix} V_f \tag{13-111}$$

There are three state equations associated with the three independent energy-storing elements.

13-5 DC MOTOR TRANSFER FUNCTION

Two forms of the dc motor transfer function are of interest. One is the armature-controlled mode with fixed field current; the other is the field-controlled mode with constant armature current flowing.

Armature-Controlled DC Motor

Figure 13-19 depicts the armature-controlled dc motor. The motor speed is made to respond to variations in the applied motor armature voltage V_t. For simplicity, armature winding leakage inductance is assumed to be negligible. Our concern here is to develop a transfer function that relates motor speed to applied armature voltage. Let J denote the total inertia of the load and the rotor of the motor. Let

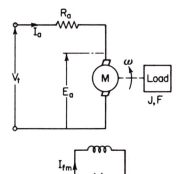

Figure 13-19 Armature-controlled dc motor, field current maintained constant.

F denote the total viscous friction of the motor and load. A good starting point in the development is to write the torque equation as it applies at the motor shaft. Accordingly,

$$T = K_t I_a$$

$$= J \frac{d\omega}{dt} + F\omega \tag{13-112}$$

where T is the motor electromagnetic torque and ω is the motor speed. The expression for the armature current is given by

$$I_a = \frac{V_t - E_a}{R_a}$$

$$= \frac{V_t - K_\omega \omega}{R_a} \tag{13-113}$$

where E_a is the motor counter-emf, which is also expressed as a speed voltage constant K_ω in volts per radian per second. In the interest of leading to the appropriate block diagram representation, it is worthwhile to write Eq. (13-113) in Laplace notation. Keeping in mind that both I_a and ω are the variables, we have

$$I_a(s) = [V_t(s) - K_\omega \omega(s)] \frac{1}{R_a} \tag{13-114}$$

The right-hand side has been written as indicated to emphasize that the reference input quantity $V_t(s)$ and a quantity derived from the output variable, $K_\omega \omega(s)$, are being compared and that the difference is then multiplied by $1/R_a$ to yield the armature current variable. Hence the symbol for a summing point followed by a block marked $1/R_a$ can be made to represent Eq. (13-114). It is interesting to note that the counter-emf of a motor can always be treated as a negative feedback quantity.

Once $I_a(s)$ is available, the transfer function relating it to the electromagnetic torque is

$$\frac{T(s)}{I_a(s)} = K_t \tag{13-115}$$

Finally, the transfer function that describes the variation of the motor speed with torque is obtained from Eq. (13-112) after it is Laplace transformed. Thus

$$T(s) = Js\omega(s) + F\omega(s) \tag{13-116}$$

or

$$\frac{\omega(s)}{T(s)} = \frac{1}{F(1 + s\tau_m)} \tag{13-117}$$

where

$$\tau_m = \text{mechanical time constant} = \frac{J}{F} \qquad (13\text{-}118)$$

The successive block-diagram notation of Eqs. (13-114), (13-115), and (13-117) leads to the complete block diagram depicted in Fig. 13-20.

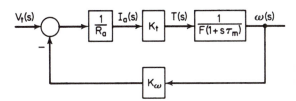

Figure 13-20 Block diagram representation of Fig. 13-19. The output variable is motor velocity.

The desired transfer function between the motor speed and the motor applied voltage is found by applying Eq. (13-16) to the block diagram of Fig. 13-20. Accordingly,

$$\frac{\omega(s)}{V_t(s)} = \frac{(K_t/R_aF)[1/(1 + s\tau_m)]}{1 + (K_\omega K_t/R_aF)[1/(1 + s\tau_m)]}$$

or

$$\boxed{\frac{\omega(s)}{V_t(s)} = \frac{K_t}{K_t K_\omega + R_a F} \frac{1}{1 + s\tau'_m}} \qquad (13\text{-}119)$$

where

$$\tau'_m = \tau_m \frac{R_a F}{R_a F + K_t K_\omega} = \frac{J R_a}{R_a F + K_t K_\omega} \qquad (13\text{-}120)$$

Equation (13-120) discloses that the feedback path that appears around the direct transmission path of Fig. 13-20 is responsible for reducing the motor time constant over the nominal value of J/F. The practical aspect of this observation is that the motor responds dynamically more quickly than one would expect by looking at τ_m alone.

If it is desirable to derive the transfer function of Eq. (13-119) without resorting to the block diagram representation followed by the application of the feedback formula, this may be readily accomplished by inserting Eq. (13-113) into Eq. (13-112) and then rearranging terms. Thus

$$\frac{V_t - K_\omega \omega}{R_a} K_t = J \frac{d\omega}{dt} + F\omega \qquad (13\text{-}121)$$

or

$$V_t = \frac{R_a F}{K_t}\left[\frac{J}{F}\frac{d\omega}{dt} + \frac{R_a F + K_\omega K_t}{R_a F}\omega\right]$$

$$= \frac{R_a F + K_\omega K_t}{K_t}\left[\tau_m \frac{R_a F}{R_a F + K_\omega K_t}\frac{d\omega}{dt} + \omega\right] \qquad (13\text{-}122)$$

Laplace transforming and formulating the ratio of output to input yields a result that is identical to Eq. (13-119).

Sometimes the dc motor Fig. 13-19 is used as the output actuator in feedback control systems where the controlled variable may be position rather than velocity. Since velocity is the derivative of the position variable c, the transfer function relating the two is obtained as follows:

$$\frac{dc}{dt} = \omega \qquad (13\text{-}123)$$

Laplace transforming for zero initial conditions gives

$$c(s) = \frac{1}{s}\,\omega(s) \qquad (13\text{-}124)$$

The complete block diagram now becomes that shown in Fig. 13-21. Moreover, the complete transfer function between the motor displacement variable and the applied armature voltage becomes

$$\boxed{\frac{c(s)}{V_t(s)} = \frac{K_t}{K_\omega K_t + R_a F}\frac{1}{s(1 + s\tau'_m)}} \qquad (13\text{-}125)$$

where τ'_m is specified by Eq. (13-120).

Appearing in Fig. 13-22 is the form of the block diagram of Fig. 13-21 that allows convenient identification of the state variables, which in this case are

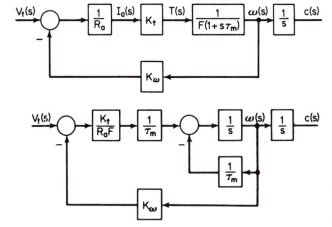

Figure 13-21 Block diagram of the motor of Fig. 13-19 when the output variable is motor displacement.

Figure 13-22 Alternative form of Fig. 13-21, drawn to facilitate writing the appropriate state equations.

clearly the motor velocity and the motor displacement. The first state equation in transformed form is found by taking the input to the first integrator and setting it equal to $s\omega(s)$. Hence

$$s\omega(s) = \frac{K_t V_t(s)}{R_a F\tau_m} - \frac{K_\omega K_t}{R_a F\tau_m}\,\omega(s) - \frac{1}{\tau_m}\,\omega(s)$$

$$= \frac{K_t}{R_a F\tau_m}\,V_t(s) - \left(\frac{K_\omega K_t}{R_a F\tau_m} + \frac{1}{\tau_m}\right)\omega(s) \qquad (13\text{-}126)$$

The second state equation is given by Eq. (13-124). The corresponding time-domain expressions are

$$\frac{d\omega}{dt} = \frac{K_t}{R_a F\tau_m}\,V_t - \left(\frac{K_\omega K_t}{R_a F\tau_m} + \frac{1}{\tau_m}\right)\omega \qquad (13\text{-}127)$$

$$\frac{dc}{dt} = \omega \qquad (13\text{-}128)$$

The matrix notation in normal form is then

$$
\begin{bmatrix} \dfrac{d\omega}{dt} \\[2mm] \dfrac{dc}{dt} \end{bmatrix}
=
\begin{bmatrix} -\left(\dfrac{K_\omega K_t}{R_a F\tau_m} + \dfrac{1}{\tau_m}\right) & 0 \\[2mm] 1 & 0 \end{bmatrix}
\begin{bmatrix} \omega \\[2mm] c \end{bmatrix}
+
\begin{bmatrix} \dfrac{K_t}{R_a F\tau_m} \\[2mm] 0 \end{bmatrix} V_t \qquad (13\text{-}129)
$$

Field-Controlled DC Motor

The circuit configuration for this mode of operation is shown in Fig. 13-23. The armature is supplied with a constant current from a suitable current source. In light of this fact the variation of electromagnetic torque is dependent upon the

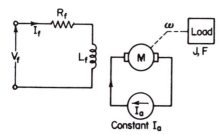

Figure 13-23   Field-controlled dc motor.

variation in field current. This is apparent upon recalling that $T = K_T \Phi I_a$ and that the air-gap flux Φ is directly related to the field current for the assumed condition of no saturation. Therefore, the torque in this instance may be written simply as

$$T = K_{tf} I_f \qquad (13\text{-}130)$$

where K_{tf} is a motor constant expressed in units of torque per field ampere. The torque equation at the motor shaft then becomes

$$K_{tf} I_f = J \frac{d\omega}{dt} + F\omega \qquad (13\text{-}131)$$

where J and F have the same meanings as in the preceding. Laplace transforming and formulating the ratio of output to input yields the transfer function that relates motor velocity to field current. Thus

$$\frac{\omega(s)}{I_f(s)} = \frac{K_{tf}}{F(1 + s\tau_m)} \qquad (13\text{-}132)$$

In turn the field current is related to the applied field voltage by the usual transfer function. Accordingly,

$$\frac{I_f(s)}{V_f(s)} = \frac{1}{R_f(1 + s\tau_f)} \qquad (13\text{-}133)$$

where $\tau_f = L_f/R_f$. By combining the last two expressions, the transfer function between speed and field voltage results:

$$\boxed{\frac{\omega(s)}{V_f(s)} = \frac{I_f(s)}{V_f(s)} \frac{\omega(s)}{I_f(s)} = \frac{K_{tf}}{R_f F} \frac{1}{1 + s\tau_f} \frac{1}{1 + s\tau_m}} \qquad (13\text{-}134)$$

Finally, to obtain the relationship between motor displacement and applied field voltage, it is merely necessary to multiply the last equation by Eq. (13-124) expressed in the form $[c(s)/\omega(s)] = 1/s$. This yields

$$\boxed{\frac{c(s)}{V_f(s)} = \frac{K_{tf}}{R_f F} \frac{1}{s(1 + s\tau_f)(1 + s\tau_m)}} \qquad (13\text{-}135)$$

A glance at the denominator of Eq. (13-135) discloses that now the characteristic equation is third order. Figure 13-24 depicts the block diagram of Eq. (13-135).

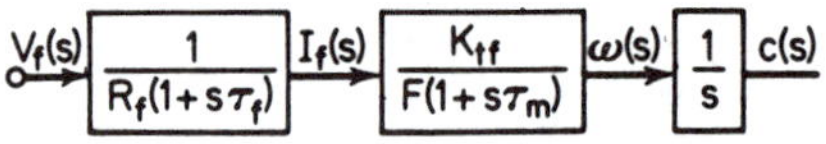

Figure 13-24   Block diagram of the system of Fig. 13-23.

Consistent with the third-order nature of the characteristic equation, there are three state equations which describe the complete dynamic behavior of this system. These equations are more easily identified by referring to the alternative block diagram illustrated in Fig. 13-25. Clearly, the three state variables are field current, motor speed, and motor displacement, which are respectively the output quantities of each of the integrators in Fig. 13-25. The transformed versions of the three state equations as associated with the three integrators are then

$$s I_f(s) = \frac{1}{R_f \tau_f} V_f(s) - \frac{1}{\tau_f} I_f(s) \qquad (13\text{-}136)$$

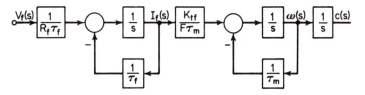

Figure 13-25   Alternative form of Fig. 13-24.

$$s\omega(s) = \frac{K_{tf}}{F\tau_m} I_f(s) - \frac{1}{\tau_m} \omega(s) \qquad (13\text{-}137)$$

$$sc(s) = \omega(s) \qquad (13\text{-}138)$$

The corresponding first-order differential equations then become

$$\frac{dI_f}{dt} = \frac{1}{R_f\tau_f} V_f - \frac{1}{\tau_f} I_f \qquad (13\text{-}139)$$

$$\frac{d\omega}{dt} = \frac{K_{tf}}{F\tau_m} I_f - \frac{1}{\tau_m} \omega \qquad (13\text{-}140)$$

$$\frac{dc}{dt} = \omega \qquad (13\text{-}141)$$

Finally the matrix formulation is

$$\begin{bmatrix} \dfrac{dI_f}{dt} \\[2ex] \dfrac{d\omega}{dt} \\[2ex] \dfrac{dc}{dt} \end{bmatrix} = \begin{bmatrix} -\dfrac{1}{\tau_f} & 0 & 0 \\[2ex] \dfrac{K_{tf}}{F\tau_m} & -\dfrac{1}{\tau_m} & 0 \\[2ex] 0 & 1 & 0 \end{bmatrix} \begin{bmatrix} I_f \\[2ex] \omega \\[2ex] c \end{bmatrix} + \begin{bmatrix} \dfrac{1}{R_f\tau_f} \\[2ex] 0 \\[2ex] 0 \end{bmatrix} V_f \qquad (13\text{-}142)$$

13-6 TWO-PHASE SERVOMOTOR TRANSFER FUNCTIONS

The steady-state characteristics of the two-phase servomotor are discussed in Chapter 10. Attention is directed here to the dynamic behavior of these motors as described by their transfer functions.

Although the servomotor transfer functions could be derived by following the procedure outlined for the dc motor in the preceding section, in the interest of versatility an alternative approach is employed using the external torque-speed characteristic. Typical torque-speed characteristics such as those depicted in Fig. 10-13 are assumed to be available. Furthermore, as indicated in Fig. 13-26, the servomotor is assumed to be operated at a fixed reference voltage. The input variable in this arrangement is the control winding voltage V_c. What quantity then

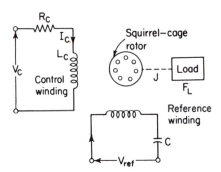

Figure 13-26 Two-phase servomotor with fixed reference winding voltage.

shall we consider to be the output variable? Shall it be velocity ω or displacement c? Examination of Fig. 13-26 certainly leaves no doubt that the application of a given control winding voltage manifests itself as an output velocity. However, this alone is not necessarily the determining factor in establishing what the output variable shall be. Keep in mind that servomotors are used in feedback control systems so that a more important consideration is the function which the servomotor will fulfill. If the output of the control system is to be velocity, then it is appropriate to derive a transfer function that relates velocity to the control winding voltage. On the other hand, if the motor output is to be used to drive a *position-sensitive* transducer such as a potentiometer or a resolver or a synchro, then clearly the motor displacement should be considered the output variable.

The primary concern at the moment is the general form that the servomotor transfer function assumes. Consequently, the torque-speed curves shown in Fig. 10-13 are assumed to be linear for simplicity. A study of these curves reveals that an increased motor torque comes about either by an increase in control winding voltage or by a decrease in speed for a fixed V_c. Expressed mathematically the developed motor torque is

$$T = K_M V_c - F_M \frac{dc}{dt} \tag{13-143}$$

where K_M = motor torque constant in units of torque/volt

F_M = motor equivalent viscous-friction constant in units of torque/rad/sec

It is helpful to note that F_M is nothing more than the slope of the torque-speed curve at constant V_c. Also, once the torque-speed curves are known, the parameter K_M is merely the change in torque per unit of change in control voltage at constant speed. Equating this torque to the sum of the opposing load torques yields the governing differential equation that relates motor displacement to control voltage. Thus

$$K_M V_c - F_M \frac{dc}{dt} = J \frac{d^2c}{dt^2} + F_L \frac{dc}{dt} \tag{13-144}$$

where J = motor inertia + load inertia referred to motor shaft

F_L = viscous friction of load referred to motor shaft

Rearranging and Laplace transforming lead to

$$K_M V_c(s) = (Js^2 + sF)c(s) \tag{13-145}$$

where

$$F = F_M + F_L \tag{13-146}$$

Hence the desired transfer function becomes

$$\boxed{\frac{c(s)}{V_c(s)} = \frac{K_m}{s(1 + s\tau_m)}} \tag{13-147}$$

where

$$K_m = \frac{K_M}{F} \quad \text{(V-s)}^{-1} \tag{13-148}$$

$$\tau_m = \frac{J}{F} \quad \text{s} \tag{13-149}$$

Frequently the motor time constant τ_m plays an important role in determining the character of the dynamic response of the control system of which it is a part. This is because the magnitude of τ_m is established in terms of the inertial properties of physical elements such as the size of the servomotor rotor and the reflected load inertia. These elements are such that they cannot be reduced to negligence primarily because the size of J goes hand in hand with the required level of output power. Accordingly, in many cases the motor inertia is the determining factor that establishes the extent to which the system will be capable of following variations of the command signal with time.

Effect of the Control Winding Time Constant

The result of Eq. (13-147) is approximate because it neglects the time lag caused by the inductance of the control winding. A more exact description of the servomotor transfer function can be obtained by recognizing that the control winding voltage really produces a control winding current I_c, which in turn produces the flux that enables the motor to develop torque. The torque equation as it applies at the motor shaft then becomes

$$K_{ct}I_c - F_M \frac{dc}{dt} = J \frac{d^2c}{dt^2} + F_L \frac{dc}{dt} \tag{13-150}$$

where K_{ct} denotes a torque constant expressed in units of torque per field ampere. The corresponding transfer function is given by

$$\frac{c(s)}{I_c(s)} = \frac{K_{ct}}{F} \frac{1}{s(1 + s\tau_m)} \tag{13-151}$$

Next compare Fig. 13-27 with Fig. 10-13. At a fixed speed the input impedance at the control winding terminals remains constant. If operation is say at V_{c2},

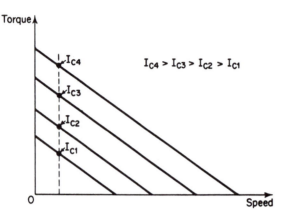

Figure 13-27 Linearized torque-speed curves showing variation of control winding current with control winding voltage at constant speed. Each straight line corresponds to a fixed control voltage: the higher the line, the larger the control voltage.

then a sudden increase in control voltage to V_{c3} at the same speed increases the torque along an ordinate line. If the field winding time constant is not negligible then, as V_{c2} increases to V_{c3}, so too does the control current increase from I_{c2} (in Fig. 13-27) to I_{c3}, but with a time delay.

The field current is therefore related to the control voltage by the usual transfer function for an R-L circuit. Thus

$$\frac{I_c(s)}{V_c(s)} = \frac{1}{R_c(1 + s\tau_c)} \tag{13-152}$$

where $\tau_c = L_c/R_c$. Upon combining Eqs. (13-151) and (13-152), the desired transfer function results:

$$\frac{c(s)}{V_c(s)} = \frac{K_{ct}}{FR_c} \frac{1}{s(1 + s\tau_m)(1 + s\tau_c)}$$

or

$$\boxed{\frac{c(s)}{V_c(s)} = \frac{K_m}{s(1 + s\tau_m)(1 + s\tau_c)}} \tag{13-153}$$

It is interesting to note in this development that

$$\frac{K_{ct}}{R_c} = K_M \tag{13-154}$$

A comparison of Eq. (13-153) with Eq. (13-147) indicates that the difference does indeed lie in the time constant of the control field winding.

Figure 13-28 depicts the block diagram representation of Eq. (13-153). Note again that, whenever the motor output variable is taken to be displacement and

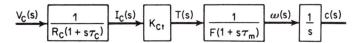

Figure 13-28   Block diagram of the system of Fig. 13-26.

the motor field winding time constant is included in the analysis, the result is a third-order characteristic equation. Of course this means that there are three state variables and three state equations. Reference to Fig. 13-29 reveals these state equations in transformed form to be as follows:

$$sI_c(s) = \frac{1}{R_c \tau_c} V_c(s) - \frac{1}{\tau_c} I_c(s) \tag{13-155}$$

$$s\omega(s) = \frac{K_{ct}}{F\tau_m} I_c(s) - \frac{1}{\tau_m} \omega(s) \tag{13-156}$$

$$sc(s) = \omega(s) \tag{13-157}$$

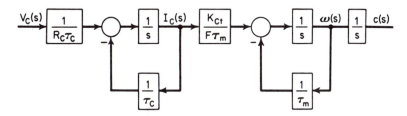

Figure 13-29   Alternative form of Fig. 13-28, modified to show explicitly the pure integrators, the outputs of which identify the state variables of the system.

The corresponding first-order differential equations can be written as

$$\frac{dI_c}{dt} = \frac{1}{R_c \tau_c} V_c - \frac{1}{\tau_c} I_c \tag{13-158}$$

$$\frac{d\omega}{dt} = \frac{K_{ct}}{F\tau_m} I_c - \frac{1}{\tau_m} \omega \tag{13-159}$$

$$\frac{dc}{dt} = \omega \tag{13-160}$$

The matrix formulation is then:

$$
\begin{bmatrix} \dfrac{dI_c}{dt} \\[2ex] \dfrac{d\omega}{dt} \\[2ex] \dfrac{dc}{dt} \end{bmatrix} = \begin{bmatrix} -\dfrac{1}{\tau_c} & 0 & 0 \\[2ex] \dfrac{K_{ct}}{F\tau_m} & -\dfrac{1}{\tau_m} & 0 \\[2ex] 0 & 1 & 0 \end{bmatrix} \begin{bmatrix} I_c \\[2ex] \omega \\[2ex] c \end{bmatrix} + \begin{bmatrix} \dfrac{1}{R_c\tau_c} \\[2ex] 0 \\[2ex] 0 \end{bmatrix} V_c \qquad (13\text{-}161)
$$

13-7 AMPLIDYNE GENERATOR

The amplidyne generator is a power-amplifying device capable of generating high output levels, usually in the range of 1 to about 50 kW. In control system applications it often serves as a power stage that drives a dc motor. The control field winding of the amplidyne (of which there are several sections) is usually driven from the output stage of an electronic amplifier.

The amplidyne is basically a dc generator. It is driven at constant speed by a suitable motor which serves as a source of energy for the unit. The magnitude of its output voltage is controlled by the amount of field current flowing through the control winding. Since the armature winding of a dc motor is frequently placed across the output terminals of the amplidyne, it is important to account for the motor loading effect whenever the transfer function is derived.

The principle of operation of the amplidyne is straightforward. A small current flowing in the control winding is made to create a flux Φ_d directed as illustrated in Fig. 13-30. Because the armature winding is being driven at full speed, a voltage E_q is induced and it is collected at the brushes marked q-q. By short-circuiting these brushes and designing an armature winding of low resis-

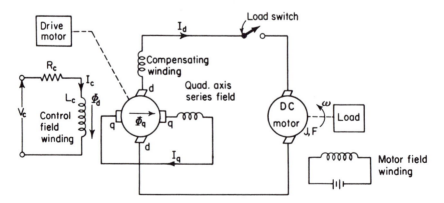

Figure 13-30 Amplidyne generator, showing windings and connected dc motor and load.

tance, very large armature currents can be made to flow. For example, if E_q is 1 V and the armature winding resistance is 0.01 Ω, an armature current of 100 A results. As a consequence of the huge armature reaction associated with such a large current, a very strong flux field Φ_q is produced and directed in quadrature to the original field Φ_d. By placing a second set of brushes, *d-d*, normal to the Φ_q field, a high-level voltage source is made available that can be applied to the dc motor, which in turn provides the desired output at the mechanical load. In practice it is customary to short-circuit the brushes *q-q* through a series field which enables the required values of Φ_q to be established with smaller quadrature-axis armature current. Another modification involves the inclusion of a compensating winding placed in the direct-axis circuit. Its purpose is to prevent the armature reaction flux produced by the armature current I_d from diminishing the original field Φ_d. The compensating winding is designed to furnish 100% compensation, a condition that is essential in view of the small values that Φ_d assumes during normal operation.

The transfer function of the amplidyne as an isolated unit—without its connected load—is found by considering the direct-axis induced voltage E_d as the output quantity and the control voltage V_c as the input quantity. The derivation that follows is based on the assumptions that there is zero coupling between the flux in the direct axis and the flux in the quadrature axis and that the magnetization curve is linear. The differential equation for the field winding is as usual

$$V_c = I_c R_c + L_c \frac{dI_c}{dt} \tag{13-162}$$

where R_c = control winding resistance, Ω
$\qquad L_c$ = control winding inductance, H

Expressed in transformed form, this equation becomes

$$\frac{I_c(s)}{V_c(s)} = \frac{1}{R_c(1 + s\tau_c)} \tag{13-163}$$

where $\tau_c = L_c/R_c$ and denotes the control winding time constant.

The presence of the control winding current produces the direct-axis flux Φ_d which in turn causes an emf to appear across the *q-q* brushes and which specifically is expressed as

$$E_q = K_1 \Phi_d \tag{13-164}$$

where K_1 includes winding factors as well as the speed of rotation of the armature. A more convenient form of Eq. (13-164) results by recalling that Φ_d is proportional to I_c so that we can write

$$E_q = K_q I_c \tag{13-165}$$

where K_q is a proportionality factor for the number of volts generated per control-winding ampere. The Laplace transformed version of this expression is simply

$$\frac{E_q(s)}{I_c(s)} = K_q \tag{13-166}$$

The governing differential equation for the quadrature-axis circuit is readily seen from Fig. 13-30 to be

$$E_q = I_q R_q + L_q \frac{dI_q}{dt} \tag{13-167}$$

where R_q = total resistance in the quadrature axis circuit, Ω
       L_q = total inductance in the quadrature axis circuit, H
       I_q = current flowing in the quadrature axis, A

The transformed version of the last equation expressed as a ratio then becomes

$$\frac{I_q(s)}{E_q(s)} = \frac{1}{R_q(1 + s\tau_q)} \tag{13-168}$$

where

$$\tau_q = \frac{L_q}{R_q} \tag{13-169}$$

Finally, with the amplidyne output circuit open it follows that the output voltage appearing at the direct-axis brushes is simply

$$E_d = K_d I_q \tag{13-170}$$

where K_d denotes the number of volts induced in the direct axis per quadrature-axis ampere. The Laplace transformed expression is

$$\frac{E_d(s)}{I_q(s)} = K_d \tag{13-171}$$

The transfer function that applies for the unloaded amplidyne generator readily follows by multiplying Eqs. (13-163), (13-166), (13-168), and (13-171). Thus

$$\boxed{\frac{E_d(s)}{V_c(s)} = \frac{K_q K_d}{R_c R_q} \frac{1}{(1 + s\tau_c)(1 + s\tau_q)}} \tag{13-172}$$

The characteristic equation is second order in this mode of operation. Clearly the usefulness of this result is limited to those applications where the load current of the amplidyne, I_d, is negligible. This is certainly not the case in the situation depicted in Fig. 13-30 when the load switch is closed.

Amplidyne Transfer Function with Connected Load

The transfer function for this arrangement must be found by considering the dc motor as part of the direct-axis circuitry of the amplidyne and by this means accounting for the loading effect. The resulting transfer function then provides the correct description of the dynamics of the amplidyne-motor combination. For this derivation the additional assumption is made that the output circuit is compensated perfectly, which ensures that the load current I_d has no effect on the control flux Φ_d.

A few more equations are needed to complete the analysis. The first is obtained by applying Kirchhoff's voltage law to the direct-axis circuit. Thus

$$E_d = K_q I_c = I_d R_d + K_\omega \omega \qquad (13\text{-}173)$$

where R_d is the total resistance found in the direct-axis path, ω is the motor speed, and K_ω denotes the number of volts per radian per second generated by the dc motor. For simplicity the inductance in the direct axis is assumed negligibly small. Expressed in the Laplace transform notation this equation becomes

$$\frac{E_d(s) - K_\omega \omega(s)}{R_d} = I_d(s) \qquad (13\text{-}174)$$

Keep in mind that $E_d(s)$ is the Laplace transform of E_d, $I_d(s)$ is the Laplace transform of I_d, and $\omega(s)$ is the Laplace transform of ω. Also, Eq. (13-174) is written as indicated on the left-hand side in order to emphasize the need for a summing point in the block diagram representation. Recall that $\omega(s)$ is treated as a feedback quantity in the formulation of Eq. (13-174).

A second required equation relates the motor-developed torque to the direct-axis current through the torque constant K_t. In Laplace transform notation we have

$$\frac{T(s)}{I_d(s)} = K_t \qquad (13\text{-}175)$$

Moreover, the relationship between torque and speed follows from writing the torque equation as it applies to the motor shaft. Thus

$$T = J\frac{d\omega}{dt} + F\omega \qquad (13\text{-}176)$$

Here again J denotes the total inertia reflected to the motor shaft and F denotes the total viscous friction reflected to the motor shaft. The transfer function for Eq. (13-176) thus becomes

$$\frac{\omega(s)}{T(s)} = \frac{1}{F(1 + s\tau_m)} \qquad (13\text{-}177)$$

where $\tau_m = J/F$. Finally, in those applications where displacement rather than velocity is of primary interest, use is made of

$$\frac{c(s)}{\omega(s)} = \frac{1}{s} \tag{13-178}$$

The total system block diagram can now be drawn by employing a successive block representation of Eqs. (13-163), (13-166), (13-168), (13-170), (13-174), (13-175), (13-177), and (13-178). The result is depicted in Fig. 13-31. The one feedback path exists because of the presence of the dc motor and its associated counter-emf during running conditions. By employing Eq. (13-16) the single

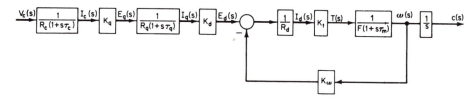

Figure 13-31 Block diagram representation of the amplidyne motor system of Fig. 13-30.

transfer function that relates motor speed to the amplidyne induced emf in the direct axis is obtained. Accordingly,

$$\frac{\omega(s)}{E_d(s)} = \frac{K_t/[R_d F(1 + s\tau_m)]}{1 + [K_\omega K_t/R_d F(1 + s\tau_m)]}$$

$$= \frac{K_t}{FR_d[1 + (K_\omega K_t/FR_d)]} \frac{1}{1 + s\tau_m'} \tag{13-179}$$

where

$$\tau_m' = \tau_m \frac{1}{1 + (K_\omega K_t/FR_d)} \tag{13-180}$$

The complete amplidyne-motor-system transfer function can now be written by formulating the product of Eqs. (13-163), (13-166), (13-168), (13-170), (13-179), and (13-178). The result is

$$\boxed{\frac{c(s)}{V_c(s)} = \frac{K_q K_d}{R_c R_q} \frac{K_t}{FR_d[1 + (K_\omega K_t/FR_d)]} \frac{1}{s(1 + s\tau_c)(1 + s\tau_q)(1 + s\tau_m')}} \tag{13-181}$$

A comparison of this expression with Eq. (13-172) reveals the precise manner in which the presence of the motor modifies the transfer function. It introduces a change in direct transmission gain along with pure integration and a simple time lag as represented by τ'_m. If the loading effect that the motor has on the amplidyne generator had been neglected, the result would have been a larger mechanical time constant and a different direct gain.

The magnitudes of the amplidyne time constants τ_c and τ_q vary with the size of the unit, but often they lie in a range of 0.05 s for smaller units to 0.5 s for the larger ones. The value of τ'_m of course depends upon the size of the motor and the nature of the connected mechanical load, but it easily can be of the order of several seconds. The use of such a combination in control systems leads to stability problems because of the time lags associated with the composite transfer function.

A glance at Eq. (13-181) discloses the characteristic equation of the amplidyne motor system to be of fourth degree. Accordingly, there are also four state variables and four state equations. Figure 13-32 reveals the state variables to be the control winding current I_c, quadrature axis current I_q, motor speed ω, and motor displacement c. The state equations in Laplace notation are found by

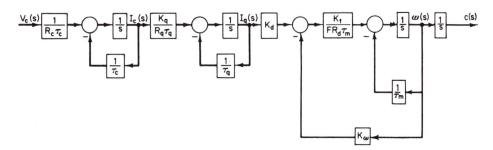

Figure 13-32   Alternate form of the block diagram of the amplidyne motor combination. The output of the pure integrators discloses the state variables.

setting the input signal to each of the four integrators equal to s times the output of the respective integrator. This leads to

$$sI_c(s) = \frac{1}{R_c \tau_c} V_c(s) - \frac{1}{\tau_c} I_c(s) \tag{13-182}$$

$$sI_q(s) = \frac{K_q}{R_q \tau_q} I_c(s) - \frac{1}{\tau_q} I_q(s) \tag{13-183}$$

$$s\omega(s) = \frac{K_d K_t}{FR_d \tau_m} I_q(s) - \left(\frac{K_\omega K_t}{FR_d \tau_m} + \frac{1}{\tau_m} \right) \omega(s) \tag{13-184}$$

$$sc(s) = \omega(s) \tag{13-185}$$

The corresponding time-domain equations expressed in the matrix formulation are

$$
\begin{bmatrix} \dfrac{dI_c}{dt} \\[2ex] \dfrac{dI_q}{dt} \\[2ex] \dfrac{d\omega}{dt} \\[2ex] \dfrac{dc}{dt} \end{bmatrix} =
\begin{bmatrix}
-\dfrac{1}{\tau_c} & 0 & 0 & 0 \\[2ex]
\dfrac{K_q}{R_q\tau_q} & -\dfrac{1}{\tau_q} & 0 & 0 \\[2ex]
0 & \dfrac{K_dK_t}{FR_d\tau_m} & -\left(\dfrac{K_\omega K_t}{FR_d\tau_m} + \dfrac{1}{\tau_m}\right) & 0 \\[2ex]
0 & 0 & 1 & 0
\end{bmatrix}
\begin{bmatrix} I_c \\[2ex] I_q \\[2ex] \omega \\[2ex] c \end{bmatrix} +
\begin{bmatrix} \dfrac{1}{R_c\tau_c} \\[2ex] 0 \\[2ex] 0 \\[2ex] 0 \end{bmatrix} V_c
\qquad (13\text{-}186)
$$

13-8 A FEEDBACK VOLTAGE CONTROL SYSTEM

To illustrate how knowledge of the steady-state and dynamic performance of a dc generator can be applied in a general way, attention is directed to the elementary voltage control system depicted in Fig. 13-33. The purpose of this control system is to provide a terminal voltage across the load that will remain essentially invariant in spite of changes in load current. Here the load current and the armature

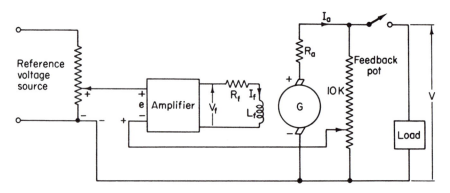

Figure 13-33 Elementary feedback voltage system.

current are assumed to be the same because the current drawn by the 10-kΩ potentiometer is negligible for all practical purposes. Also, assume that the generator external characteristics are properly represented by Fig. 13-34. The curves are linearized for simplicity. These characteristics state, e.g., that, if the generator were operating isolated and at a constant field current of I_{f3}, a 20-V drop would occur in terminal voltage for a load demand corresponding to 20 A. This is consistent with an armature circuit resistance R_a of 1 Ω for the generator. Of course the purpose of the arrangement of Fig. 13-33 is to prevent this drop from occurring and to maintain stability in doing so.

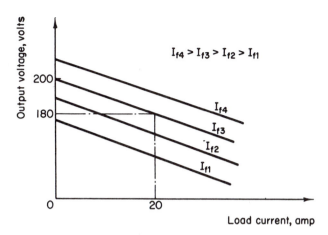

Figure 13-34 Generator output characteristics.

Before proceeding with a description of how this automatic regulating action is achieved, let us first calculate the reference voltage needed to make available a specified no-load voltage. Assume the remaining system parameters are as follows:

$$K_a = \text{amplifier gain constant} = 2000 \text{ V/V}$$

$$R_f = \text{generator field winding resistance} = 200 \text{ }\Omega$$

$$K_f = \text{generated emf constant} = 300 \text{ V/field ampere}$$

$$h = \text{feedback factor} = 0.1 \text{ numeric}$$

The generated emf constant is valid for a specified constant prime mover speed. The feedback factor refers to the slider-arm setting of the feedback potentiometer. Assuming the desired output at no-load to be 200 V, an amplifier input signal e_o is needed which satisfies the equation

$$\frac{e_o K_a}{R_f} K_f = 200 \text{ V} \tag{13-187}$$

or

$$\frac{e_o 2000}{200} 300 = 200$$

Hence

$$e_o = \tfrac{1}{15} = 0.065 \text{ V}$$

It should be apparent that Eq. (13-187) merely identifies how the amplifier input signal at no-load is operated upon to yield the desired no-load terminal voltage in steady state. Thus $e_o K_a$ denotes the amplifier output voltage that is applied to the field circuit to produce a field current. The steady-state value of this field current is found by dividing the field voltage by R_f, which yields $e_o K_a/R_f$. Finally, by multiplying this quantity by K_f the desired no-load voltage is obtained. The required reference voltage is found from the equation that indicates how the amplifier input voltage is obtained. A glance at Fig. 13-33 discloses that it is the *difference* between the reference voltage and the portion of the output voltage that is fed back. Expressing this mathematically we have generally

$$E_R - hV = e \qquad (13\text{-}188)$$

Therefore for the case at hand

$$E_R = e_o + hV = 0.065 + (0.1)200 = 20.065 \text{ V} \qquad (13\text{-}189)$$

This expression states that, if the slider arm of the potentiometer at the reference source is set at 20.065 V, the generated output voltage at no-load will be 200 V. Obviously, as a matter of convenience, the reference potentiometer can be calibrated to convey this information.

Let us now describe how the closed-loop system of Fig. 13-33 automatically furnishes a self-correcting action as load current to the extent of 20 A is permitted to flow. From Eq. (7-12) we know that the terminal voltage is related to the generated emf by

$$V = E_a - I_a R_a \qquad (13\text{-}190)$$

Hence, for $I_a = 20$ and $R_a = 1$, the terminal voltage attempts to fall to a value of 180 V. However, as it moves in this direction the feedback quantity hV becomes correspondingly smaller, which increases the amplifier input signal as indicated by Eq. (13-188). The increased amplifier input means a greater field current, which in turn generates an increased E_a to offset the drop in terminal voltage. Complete neutralization does not occur in this system because some small drop in terminal voltage must be allowed in order for the corrective action to exist.

The block diagram of this system may be developed by following the procedure employed in the preceding sections. The transfer function of each part of the system is found and these are then combined to yield the total system transfer function and block diagram. The Laplace transformed version of the equation that provides the actuating signal for the amplifier is obtained from Eq. (13-188). Thus

$$E_R(s) - hV(s) = e(s) \qquad (13\text{-}191)$$

Since $V(s)$ is the output variable, a summing point is used to identify this equation in the block diagram with $hV(s)$ coming from a feedback path. The transfer function for the amplifier is clearly just the associated gain factor. Hence

$$\frac{V_f(s)}{e(s)} = K_a \qquad (13\text{-}192)$$

where $V_f(s)$ denotes the amplifier output voltage which is applied to the field circuit. Next, because of the *R-L* nature of the field circuit, our previous experience† tells us that field current and field voltage are related by

$$\frac{I_f(s)}{V_f(s)} = \frac{1}{R_f(1 + s\tau_f)} \tag{13-193}$$

where $\tau_f = L_f/R_f$. Proceeding on through the system, we next note that the field current of the last equation yields a generated emf described by the transfer function

$$\frac{E_a(s)}{I_f(s)} = K_f \tag{13-194}$$

The final step in the development is the Laplace transformed version of Eq. (13-190), which provides the relationship between the $E_a(s)$ of the last equation and the output quantity $V(s)$. Thus

$$E_a(s) - I_a(s)R_a = V(s) \tag{13-195}$$

A successive block interpretation of Eqs. (13-191) to (13-195) leads to the complete block diagram depicted in Fig. 13-35. Note that the increase in armature current is treated as a negative input disturbance that acts to reduce the output voltage.

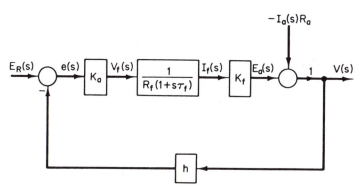

Figure 13-35 Block diagram for the system of Fig. 13-34.

It is important to understand that from the system block diagram it is now possible to write conveniently expressions which describe the manner in which the output voltage responds to changes in either the reference signal or load disturbances or in both. Moreover, these expressions can be made to reveal information about dynamic behavior as well as steady-state performance. To illustrate, let us first write the transfer function relating output voltage to reference input. Application of Eq. (13-16) to the block diagram yields

$$\frac{V(s)}{E_R(s)} = \frac{[K_aK_f/R_f(1 + s\tau_f)]}{1 + [hK_aK_f/R_f(1 + s\tau_f)]} = \frac{K}{1 + hK}\frac{1}{1 + s\tau_f'} \tag{13-196}$$

†See derivation of Eq. (13-57).

where

$$K = \frac{K_a K_f}{R_f} = \text{direct transmission gain} \qquad (13\text{-}197)$$

$$\tau'_f = \frac{\tau_f}{1 + hK} \qquad (13\text{-}198)$$

In a similar fashion the transfer function between the output voltage and the armature voltage drop can be shown to be given by

$$\frac{V(s)}{-R_a I_a(s)} = \frac{1}{1 + [hK_a K_f / R_f (1 + s\tau_f)]} = \frac{1}{1 + hK} \frac{1 + s\tau_f}{1 + s\tau'_f} \qquad (13\text{-}199)$$

The complete expression for the output voltage corresponding to E_R and I_a is found by superposing the results obtained from Eqs. (13-196) and (13-199). Accordingly, we have

$$\boxed{V(s) = \frac{K}{1 + hK} \frac{1}{1 + s\tau'_f} E_R(s) - \frac{R_a I_a(s)}{1 + hK} \frac{1 + s\tau_f}{1 + s\tau'_f}} \qquad (13\text{-}200)$$

To illustrate the use of the last equation, let it be desirable to find the value of the output voltage for a reference input of 20.065 V and zero armature current. The Laplace transform of the reference input is given by

$$E_R(s) = \frac{20.065}{s} \qquad (13\text{-}201)$$

Hence

$$V(s) = \frac{1}{1 + hK} \frac{1}{1 + s\tau'_f} \frac{20.065}{s} \qquad (13\text{-}202)$$

Keep in mind that Eq. (13-202) furnishes total information about the response, dynamic as well as steady-state. Since for the moment our interest is focused on the steady-state value, applying the final value theorem to Eq. (13-202) yields the desired result. Thus

$$V = \lim_{s \to 0} sV(s) = \frac{K}{1 + hK} 20.065 = \frac{3000}{301} 20.065 = 200 \text{ V} \qquad (13\text{-}203)$$

This, of course, is exactly the value used previously in connection with $E_R = 20.065$ V. Refer to Eq. (13-189).

If, in addition to E_R, it is further assumed that a current of 20 A is delivered to the load, then application of the final value theorem to Eq. (13-200) discloses that the new steady-state value is

$$V = \frac{K}{1 + hK} 20.065 - \frac{20}{1 + hK} = 200 - \frac{20}{301} = 200 - 0.0655 = 199.9345 \text{ V}$$

$$(13\text{-}204)$$

Clearly, only a very slight drop in output voltage occurs in the closed-loop system. Note that the placement of the generator in a closed-loop system caused the normal 20-V drop in terminal voltage to be reduced to 0.0655, a reduction by the factor $1/(1 + hK)$ where hK is the loop gain factor. What is more, all of this has been achieved automatically.

If it should be desirable to describe the dynamic behavior of the system in response to $E_R(s)$ and $I_a(s)$, the usual procedure must be applied. First, the appropriate expressions for $E_R(s)$ and $I_a(s)$ are introduced into Eq. (13-200). This is then followed by a partial fraction expansion and subsequent evaluation of the associated coefficients. Evaluation of the inverse Laplace transforms by means of a table of Laplace transform pairs furnishes the complete time solution for the output voltage. In this instance the output voltage will be seen to rise in accordance with the usual buildup that takes place in any system containing a single energy-storing element. See Eq. (13-12) for example.

PROBLEMS

13-1. Obtain the transfer function relating output voltage to input voltage for each circuit shown in Fig. P13-1. Put the result in the form that involves factors such as $(1 + s\tau)$.

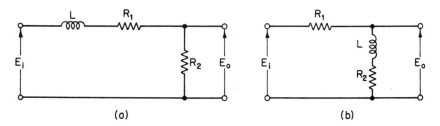

Figure P13-1

13-2. The values of the parameters in the circuit of Fig. P13-1 are as follows: $L = 0.5$ H, $R_1 = 2 \ \Omega$, and $R_2 = 3 \ \Omega$. Find the complete expression for the output voltage for a unit step input voltage.

13-3. Find the transfer function relating the output voltage to the input voltage for each circuit depicted in Fig. P13-3. Manipulate the result into the standard form.

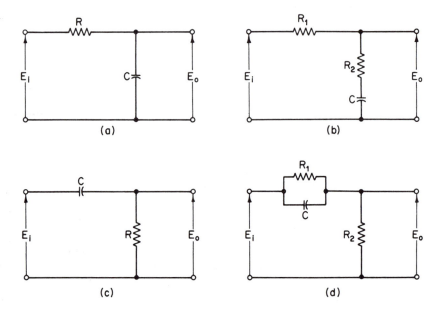

Figure P13-3

13-4. Refer to Fig. P13-4.
  (a) Derive the transfer function of this circuit relating E_o to E_i.
  (b) Obtain the transfer function that relates the voltage across capacitor C_2 to the input voltage E_i.
  (c) Represent the transfer function between E_o and E_i by a block diagram that brings into evidence a pure integration function in the direct transmission path.
  (d) Identify the state variable for this network and indicate its location in the block diagram.

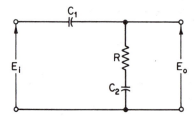

Figure P13-4

13-5. The block diagram relating output to input of an electric circuit is shown in Fig. P13-5.
  (a) Find the transfer function between input E_i and output E_o.
  (b) Can you suggest a circuit for which this transfer function applies? The C's denote capacitance and R resistance.

13-6. Refer to Fig. P13-3(d).
  (a) Derive the transfer function $E_o(s)/E_i(s)$.
  (b) Find the transfer function relating the voltage across the capacitor to the input voltage.

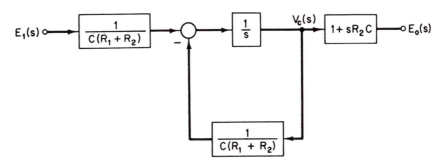

Figure P13-5

(c) Represent the transfer function between input and output in terms of a pure integration in the direct transmission path.

(d) Identify the state variable in the block diagram of part (c).

13-7. Determine the transfer function that relates the voltage E_2 to the input voltage E_i in the system depicted in Fig. P13-7. The system parameters are as follows:

Generator: $L_f = 50$ H, $R_f = 50\ \Omega$, $R_a = 1\ \Omega$, $L_a = 1$ H

emf constant: $K_g = 100$ V/field ampere

Low-pass filter: $L = 1$ H
$R = 1\ \Omega$

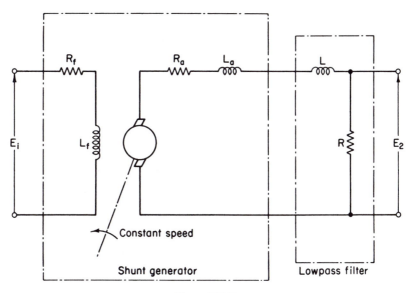

Figure P13-7

13-8. The motor of Fig. 13-5 is operated at constant field current. The motor parameters are: $R_a = 0.5\ \Omega$, $L_a = 0$, $K_\omega = 2$ V/rad/s, $K_t = 5$ lb-ft/A, and $F = 1$ lb-ft/rad/s. The equivalent inertia is 42 lb-ft-s^2.

(a) What is the numerical expression for the transfer function that relates motor speed in radians per second to the voltage applied to the armature circuit?

(b) If the applied armature voltage is a step of 200 V magnitude, find the steady-state speed of the motor.

(c) Obtain the complete expression for the motor speed from the instant the voltage of part (b) is applied to the time that virtual steady state is reached.

(d) Show the block diagram of this system, using the numerical values.

(e) Write the state equation for the system.

13-9. The application of 100 V to the armature of the machine of Prob. 13-8 at no-load generates an output velocity of 47.6 rad/s.

(a) If a load torque of 40 lb-ft is applied to the motor shaft, find the new operating speed.

(b) How long does it take for the new speed to be reached? Use three time constants as a satisfactory indication of settling time.

13-10. Repeat Prob. 13-8, taking into account the motor armature leakage inductance. Assume that $L_a = 0.05$ H.

13-11. The parameters of the system of Fig. P13-11 are as follows:

$$K_t = 0.0234 \text{ lb-ft/A} = \text{torque constant}$$

$$K_n = 0.2 \text{ V/rev/s} = \text{speed constant}$$

$$K_A = 0.15 \text{ V/V} = \text{amplifier gain}$$

$$R_a = 0.2 \text{ }\Omega$$

$$L_a = 0.0$$

$$J = 2 \times 10^{-6} \text{ slug-ft}^2$$

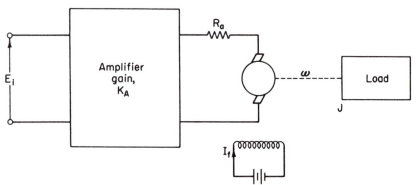

Figure P13-11

The viscous friction coefficient of the load is negligible.

(a) Develop the block diagram for this system, relating output velocity to amplifier input.

(b) Find the expression for the closed-loop transfer function, using the numerical values.

(c) Write the differential equation for this system.

(d) Is there an equivalent viscous friction (i.e., damping coefficient) in this system? Explain.

13-12. Find the complete solution for the motor velocity in the system of Fig. P13-11 corresponding to a unit step input to the amplifier.

13-13. Determine the differential equation of the system shown in Fig. P13-13 in terms of the letter symbols denoting the system parameters. Consider that the armature leakage inductance of the generator and motor are both negligible. Also, take the generator voltage constant as K_g, in volts per field ampere, the motor speed constant as K_ω in volts per radian per second, and the motor torque constant as K_t in lb-ft per armature ampere.

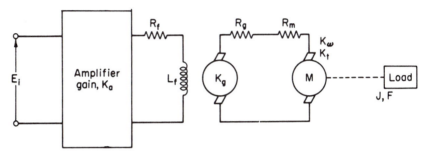

Figure P13-13

13-14. Express the dynamical behavior of the system of Fig. P13-13 in terms of the state equations expressed in normal matrix form.

13-15. A speed control system is depicted in Fig. P13-15. The generator field winding has resistance R_f and inductance L_f. Let R denote the armature circuit resistance of

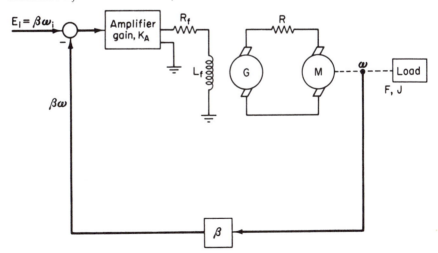

Figure P13-15

both the motor and generator. The armature winding leakage inductances are negligible and zero-load torque is applied. Also, the output voltage of the generator is described by

$$K_g I_f = I_a R + K_\omega \omega \qquad \text{where } K_g = \text{V/field amp}$$

and the output speed and developed motor torque are related by

$$K_t I_a = \frac{d\omega}{dt} + F\omega \qquad \text{where } K_t = \text{lb-ft/arm. amp}$$

Moreover, β is a tachometer constant expressed in volts per radian per second.
(a) Develop the block diagram showing the transfer function of each block.
(b) What is the characteristic equation of this system?
(c) Write the state equations in the standard matrix form.

13-16. A system is represented by the block diagram shown in Fig. P13-16.
(a) Find the expression for the closed loop transfer function.
(b) What is the characteristic equation?
(c) What influence does the parameter K_e exert on the system dynamic response?

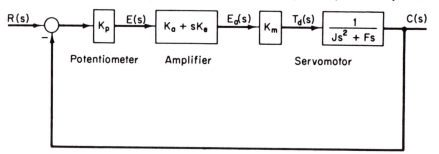

Figure P13-16

13-17. A system is represented by the block diagram depicted in Fig. P13-17.

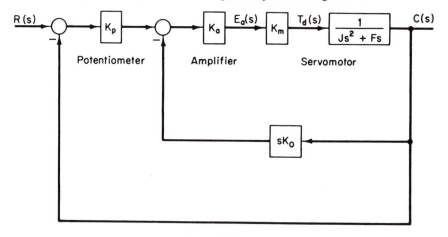

Figure P13-17

(a) Determine the closed loop transfer function.

(b) What is the expression for the characteristic equation?

(c) Write the defining differential equation for this system.

(d) What role does K_0 play in determining the dynamic response?

13-18. A system is represented by the block diagram depicted in Fig. P13-18. The amplifier section is designed to provide an output signal that contains a proportional as well as an integral component of the input signal.

(a) Find the closed loop transfer function.

(b) What is the characteristic equation?

(c) How does the result of part (b) compare with part (b) of Probs. 13-16 and 13-17?

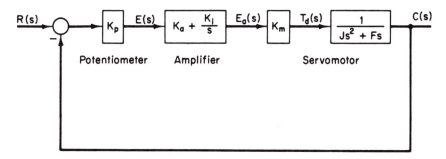

Figure P13-18

13-19. In the system of Fig. P13-19 assume that the reference and feedback tachometer generators are identical. In addition, the system has the following parameters:

$$K_a = \text{amplifier gain, V/V}$$

$$R_f = \text{generator field winding resistance, } \Omega$$

$$L_f = \text{generator field winding inductance, H}$$

$$K_f = \text{generator emf factor, V/field amp}$$

$$R = \text{armature circuit resistance of motor and generator}$$

$$K_t = \text{motor torque constant, lb-ft/arm. amp}$$

$$K\omega = \text{motor speed constant, V/rad/s}$$

$$\beta = \text{dc tachometer speed constant, V/rad/s}$$

$$J = \text{system inertia in consistent units}$$

$$F = \text{system viscous friction coefficient in consistent units}$$

(a) Draw the complete block diagram of this system, indicating clearly the transfer function of each block.

(b) Identify the state variables.

(c) Write the state equations in matrix form.

(d) Determine the characteristic equation of this system.

(e) Are oscillations possible in the dynamic state? Explain.

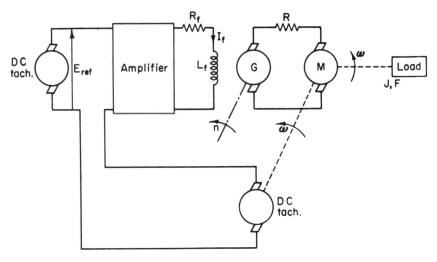

Figure P13-19

13-20. A 10-hp armature-controlled dc motor drives a load whose viscous friction coefficient is 2 lb-ft/rad/s and whose angular inertia is 17.5 lb-ft-s². The field winding is separately excited and maintained fixed. The corresponding motor parameters are then as follows:

$$R_a = 0.3 \ \Omega$$

$$K_\omega = 1.0 \ \text{V/rad/s}$$

$$K_t = 1.5 \ \text{lb-ft/arm. amp}$$

The armature winding inductance is negligible.
(a) Compute the steady-state speed corresponding to a step applied armature voltage of 210 V.
(b) How long does the motor take to reach within 95% of the steady-state speed of part (a)?
(c) What is the value of the total effective viscous damping coefficient of this motor-load configuration?

13-21. The motor of Prob. 13-20 is now used as a field-controlled dc machine in the manner illustrated in Fig. 13-23. The armature is assumed to be energized from a constant current source. The field winding parameters are: $R_f = 50 \ \Omega$ and $L_f = 20 \ \text{H}$. Moreover, the motor torque constant now expressed in lb-ft per field amperes (K_{tf}) is 60.
(a) Draw the block diagram.
(b) Calculate the value of steady-state speed for a step applied field voltage of 100 V.
(c) Approximately how long does it take the motor to reach within 95% of the final speed of part (b)?
(d) Write the state equations in matrix form, using numerical values.

13-22. A two-phase servomotor has a standstill torque at full control winding voltage of 4 oz-in. Also, at full control voltage, the no-load speed is known to be 100 rad/s. The motor torque constant is 0.1 oz-in. per V, or 4 oz-in. per control ampere. The control winding inductance is 6 H. Assume linear torque-speed curves and a rotor inertia of 0.5 oz-in.^{2}

(a) Determine the complete numerical expressions relating motor displacement to control winding voltage.

(b) If the result of part (a) is assumed driven from an amplifier having a voltage gain of 200, write the total expression for the direct transmission function.

(c) Place a unity feedback loop around the direct transmission function of part (b), and find the closed loop transfer function.

(d) What is the characteristic equation of the system of part (c)?

(e) What is the characteristic equation of the system of part (b)?

13-23. An amplidyne generator is used as a power stage that drives the dc motor of Prob. 13-20. The amplidyne has a control winding resistance of 400 Ω, a quadrature-axis circuit resistance of 0.1 Ω, and a direct-axis circuit resistance of 0.5 Ω. Moreover, the voltage induced in the cross-axis for each milliampere of control current is 0.1 V, and the voltage induced in the direct axis for each ampere in the quadrature axis is 5 V. The inductance of the control winding is 50 H and that of the quadrature axis is 2 mH.

(a) Determine the transfer function that relates motor output speed to amplidyne control winding voltage.

(b) For a control voltage of 5 V compute the steady-state speed of the motor.

(c) Identify the predominant factor in establishing the dynamic response of this amplidyne-motor system.

13-24. A unity feedback loop is placed around the amplidyne-motor configuration of Prob. 13-23. The feedback element is a position-sensitive device that has a transducer constant of 1 volt per radian of motor displacement.

(a) Find the closed-loop transfer function.

(b) Identify the characteristic equation.

(c) How many transient modes are there?

(d) Compute the time constant of each transient mode.

13-25. Refer to the system of Fig. 13-33, and assume that the parameters are those listed on p. 547.

(a) Calculate the no-load terminal voltage corresponding to a reference input voltage of 30.1 V.

(b) What is the value of the closed loop *gain*?

(c) Determine the complete solution for the output voltage corresponding to a step input of 30.1 V. Assume no-load.

(d) What is the time constant of this closed loop system?

13-26. In the system configuration of Fig. 13-33 find the complete solution for the output voltage for a step reference input of 30.1 V and the simultaneous load current demand of 25 A.

13-27. A speed control system is designed to have the configuration shown in Fig. P13-27. The system parameters have the following identification:

$$K_a = \text{amplifier gain, V/V}$$

$$R_f = \text{generator field winding resistance, } \Omega$$

$$L_f = \text{generator field winding inductance, H}$$

K_f = emf constant of generator, V/field amp

R = combined armature circuit resistance of generator and motor, Ω

K_t = motor torque constant, units of torque/arm. amp

J = equivalent inertia of motor and load

F = viscous friction coefficient of the load, assume negligible

K_ω = motor speed-voltage constant, V/rad/s

K_0 = dc tachometer speed-voltage constant, V/rad/s

Develop the complete block diagram for this system, and clearly show the transfer function for each block.

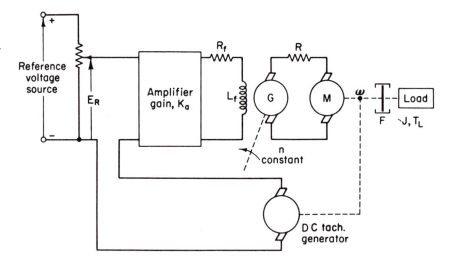

Figure P13-27

13-28. Refer to the system of Prob. 13-27, and find the expression for the no-load steady-state speed in terms of the system parameters for a given value of reference voltage.

13-29. Refer to the system of Prob. 13-27, and determine the expression for the final change in speed in terms of the system parameters corresponding to an applied load torque of T_L units of torque.

13-30. Determine the state equations in matrix form for the system of Fig. P13-27. Assume that the system is operating in response to an applied reference voltage E_R and an applied load torque of T_L.

Appendix
to Chapter 3

3A DEVELOPED TORQUE FOR SINUSOIDAL B AND NI: THE CASE OF AC MACHINES

The expression for electromagnetic torque as it appears in Eq. (3-9) has limited application because B is assumed constant and all conductors are assumed to produce equal torques in the same direction—a situation which rarely occurs in practical machines. In fact in most ac machines the field winding is intentionally designed to produce an almost sinusoidally distributed flux density, and the armature winding is similarly arranged to yield an ampere-conductor distribution which is also almost sinusoidally distributed along the periphery of the rotor structure. For this reason we shall now consider the development of an electromagnetic torque equation that applies for such cases.

The armature winding ampere-conductor distribution refers to a point-by-point plot of the current flowing through the individual conductors of the armature as depicted in Figs. 3-12(a) and (b). Note that the instantaneous value of the current existing in the conductors is assumed to be sinusoidally distributed. This situation is readily achieved in practical machines. For the 12 conductors shown in Fig. 3-12(a), 6 carry current into the plane of the paper and the other 6 carry current out of the paper, thus giving the appearance of two poles. Actually the number of poles associated with the ampere-conductor distribution of the arma-

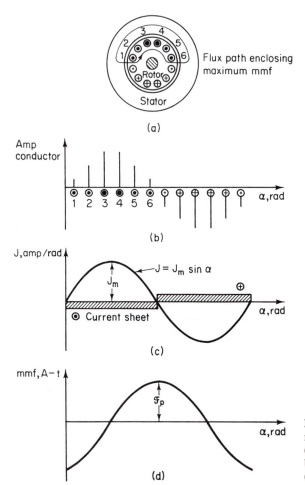

Figure 3-12 Development of the current sheet of sinusoidal density: (a) and (b) show finite ampere-conductor distribution; (c) current sheet representation; (d) corresponding mmf wave.

ture winding is always equal to the number of field poles produced by the field winding.

Note too that the greater the number of conductors the more nearly sinusoidal will be the ampere-conductor distribution. In fact, with a truly large number of conductors a current sheet of sinusoidal density as shown in Fig. 3-12(c) can be used to represent the ampere-conductor distribution. Of course this means that the latter now takes on a purely sinusoidal variation. Moreover, the armature winding can be considered a single sheet of one turn having the units of ampere-turns per radian. A per-unit radian notation is used because in Fig. 3-12(c) the distribution is no longer discrete.

A study of Fig. 3-12(a) reveals that for the ampere-conductor distribution shown, the armature winding taken by itself looks like a solenoid, the mmf of which is directed towards the right. For any specified armature-winding distribution the total value of this mmf is found by applying Ampere's circuital law. The essence of Ampere's circuital law is embodied in the statement that any closed flux path can be said to exist if it encloses an mmf that is not equal to zero. Thus, if a closed flux path is taken so that it encloses just conductor 1 in Fig. 3-12(a) as it crosses the air gap twice, the associated mmf is equal to the magnitude of the current in this single conductor. If the path is extended to embrace conductors 1 and 2, the total enclosed mmf is now the sum of the currents flowing in these two conductors. It follows then that Ampere's circuital law involves an integration (or summation) of the current sheet in order to obtain the mmf characteristic of the machine. In fact, for the closed flux path depicted in Fig. 3-12(a) showing all conductors carrying current in the same direction (which is the same as enclosing the current sheet over the displacement range from 0 to π radians), the maximum mmf associated with this current sheet is given by

$$\mathscr{F} = \int_0^\pi J_m \sin \alpha \, d\alpha = 2J_m \tag{3-49}$$

where J_m denotes the peak value of the current sheet density and $\mathscr{F}$ denotes the mmf per pole pair. If the upper limit in Eq. (3-12) is left unspecified, it follows that the ensuing integration yields the negative cosine function that is illustrated in Fig. 3-12(d). Whenever such a plot is generated from Ampere's circuital law, the abscissa axis is always placed so that the areas of the mmf curve above and below this axis are the same, in recognition of the fact that the flux that leaves the stator must also be the flux that returns to the stator.

Because the flux path used in obtaining the result of Eq. (3-49) involves two air gaps, it follows that the mmf per pole (or per air-gap crossing) is

$$\boxed{\mathscr{F}_p = \frac{\mathscr{F}}{2} = J_m} \tag{3-50}$$

This result will be useful later in converting the basic electromagnetic torque equation to its practical forms for ac machines.

Example 3-4

A two-pole machine has the stator construction shown in Fig. 3-13(a). Conductors 1, 2, 3, and 4 carry a current of 15 A directed out of the page. Correspondingly, conductors 5 to 8 carry the same current directed into the page.

(a) Draw the magnetomotive force as a function of the displacement along the air gap starting at point A and progressing clockwise. Use Ampere's circuital law.

(b) Find the maximum value of the mmf per pole.

(c) Identify the location of the axis of the flux produced by the mmf wave of part (a).

(d) Draw the equivalent current sheet associated with this mmf distribution.

Solution   (a) Starting at point A, draw a closed path that crosses the air gap twice and encircles conductor 1 in the manner depicted in Fig. 3-13(a). By Ampere's circuital law such a flux line path does exist because the net mmf enclosed by this

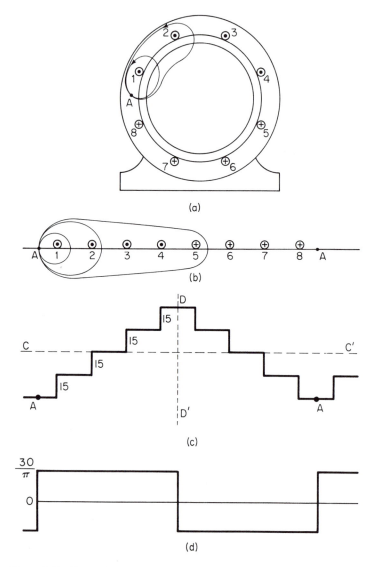

Figure 3-13 Determination of mmf distribution: (a) two-pole machine configuration; (b) plan view of the ampere-conductor distribution; (c) mmf distribution; (d) the equivalent current sheet.

closed path is not zero. In a practical machine conductor 1 is connected to a matching conductor, which in the case of a full-pitch coil would be conductor 5. Hence encirclement of conductor 1 means encirclement of the turn represented by the series connection of conductors 1 and 5. Accordingly, the number of ampere-turns that acts to produce such a flux line is here equal to 15 A-t. This situation is represented graphically in Figs. 3-13(b) and (c). Observe that until a flux path starting at point A is taken that reaches to enclose conductor 1, the enclosed mmf is zero. Just beyond the location of conductor 1, a step increase in mmf is assumed to occur on the assumption that thickness of the conductor is negligible. As additional flux lines, based at A are drawn, threading the space between conductors 1 and 2, no additional mmf is enclosed. Hence in Fig. 3-13(c), the mmf wave stays constant at 15 A-t until conductor 2 is reached. By drawing a closed flux path from A that just encloses conductors 1 and 2, there occurs an additional increase in the mmf enclosed by such a path. This is shown as the second step of the curve in Fig. 3-13(c). A continuation of this procedure with A as the starting point, yields the resultant mmf wave of Fig. 3-13(c). Several points are worth noting. When a flux path is drawn to enclose conductors 1 to 4, the enclosed mmf is a maximum because for such a flux path, all the conductors carry current in the same direction. Hence this spatial integration process carried on up to conductor 4 makes all positive contributions. As the process is extended beyond conductor 4, the additional contributions become negative since the newly added conductors now involve current flowing in the opposite direction. Thus if a flux path is drawn from A to include conductors 1 to 5, the net encircled mmf is now the same as when the path is made to enclose conductors 1 to 3. Of course, a flux line that encircles conductors 1 to 8 cannot in fact exist because the net mmf is zero.

(b) Keep in mind that the closed paths to find the mmf wave by spatial integration in part (a) were based on two crossings of the air gap, which is the situation that prevails in practical machines by design. The double crossing, therefore, implies that the mmf associated with the closed line integrals involves a *pair* of poles. Accordingly, the mmf curve of Fig. 3-13(c) is for two poles with the mmf equally divided. This point can be stressed by drawing a line CC' in Fig. 3-13(c) that divides the area between the mmf curve and line CC' into equal positive and negative halves. In this case, the negative half represents north-pole flux and the positive half south-pole flux.

It is worthwhile to note that the spatial integration process could start at any point on the stator surface. All that is required in the end is that the resulting curve be divided equally between north- and south-pole fluxes as represented by the mmf's that produce these fluxes.

Once line CC' is properly established, it follows in this case that the maximum height of the mmf wave above (or below) line CC' is 30 A-t. Hence

$$\mathscr{F}_p = 30 \text{ A-t/pole}$$

(c) On the assumption that the presence of the iron in the closed flux paths produces negligible effects, the axis of the flux produced by the mmf wave is coincident with the axis of the mmf wave itself, which is along line DD' in Fig. 3-13(c).

(d) The equivalent current sheet is depicted in Fig. 3-13(d). The distribution is rectangular because the current carried by the conductors is the same and fixed in magnitude. Take note of the quadrature relationship that the ampere-conductor distribution (represented by the current sheet here) bears to the mmf distribution along the air gap of the machine.

We are now in a position to develop a basic torque formula as it applies to a device that has a sinusoidally distributed flux density along the periphery of the air gap and an ampere-conductor distribution which for convenience is represented

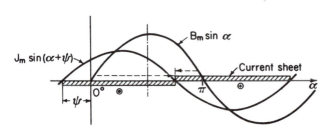

Figure 3-14 Configuration for development of the torque equation, showing the sinusoidally distributed flux density produced by the field winding and the current sheet, which represents the armature winding. The dashed arrows denote relative magnitude and direction of the electromagnetic torque developed per pole. The magnetic field and the current sheet are stationary relative to each other.

by a current-sheet density. Also, for generality, it is assumed that the current-sheet density is out of phase with the B-distribution by ψ degrees as depicted in Fig. 3-14. The procedure is really not much different from that which leads to Eq. (3-9), but several modifications are needed. We can again start with the fundamental force expression Eq. (3-7) and replace B with its sinusoidal variation $B_m \sin \alpha$. Furthermore, in place of the conductor and the current flowing through it, we now deal with a current element which is a differential strip of width $d\alpha$ of the current sheet. Accordingly, for i in Eq. (3-7) we may use

$$i = J \, d\alpha = J_m \sin (\alpha + \psi) \, d\alpha \qquad (3\text{-}51)$$

The force appearing on the elemental current strip is then expressed as

$$F_e = Bli = (B_m \sin \alpha)l[J_m \sin (\alpha + \psi) \, d\alpha] \qquad (3\text{-}52)$$

where the subscript e denotes elemental current strip. Of course the corresponding torque is obtained by multiplying by the rotor radius r. Hence

$$T_e = B_m J_m lr \sin \alpha \sin (\alpha + \psi) \, d\alpha \qquad (3\text{-}53)$$

To determine the total torque developed on the current sheet beneath one pole of the field distribution, we must take the sum of the elemental torques produced by each element of the current sheet lying beneath the pole. In other words, it is necessary to integrate Eq. (3-53) over π electrical radians, which is the span of one pole. Hence

$$T_p = \sum \text{torques on current elements beneath one pole}$$

$$= B_m J_m lr \int_0^\pi \sin \alpha \sin (\alpha + \psi) \, d\alpha \qquad (3\text{-}54)$$

where T_p denotes the torque developed per pole. Clearly for a field distribution which has p poles the total resultant electromagnetic torque is

$$T = p B_m J_m lr \int_0^\pi \sin \alpha \sin (\alpha + \psi) \, d\alpha \qquad (3\text{-}55)$$

By introducing the trigonometric identity

$$\sin \alpha \sin (\alpha + \psi) = \tfrac{1}{2} \cos \psi - \tfrac{1}{2} \cos (2\alpha + \psi) \qquad (3\text{-}56)$$

into Eq. (3-55) we obtain

$$T = pB_mJ_mlr \left[\int_0^\pi \tfrac{1}{2} \cos \psi \, d\alpha - \int_0^\pi \tfrac{1}{2} \cos (2\alpha + \psi) \, d\alpha \right] \qquad (3\text{-}57)$$

The second term in brackets has a value of zero, so the final expression for the developed electromagnetic torque is

$$T = \pi \frac{p}{2} J_m B_m lr \cos \psi \qquad (3\text{-}58)$$

Keep in mind that this equation is valid for sinusoidal distributions of the flux density and ampere conductors (or current sheet). Also note that for the indicated directions for the current sheet and magnetic flux, the developed torque is to the left and that its direction can be reversed when $\psi > 90°$.

An inspection of Eq. (3-58) reveals that the torque is dependent upon the space-displacement angle ψ. Thus when $\psi = 0°$ the best *field pattern* for developing torque prevails because the entire current sheet beneath a pole has a unidirectional torque. When $\psi = 45°$, as depicted in Fig. 3-14, note that a portion of the current sheet beneath a pole will produce an opposing torque, which results in a reduced torque if all other things are equal. It should not be inferred, however, that because $\psi \neq 0°$ the developed torque cannot be a maximum. In some ac machines, operation at $\psi \neq 0°$ causes a manifold increase in J_m, which gives a considerable increase in developed torque in spite of a nonoptimum field pattern.

Often it is convenient to express Eq. (3-58) in terms of the total flux per pole, which is the flux associated with any one pole of the flux-density curve. One such pole is shown in Fig. 3-15. But before proceeding further it is necessary to distinguish between mechanical and electrical degrees. It has already been pointed out that, regardless of the number of poles in the flux-density curve, each pole spans π electrical radians. Now, since there are always 2π mechanical radians in a configuration such as appears in Fig. 3-12(a), it follows that when the flux-density curve has two poles the mechanical and electrical degrees are the same. However, when the flux-density curve has four poles, then clearly the number of mechanical degrees is one-half the number of electrical degrees. For a p-pole flux-density

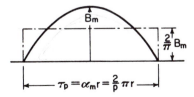

Figure 3-15 Relationships for computing flux per pole.

curve the relationship is

$$\boxed{\alpha_m = \frac{2}{p} \alpha} \qquad (3\text{-}59)$$

where α_m denotes mechanical degrees and α denotes electrical degrees. Therefore, in Fig. 3-15 the pole pitch expressed in meters is

$$\tau_p \equiv \text{pole pitch} = \alpha_m r = \frac{2}{p} \pi r \qquad (3\text{-}60)$$

where r is the radius of the rotor in meters. Moreover, for a rotor axial length of l the area associated with a pole of the flux-density curve is

$$A_p = l\tau_p = \frac{2}{p} \pi l r \qquad (3\text{-}61)$$

A further simplification in the computation leading to the expression for the flux per pole results upon recalling that the average value of the positive (or negative) half of a sine wave is

$$B_{av} = \frac{2}{\pi} B_m \qquad (3\text{-}62)$$

Thus the sine wave of Fig. 3-15 may be replaced with a rectangular wave having a height of B_{av} and spanning the entire pole pitch as indicated by the broken line in Fig. 3-15. The expression for the flux per pole then follows from

$$\Phi = B_{av} A_p \qquad (3\text{-}63)$$

Inserting Eqs. (3-61) and (3-62) into Eq. (3-63) yields

$$\Phi = \frac{4}{p} B_m l r \qquad (3\text{-}64)$$

Again keep in mind that this result is valid for sinusoidal variations of flux density.

By means of Eq. (3-64) the equation for the developed electromagnetic torque may be expressed in terms of the flux per pole. Insertion of Eq. (3-64) into Eq. (3-58) yields

$$\boxed{T = \frac{\pi}{8} p^2 \Phi J_m \cos \psi} \qquad \text{N-m} \qquad (3\text{-}65)$$

A study of Eq. (3-65) points out more clearly the three conditions that must be satisfied for the development of torque in conventional electromechanical energy-conversion devices. There must be a field distribution represented by Φ. There must be an ampere-conductor distribution represented by J_m. Finally, there must exist a favorable space-displacement angle between the two distributions. Note that it is possible to have both a field and an ampere-conductor distribution and still the torque can be zero if the field pattern is such that equal and opposite torques are developed by the conductors, i.e., $\psi = 90°$.

It is also worthwhile to note that no really new concepts have been used to derive the torque formula of Eq. (3-65). It is merely an extension of Ampere's law.

3B TORQUE FOR NONSINUSOIDAL B AND UNIFORM AMPERE-CONDUCTOR DISTRIBUTION: THE CASE OF DC MACHINES

The typical distributions are depicted in Fig. 3-16. Again for generality the current sheet that is representative of the ampere-conductor distribution is shown displaced in phase by the angle ψ. These distributions are commonly found in dc machines, as is described in Sec. 7-3. The starting point of the derivation is Ampere's law, again as expressed in Eq. (3-7). Of course modifications are needed consistent with the assumed distributions. Because the armature winding is represented by a current sheet, we deal with an elemental strip of the current sheet rather than with a particular conductor and the current it carries. Thus in Eq. (3-7) we introduce

$$i = J \, d\alpha \tag{3-66}$$

and

$$B = B_\alpha \tag{3-67}$$

where B_α denotes the nonsinusoidal variation of flux density along the air gap. An exact analytical expression for B_α is difficult to obtain, but this is not disturbing because one deals rather with the flux per pole, which is readily available. Hence the force developed by an elemental strip of the current sheet is

$$F_e = B_\alpha l J \, d\alpha \tag{3-68}$$

The corresponding torque for a p-pole machine is then

$$T_e = pJ \, B_\alpha l r \, d\alpha \qquad \text{N-m} \tag{3-69}$$

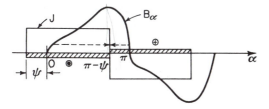

Figure 3-16 Current-sheet and flux-density curves characteristic of dc machines.

The total resultant torque is obtained upon integrating Eq. (3-69) over one pole pitch. Thus

$$T = pJ \left[\int_0^{\pi-\psi} B_\alpha lr \, d\alpha - \int_{\pi-\psi}^{\pi} B_\alpha lr \, d\alpha \right] \qquad (3\text{-}70)$$

The second term in brackets carries the minus sign to account for the fact that J is negative beneath the positive pole of the B_α curve in the region from $\pi - \psi$ to π. Clearly, without an analytical expression for B_α, a closed-form solution for the torque is not readily available. However, in electromechanical energy-conversion devices where these distributions apply, the designer by intention makes $\psi = 0$. Unlike the preceding case this ensures that no conductors contribute negative torques beneath the poles. Introducing $\psi = 0$ simplifies the expression for the electromagnetic torque to

$$T = pJ \int_0^{\pi} B_\alpha lr \, d\alpha \qquad \text{N-m} \qquad (3\text{-}71)$$

Now in the interest of avoiding B_α we can express the torque in terms of the flux per pole—a quantity which is available from other considerations. In terms of B_α the flux per pole Φ is

$$\Phi = \int_0^{\pi} B_\alpha \, dA_p \qquad (3\text{-}72)$$

But the area per pole can be expressed as

$$A_p = lr\alpha_m = lr \frac{2}{p} \alpha \qquad (3\text{-}73)$$

Hence Eq. (3-72) becomes

$$\Phi = \int_0^{\pi} B_\alpha lr \frac{2}{p} \, d\alpha = \frac{2}{p} \int_0^{\pi} B_\alpha lr \, d\alpha \qquad (3\text{-}74)$$

Therefore

$$\int_0^{\pi} B_\alpha lr \, d\alpha = \frac{p}{2} \Phi \qquad (3\text{-}75)$$

Substituting Eq. (3-75) into Eq. (3-71) yields the final form for the electromagnetic torque developed when the distributions are as depicted in Fig. 3-16 with $\psi = 0$. Thus

$$\boxed{T = \frac{p^2}{2} J\Phi} \qquad \text{N-m} \qquad (3\text{-}76)$$

It is worthwhile to note again that electromagnetic torque is dependent upon a field distribution represented by Φ, an ampere-conductor distribution represented by J, and the angular displacement between the two distributions. Of course ψ does not appear in Eq. (3-76) because for simplicity the current sheet was assumed in phase with the flux-density curve, thereby ensuring that all parts of the current sheet experience a unidirectional torque. A glance at Fig. 3-16 should make it apparent that for $\psi = 90°$ the resultant electromagnetic torque is zero.

Example 3-5

A p-pole machine has the sinusoidal field distribution $B_p \sin(\alpha - \psi)$, where α and ψ are in electrical degrees. The armature consists of a uniform current sheet of value J, which has associated with it a triangular mmf having an amplitude per pole of F_A. Refer to Fig. 3-17(a). The machine has an axial length l and a radius r.

    (a) Determine the expression for the maximum flux/pole.

    (b) Obtain the expression for the developed electromagnetic torque in terms of B_p, J, and Δ.

    (c) Express the torque found in part (b) in terms of F_A and the flux per pole Φ. Assume that $\Delta = \pi/2$.

    (d) Is the torque found in part (b) constant at every point in the air gap? Explain.

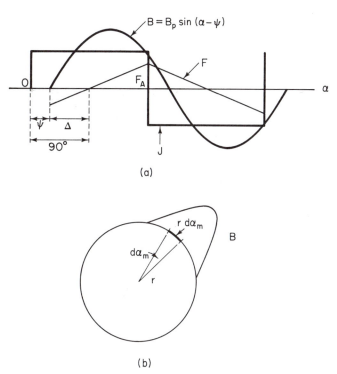

(a)

(b)

Figure 3-17 Configuration for Example 3-5: (a) relationship existing between flux density, mmf, and current sheet; (b) a view of the rotor showing one pole and the differential arc length $r \, d\alpha_m$.

Solution (a) The expression for the differential flux is given by

$$d\phi = B \, dA$$

where dA denotes a differential area beneath the flux density curve. This area is given by the product of the effective axial length l and the arc length, which is denoted by $r \, d\alpha_m$. Here use is made of the mechanical angle α_m because we are dealing with a physical differential arc length. See Fig. 3-17(b). By using Eq. (3-59), we can express the differential arc length in terms of electrical degrees as $(2r/p) \, d\alpha$. Accordingly, the differential flux can then be written as (for $\psi = 0$ for convenience in this part)

$$d\phi_p = \frac{2lrB}{p} \, d\alpha = \frac{2lrB_p}{p} \sin \alpha \, d\alpha$$

The total flux per pole can now be determined by integrating the last expression over one pole pitch which is equal to π electrical radians. Thus

$$\Phi_p = \frac{2B_p lr}{p} \int_0^\pi \sin \alpha \, d\alpha = \frac{4B_p lr}{p}$$

Note that this result entirely agrees with Eq. (3-64).

(b) Let i denote the differential current element, which for a uniform current sheet of value J can be written as

$$i = J \, d\alpha$$

Associated with this differential current element under the influence of the pole flux density is a differential force $Bli = BJl \, d\alpha$. When acting on a torque arm of the machine radius r, the differential torque expression becomes

$$BJlr \, d\alpha$$

By integrating this quantity over one pole pitch (i.e., π electrical radians), the net torque produced by the interaction of the field distribution and the current sheet is determined for one pole. The total torque results upon multiplying by the number of poles. Thus the expression for the total differential torque is

$$dT = p(BJlr \, d\alpha)$$

Or the total torque produced becomes

$$T = plrJ \int_0^\pi B \, d\alpha = plrJB_p \int_0^\pi \sin (\alpha - \psi) \, d\alpha$$

Upon integrating, inserting the limits and simplifying, the result is

$$T = 2B_p Jlr \cos \psi$$

Because ψ and Δ are complementary angles, we get

$$\psi = 90° - \Delta$$

which when introduced into the preceding equation and simplified yields

$$T = 2plrB_p J \sin \Delta$$

(c) For $\Delta = \pi/2$ the last equation simplifies to

$$T = p2B_plrJ$$

Algebraic manipulation of this last expression leads to expressions which can be identified as the flux per pole and the amplitude of the mmf per pole. The flux per pole is readily identified by rearranging the last expression to read as follows:

$$T = p^2 \left(\frac{4}{p} B_plr\right) \frac{J}{2} = p^2\Phi_p \frac{J}{2}$$

Clearly, the quantity in parentheses is the flux per pole.

To get F_A into this formulation, it is necessary to recognize that when $\Delta = \pi/2$, the total mmf per pair of poles is given by [See Fig. 3-11(a)]

$$\mathcal{F} = \pi J$$

Hence the amplitude per pole is then

$$F_A = \frac{\mathcal{F}}{2} = \frac{\pi J}{2}$$

Accordingly, by rewriting the torque equation in the further manipulated form

$$T = p^2\phi_p \frac{1}{\pi} \left(\frac{\pi J}{2}\right)$$

We get finally

$$T = \frac{p^2}{\pi} \Phi_p F_A$$

(d) No. Although the current sheet is of constant value, the flux density is not. Therefore, the instantaneous value of torque varies with the value of B.

3C THE GENERAL TORQUE EQUATION

The formulas for the electromagnetic torque that are derived in Secs. 3-1, 3-4, and 3A apply in situations where the field distribution in one part of an electric machine interacts with the ampere-conductor distribution in the other part. Torque is produced in the machine because of the mutual action existing between the windings to bring about an alignment of their associated fields. As long as currents exist in each winding and the field pattern is proper, there exists a torque owing to this mutual coupling. Throughout the development the assumption was made that currents exist in *each* winding. However, there are machines in which it is possible to produce torque when only one part of the machine is excited. Surely this statement is not so surprising in light of the results of Sec. 1-10, where it was shown that a force can exist whenever a flux flows in a geometrical configuration characterized by a reluctance that varies with position. Refer to Eq. (1-55). What is important to us here, however, is to recognize that the torque associated with reluctance variation cannot be computed from the results derived in Secs. 3-1 and 3A. It seems worthwhile therefore at this point in our study to obtain a general expression for the torque developed by an electromechanical energy-conversion device—one that permits calculation of the *reluctance torque* as well as the *electromagnetic torque* of the individually excited, mutually coupled windings.

The general torque equation can be derived by analyzing the configuration shown in Fig. 3-18. A current i_1 is assumed to flow in the rotor winding producing the air-gap flux field. The current i_2 is assumed to flow in the stator winding. This current may be direct current, single-phase alternating current, or the phase current of a polyphase system applied to a polyphase stator winding; in any event, irrespective of the form of i_2 it is assumed that the stator current is responsible for the ampere-conductor distribution shown. Next consider that the rotor is allowed to move by some differential angle $d\alpha$. In turn this causes a change in coupling between the rotor field distribution and the stator ampere-conductor distribution. Moreover, the induced emf's of both windings will be affected, which will influence the energy input. If the input energy originates solely with the electrical sources, it follows that corresponding to the displacement $d\alpha$ the change in input energy is described by

$$dW_e = e_1 i_1 \, dt + e_2 i_2 \, dt \tag{3-77}$$

But by Faraday's law

$$e_1 = \frac{d\lambda_1}{dt} \quad \text{and} \quad e_2 = \frac{d\lambda_2}{dt} \tag{3-78}$$

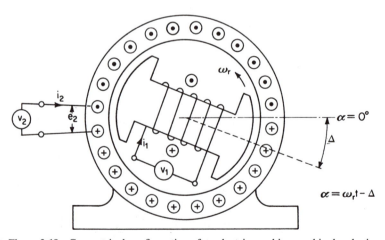

Figure 3-18 Geometrical configuration of an electric machine used in developing the general torque equation. Current i_1 flows in the rotor winding and current i_2 flows in the stator winding; together these produce the electromagnetic torque. The nonuniform air gap gives rise to a reluctance torque.

where λ_1 and λ_2 denote the total flux linkages of the rotor and stator windings respectively. Specifically, these quantities may be expressed in terms of the self- and mutual inductances provided that nonlinearities and hysteresis are neglected. In large measure this is justified by the presence of air gaps. Thus

$$\lambda_1 = L_{11}i_1 + M_{12}i_2 \qquad (3\text{-}79)$$

$$\lambda_2 = M_{21}i_1 + L_{22}i_2 \qquad (3\text{-}80)$$

Here L_{11} is the self-inductance of the rotor winding, L_{22} is the self-inductance of the stator winding, and M_{12} is the mutual inductance between rotor and stator windings. The positive sign is used for M_{12} because i_1 and i_2 produce fluxes in the same direction in the configuration of Fig. 3-18.

Upon substituting Eqs. (3-78) in Eq. (3-77) there results

$$dW_e = i_1\,d\lambda_1 + i_2\,d\lambda_2 \qquad (3\text{-}81)$$

This expression emphasizes that any change in electrical input energy comes about through changes in flux linkages. Furthermore, the energy balance condition requires that this change in electrical input energy be equal to the change in field energy or the mechanical work done by the field or both. Mathematically expressed,

$$dW_e = dW_f + T\,d\alpha \qquad (3\text{-}82)$$

where T denotes the total torque developed as a consequence of the rotor displacement $d\alpha$ and W_f represents the field energy associated with the currents i_1 and i_2. For the doubly excited configuration of Fig. 3-18 the expression for this field energy is given by

$$W_f = \tfrac{1}{2}L_{11}i_1^2 + \tfrac{1}{2}L_{22}i_2^2 + Mi_1i_2 \qquad (3\text{-}83)$$

A word is in order at this point concerning the variational nature of the self- and mutual inductances as a function of the rotor displacement angle α. In accordance with Eq. (1-40) we can express the self-inductance of the rotor as

$$L_{11} = \frac{N_1^2}{\mathcal{R}_1} \qquad (3\text{-}84)$$

where N_1 denotes the number of turns of the rotor winding and $\mathcal{R}_1$ denotes the reluctance of the flux produced by the rotor winding. It is interesting to note here that irrespective of the position taken by the rotor the reluctance $\mathcal{R}_1$ remains invariant for a symmetrical and concentrically located rotor structure. Hence for this geometry L_{11} is a constant.

However, for the stator winding, the situation is different. Although a similar expression for the stator self-inductance may be written, i.e.,

$$L_{22} = \frac{N_2^2}{\mathcal{R}_2} \qquad (3\text{-}85)$$

examination of the geometry of Fig. 3-18 indicates that the reluctance $\mathcal{R}_2$ varies with the rotor position. This is readily understood by considering for a moment that i_2 is made to flow alone (i_1 assumed zero) in the stator winding to produce the ampere-conductor distribution depicted in Fig. 3-18. The stator thus behaves like a solenoid producing flux on an axis coincident with the horizontal. Clearly, then, when the rotor axis lies along this horizontal axis (i.e., $\alpha = 0$ degrees), the reluctance seen by this flux will be a minimum because the magnetic path consists of two small air gaps and a large amount of high-permeability iron. This path is called the *direct axis* of the machine and leads to a minimum reluctance $\mathcal{R}_d$. As the rotor iron is displaced towards $\alpha = \pi/2$, the reluctance increases because the stator flux is still horizontally directed but the rotor iron is advancing towards the vertical position as α increases which increases the air gap on the horizontal axis. At $\alpha = \pi/2$ the reluctance seen by the stator-produced flux is a maximum because it now consists of two very large air gaps. The rotor alignment with respect to the stator-produced flux field is now said to be in its *quadrature* axis. Call this maximum value of reluctance $\mathcal{R}_q$. As a result of the shaping of the pole pieces and the distributed nature of the stator winding,† the reluctance variation $\mathcal{R}_2$ is a cosinusoidal function of α. Since for each full revolution of the rotor structure there are two positions each of minimum and maximum reluctance, it follows that L_{22} is a double frequency function of α. The complete variation of the reluctance is depicted graphically in Fig. 3-19. The analytical expression for this reluctance is given by

$$\mathcal{R}_2 = \frac{\mathcal{R}_q + \mathcal{R}_d}{2} - \frac{\mathcal{R}_q - \mathcal{R}_d}{2} \cos 2\alpha \qquad (3\text{-}86)$$

Insertion of this expression into Eq. (3-85) yields the corresponding equation that describes the variation of the stator self-inductance L_{22} with rotor position.

There now finally remains to be determined the manner in which the mutual inductance M_{12} varies with rotor displacement, if at all. A simple and direct way to obtain this information is to perform a straightforward test on the machine of Fig. 3-18. Apply a fixed sinusoidal voltage to the rotor winding. Remove the source V_2 from the stator winding, and replace it with a suitable rms-reading voltmeter. The voltmeter reads the voltage induced through the mutual coupling with the changing current in the rotor winding. Specifically, this induced emf is related to the rotor current by

$$e_2 = M_{12} \frac{di_1}{dt} \qquad (3\text{-}87)$$

For a sinusoidal variation of i_1 the magnitude of emf induced in the stator winding, E_2, by the constant rms rotor current I_1 is

$$E_2 = \omega M_{12} I_1 \qquad (3\text{-}88)$$

---

†The stator winding is not concentrated as a coil but rather is distributed in slots along the entire inner periphery of the stator structure.

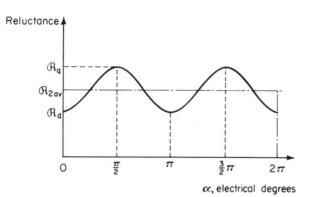

Figure 3-19 Variations of reluctance as a function of rotor displacement, as viewed from the stator winding. To a different scale this plot also represents the variation of L_{22} versus α.

or

$$M_{12} = \frac{E_2}{\omega I_1} = \frac{E_{2\,\text{max}}\cos\alpha}{\omega I_1} \qquad (3\text{-}89)$$

The value of E_2 in this equation is directly a cosinusoidal function of the coupling between the rotor flux field and the stator winding. A glance at Fig. 3-18 shows that at $\alpha = 0$ degrees the rotor flux has maximum linkage with the stator winding, which registers a maximum reading on the voltmeter, $E_{2\text{max}}$. At $\alpha = \pi/2$ the net flux linkage is zero (ideally). Hence E_2 is zero. At $\alpha = \pi$ the flux linking the stator winding is once again a maximum, but because the flux field is reversed the stator-induced emf undergoes a reversed polarity. Hence the mutual inductance is at its negative maximum value. The variation of M_{12} between its extreme values occurs in accordance with a cosinusoidal function owing to the inherent construction features of the machine. Appearing in Fig. 3-20 is the complete variation of the mutual inductance for one full revolution of the rotor. Note that, unlike the stator self-inductance, the mutual inductance has a fundamental variation with the rotor displacement. If the maximum value of the mutual inductance is called M, then

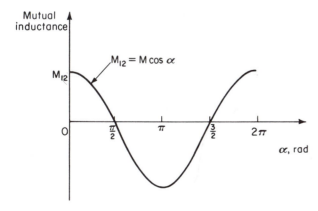

Figure 3-20 Variation of mutual inductance between the rotor and stator windings of the machine depicted in Fig. 3-18.

$$\boxed{M_{12} = M \cos \alpha} \tag{3-90}$$

The foregoing results demonstrate that, for the geometrical configuration of Fig. 3-18, L_{11} is independent of rotor displacement while L_{22} and M_{12} are both functions of α. Keeping these facts in mind, we can now return to Eq. (3-81) and compute the change in the electrical energy input in terms of the self- and mutual inductances. Inserting Eqs. (3-79) and (3-80) into Eq. (3-81) leads to

$$dW_e = i_1 \, d(L_{11}i_1 + M_{12}i_2) + i_2 \, d(M_{12}i_1 + L_{22}i_2) \tag{3-91}$$

Since L_{11} is a constant, its differential is zero. This is not so for L_{22} and M_{12}. Accordingly, Eq. (3-91) becomes

$$dW_e = i_1 L_{11} \, di_1 + i_1 M_{12} \, di_2 + i_1 i_2 \, dM_{12} + i_2 M_{12} \, di_1 + i_1 i_2 \, dM_{12}$$
$$+ i_2 L_{22} \, di_2 + i_2^2 \, dL_{22} \tag{3-92}$$

Also, the change that takes place in the field energy is readily found by differentiating Eq. (3-83). Thus

$$dW_f = i_1 L_{11} \, di_1 + i_2 L_{22} \, di_2 + \tfrac{1}{2} i_2^2 \, dL_{22} + M_{12} i_1 \, di_2 + M_{12} i_1 \, di_2$$
$$+ M_{12} i_2 \, di_1 + i_1 i_2 \, dM_{12} \tag{3-93}$$

When Eqs. (3-92) and (3-93) are introduced into Eq. (3-82)—the energy balance equation—all the terms of dW_e that involve differential currents have matching terms in the expression for dW_f and so cancel out. This is by way of stating that differential changes in the currents have no bearing on the development of mechanical forces. The remaining terms in the energy balance equation are then as follows:

$$i_2^2 \, dL_{22} + 2i_1 i_2 \, dM_{12} = T \, d\alpha + \tfrac{1}{2} i_2^2 \, dL_{22} + i_1 i_2 \, dM_{12} \tag{3-94}$$

The left side of Eq. (3-94) represents the change in electrical input energy corresponding to the rotor displacement $d\alpha$. The last two terms on the right side denote the change in the field energy. The remainder of the input energy goes to do mechanical work as represented by $T \, d\alpha$. It follows from Eq. (3-94) that the desired general expression for the developed torque in the machine of Fig. 3-18 is

$$\boxed{T = \frac{1}{2} i_2^2 \frac{dL_{22}}{d\alpha} + i_1 i_2 \frac{dM_{12}}{d\alpha}} \tag{3-95}$$

This is the expression for the instantaneous total developed torque.† It depends

---

†A little thought should make it apparent that, if the machine configuration were such that L_{11} were also a function of α, the general torque equation would become

$$T = \frac{1}{2} i_1^2 \frac{dL_{11}}{d\alpha} + \frac{1}{2} i_2^2 \frac{dL_{22}}{d\alpha} + i_1 i_2 \frac{dM_{12}}{d\alpha} \tag{3-95a}$$

upon the instantaneous value of i_1 and i_2 as well as on the change of the stator self-inductance and the mutual inductance with rotor position.

A study of Eq. (3-94) leads to a useful alternative form of the general torque equation. Note that, if i_1 and i_2 are held constant, the change in field energy is exactly equal to half the change in electrical energy input. Therefore the mechanical work done is equal to the change in field energy. Expressed mathematically, this is

$$T = \left(\frac{\partial W_f}{\partial \alpha}\right)_{i_1, i_2 \text{ const}} \tag{3-96}$$

The partial derivative is employed here to emphasize that i_1 and i_2 are held constant in this operation. Equation (3-96) can be useful in several ways. Once particular application is to reveal that the positions of maximum or minimum field energy corresponding to zero developed torque.

Finally, using Eq. (3-95), it can be shown that the torque of mutual coupling can be expressed in the following two alternative forms

$$T = \frac{\pi}{8} p^2 \Phi \mathscr{F}_p \sin \Delta = \frac{\pi}{8} p^2 \Phi J_m \cos \psi \tag{3-97}$$

where

$$\Delta = \frac{\pi}{2} - \psi \tag{3-98}$$

Moreover, the associated reluctance torque becomes

$$T_r = \frac{1}{4} \frac{V^2}{\omega} \frac{x_d - x_q}{x_d x_q} \sin 2\Delta \tag{3-99}$$

where

$$x_d = \omega L_d = \omega \frac{N_2^2}{R_d} \quad \text{and} \quad x_q = \omega L_q = \omega \frac{N_2^2}{R_q} \tag{3-100}$$

Example 3-6

Two coils are mounted in a machine with a uniform air gap in the manner illustrated in Fig. 3-21. The self- and mutual inductances are given by

$$L_1 = 5 \text{ H} \qquad M = 3 \cos \alpha$$

$$L_2 = 2 \text{ H}$$

The winding resistances are negligible. Coil 2-2' is initially placed at an angle α from coil 1-1'. Coil 1-1' is made to carry a current of $10 \sin \omega t$, where ω is the frequency of the voltage source that energizes coil 1-1'.

(a) Find the expression for the torque when coil 2-2' is supplied with a direct current of -5 A from an ideal current source. Will a displacement of the rotor occur? Explain.

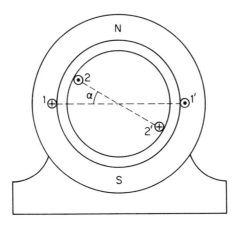

Figure 3-21   Configuration for Example 3-6.

(b) Now assume that the current that flows through coil 2-2' is $-5 \sin \omega t$. Repeat part (a).

(c) At what value of α will the torque be a maximum?

Solution   (a) Equation (3-95a) may be applied here. However, note that dL_1/dt and dL_2/dt are zero because L_1 and L_2 are given as constants. Hence the electromagnetic torque is attributable solely to the change in mutual inductance. Thus

$$T = i_1 i_2 \frac{dM}{d\alpha} = (10 \sin \omega t)(-5) \frac{d}{d\alpha} (3 \cos \alpha)$$

$$= 150 \sin \alpha \sin \omega t$$

Clearly, the average torque over one cycle of the source is zero and so no rotation occurs in this case. This statement is based on the assumption that the source frequency is rapid enough so that the rotor cannot make an instantaneous response.

(b) In this case we get

$$T = (10 \sin \omega t)(-5 \sin \omega t)(-3 \sin \alpha)$$

$$= 150 \sin \alpha \sin^2 \omega t$$

Introducing the trigonometric substitution

$$\sin^2 \omega t = \tfrac{1}{2} - \tfrac{1}{2} \cos 2\omega t$$

then yields

$$T = 75 \sin \alpha - 75 \cos 2\omega t$$

In this case, there does exist an average torque which is different from zero. In fact, specifically it is equal to the first term on the right side of the last equation. Of course, the second term is zero over one cycle of the source frequency.

(c) A glance at the last equation makes it obvious that the maximum torque occurs at $\alpha = 90°$. Accordingly, when coil 2-2' is energized while at position α in Fig. 3-21, the rotor will rotate in a clockwise position until it reaches a position where $\alpha = 90°$, provided that a restraining force in excess of $75 \sin \alpha$ does not appear on the rotor shaft.

PROBLEMS (cont.)

3-19. A two-pole machine is equipped with a total of 12 conductors uniformly distributed in slots along the inner surface of the stator periphery. The conductors are paired so that conductor 1 is series connected to conductor 7 by an appropriate end connection. Similarly, conductor 2 is joined to conductor 8 and so on until all conductors are series connected and energized by a dc source which forces the flow of a 10-A current.

(a) Draw the mmf distribution as a function of displacement along the air gap.

(b) Determine the number of A-t per pole.

(c) Draw the equivalent current sheet of the ampere-conductor distribution.

3-20. An electric motor is designed to have a triangular field distribution as shown in Fig. P3-20. The ampere-conductor distribution is uniform at value J. Assume that the machine has p poles. A rotor radius of r meters and air axial length of l meters.

(a) Find the expression for the flux per pole.

(b) Derive the equation for the developed torque in terms of the flux per pole, ψ, and the peak mmf per pole $\mathscr{F}_p$.

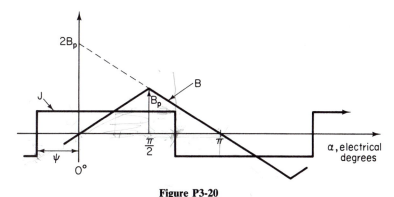

Figure P3-20

3-21. A p-pole machine has the sinusoidal current sheet variation and constant field distribution depicted in Fig. P3-21. The machine has an axial length l and radius r.

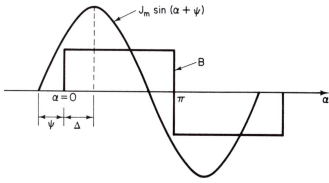

Figure P3-21

(a) Find an expression for the maximum flux per pole.
(b) Determine the expression for the developed electromagnetic torque in terms of B, J_m, and Δ.
(c) Is the torque constant at every point along the air gap? Explain.

3-22. Two windings are mounted in a machine with a uniform air gap as shown in Fig. P3-22. The self- and mutual inductances are

$$L_{11} = 2.2 \text{ H} \qquad M_{12} = \sqrt{2} \cos \alpha \text{ H}$$

$$L_{22} = 1.0 \text{ H}$$

The winding resistances are negligible. When winding 2-2′ is short-circuited with voltage applied to winding 1-1′, the current in winding 1-1′ is known to be $i_1 = 10 \sqrt{2} \sin \omega t$, where ω is the frequency of the source.

(a) If the rotor is stationary, derive an expression for the instantaneous torque, in newton-meters, in terms of the space angle α.
(b) Compute the average torque in newton-meters when $\alpha = 45°$.
(c) If the rotor is permitted to move, will it rotate continuously or will it tend to come to rest? If the latter, at what value of α?

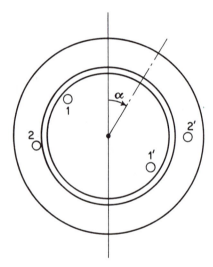

Figure P3-22

3-23. In the machine configuration of Fig. P3-23, the self- and mutual inducíances have the following identification:

$$L_{11} = \frac{N_1^2}{\mathcal{R}} \qquad M_{12} = \frac{2N_1 N_2 \alpha}{\pi \mathcal{R}} \qquad \text{for } -\frac{\pi}{2} < \alpha < \frac{\pi}{2}$$

$$L_{22} = \frac{N_2^2}{\mathcal{R}}$$

where N_1 denotes the stator turns and N_2 denotes the rotor turns. The machine is equipped with a uniform air gap. Through use of the rate of change of flux linkage with angular displacement, the torque is found to be

$$T = i_1 \frac{d\lambda_{12}}{d\alpha}$$

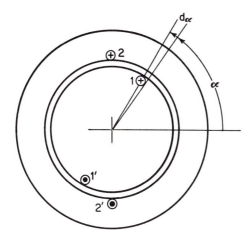

Figure P3-23

Moreover, through the use of the *Blri* relationship, the expression for developed torque is found to be $T = 2N_1 N_2 B_2 lri_2$. Finally, from the conservation of energy, the torque equation can be written as

$$T = \frac{1}{2} i_1^2 \frac{dL_{11}}{d\alpha} + \frac{1}{2} i_2^2 \frac{dL_{22}}{d\alpha} + i_1 i_2 \frac{dM_{12}}{d\alpha}$$

Show that the torques computed by all three methods are equal.

3-24. In the configuration of Fig. P3-23 let $\alpha = \omega_0 t$. By means of the flux linkage relationship the emf induced in winding 1-1' can be expressed by $e_1 = d\lambda_{12}/dt$. Moreover, through the use of the *Blv* relationship the emf equation becomes $e_1 = 2N_1 N_2 B_2 lv$. Finally, the expression for the speed voltage can be written as

$$e_{1s} = i_1 \frac{dL_{11}}{dt} + i_2 \frac{dM_{12}}{dt} = \left(i_1 \frac{dL_{11}}{d\alpha} + i_2 \frac{dM_{12}}{d\alpha} \right) \omega_0$$

Show that the speed voltages computed by all three methods are equal.

3-25. In the machine configuration of Fig. P3-23 the positive directions for the currents i_1 and i_2 are defined by the dots and crosses appearing in the diagram. If both currents are of constant magnitude with $i_1(t) = I$ and $i_2(t) = -(N_1/N_2)I$, and if $\alpha = 50°$, as depicted, in what direction is the torque on the rotor? Explain. With the same currents, what should be the value of α for maximum energy to be stored in the magnetic field? Explain.

3-26. In the idealized uniform air-gap machine of Fig. P3-26, excitation of either winding produces a radial field. Coil 1 lies at radius r_1 from the axis of rotation. Coil 2 lies at radius r_2 from the axis of rotation. Given that

$$T_1 = 2N_1 B_2 lr_1 i_1$$

$$T_2 = 2N_2 B_1 lr_2 i_2$$

where B_1 is the flux density at coil 2 produced by i_1 in coil 1, and B_2 is the flux density at coil 1 produced by i_2 in coil 2.
(a) Show that T_1 and T_2 have equal magnitudes.
(b) Show that T_1 and T_2 act in opposite directions.

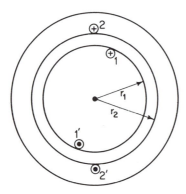

Figure P3-26

3-27. (a) Prove the validity of Eq. (3-83) for the doubly excited configuration.

   (b) Demonstrate how Eq. (3-88) follows from Eq. (3-87).

3-28. In the machine configuration of Fig. 3-18 the rotor winding is unexcited and the stator winding is energized with the direct current I_{dc}.

   (a) Can a reluctance torque exist? Why?

   (b) At what rotor speed, if any?

   (c) Derive an expression for the reluctance torque in terms of the direct-axis self-inductance L_d and the quadrature-axis self-inductance L_q.

3-29. The machine configuration of Prob. P3-23 is doubly excited with currents i_1 and i_2. Are the positions in space at which the maximum torque occurs affected by the values of i_1 and i_2? Explain.

3-30. A reluctance torque motor has the configuration of Fig. 3-18 with a sinusoidal reluctance variation along the periphery of the air gap. The motor is energized from a 120-V, 60-Hz source. With the rotor fixed in the minimum reluctance position, a line ammeter reads 0.5 A rms. In the maximum reluctance position it reads 2 A.

   (a) Determine the power developed in watts when the motor is operating with a power angle of 10 electrical degrees.

   (b) What is the speed of rotation of the rotor?

   (c) What is the maximum developed power in watts for this motor?

3-31. A doubly excited system with a nonuniform air gap has inductances described by

$$L_{11} = 6 + 1.5 \cos 2\alpha \qquad L_{22} = 4 + \cos 2\alpha \qquad M = 6 \cos \alpha$$

The coils are excited by dc currents with $i_1 = 0.5$ A and $i_2 = 0.4$ A.

   (a) Determine the developed torque as a function of α.

   (b) Find the stored energy as a function of α.

   (c) For $\alpha = 30°$ find the magnitude and direction of the torque on the rotor.

3-32. The dynamometer (moving-coil) instrument is distinguished from other measuring instruments by its linear scale. In construction the moving coil is restrained by a linear spring to have limited movement in flux field of constant value produced by a permanent magnet. Derive an expression to show that the deflection of an indicator needle attached to the moving coil is directly proportional to the coil current.

3-33. A singly excited reluctance motor has a variation of reluctance over the periphery of the air gap that is as depicted in Fig. P3-33(a) for an assumed power angle displace-

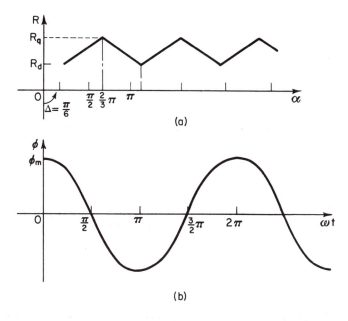

Figure P3-33

ment Δ. At the same time the variation of flux with time is described in Fig. P3-33(b).

(a) Can a net torque different from zero be produced through the cooperation of the given flux and reluctance variation? Explain.

(b) Sketch the variations of the instantaneous torque for $\Delta = 30°$.

(c) Compute the average value of torque associated with this instantaneous variation.

3-34. A machine configuration is such that the variation of mutual and self-inductances can be described by

$$M_{12} = 1 - \sin \alpha \qquad \text{H}$$

$$L_{11} = 1 + \sin \alpha \qquad \text{H}$$

$$L_{22} = 2(1 + \sin \alpha) \qquad \text{H}$$

(a) Find the value and direction of the developed torque when the current in coil 1 is maintained constant at 10 amperes and that in coil 2 is fixed at -20 A. Take $\alpha = 60°$.

(b) For the condition of part (a), compute the amount of energy supplied by each electrical source.

(c) If the current of coil 1 of part (a) is changed to a sinusoidal current of 10 amperes rms and coil 2 is short-circuited, find the rms value of coil 2 current. Assume a 60-Hz, 120-V source.

(d) Determine the expression for the instantaneous torque produced in part (c).

(e) What is the average value of the torque in part (c)?

Appendices

A. UNITS AND CONVERSION FACTORS

TABLE A-1 UNITS

Quantity	Symbol	Name of unit	Dimension
Fundamental			
Length	l, L	meter	L
Mass	m, M	kilogram	M
Time	t	second	T
Current	i, I	ampere	I
Mechanical			
Force	F	newton	MLT^{-2}
Torque	T	newton-meter	ML^2T^{-2}
Angular displacement	θ	radian	—
Velocity	v	meter/second	LT^{-1}
Angular velocity	ω	radian/second	T^{-1}
Acceleration	a	meter/second^2	LT^{-2}
Angular acceleration	α	radian/second^2	T^{-2}
Spring constant (translation)	K	newton/meter	MT^{-2}
Spring constant (rotational)	K	newton-meter	ML^2T^{-2}

TABLE A-1 CONTINUED

Damping coefficient (translational)	D, F	newton-second/meter	MT^{-1}
Damping coefficient (rotational)	D	newton-meter-second	ML^2T^{-1}
Moment of inertia	J	kilogram-meter^2	ML^2
Energy	W	joule (watt-second)	ML^2T^{-2}
Power	P	watt	ML^2T^{-3}
Electrical			
Charge	q, Q	coulomb	TI
Electric potential	v, V, E	volt	$ML^2T^{-3}I^{-1}$
Electric field intensity	ε	volt/meter (or newton/coulomb)	$MLT^{-3}I^{-1}$
Electric flux density	D	coulomb/meter^2	$L^{-2}TI$
Electric flux	Q	coulomb	TI
Resistance	R	ohm	$ML^2T^{-3}I^{-2}$
Resistivity	ρ	ohm-meter	$ML^3T^{-3}I^{-2}$
Capacitance	C	farad	$M^{-1}L^{-2}T^4I^2$
Permittivity	ϵ	farad/meter	$M^{-1}L^{-3}T^4I^2$
Magnetic			
Magnetomotive force	$\mathscr{F}$	ampere(-turn)	I
Magnetic field intensity	H	ampere(-turn)/meter	$L^{-1}I$
Magnetic flux	ϕ	weber	$ML^2T^{-2}I^{-1}$
Magnetic flux density	B	tesla	$MT^{-2}I^{-1}$
Magnetic flux linkages	λ	weber-turn	$ML^2T^{-2}I^{-1}$
Inductance	L	henry	$ML^2T^{-2}I^{-2}$
Permeability	μ	henry/meter	$MLT^{-2}I^{-2}$
Reluctance	$\mathscr{R}$	ampere/weber	$M^{-1}L^{-2}T^2I^2$

TABLE A-2 CONVERSION FACTORS

Quantity	Multiply number of	by	to obtain
Length	meters	100	centimeters
	meters	39.37	inches
	meters	3.281	feet
	inches	0.0254	meters
	inches	2.54	centimeters
	feet	0.3048	meters
Force	newtons	0.2248	pounds
	newtons	10^5	dynes
	pounds	4.45	newtons
	pounds	4.45×10^5	dynes
	dynes	10^{-5}	newtons
	dynes	2.248×10^{-6}	pounds
Torque	newton-meters	0.7376	pound-feet
	newton-meters	10^7	dyne-centimeters
	pound-feet	1.356	newton-meters
	dyne-centimeters	10^{-7}	newton-meters

TABLE A-2 CONTINUED

Energy	joules (watt-seconds)	0.7376	foot-pounds
	joules	2.778×10^{-7}	kilowatt-hours
	joules	10^7	ergs
	joules	9.480×10^{-4}	British thermal units
	foot-pounds	1.356	joules
	joules	1.6×10^{-19}	electron-volts
Power	watts	0.7376	foot-pounds/second
	watts	1.341×10^{-3}	horsepower
	horsepower	745.7	watts
	horsepower	0.7457	kilowatts
	foot-pounds/second	1.356	watts

B. PITCH AND DISTRIBUTION FACTORS FOR MACHINE WINDINGS

In the interest of making more effective use of the copper and iron as well as to discriminate severely against the presence of space harmonics in the air-gap field distribution, the windings of ac machines are intentionally distributed and pitched so that all the phase turns do not fully link the total air-gap flux. In a three-phase machine, the distribution of the winding is such that each phase is made to occupy a region of 60 electrical degrees beneath each pole. Moreover, each phase coil is made to span less than 180 electrical degrees; i.e., the span is less than a pole pitch. Such a coil is referred to as a *fractional-pitch coil*. Although the use of the fractional-pitch coil means some reduction in the value of emf induced by the fundamental component of the field, the attenuation of the space harmonics is so much greater that it easily justifies the cost.

The Pitch Factor

This factor serves as a measure of the effectiveness of a selected coil pitch in reducing the influence of field harmonics on the emf induced in the armature winding of the machine. To understand how this comes about, attention is directed first to the reduction caused in the fundamental value of the induced emf compared to the case of a full-pitch coil. Refer to Fig. B-1, and assume that the coil consists of the two coil sides 1 and 2 which are moving toward the left relative to the stationary north- and south-pole fundamental field component. The voltage induced in coil side 1 is therefore out of the paper, and that induced in 2 is into the paper. The instantaneous induced emf in these coil sides can conveniently be represented graphically by means of Fig. B-2. The voltages E_1 and E_2 denote the maximum values of emf associated with each coil side. Accordingly, the instantaneous values are found by obtaining the projection of these peak voltages on the

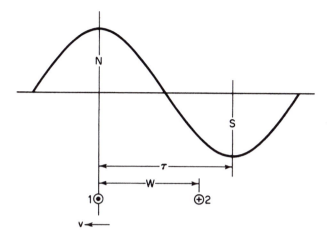

Figure B-1 Fractional-pitch coil in the fundamental field.

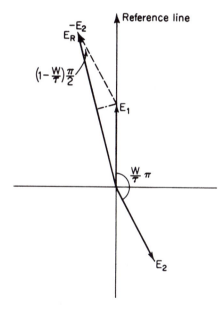

Figure B-2 Phasor diagram for the fractional-pitch coil.

vertical reference line. Of course, each coil side has the same peak voltage. Placing E_1 coincident with the reference line implies that coil side 1 is instantaneously under the influence of the maximum flux density. Figure B-2 is the phasor diagram representation of the situation depicted in Fig. B-1. The projection of E_2 onto the reference line is negative, which is consistent with the fact that coil side 2 is under the influence of a pole of reversed polarity from that of coil side 1. Because the end connection that joins coil sides 1 and 2 provides addition of the coil voltages, the resultant coil voltage is found in Fig. B-2 by adding $-E_2$ to E_1. It is worthwhile to keep in mind here that the maximum value of this resultant coil voltage remains fixed. The instantaneous value, which varies with the position of the coil relative to the flux field, is readily found as the vertical projection of the E_R phasor. It is important to note that, because of the use of a fractional-pitch coil, the resultant voltage can never be equal to the algebraic sum of the coil side voltages. As a matter of fact, the pitch factor is simply defined as the *ratio of the resultant coil voltage to the absolute sum of the voltages induced per coil side*.

A convenient analytical expression for the pitch factor for the fundamental induced voltage can be derived easily from the geometry of Fig. B-2. Let W denote the coil width in suitable units, and let τ denote the pole pitch in the same units. The angle between E_1 and E_2 is then $W\pi/\tau$. Drawing the perpendicular bisector from the tip of E_1 onto E_R reveals two right triangles. The resultant coil voltage can then be expressed in terms of the coil side voltage by

$$E_R = 2E_1 \cos \left(1 - \frac{W}{\tau} \right) \frac{\pi}{2} \qquad\qquad \text{(B-1)}$$

The expression for the pitch factor for the fundamental is therefore

$$K_{p1} = \frac{E_R}{2E_1} = \cos \left(1 - \frac{W}{\tau} \right) \frac{\pi}{2} = \sin \frac{\pi}{2} \sin \frac{W}{\tau} \frac{\pi}{2}$$

$$= \sin \frac{W}{\tau} \frac{\pi}{2} \qquad\qquad \text{(B-2)}$$

For a machine that has nine slots per pole, the use of a coil width of seven slots leads to a value of $K_{p1} = 0.940$. Thus there is just a 6% reduction in fundamental induced emf. However, the corresponding reduction in the harmonics is much greater. This is demonstrated next.

If n denotes the order of a space field harmonic, a little thought discloses that the phasor diagram differs by the amount of rotation associated with one coil side voltage relative to the other. Thus in Fig. B-2 the angle that $-E_2$ makes with the vertical is increased by n times. Accordingly, the expression for the nth harmonic pitch factor becomes

$$K_{pn} = \frac{E_{Rn}}{2E_{cn}} = \cos n \left(1 - \frac{W}{\tau} \right) \frac{\pi}{2} = \sin n \frac{\pi}{2} \sin n \frac{W}{\tau} \frac{\pi}{2} \qquad \text{(B-3)}$$

For the same machine with nine slots per pole, the value of fifth harmonic pitch factor becomes

$$K_{p5} = -0.174 \qquad\qquad \text{(B-4)}$$

Note that the emf induced by the fifth harmonic field component is reduced by approximately 83% compared to 6% for the fundamental. The minus sign in Eq. (B-4) indicates a phase reversal for the seventh harmonic emf, brought about by a negative flux linkage caused by the chording of the winding.

The Distribution, or Breadth, Factor

In order to focus solely upon the effect of distributing the winding of an ac machine over sixty electrical degrees, it is assumed that only full-pitch coils are used. Once again the derivation is carried out first for the fundamental field wave. In this connection refer to Fig. B-3, which shows the four coils that are assumed to constitute all of phase I of a single-layer, two-pole, three-phase machine. It is helpful to note that the instant is shown which corresponds to a net phase voltage of zero. This means that in the phasor diagram of Fig. B-4 the phasor denoting the phase voltage (AE) must have a zero projection on the reference line. In Fig. B-4 the quantity AB denotes the peak value of the voltage induced in coil 1-13. Note that this coil is displaced from the position of maximum voltage by the angle α. At the same time, because of the distribution pattern, the voltage of coil 2-14 (denoted by BC) is away from its peak value by the angle $(\alpha + \beta)$, where β denotes the angle between slots. It is interesting to observe that the projection of BC onto the reference line is smaller than that of AB. A glance at Fig. B-3 indicates consistency with the physical relationship these coils bear to the field distribution. Coil 3-15 is represented by CD in Fig. B-4. Its projection on the

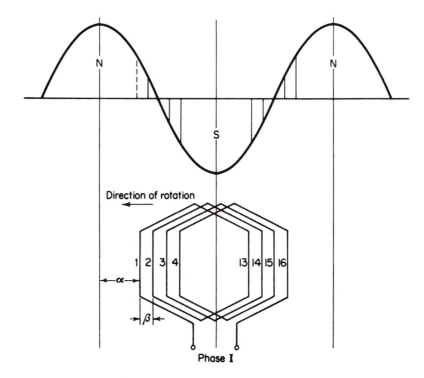

Figure B-3 Showing a distributed winding relative to the fundamental field flux at the instant of zero net phase voltage.

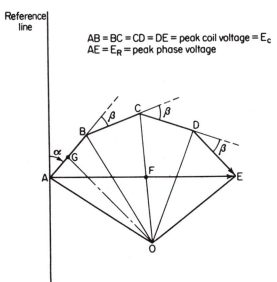

Reference line

AB = BC = CD = DE = peak coil voltage = E_c
AE = E_R = peak phase voltage

Figure B-4 Derivation of the distribution factor. Illustration is for four slots per pole per phase.

time line is negative, thus indicating that coil sides 3 and 15 are under the influence of poles of reversed respective polarity than those for coils 1-13 and 2-14. Coil 4-16 is represented by *DE* in Fig. B-4. Each successive coil voltage is displaced by an additional amount (equal to the slot angle β) from the preceding coil. The effect is that a resultant phase voltage is obtained that is *less* than the absolute sum of the individual coil voltage. In fact, the distribution factor is defined as the *ratio of the resultant voltage to the absolute sum of the coil voltages comprising the resultant.*

An examination of Fig. B-4 leads to a convenient analytical expression for the distribution factor. By drawing the perpendicular bisectors of each of the coil voltages represented by *AB*, *BC*, and so on, the center of a circle passing through points *A, B, C, D,* and *E* can be found. The quantities *AB, BC, CD,* and *DE* are then also equal chords of this circle. Accordingly, it follows that

$$\sphericalangle\, AOE = Q\beta \tag{B-5}$$

where Q denotes the number of slots per pole per phase. Moreover,

$$\sphericalangle\, AOB = \beta \tag{B-6}$$

$$\sphericalangle\, AOG = \frac{\beta}{2} \tag{B-7}$$

and

$$\sphericalangle\, AOF = \frac{Q\beta}{2} \tag{B-8}$$

The expression for the peak value of the resultant phase voltage becomes

$$E_R = AE = 2(AF) = 2(OA) \sin \frac{Q\beta}{2} \qquad \text{(B-9)}$$

By definition, the distribution factor is then

$$K_d \equiv \frac{E_R}{QE_c} = \frac{2(OA) \sin \frac{Q\beta}{2}}{Q2(AG)} = \frac{2(OA) \sin \frac{Q\beta}{2}}{Q2(OA) \sin \beta/2}$$

$$= \frac{\sin \frac{Q\beta}{2}}{Q \sin \beta/2} \qquad \text{(B-10)}$$

For the machine shown in Fig. B-4 the values of the quantities in Eq. (B-10) are $Q = 4$, $\beta = 15°$. Hence the value of the distribution factor for the fundamental is 0.958, thus representing less than a 5% reduction. The value of the distribution factor for the fifth harmonic for the same machine is found to be 0.205, thus representing an 80% reduction.

The net winding factor for a winding that is pitched, as well as distributed, is equal to the product of the pitch and distribution factors. Expressed mathematically,

$$K_{wn} = K_{pn} K_{dn} \qquad \text{(B-11)}$$

where the second subscript n denotes the nth harmonic. When K_{pn} and K_{dn} are low for a given harmonic, the net winding factor assumes such small values as to render that particular harmonic inconsequential.

C. DERIVATION OF THE PEAK MMF PER POLE OF A POLYPHASE WINDING

To begin simply, consider a two-pole machine with three phases and a total of six slots. Each phase therefore consists of a single coil, which for a full-pitch winding means that the coil spans from slot 1 to slot 4, as depicted in Fig. C-1. For a uniform air-gap machine, any current that flows through this coil produces a rectangular mmf distribution. The total coil mmf is distributed equally between slots 1 and 4 (say for flux entering the air gap) and slots 4 and 1 (where the flux is assumed leaving the air gap for the same instant). Attention for the moment is directed to the mmf distribution of one phase. The effect of the remaining phases is considered presently.

If the phase coil is assumed to be carrying a sinusoidally varying current of rms value I, then the height of the rectangular distribution alternates between a

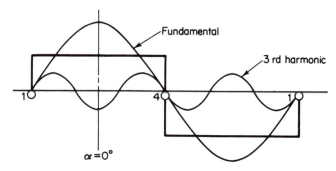

Figure C-1 Magnetomotive force distribution of a full-pitch, single-coil phase winding in a uniform air-gap machine.

positive maximum and a negative maximum value, in accordance with the following expression:

$$\mathcal{F} = \frac{\sqrt{2}}{2} z_s I \sin \omega t \qquad \text{(C-1)}$$

where z_s denotes the conductors per slot. Note that when the argument is $\pi/2$, the coil mmf has a positive maximum value. In the interest of obtaining a result that is applicable generally, i.e., to the harmonics as well as to the fundamental contained in the rectangular distribution of Fig. C-1, a Fourier analysis is applied to this wave. The equation for the mmf distribution for phase I, expressed as a function of time and space displacement from the indicated reference position $\alpha = 0°$, then becomes

$$\mathcal{F}_I(\alpha) = \sum_{n=1,3}^{\infty} a_n \cos n\alpha \qquad \text{(C-2)}$$

where

$$a_n = \frac{1}{\pi} \int_0^{2\pi} \mathcal{F} \cos n\alpha \, d\alpha \qquad \text{(C-3)}$$

and $\mathcal{F}$ is given by Eq. (C-1).

Inserting Eq. (C-1) into Eq. (C-3) and noting that the given wave possesses quarter-wave symmetry, we may write

$$a_n = \frac{4}{\pi} \int_0^{\pi/2} \frac{\sqrt{2}}{2} I z_s \sin \omega t \cos n\alpha \, d\alpha = 0.9 z_s I \sin \omega t \left(\frac{1}{n} \sin n \frac{\pi}{2} \right) \qquad \text{(C-4)}$$

Substituting this result into Eq. (C-2) then yields

$$\mathcal{F}_I(\alpha) = 0.9 z_s I \sin \omega t \, (\cos \alpha - \tfrac{1}{3} \cos 3\alpha + \tfrac{1}{5} \cos 5\alpha - \tfrac{1}{7} \cos 7\alpha + \cdots) \qquad \text{(C-5)}$$

Now, if instead of one slot per pole per phase the machine is equipped with Q slots per pole per phase, the foregoing equation can be made to apply by introducing two modifications: replacing z_s by $z_s Q$ and including the distribution factor of each harmonic. Accordingly, we get

$$\mathcal{F}_I(\alpha) = 0.9 z_s Q I \sin \omega t \left(K_{d1} \cos \alpha - \frac{K_{d3}}{3} \cos 3\alpha + \frac{1}{5} K_{d5} \cos 5\alpha - \cdots \right) \quad \text{(C-6)}$$

Furthermore, if each coil is assumed to be fractional pitch, then each term of Eq. (C-6) must be further modified by the pitch factor for each harmonic. Thus the complete expression that describes the mmf distribution of one phase of a fractional-pitch distributed winding is

$$\mathcal{F}_I(\alpha) = 0.9 z_s Q I \sin \omega t \left(K_{w1} \cos \alpha - \frac{K_{w3}}{3} \cos 3\alpha + \frac{K_{w5}}{5} \cos 5\alpha - \cdots \right) \quad \text{(C-7)}$$

where $K_{wn} = K_{dn} K_{pn}$ is the winding factor obtained as the product of the individual distribution and pitch factors.

Since our interest is chiefly in the resultant mmf produced by three-balanced windings cooperating to produce a revolving mmf in the manner described in Sec. 4-1, we need first the expressions for the mmf of phases II and III. These are readily obtained from Eq. (C-7) by introducing the required 120° displacement in time and space. Hence

$$\mathcal{F}_{II}(\alpha) = 0.9 z_s Q I \sin (\omega t - 120°)\left[K_{w1} \cos (\alpha - 120°) - \frac{K_{w3}}{3} \cos (3\alpha - 360°) \right.$$

$$\left. + \frac{K_{w5}}{5} \cos (5\alpha + 600°) - \cdots \right] \quad \text{(C-8)}$$

$$\mathcal{F}_{III}(\alpha) = 0.9 z_s Q I \sin (\omega t - 240°)\left[K_{w1} \cos (\alpha - 240°) - \frac{K_{w3}}{3} \cos 3(\alpha - 240°) \right.$$

$$\left. + \frac{K_{w5}}{5} \cos 5(\alpha + 240°) - \cdots \right] \quad \text{(C-9)}$$

It is helpful to note that the mmf of each individual phase has an alternating character. This is true of the harmonic as well as the fundamental components. However, the resultant fundamental or harmonic waves are found to revolve with time. For any particular harmonic this is readily corroborated by summing the contributions for each phase. Thus for the fundamental the result is found to be

$$\mathcal{F}_{r1}(\alpha) = \tfrac{3}{2}(0.9) z_s Q I K_{w1} \sin (\omega t - \alpha) \quad \text{(C-10)}$$

As already demonstrated in Sec. 4-1, Eq. (C-10) is the equation of a traveling sine wave moving in the positive α direction. It follows therefore that the *amplitude* of this resultant wave is the coefficient of the trigonometric term. Hence

$$\mathcal{F}_A = \tfrac{3}{2}(0.9) z_s Q K_{w1} I \quad \text{(C-11)}$$

The factor 3 appears because we are dealing with three phases. For a q-phase machine, Eq. (C-11) becomes

$$\mathscr{F}_A = \frac{q}{2}(0.9)z_s QK_{w1}I \tag{C-12}$$

It is often desirable to express this quantity in terms of the total number of turns per phase (N). This is possible to do upon recognizing that

$$N\left(\frac{\text{turns}}{\text{phase}}\right) = \frac{z_s}{2}\left(\frac{\text{turns}}{\text{slot}}\right) Q\left(\frac{\text{slots}}{\text{pole} \times \text{phase}}\right) p\,(\text{poles}) \tag{C-13}$$

Accordingly,

$$\frac{z_s Q}{2} = \frac{N}{p} \tag{C-14}$$

Insertion of the last expression into Eq. (C-12) yields the final and desired form of the mmf amplitude of a polyphase winding, namely,

$$\mathscr{F}_A = 0.9q\,\frac{NK_{w1}}{p}I \tag{C-15}$$

A similar procedure can be followed to find the expression of the peak mmf of the harmonics in the resultant revolving field. Adding the third harmonic components of each of the three phases yields zero—a result that is obvious from an inspection of Eqs. (C-7), (C-8), and (C-9). This conclusion also applies to all multiples of the third harmonic. Accordingly, there is no resultant third harmonic component in the traveling mmf wave of a balanced three-phase winding.

The addition of the fifth harmonic contributions in each of the three phases leads to the following result:

$$\mathscr{F}_{r5}(\alpha) = \frac{3}{2}(0.9)z_s QI\,\frac{K_{w5}}{5}\sin\,(\omega t + \alpha) \tag{C-16}$$

Note the form of the argument is such as to indicate that the fifth harmonic component of the resultant wave moves in the negative α direction. The amplitude of the fifth harmonic wave is then clearly

$$\mathscr{F}_{A5} = 3(0.9)\frac{Qz_s}{2}\frac{K_{w5}}{5}I = 3(0.9)\frac{N}{p}\frac{K_{w5}}{5}I \tag{C-17}$$

For a q-phase machine this becomes

$$\mathscr{F}_{A5} = 0.9q\,\frac{NK_{w5}}{5p}I \tag{C-18}$$

A little thought concerning the derivation of the last expression readily reveals that the general expression for the amplitude of the resultant traveling wave of the nth harmonic of a q-phase winding is

$$\mathscr{F}_{An} = 0.9q \frac{NK_{wn}}{np} I \qquad\qquad \text{(C-19)}$$

Upon inserting $n = 1$, Eq. (C-15) is obtained.

D. REDUCTION FACTORS FOR POLYPHASE WINDINGS

A *reduction factor* is a numerical quantity which, when applied to a rotor quantity such as voltage, current, impedance, and so on, refers that quantity to the stator. This means that the effect produced by the referred quantity in the stator is identical to the effect caused by the original quantity in the rotor. The reduction factors are different for the various electrical quantities, and they also differ depending upon whether the rotor is of the wound or squirrel-cage type.

The Wound Rotor

There are three reduction factors of interest: one to handle voltage, one to handle current, and the third to deal with resistance, reactance, and impedance. As described in Chapter 2, a reduction factor is essentially a transformation ratio that converts the actual rotor winding into an equivalent rotor winding having the same number of turns as the stator winding. If we let E_1 denote the standstill emf induced per phase in the stator winding by the air-gap flux, and E_2 denote the standstill emf induced per phase in the rotor winding by the same air-gap flux, then by Eqs. (4-9) and (4-10) the ratio of these two quantities is

$$\frac{E_1}{E_2} = \frac{N_1 K_{w1}}{N_2 K_{w2}} = a_v \qquad\qquad \text{(D-1)}$$

Equation (D-1) is the reduction factor for referring *any* voltage associated with the rotor winding back to the stator. The convention used in this book to denote the referred quantity is the *prime* notation. Accordingly, to refer the quantity E_2 to the stator, we write

$$E_2' = a_v E_2 \qquad\qquad \text{(D-2)}$$

A comparison with Eq. (D-1) shows E_2' is identical to E_1, and they should be since both quantities are related to a winding having the same effective turns cut by the same gap flux.

The reduction factor for current can be derived by recalling how the rotor current is made to flow. Ideally, at no-load and for an induction motor with negligible rotational losses, the rotor current is zero. The slip correspondingly is also zero. As load is applied to the rotor shaft, the rotor structure slows down, thus causing an emf to be induced in the rotor winding, which in turn produces enough rotor current to supply the required energy to the load. As rotor current flows through the q_2-phases of the rotor, it presents a total rotor mmf of magnitude $q_2 N_2 K_{w2} I_2$ that acts in a direction to oppose the original stator mmf, $q_1 N_1 K_{w1} I_m$. The latter quantity is the effective stator mmf that produces the air-gap flux, the value of which is essentially determined by the applied stator voltage. To offset this opposing action of the rotor mmf and thus preserve the value of air-gap flux demanded by the applied stator voltage, there must occur an increase in stator mmf equal and opposite to the total rotor mmf. Calling the increased stator current I_2', at equilibrium we can then write

$$q_1 I_2' N_1 K_{w1} = q_2 I_2 N_2 K_{w2} \tag{D-3}$$

Or, the required increase in stator current becomes

$$I_2' = \frac{q_2 N_2 K_{w2}}{q_1 N_1 K_{w1}} I_2 = a_i I_2 \tag{D-4}$$

where

$$a_i \equiv \frac{q_2 N_2 K_{w2}}{q_1 N_1 K_{w1}} = \text{current reduction factor} \tag{D-5}$$

The application of a_i to any rotor phase current gives the equivalent effect of that current in the stator winding. It is interesting to note that when the rotor and stator phases are the same, the reduction factor for current is the inverse of the reduction factor for voltage.

The manner of transferring the effect of a given rotor resistance to the stator can be found from the relationship that preserves either copper loss or per-unit resistance drop. For example, for a rotor winding resistance r_2 and rotor current I_2 the per-unit resistance drop is $I_2 r_2 / E_2$. When placed in the stator, the effect of this rotor resistance (now called r_2') must be the same as it produces in the rotor itself, in spite of the fact that in the stator the r_2' carries the current I_2' and must be associated with the stator voltage E_1 (or E_2'). Equivalence is preserved by satisfying the following equation:

$$\frac{I_2' r_2'}{E_2'} = \frac{I_2 r_2}{E_2} \tag{D-6}$$

It then follows that the rotor resistance as it must appear in the stator is

$$r_2' = \frac{I_2}{I_2'} \frac{E_2'}{E_2} r_2 \tag{D-7}$$

Inserting Eqs. (D-1), (D-2), and (D-4) yields

$$r_2' = \frac{q_1}{q_2} \left(\frac{N_1 K_{w1}}{N_2 K_{w2}}\right)^2 r_2 = a_z r_2 \tag{D-8}$$

where

$$a_z = \frac{q_1}{q_2}\left(\frac{N_1 K_{w1}}{N_2 K_{w2}}\right)^2 = \frac{\text{reduction factor for resistance,}}{\text{reactance and impedance}} \qquad \text{(D-9)}$$

It is a simple matter to show that Eq. (D-9) applies with equal validity to reactance and impedance quantities. This is left as an exercise for the reader.

The Squirrel-Cage Rotor

It is helpful to keep in mind that the squirrel-cage rotor consists of solid copper bars that are often driven through the slots and connected at both ends by means of copper end rings. As a result of this construction each bar can be considered to be a separate and individual phase of a q-phase system, reckoned in terms of the traditional definition of what constitutes a phase. Since each bar (or conductor) is a phase, it follows that the number of turns per phase must be $\frac{1}{2}$ (it takes two conductors to make one turn), and the winding factor must be identically unity.

The voltage reduction factor for the squirrel-cage rotor is readily determined by inserting $N_2 = \frac{1}{2}$ and $K_{w2} = 1$ into Eq. (D-1), which is a general result. Thus

$$a_v = 2N_1 K_{w1} \qquad \text{(D-10)}$$

When the number of rotor bars is not divisible by the pole pairs through an integer, the number of rotor phases is equal to the number of bars. Hence, under these circumstances for the squirrel-cage rotor, Eq. (D-3) becomes

$$q_1 I_2' N_1 K_{w1} = S_2(\tfrac{1}{2})(1)I_2 \qquad \text{(D-11)}$$

or

$$I_2' = \left(\frac{S_2}{2q_1 N_1 K_{w1}}\right) I_2 \qquad \text{(D-12)}$$

where S_2 denotes the number of rotor slots, and the quantity in parentheses is the current reduction factor.

The reduction factor for resistance is found again by employing Eq. (D-6). For the squirrel-cage rotor this leads to

$$r_2' = \frac{E_2'}{E_2}\frac{I_2}{I_2'}r_2 \qquad \text{(D-13)}$$

Introducing the appropriate current and voltage reduction factors into the last equation yields

$$r_2' = \left[\frac{4q_1(N_1 K_{w1})^2}{S_2}\right] r_2 \qquad \text{(D-14)}$$

where the quantity in brackets identifies the desired result. As before, this reduction factor is applicable to reactance and impedance as well as resistance.

E. SATURATED SYNCHRONOUS REACTANCE METHOD BASED ON RESULTANT FLUX

Frequently, the design of synchronous machines is such that when rated power is delivered at rated voltage the operating point on the magnetization curve is located somewhere along the knee of the curve. Accordingly, the machine can be described as operating in a partially saturated state. A good indication of the degree of saturation of the magnetic circuit is given by the resultant flux Φ_r or its associated emf E_r. Thus, if a synchronous generator is assumed to be delivering rated power at rated voltage and a specified power factor, the actual armature induced emf $\bar{E}_r$ can be found as illustrated by Eq. (5-22) and then located on the magnetization curve. This is illustrated in Fig. E-1, where E_r is represented by point a on the saturation curve and is produced by the field current R. It is worthwhile to note here that if there were no saturation of the iron, the voltage E_r would be induced by a smaller excitation R_{ag}. In essence, then, it can be said that the degree of saturation prevailing under these conditions is represented by the

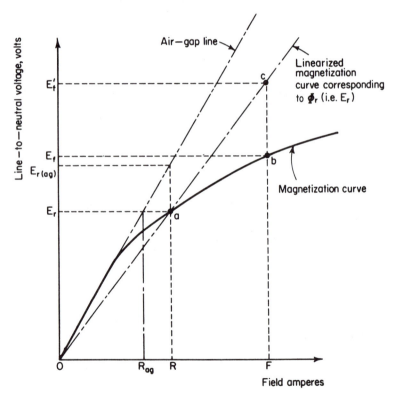

Figure E-1 Illustrating the saturated synchronous reactance method based on E_r.

ratio of these two quantities. That is, the reluctance of the magnetic circuit has been increased by a saturation factor k where

$$k = \frac{R}{R_{ag}} = \frac{E_{r(ag)}}{E_r} \tag{E-1}$$

The second form of this equation follows from the geometry of similar triangles in Fig. E-1.

How would this synchronous generator perform if it were now assumed that the degree of saturation remained fixed at Φ_r and the armature current were reduced to zero while the field current remained constant? A study of Fig. E-1 discloses that the operating point would move from a to c along a linearized magnetization curve that corresponds to the resultant flux. Point c is found as the point of intersection of a line drawn from the origin through a and the ordinate line placed at the constant field excitation (F). The no-load voltage that corresponds to c is identified as E_f' and represents the no-load terminal voltage that would be produced if the air-gap flux remained invariant as the armature current diminished to zero. Since this does not actually happen, clearly E_f' is entirely a fictitious quantity. On the other hand, the field current F (that is associated with E_f') is not fictitious, but real. In fact, F is the quantity that we are seeking, and the foregoing construction has been introduced simply to facilitate the determination of F. The determination immediately follows from the similar triangles OaR and OcF. Thus

$$F = \frac{E_f'}{E_r} R \tag{E-2}$$

where these quantities are defined as illustrated in Fig. E-1.

In applying Eq. (E-2), the quantity E_r is found by means of Eq. (5-22); R is found from the actual magnetization curve. E_f' is then computed by the synchronous reactance method in the manner illustrated in Example 5-2 but with the necessary modification that that value of synchronous reactance be used that corresponds to the degree of saturation represented by the resultant flux. If x_s were entirely associated with the iron structure of the machine, the saturated synchronous reactance could be found by dividing the unsaturated x_s by the saturation factor k. However, as indicated by Eq. (5-16), the synchronous reactance also includes the effect of leakage reactance, which remains essentially unaffected by the changes in the degree of saturation of the main iron paths of the machine. Accordingly, the saturation factor must be applied to x_ϕ only in Eq. (5-16). This leads to the following expression for the saturated synchronous reactance x_{ss}:

$$x_{ss} = x_l + \frac{x_s - x_l}{k} \tag{E-3}$$

where x_l denotes leakage reactance, k is the saturation factor given by Eq. (E-1), and x_s is the unsaturated synchronous reactance given by Eq. (5-18). The following example illustrates the effectiveness of this method.

Example E-1

For the machine of Example 5-1, find the field current that produces rated terminal voltage when rated armature current is delivered to a 0.8 lagging pf three-phase balanced load. Use the saturated synchronous reactance method.

Solution The values of E_r and R are given by Eqs. (5-22) and (5-23), which are repeated here for convenience.

$$E_r = 8444.4 \text{ V} \quad \text{and} \quad R = 222 \text{ A}$$

The saturation factor is found to be

$$k = \frac{R}{R_{ag}} = \frac{222}{157} = 1.41$$

Hence by Eq. (E-3) the saturated synchronous reactance is

$$x_{ss} = x_l + \frac{x_s - x_l}{k} = 1.9 + \frac{22 - 1.9}{1.41} = 16.2 \ \Omega$$

Accordingly, we get

$$\bar{E}_f' = V_t \underline{/0^\circ} + \bar{I}_a(jx_{ss})$$

$$= 7967.4 + j418.4(0.8 - j0.6)16.2$$

$$= 7967.4 + 4042 + j5423$$

$$= 12,009 + j5423 = 13,177 \underline{/24.3^\circ}$$

Finally, by Eq. (E-2) the required excitation current is found to be

$$F = \frac{13,177}{8444.4} (222) = 346 \text{ A}$$

which is identical to the result found by the general method.

When it is desirable to compute voltage regulation by the saturated synchronous reactance method, a glance at Fig. E-1 reveals that the no-load terminal voltage to be used in Eq. (5-26) is to be found on the actual magnetization curve corresponding to the value of F computed by Eq. (E-2). This is denoted by point b in Fig. E-1.

F. SLIP TEST FOR SALIENT-POLE SYNCHRONOUS MACHINES

Information about the direct-axis synchronous reactance and the quadrature-axis synchronous reactance of synchronous machines can be obtained from a relatively simple no-load test known as the *slip test*. The reason for the name will be apparent presently. Figure F-1 depicts the wiring arrangement for the test, including the instruments that permit calculations of the reactances.

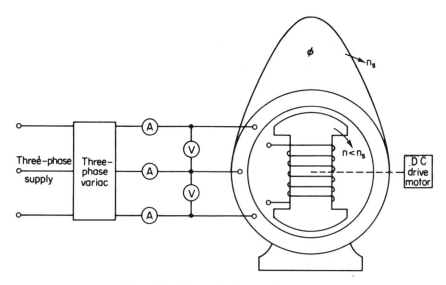

Figure F-1 Schematic diagram for the slip test.

The three-phase variac is needed to apply an adjustable voltage of rated frequency to the three-phase stator winding, which is set appreciably below the rated machine voltage. Application of this voltage creates a revolving set of magnetic poles in the air gap of the synchronous generator, traveling at synchronous speed n_s. At the same time, a separate source is used to drive the salient-pole rotor structure at a speed slightly different from synchronous speed and in the *same* direction as the revolving stator mmf. A quick and easy way to ascertain that the direction is proper is to put a voltmeter across the open-field winding terminals. If a small reading is registered the direction is proper. (Why?) The use of a dc motor is convenient as the separate drive source because of the ease with which the speed may be adjusted.

By driving the rotor structure at a speed either slightly below or slightly above synchronous speed, the rotor appears to be slipping either behind or ahead of the rotating magnetic poles of the stator. Consequently, there will be a period of time when the stator magnetic poles are in a position of alignment with the rotor iron pole structure, thus presenting a minimum reluctance to the stator mmf and resulting in a minimum magnetizing current, as indicated by the line ammeters, A, in Fig. F-1. At a slightly later instant the direct axis of the rotor pole structure will be in a position of quadrature with respect to the axes of the stator magnetic poles. Accordingly, the air gap is very large, and so a large magnetizing current must flow to establish essentially the same air-gap flux demanded by the applied voltage. The increase in current is quite appreciable and is easily registered by the line ammeters. Keep in mind that this variation of the current from a minimum to a maximum level occurs at a frequency that is dependent directly upon $n_s - n$.

Obviously, one technique required in this test is to keep this frequency small enough so that the true maximum and minimum values can be recorded. However, at the same time, this frequency must not be made so small that the rotor structure is pulled into synchronism with the stator magnetic poles through the action of the reluctance power. As a matter of fact, the reason for performing this test at reduced voltage is precisely to minimize the influence of the reluctance torque to pull the rotor into synchronism. Unfortunately, a lower limit must be placed on this applied voltage; otherwise, the computed value of synchronous reactance will be nearer to the unsaturated value, which we already know can be misleading.

Because the internal impedance of the variac is not likely to be negligible, the wide swing in the minimum and maximum values of the line ammeters will cause some fluctuations in the voltmeter reading. These too should be recorded. Also note that the maximum voltage reading is associated with the minimum line current, and the minimum voltage reading goes with the maximum current.

The direct-axis synchronous reactance is found by using V_{max} and I_{min} since these are the readings that are registered when the rotor-pole axis and the magnetic poles of the stator are in alignment. Assuming a Y-connected stator winding, the expression for the direct-axis synchronous reactance thus becomes

$$x_d = \frac{V_{max}}{\sqrt{3}\,I_{min}} \qquad \text{(F-1)}$$

Average values of the three ammeters and two voltmeters should be used.

By a similar argument, it should be clear that the quadrature-axis synchronous reactance can be computed from

$$x_q = \frac{V_{min}}{\sqrt{3}\,I_{max}} \qquad \text{(F-2)}$$

The reliability of the results obtained from Eqs. (F-1) and (F-2) can be improved by employing oscillographic methods to record and measure the variations in line currents and line voltages. This permits a larger slip frequency to be used, which in turn allows a higher applied voltage to be used without the danger of pull-in. Furthermore, the maximum and minimum values can be measured much more accurately because the inertia effects of the indicating ammeters and voltmeters are eliminated entirely.

G. A SHORT BIBLIOGRAPHY

The following is a short selected list of books that the reader can find helpful as a supplementary source of information in the study of this textbook.

CHAPMAN, STEPHEN, J., *Electric Machinery Fundamentals* (New York: McGraw-Hill Book Company, 1985).

EL-HAWARY, M.E., *Principles of Electric Machines with Power Electronic Applications* (Englewood Cliffs, NJ: Prentice-Hall, Inc., 1986).

FITZGERALD, A.E., C. KINGSLEY, JR., AND S.D. UMANS, *Electric Machinery,* 4th ed. (New York: McGraw-Hill Book Company, 1983).

LIWSHITZ-GARIK, M. AND C.C. WHIPPLE, *Electric Machinery,* vols I and II (Princeton, NJ: D. Van Nostrand Co. Inc., 1946).

MCPHERSON, GEORGE, *An Introduction to Electrical Machines and Transformers* (New York: John Wiley & Sons, Inc., 1981).

NASAR, S.A. AND L.E. UNNEWEHR, *Electromechanics and Electric Machines* (New York: John Wiley & Sons, Inc., 1979).

STEIN, ROBERT AND WILLIAM T. HUNT, *Electric Power System Components: Transformers and Rotating Machines* (New York: Van Nostrand Reinhold, 1979).

Answers to Selected Problems

1-1. (a) $4/\pi$ A/in   **(b)** 32×10^{-7} N/m   **(c)** 0.32 N/m

1-3. (a) 0   **(b)** -0.1 N/m, repulsion   **(c)** same as (b)

1-5. 20 A

1-7. (a) 98 A-t   **(b)** 1.225 A

1-9. 183.75 A-t

1-11. 0.00475 Wb

1-13. 357.4 turns

1-15. 158.8

1-17. (a) 0.5 H   **(b)** 0.0025 J

1-19. 2.41 W

1-21. 0.302 J/cycle

1-23. Hysteresis: 1400 W @ 60 Hz, 2100 W @ 90 Hz; Eddy: 400 W @ 60 Hz, 900 W @ 90 Hz

1-25. $F = (B^2/\mu)bg$

1-27. (a) 0.53 A   **(b)** 0.4975 J   **(c)** 0.165 J   **(d)** 4.72 H

2-1. (a) 0.00108 Wb   **(b)** 1.44 in

2-3. ϕ decreases, I_m remains constant

2-5. (a) 0.01126 Wb   **(b)** 1.5   **(c)** Increase exceeds 50%   **(d)** 90 Hz

2-7. **(a)** 97.65%   **(b)** 2453.2$\underline{/0.33°}$ **(c)** 2.216%

2-9. 96.68%

2-11. 4.82%

2-13. 0.7055 A

2-15. 206 A

2-17. Purely capacitive

2-20. 87.8 V

2-21. 0.8566 + j0.8072

2-23. Leading load with pf angle of 28.6°

2-25. 1 A

2-27. 10 H

2-30. **(a)** 1.9 H   **(b)** 0.2 H

2-33. **(a)** 1.48%   **(b)** 0.575 Ω **(c)** 55 kW, 16.5 kW   **(d)** 22 kVA, 440/330 V

2-35. 621$\underline{/-68.75°}$ A, high side

2-37. 48.4$\underline{/-60.46°}$ A, 50$\underline{/-36.9°}$

2-39. **(a)** 225 kVA   **(b)** 25 kVA   **(c)** 200 kVA

2-42. **(a)** 0.0206, 0.0206   **(b)** 95.63%

2-43. 0.0256 pu, 0.016 pu

2-45. 2 ohms

2-47. 0.866

2-49. **(a)** 120.09 V   **(b)** 126.6 V

3-1. **(a)** 0.09425 Wb   **(b)** 39.27 V   **(c)** 3.927 Wb/s

3-3. **(a)** 4.8 N-m   **(b)** 50 V   **(c)** 400 W

3-5. **(a)** 0.0024 Wb   **(b)** 0.144 Wb-t   **(c)** 0.144 cos $5\pi t$ **(d)** 2.26
      (e) 0   **(f)** 1.96 V

3-7. **(a)** $\phi = \text{Bl}r(\pi - 2\theta_0)$ **(b)** $e = 2NBlv$ **(c)** 7.54 V   **(d)** 4.8 V

3-9. **(a)** $\lambda = 2.4 \cos \omega_0 t$ **(b)** 130.6 V   **(c)** 130.6 V   **(d)** 150.8 V   **(e)** 0

3-12. 532 A-t/pole

3-15. **(a)** 8620 A-t   **(b)** 53.1 N-m   **(c)** 104 V   **(d)** 0.00122 Wb

4-2. 900 rpm, 1.8 Hz

4-5. **(a)** 1.50ϕ_m directed CW 60° from vertical in 1st quad.   **(b)** 1.50ϕ_m @ 120 CW
      4th quad.

4-9. $F_m \cos (\omega t - \alpha)$

4-10. **(a)** 1800 rpm   **(b)** 3 Hz   **(c)** 90 rpm   **(d)** 1800 rpm   **(e)** 0   **(f)** yes

4-13. **(a)** 700 rpm   **(b)** 1700 rpm

4-16. 87.45%

4-17. 99.3 A

4-19. **(a)** 21.68$\underline{/-25.3°}$ **(b)** 8.38 hp   **(c)** 80.3%   **(d)** 36.32 N-m
      (e) 56.73$\underline{/-78.4°}$ A

4-21. (a) 0.288 lag   (b) 0.98 lag   (c) 5.41 kW

4-23. (a) $160.5 \underline{/-27.53°}$ (b) 90.49%

4-25. (a) 0.9 lag   (b) 87.35%   (c) 41 hp

4-28. (a) yes   (b) 5 hp   (c) slightly less   (d) $\frac{1}{2}$

4-30. 16.7% drop

4-33. (a) 0.096   (b) 45.55 N-m

4-35. 0.0333 ohm

4-37. (a) 40.033 A, 13.52 kW   (b) 593.5 V, 71.75 kW

4-40. 1.215 ohms

4-41. $r_1 = 0.1174 \ \Omega$, $r_2' = 0.13 \ \Omega$, $x_1 + x_2' = 0.9 \ \Omega$, $r_t = 136.6 \ \Omega$, $x_\phi = 15.7 \ \Omega$.

4-43. (a) 6 A, 0.301 lag   (b) 534 W

4-45. (a) 160 V

4-47. (a) 900 rpm   (b) 945 rpm

4-49. (a) −0.2   (b) 1440 rpm   (c) 60 kW

4-51. (a) 32 V   (b) 6 V in quadrature with sE_2

4-53. (a) 10 V   (b) 3 kW

4-56. 1.1 ohms

4-57. 25.87 ohms

5-1. 1.8 pu

5-3. 17.8 A

5-5. (a) $340 \underline{/14.75°}$ V   (b) 329 V   (c) 233 kW

5-7. (a) 10.56 A   (b) 12.68 A

5-9. (a) 1.58 Ω (b) 1.284 Ω (c) 33.06 A

5-11. 11.3 t/p

5-13. 38.6 A

5-15. (a) 31.5%   (b) −18.38%

5-17. 40.55%

5-19. 2 ohms

5-21. (a) 192.25 V   (b) 35.9°   (c) 35.9°

5-23. (a) 112.78 V   (b) 8.74°   (c) 45.64°

5-25. (a) 101 V   (b) 3.053 A

5-27. $275 \underline{/24.42°}$ V

5-29. (a) 6000   (b) 164 A   (c) 50.9°

6-1. (a) 0.605 leading   (b) 1000 V/phase

6-4. (a) $293 \underline{/9.1°}$

6-5. (a) 0.778 leading   (b) $10,000 \underline{/-14.7°}$ (c) 1517.2 kW

6-7. (a) 2/3 pu   (b) 2.6 pu

6-8. (a) motor   (b) −30°

6-11. (a) 24.85 A, 0.79 lagging   (b) $280.3 \underline{/-25.03°}$

6-13. 627 N-m

6-15. (a) 2.58 pu   **(b)** 0.375 pu

6-18. $x_d = 0.7$ pu, $x_q = 0.4$ pu

6-19. 1.805 pu, 0.72 leading

6-20. (a) 72°   **(b)** 1.926 pu

6-22. 481

6-24. 0.797 lag

7-2. (b) $3.73E_m$

7-3. (b) $3.86E_m$

7-8. (a) 366.7 ohms   **(b)** 12 V   **(c)** 176.5 ohms   **(d)** 0.068 A   **(e)** 436 rpm

7-9. 190 V

7-11. (a) 0.01 Wb   **(b)** 1.24 Wb   **(c)** 1.16 Wb

7-13. (a) 366.7 ohms   **(b)** 17.3 V   **(c)** 1.48 A

7-15. (a) 258 V   **(b)** 58.3 A   **(c)** 1.54 A   **(d)** 236 V

7-17. (a) 8.125 ohms   **(b)** 210 V   **(c)** 84.1%, 454 N-m

7-19. (a) 11.2 ohms   **(b)** 552 V

7-21. (a) 250 V   **(b)** 270 V   **(c)** 8.56 A   **(d)** 276 V

7-24. 38.7 ohms

8-1. (a) 1192.27 rpm   **(b)** 6.8%

8-3. 998.2 rpm

8-5. (a) 430.6 rpm   **(b)** 672.6 N-m   **(c)** 40.7 hp

8-7. (b) $\omega = 2847/\sqrt{T}$ **(c)** 47.6 rad/s   **(d)** 50.6 rad/s   **(e)** 6.3%

8-9. (a) 40.16 A, 42.16 A   **(b)** 73.2 N-m   **(c)** 10.6 hp

8-11. (a) 1112 rpm   **(b)** 93.4%   **(c)** 119.6%

8-13. (a) 134 ohms   **(b)** 880 rpm

8-15. (a) 0.009 Wb   **(b)** 129 N-m   **(c)** 0.58 kW   **(d)** 15.35%

8-17. (a) 229 A   **(b)** 1413 rpm

8-19. (a) 260 A   **(b)** 576.2 N-m   **(c)** 86.76%

8-21. (a) 72.92 N-m   **(b)** 1342.3 rpm   **(c)** 81.3%

8-23. 1432.4 rpm or 150 rad/s

8-25. 69 ohms

8-27. (a) 2.045 ohms   **(b)** 10.226   **(c)** 62.06%

8-29. (a) 24.355 rad/s (232.57 rpm)   **(b)** 80.59%

8-31. 67.5 rad/s (644.6 rpm)

8-34. (a) 1. external armature resistance, 2. field control, 3. armature voltage control   **(b)** 1. 2000 rpm, 200 hp; 2. 500 rpm, 200 hp; 3. 2000 rpm, 200 hp

8-36. 880 rpm, 77.5 A

8-39. 80.6%

8-42. (a) 24.4 ohms   **(b)** 262 V   **(c)** 175 ohms

8-45. $0.585V_p$

8-47. **(a)** 393 rpm   **(b)** 379 rpm   **(c)** 3.7%

8-49. **(a)** 176.93 rpm   **(b)** 191 rpm, 7.9%   **(c)** 381.57 rpm, 3.8%

8-51. 189.44 Hz

9-1. Yes

9-5. CW: 671 A-t; CCW: 1432 A-t

9-7. $8\underline{/90°}$ A

9-9. **(a)** $95\underline{/14.7°}$ **(b)** $37\underline{/-40.6°}$

9-11. $123.9\underline{/20.9°}$, $68.7\underline{/93.19°}$

9-13. **(a)** 7.854 A, 0.57 lag   **(b)** 0.411 hp   **(c)** 62.28%

9-15. 1224 W, 20.34 A

9-17. **(a)** $r_2' = 3.34\ \Omega$, $x_1 = x_2' = 3.211\ \Omega$, $(x_\phi/2) = 49.21\ \Omega$ **(b)** 65.28 W
　　(c) 546.5 Ω

9-19. **(a)** 8 A, 0.772 lag   **(b)** 1.303 hp   **(c)** 71.49%

9-21. **(a)** 49.8% rated torque   **(b)** 213 μF   **(c)** 266.67% rated torque

10-1. **(a)** $25.5\underline{/-13.35°}$ **(b)** $104.2\underline{/25.06°}$ **(c)** $9.24\underline{/106.1°}$ **(d)** $21.52\underline{/119.4°}$

10-5. **(a)** $60\underline{/-210°}$ **(b)** $60\underline{/-120°}$

10-7. **(a)** $18.83\underline{/-143.5°}$ **(b)** No

10-9. **(a)** No   **(b)** $75\underline{/0°}$ **(c)** $25\underline{/0°}$

10-11. **(a)** $90\underline{/0°}$, $60.8\underline{/-99.46°}$, $70\underline{/-180°}$, $60.8\underline{/99.46°}$ **(b)** $70\underline{/0°}$ **(c)** Yes
　　(d) $10\underline{/0°}$ **(e)** $10\underline{/0°}$

10-13. $123.9\underline{/20.9°}$, $68.7\underline{/93.19°}$

10-15. **(a)** $100\underline{/0°}$ **(b)** $20\underline{/0°}$

10-16. 3.37 S-W or 0.0179 N-m

10-19. **(a)** $115.2\underline{/-0.68°}$, $75.52\underline{/-50.84°}$ **(b)** $61.314\underline{/-113.62°}$ mA

10-21. **(a)** 5.23 S-W (0.01386 N-m)   **(b)** 2.614 S-W (0.00693 N-m)   **(c)** 0.000039

10-24. **(a)** 13.173 S-W   **(b)** 13.7 S-W   **(c)** 9.21 S-W

10-25. 6.964

11-1. **(a)** $0.863T_m$ CCW   **(b)** 18°

11-3. 0, a condition of equilibrium

11-5. 48 bits/s

11-7.

Coil A-A'				Coil B-B'			
Q1	Q2	Q3	Q4	Q1	Q2	Q3	Q4
1	1	0	0	0	0	0	0
0	0	0	0	0	0	1	1
0	0	1	1	0	0	0	0
0	0	0	0	1	1	0	0

11-8. (a) $7.2°$ **(b)** 40

12-3. Line-to-neutral voltages: 26, 26, -52; Line-to-line: 0, 78, -78.

12-4. Line-to-neutral voltages: 26, -52, 26; Line-to-line: 78, -78, 0.

12-11. 0.06 pu

12-13. 0.275 pu

12-14.

α	$0°$	$5°$	$10°$	$15°$	$20°$	$25°$	$30°$
ε_h	0	0.023	0.31	0.022	0.007	-0.001	0

12-15. (a) $-0.1736E_1$ **(b)** $0.3472E_1$ **(c)** Yes

13-1. (a) $T(s) = \dfrac{1}{\alpha}\left(\dfrac{1}{1+s\tau}\right)$ where $\alpha = \dfrac{R_1+R_2}{R_2}$, $\tau = \dfrac{L}{R_1+R_2}$

  (b) $T(s) = \dfrac{1}{\alpha}\left(\dfrac{1+s\alpha\tau}{1+s\tau}\right)$

13-3. (a) $T(s) = \dfrac{1}{1+s\tau}$ where $\tau = RC$ **(b)** $T(s) = \dfrac{1+s\tau}{1+s\alpha\tau}$, $\tau = R_2C$, $\alpha = \dfrac{R_1+R_2}{R_2}$

  (c) $T(s) = \dfrac{s\tau}{1+s\tau}$, $\tau = RC$

  (d) $T(s) = \dfrac{1}{\alpha}\left(\dfrac{1+s\tau}{1+s\tau}\right)$, $\tau = \dfrac{R_1R_2C}{R_1+R_2}$, $\alpha = \dfrac{R_1+R_2}{R_2}$

13-5. (a) $T(s) = \dfrac{1+sR_2C}{1+s(R_1+R_2)C}$ **(b)** Same as Fig. 13-3(b)

13-7. $\dfrac{E_1}{E_2}(s) = \dfrac{1}{(s+1)^2}$

13-9. (a) 45.72 rad/s   **(b)** 6 s

13-11. (b) $\dfrac{\omega(s)}{E_1(s)} = \dfrac{4.71}{1 + 5.37 \times 10^{-4}\,s}$ **(c)** $5.37 \times 10^{-4}\,\dot\omega + \omega = 4.71e_i$

13-13. $L_fJR\ddot\omega + (JRR_f + RFL_f + K_tK_\omega L_f)\dot\omega + R_f(RF + K_tK_\omega)\omega = K_aK_gK_tE_f$
  where $R = R_m + R_g$

13-13. $\begin{bmatrix} \dot I_f \\ \dot\omega \end{bmatrix} = \begin{bmatrix} -\dfrac{1}{\tau_f} & 0 \\ \dfrac{K_gK_t}{RF\tau_m} & -\dfrac{K_tK_\omega + RF}{RF\tau_m} \end{bmatrix}\begin{bmatrix} I_f \\ \omega \end{bmatrix} + \begin{bmatrix} \dfrac{K}{R_f\tau_f} \\ 0 \end{bmatrix}E_i$
  where $\tau_f = L_f/R_f$, $\tau_m = J/F$, and $R = R_m + R_g$

13-17. (a) $M(s) = \dfrac{K_pK_aK_m}{Js^2 + (F + K_aK_mK_o)s + K_pK_aK_m}$

  (b) $Js^2 + (F + K_aK_mK_o)s + K_pK_aK_m = 0$ **(c)** provides damping of transient terms

13-20. (a) 150 rad/s   **(b)** 7.5 s   **(c)** 7 lb-ft/rad/s

13-21. (b) 60 rad/s   **(c)** 26.62s   **(d)** $\begin{bmatrix} \dot I_f \\ \dot\omega_m \end{bmatrix} = \begin{bmatrix} -2.5 & 0 \\ 24/7 & -5/7 \end{bmatrix}\begin{bmatrix} I_f \\ \omega_m \end{bmatrix} + \begin{bmatrix} 1/20 \\ 0 \end{bmatrix}V_f$

13-25. (a) 300   **(b)** 10   **(c)** $v(t) = 300\,(1 - e^{-1204t})u(t)$ **(d)** $1/1204$ s

13-28. $\omega_{0ss} = \dfrac{K_a K_g K_t}{R_f(K_t K_\omega + RF) + K_a K_g K_t K_0} E_r$

Appendix to Chapter 3

3-19. (b) 30 At/p

3-21. (a) $\phi = Blr(2\pi/p)$ **(b)** $T = p^2 \dfrac{\phi}{\pi} J_m \sin \Delta$ **(c)** T varies

3-25. $\alpha = 0°$

3-28. (a) Yes because $R_d \neq R_q$ **(b)** standstill   **(c)** $T = \dfrac{1}{4} \left(\dfrac{V}{\omega}\right)^2 \left(\dfrac{L_d - L_q}{L_d L_q}\right) \sin 2\Delta$

3-30. (a) 15.45 W   **(b)** 3600 rpm   **(c)** 45 W

Index